Electric Motor Control

Electric Motor Control
DC, AC, and BLDC Motors

Sang-Hoon Kim
Department of Electrical & Electronics Engineering,
Kangwon National University

elsevier.com

Elsevier
Radarweg 29, PO Box 211, 1000 AE Amsterdam, Netherlands
The Boulevard, Langford Lane, Kidlington, Oxford OX5 1GB, United Kingdom
50 Hampshire Street, 5th Floor, Cambridge, MA 02139, United States

Library of Congress Cataloging-in-Publication Data
A catalog record for this book is available from the Library of Congress

British Library Cataloguing-in-Publication Data
A catalogue record for this book is available from the British Library

ISBN: 978-0-12-812138-2

For Information on all Elsevier publications
visit our website at https://www.elsevier.com/books-and-journals

Publisher: Joe Hayton
Acquisition Editor: Lisa Reading
Editorial Project Manager: Edward Payne
Production Project Manager: Anusha Sambamoorthy
Cover Designer: Greg Harris

Typeset by MPS Limited, Chennai, India

Contents

Preface

Owing to rapidly increasing energy costs and global interest in reducing carbon dioxide emissions, electric motors have recently become one of the most important prime movers that produce mechanical power. Recently, in many mechanical systems, traditional prime movers such as a hydraulic system, a steam turbine, a gas turbine, and an internal combustion engine are being rapidly replaced by electric motor drive systems, which are more efficient, controllable, and environment-friendly. Currently, electric motor drive systems are playing an important role in improving convenience in many areas of our lives including home appliances, office machines, transportation, and industrial machines.

The main aim of this book is to introduce practical drive techniques of electric motors for supporting stable, efficient control of such application systems, covering basic principles to high-performance motor control techniques.

Nowadays, the most widely used motors are classic direct current (DC) and alternating current (AC) motors (induction motor and synchronous motor). Besides these motors, there is a brand new brushless direct current (BLDC) motor. To control these motors efficiently, we will do a comprehensive study of fundamental operating principles, driving methods of electric motors, related control theories, and power converters for driving a motor. Many people do not have the background knowledge in these areas and may need to obtain these contents from various books. Therefore, this book is designed optimally for studying these contents at once and easy learning of difficult control principles. Furthermore, this book provides simulation examples for key subjects using MATLAB/Simulink tool and offers practical control techniques for industrial motor drive applications currently in use.

This book is the English version of my book titled *DC, AC, BLDC Motor Control*, which was first published in August 2007, in Korean. This book is based on my wide research experience in the electric motor control field over the past 20 years. During 20 years of teaching at my university, I realized that many students struggled with mastering the control of electric motors due to the complexity and difficulty of existing textbooks. Therefore, the main goal of this textbook was to address this issue by presenting the material in a straightforward way for the students to understand. As a result, this book gained popularity, and I received many appreciation letters from students and engineers after publishing the book, expressing gratitude for presenting the material well. I have also been invited many times to speak at seminars regarding the materials presented in the book.

The contents of this book are as follows.

Chapter 1, Fundamentals of electric motors, introduces the fundamental operating principle needed to understand electric motors and describes clear differences among the motors by comparing their operating principles.

Chapter 2, Control of direct current motors, describes DC motors, which used to be the typical motor for speed control and torque control, and the concept of torque control for DC motors, which will be helpful for understanding the vector control for torque control of AC motors. Furthermore, this chapter goes into detail about the design of the current and speed controllers of DC motors, which is readily applicable to the current and speed controllers of AC motors.

Chapter 3, Alternating current motors: synchronous motor and induction motor, discusses the basic characteristics of AC motors, including synchronous and induction motors, which are widely used for industrial motors. These characteristics can provide a basis for studying high-performance motor control, which will be described later in the book.

Chapters 4–8 discuss the high-performance motor control systems of AC motors for industrial motor drive applications as follows:

Chapter 4, Modeling of alternating current motors and reference frame theory—the modeling of AC motors and $d-q$ reference frame theory for control of AC motors,

Chapter 5, Vector control of alternating current motors—the concept of the vector control and its implementation for AC motors,

Chapter 6, Current regulator of alternating current motors—the design of current controller for the vector control system,

Chapter 7, Pulse width modulation inverter—the PWM inverter and its various techniques for AC motor drives,

Chapter 8, High-speed operation of alternating current motors—the field-weakening control for high-speed operations of AC motors.

Chapter 9, Speed estimation and sensorless control of alternating current motors, describes the position/speed sensors and speed estimation required for the motor control. In additional, sensorless control, which is a state-of-the-art technique in the motor control area, is explored briefly.

Chapter 10, Brushless direct current motors) deals with BLDC motors, which are not classical motors but mostly used in small motor drive systems. The operation principle, drive methods, PWM methods, and sensorless control schemes of BLDC motors are also explored in detail.

Chapters 1–3, and 10 will be adequate for providing technical background of motor control for undergraduates in a one-semester course, while Chapters 4–9 will be suitable for graduate students and engineers with the necessary background for understanding high-performance motor control systems.

I am certain that this book will be able to equip one with complete techniques for controlling electric motors required for industrial applications.

The companion web site of the book can be found at: https://www.elsevier.com/books-and-journals/book-companion/9780128121382

Sang-Hoon Kim

Seoul, November 2016

CHAPTER

Fundamentals of electric motors

1

A moving object that has either a linear motion or rotary motion is powered by a prime mover. A prime mover is an equipment that produces mechanical power by using thermal power, electricity, hydraulic power, steam, gas, etc. Examples of the prime mover include a gas turbine, an internal combustion engine, and an electric motor. Among these, the electric motors have recently become one of the most important prime movers and their use is increasing rapidly. Nearly 70% of all the electricity used in the current industry is used to produce electric power in the motor-driven system [1].

Electric motors can be classified into two different kinds according to the type of the power source used as shown in Fig. 1.1: direct current (DC) motor and alternating current (AC) motor. The recently developed *brushless DC motor* is hard to be classified as either one of the motors since its configuration is similar to that of a permanent magnet synchronous motor (AC motor), while its electrical characteristics are similar to those of a DC motor.

The first electric motor built was inspired by Michael Faraday's discovery of electromagnetic induction. In 1831, Michael Faraday and Joseph Henry simultaneously succeeded in laboratory experiments in operating the motor for the first time. Later in 1834, M. Jacobi invented the first practical DC motor. The DC motor is the prototype of all motors from the viewpoint of torque production. In 1888, Nikola Tesla was granted a patent for his invention of AC motors, which include a synchronous motor, a reluctance motor, and an induction motor. By 1895, the three-phase power source, distributed stator winding, and the squirrel-cage rotor had been developed sequentially. Through these developments, the three-phase induction motors were finally made available for commercial use in 1896 [2].

Traditionally among the developed motors, DC motors have been widely used for speed and position control applications because of the ease of their torque control and excellent drive performance. On the other hand, induction motors have been widely used for a general purpose in constant-speed applications because of their low cost and rugged construction. Induction motors account for about 80% of all the electricity consumed by motors.

Until the early 1970s, major improvement efforts were made mainly toward reducing the cost, size, and weight of the motors. The improvement in magnetic material, insulation material, design and manufacturing technology has played a

Electric Motor Control. DOI: http://dx.doi.org/10.1016/B978-0-12-812138-2.00001-5

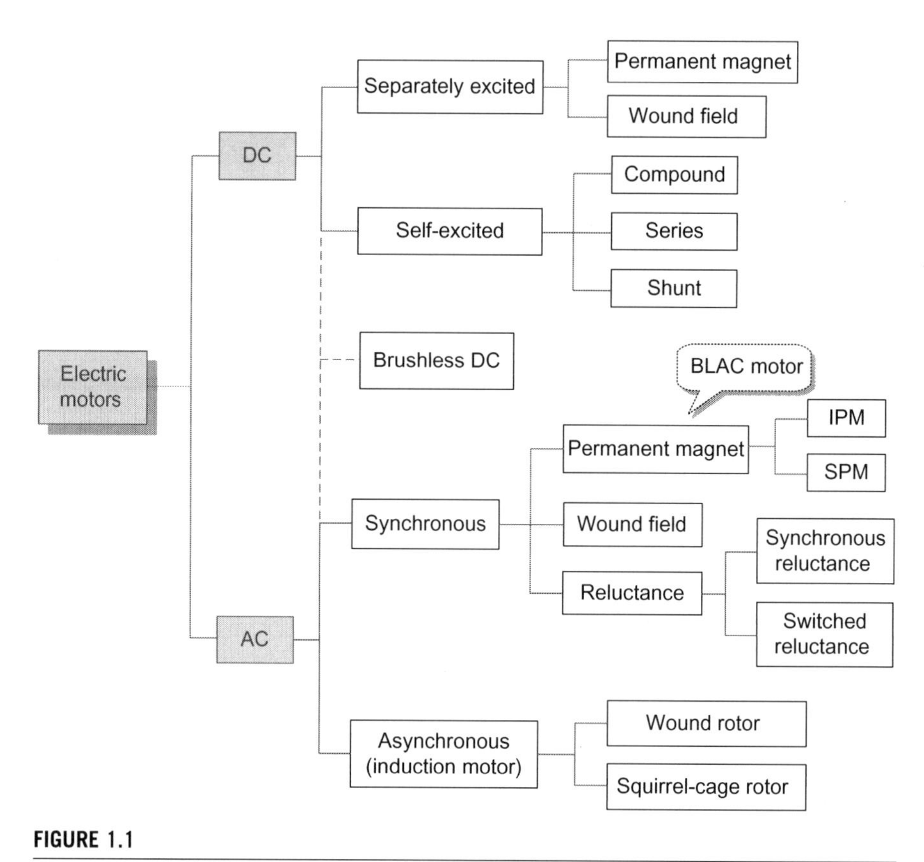

FIGURE 1.1

Classification of electric motors: AC motor and DC motor.

major role and made a big progress. As a result, a modern 100-hp motor is the same size as a 7.5-hp motor used in 1897. With the rising cost of oil price due to the oil crisis in 1973, saving the energy costs has become an especially important matter. Since then, major efforts have been made toward improving the efficiency of the motors. Recently, rapidly increasing energy costs and a strong global interest in reducing carbon dioxide emissions have been encouraging industries to pay more attention to high-efficiency motors and their drive systems [1].

Along with the improvement of motors, there have been many advances in their drive technology. In the 1960s, the advent of power electronic converters using power semiconductor devices enabled the making of motors with operation characteristics tailored to specific system applications. Moreover, using microcontrollers with high-performance digital signal processing features allowed the engineers to apply advanced control techniques to motors, greatly increasing the performance of motor-driven systems.

1.1 FUNDAMENTAL OPERATING PRINCIPLE OF ELECTRIC MOTORS

1.1.1 CONFIGURATION OF ELECTRIC MOTORS

An electric motor is composed of two main parts: a stationary part called the *stator* and a moving part called the *rotor* as shown in Fig. 1.2. The air gap between the stator and the rotor is needed to allow the rotor to spin, and the length of the air gap can vary depending on the kind of motors.

The stator and the rotor part each has both an electric and a magnetic circuit. The stator and the rotor are constructed with an iron core as shown in Fig. 1.3, through which the magnetic flux created by the winding currents will flow and which plays a role of supporting the conductors of windings. The current-carrying

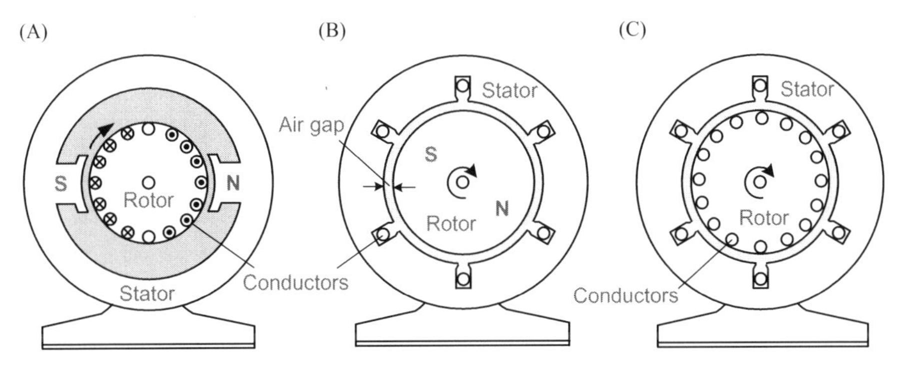

FIGURE 1.2

Configuration of electric motors. (A) DC motor, (B) AC synchronous motor, and (C) AC induction motor.

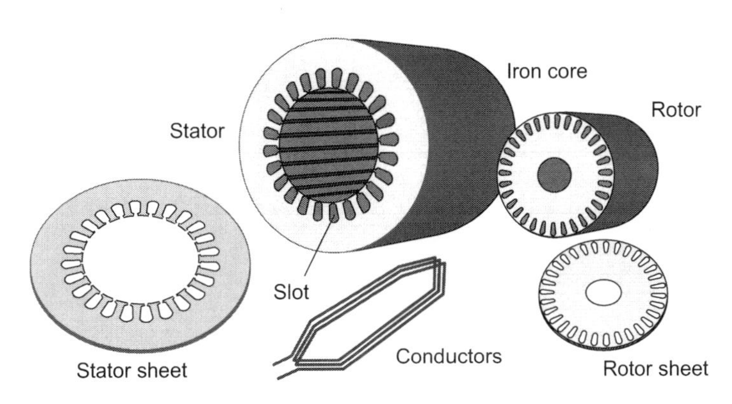

FIGURE 1.3

Electric and magnetic parts of electric motors.

conductors inserted into slots in the iron core form the electric circuit. When the current flows in these conductors, a magnetic field is created through the iron core, and the stator and the rotor each becomes an electromagnet.

To obtain a greater magnetic flux for a given current in the conductors, the iron core is usually made up of ferromagnetic material with high magnetic permeability, such as silicon steel. In some cases, the stator or the rotor creates a magnetic flux by using a permanent magnet.

1.1.2 BASIC OPERATING PRINCIPLE OF ELECTRIC MOTORS

All electric motors are understood to be rotating based on the same operating principle. As shown in Fig. 1.4A, there are generally two magnetic fields formed inside the motors. One of them is developed on the stationary stator and the other one on the rotating rotor. These magnetic fields are generated through either energized windings, use of permanent magnets, or induced currents. A force produced by the interaction between these two magnetic fields gives rise to a torque on the rotor and causes the rotor to turn. On the other hand, some motors, such as the reluctance motor, use the interaction between one magnet field and a magnetic material, such as iron, but they cannot produce a large torque (Fig. 1.4B). Most motors in commercial use today including DC, induction, and synchronous motors exploit the force produced through the interaction between two magnetic fields to produce a larger torque.

The torque developed in the motor must be produced continuously to function as a motor driving a mechanical load. Two motor types categorized according to the used power source, i.e., DC motor and AC motor, have different ways of achieving a continuous rotation. Now, we will take a closer look at these methods for achieving a continuous rotation.

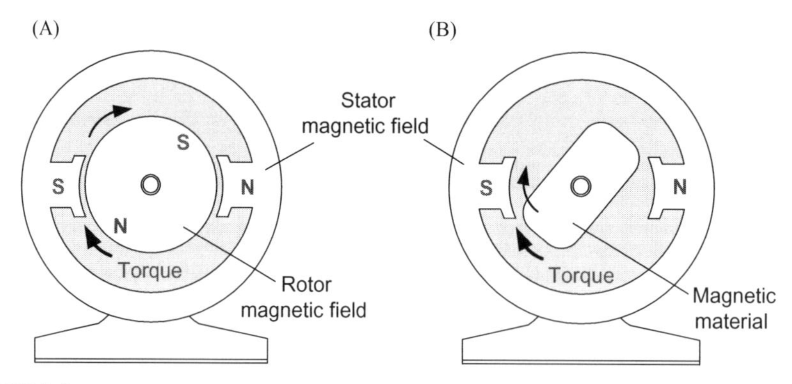

FIGURE 1.4

Rotation of electric motors. (A) Two magnetic fields and (B) one magnetic field and magnetic material.

1.1.2.1 Direct current motor

The simple concept for the rotation of a DC motor is that the rotor rotates by using the force produced on the current-carrying conductors placed in the magnetic field created by the stator as shown in Fig. 1.5A. Alternatively, we can consider the operating principle of a DC motor from the viewpoint of two magnetic fields as shown in Fig. 1.4A as follows.

There are two stationary magnetic fields in a DC motor as shown in Fig. 1.5B. One stationary magnetic field is the stator magnetic field produced by magnets or a field winding. The other is the rotor magnetic field produced by the current in the conductors of the rotor. It is important to note that the rotor magnetic field is also stationary despite the rotation of the rotor. This is due to the action of brushes and commutators, by which the current distribution in the rotor conductors is always made the same regardless of the rotor's rotation as shown in Fig. 1.5A. Thus the rotor magnetic field will not rotate along with the rotor. A consistent interaction between these two stationary magnetic fields produces a torque, which causes the rotor to turn continuously. We will study the DC motor in more detail in Chapter 2.

1.1.2.2 Alternating current motor

Unlike DC motors that rotate due to the force between two stationary magnetic fields, AC motors exploit the force between *two rotating magnetic fields*. In AC motors both the stator magnetic field and the rotor magnetic field rotate, as shown in Fig. 1.6.

As it will be described in more detail in Chapter 3, these two magnetic fields always rotate at the same speed and, thus, are at a standstill relative to each other and maintain a specific angle. As a result, a constant force is produced between them, making the AC motor is to run continually. The operating principle of the

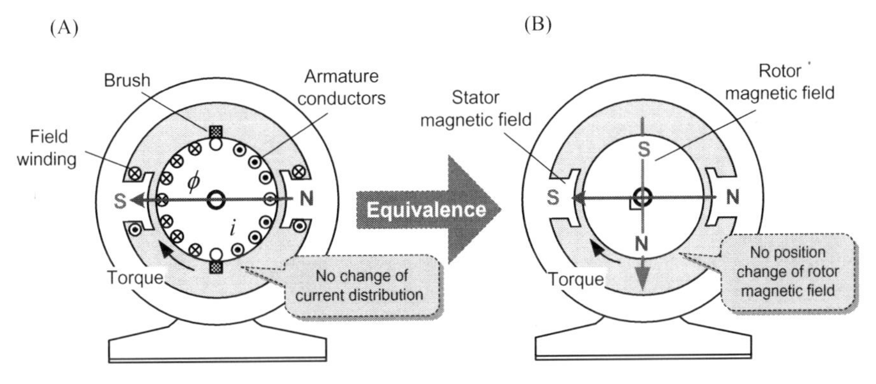

FIGURE 1.5

Operating principle of the DC motor. (A) Rotor current and stator magnetic field and (B) two magnetic fields.

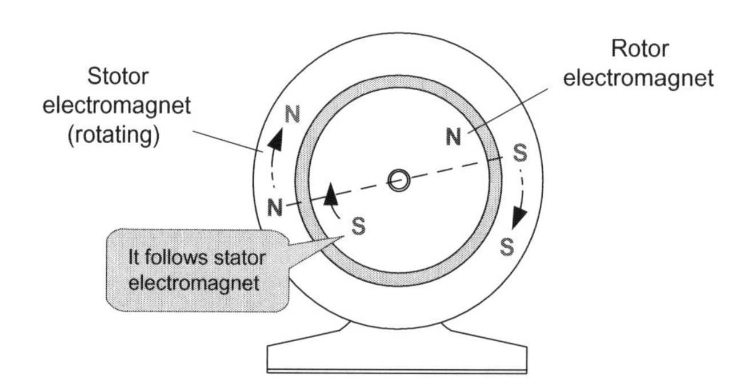

FIGURE 1.6

Operating principle of the AC motor.

AC motor is that the force produced by the interaction between the two rotating magnetic fields causes the rotor to turn.

In AC motors, the rotating magnetic field on the stator is created by three-phase currents. When a three-phase AC power source is applied to the three-phase stator windings of the AC motor, the three-phase currents flowing in these windings create a rotating magnetic field. We will examine the rotating magnetic field in more detail in Chapter 3.

There are two kinds of AC motors: a *synchronous motor* and an *induction (asynchronous) motor*. They generate the rotor magnetic field differently, whereas they generate the stator magnetic field in the same way. In a synchronous motor as depicted in Fig. 1.2B, the magnetic field on the rotor is generated either by a permanent magnet or by a field winding powered by a DC power supply separated from the stator AC power source. In this motor, the rotor magnetic field is stationary relative to the rotor. Hence, to produce a torque, the rotor should rotate at the same speed as the stator rotating magnetic field. This speed is called the *synchronous speed*. This is why this motor is referred to as the *synchronous motor*.

On the other hand, in an induction motor as shown in Fig. 1.2C, the rotor magnetic field is generated by the AC power. The AC power used in the rotor excitation is transferred from the stator by electromagnetic induction. Because of this crucial feature, this motor is referred to as the *induction motor*. In an induction motor, the rotor magnetic field rotates relative to the rotor at some speed. To produce a torque, the stator and the rotor rotating magnetic fields should rotate at the same speed. This requires that the rotor itself rotate at the speed difference between the stator and rotor rotating magnetic fields. More precisely, the rotor rotating magnetic field rotates at the speed difference between the stator rotating magnetic field and the rotor. To use the rotor excitation by electromagnetic induction, the rotor speed should always be less than the synchronous speed. Thus the induction motor is also called the *asynchronous motor*.

Among the motors, DC motors have largely been used for speed and torque control because of their simplicity. Their simplicity comes from the fact that the speed of a DC motor is proportional to the voltage, and its torque is proportional to the current. However, since DC motors require periodic maintenance of the brushes and commutators, the trend has recently moved toward employing maintenance-free AC motors as they can offer high performance at a reasonable price.

As mentioned earlier, electric motors can operate on the fundamental principle that the torque produced from the interaction between the magnetic fields generated in the stator and the rotor causes the motor to run. Now, we will take a look at the requirements that ensure continuous torque production by the motor.

1.2 REQUIREMENTS FOR CONTINUOUS TORQUE PRODUCTION [3]

An electric motor is a type of electromechanical energy conversion device that converts electric energy into mechanical energy. The principle of torque production in a motor can be understood by analyzing its energy conversion process. Electromechanical energy conversion devices normally use the magnetic field as an intermediate in their energy conversion processes. Thus an electromechanical energy conversion device is composed of three different parts: the *electric system*, *magnetic system*, and *mechanical system* as shown in Fig. 1.7.

Now, we will examine the mechanism of torque production of a motor by looking at how the force (or torque) is being produced inside the electromechanical energy conversion device.

To evaluate the force (or torque) produced inside the energy conversion devices, we will apply the law of conservation of energy: "within an isolated system, energy can be converted from one kind to another or transferred from one place to another, but it can neither be created nor destroyed." Therefore the total amount of energy is constant. Fig. 1.8 shows an application of the law of conservation of energy when the electromechanical energy conversion device is acting as a motor.

Suppose that electric energy dW_e is supplied to the energy conversion device during the differential time interval dt. The supplied electric energy dW_e is converted to field energy dW_f in the magnetic system, and to mechanical energy dW_e in the mechanical system.

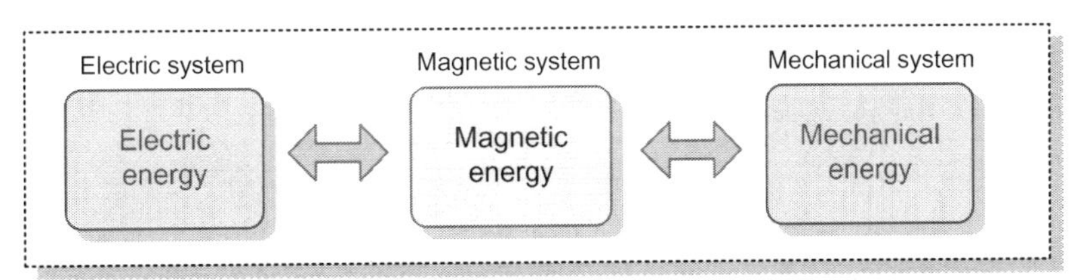

FIGURE 1.7

Electromechanical energy conversion.

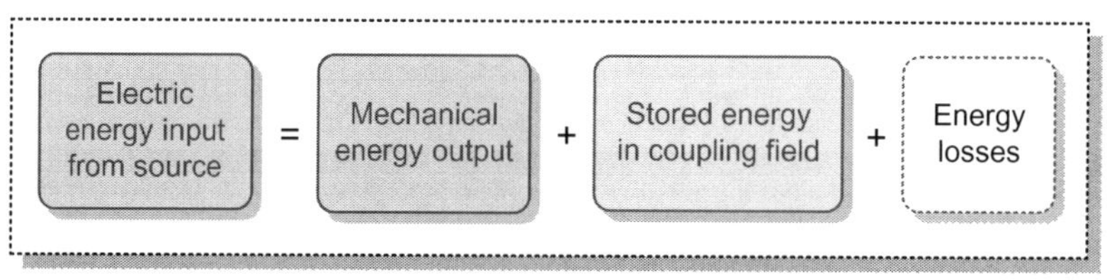

FIGURE 1.8

Equation of energy conservation for a motor.

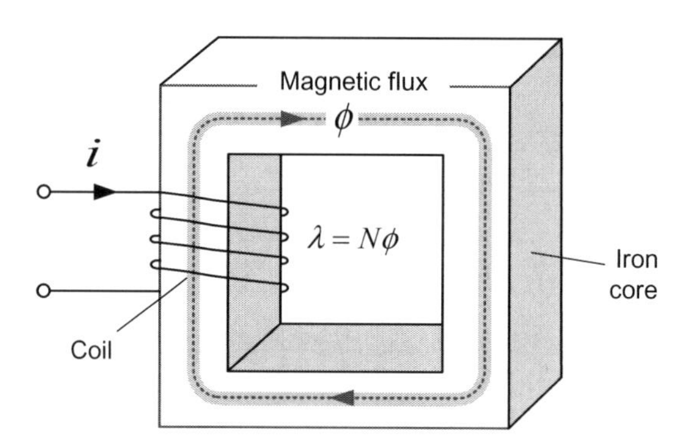

FIGURE 1.9

Coil wound on an iron core.

In addition a small amount of electric energy dW_e is converted into losses in each of the three parts: copper losses in an electric system, core losses such as hysteresis and eddy current losses in a magnetic system, and mechanical losses such as friction and windage losses. Since these losses may be dissipated in the form of heat or noise, we can leave out these losses in the process of evaluating the force or torque production in the energy conversion process. Therefore the energy conversion equation in Fig. 1.8 may be written as

$$dW_e = dW_f + dW_m \tag{1.1}$$

1.2.1 MAGNETIC ENERGY

In electric machines, including a motor, the magnetic flux is normally produced by the coil wound on an iron core as shown in Fig. 1.9. Because ferromagnetic materials such as an iron core have high permeability, much greater magnetic flux can be developed in the iron core for a given current than when the magnetic flux is developed in the air. Now, we will take a look at magnetic energy W_f stored in the magnetic system, where magnetic flux ϕ is developed in the iron core as shown in Fig. 1.9.

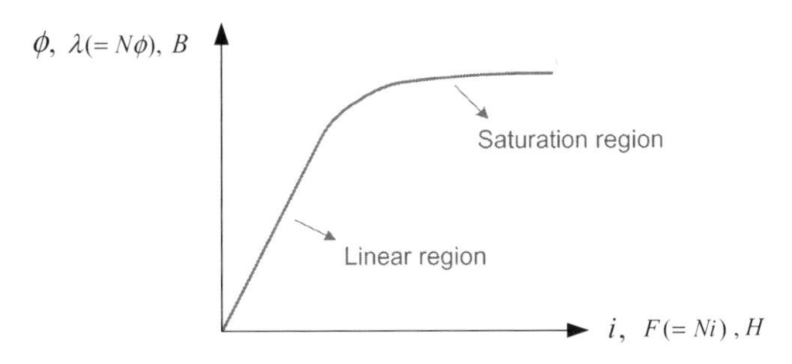

FIGURE 1.10

Flux–current characteristic.

First, we need to know the relationship between the magnetic flux ϕ produced in the core and the coil current i in this magnetic system. Fig. 1.10 demonstrates this relationship, which is the B-H characteristic called *magnetization curve*. In the beginning of the curve, the magnetic flux ϕ increases rapidly in proportion to the increase in the coil current i until it reaches a certain value. The region where the magnetic flux is linearly related to the applied current like this is called the *linear region* or *unsaturated region*.

However, above that certain value, further increases in the current produce relatively smaller increases in the magnetic flux. Eventually an increase in the current will produce almost no change at all in the magnetic flux. The region where the curve flattens out is called the *saturation region*, and the iron core is said to be saturated. When the saturation occurs, any further increase in the current will have little or no effect on the increase of the flux. In electric machines, the magnetic flux is designed to be produced as much as possible because their torque is directly proportional to the magnetic flux. Thus most electric machines normally operate near the knee of the magnetization curve, which is the transition range between the two regions.

Electric machines produce the magnetic flux by using coils, which normally consist of many turns. Thus, instead of the magnetic flux ϕ, the concept of flux linkage λ as a product of the number of turns N and the magnetic flux ϕ linking each turn is introduced. The flux linkage of N-turn coil can be given by

$$\lambda = N\phi \quad \text{(Wb-turns)} \tag{1.2}$$

The flux linkage λ can also be related to the coil current i by the definition of inductance L through the following relation.

$$L = \frac{\lambda}{i} = N\frac{\phi}{i} \quad \text{(H or Wb/A)} \tag{1.3}$$

A larger inductance value of the magnetic system implies that the same current can produce a larger flux linkage. Eq. (1.3) shows that the slope of the curve in Fig. 1.10 is the inductance of this magnetic system as shown in Fig. 1.11.

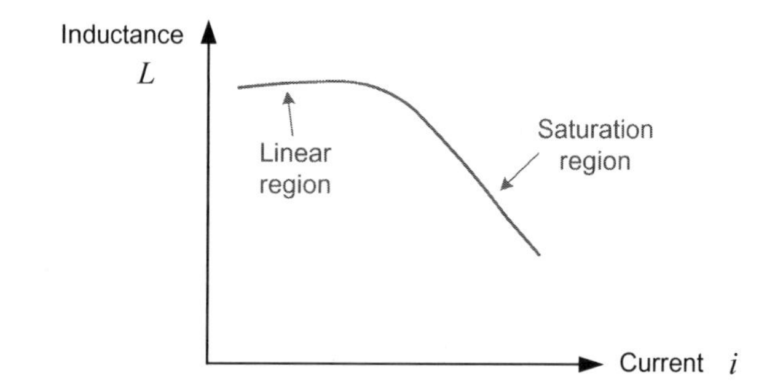

FIGURE 1.11

Inductance of the coil.

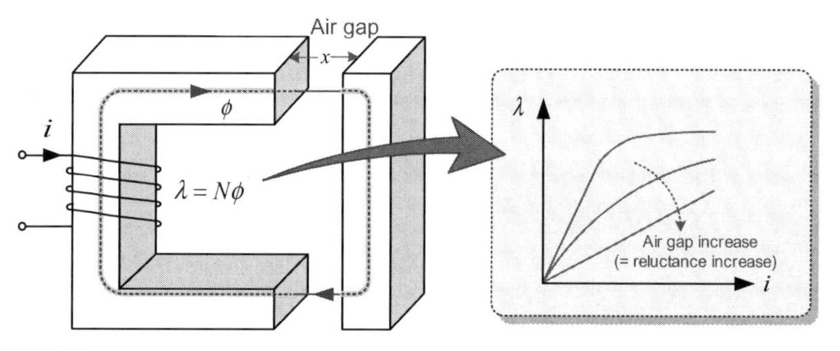

FIGURE 1.12

Flux–current characteristic according to the variation of an air-gap length.

The inductance is large and relatively constant in the unsaturated region but gradually drops to a very low value as the core becomes heavily saturated.

Now, we will take a look at magnetic energy W_f stored in the magnetic system, where magnetic flux ϕ is developed in the iron core as shown in Fig. 1.12.

The $\lambda - i$ characteristic of the magnetic system varies with the air-gap length x. Thus the inductance $L(x)$ of the coil is expressed as a function of the air-gap length x. As the air-gap length x increases, the slope of the $\lambda - i$ curve becomes smaller and results in smaller inductance as shown in Fig. 1.12.

The inductance L of the magnetic system is inversely related to its *reluctance* $\Re$, which is defined as the ratio of *magnetomotive force* (mmf) $F(=Ni)$ to magnetic flux ϕ, as in the following relation.

$$L = \frac{\lambda}{i} = \frac{N\phi}{i} = \frac{N^2}{\Re} \propto \frac{1}{\Re} \quad (F = \Re\phi = Ni) \tag{1.4}$$

Now, assume that the source voltage v is applied to the terminals of the N-turn coil on the iron core during dt as shown in Fig. 1.13. Then, the current i

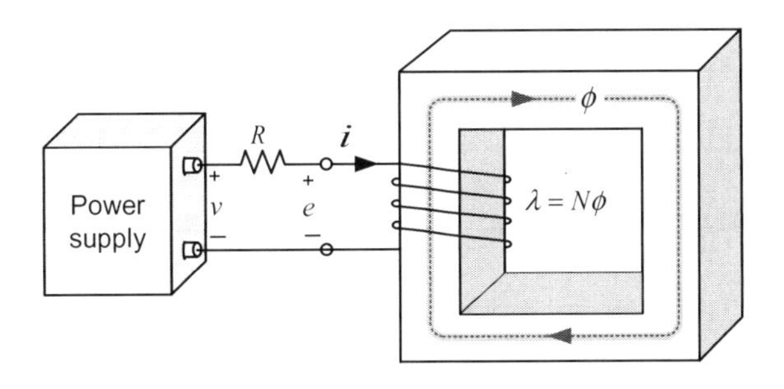

FIGURE 1.13

Magnetic system connected to source voltage.

will flow in that coil. The resulting mmf $F(=Ni)$ will produce a magnetic flux ϕ in the core and establish flux linkage $\lambda(=N\phi)$ in the coil. Now, we will discuss the energy stored in the magnetic system.

MAGNETOMOTIVE FORCE (MMF) AND MAGNETIC FLUX

Magnetic flux ϕ is indispensable in producing torque in motors. The force that is capable of producing a magnetic flux is called magnetomotive force or mmf F. In general, mmf is generated by current, so the source of magnetic flux is the current. When the current I is flowing through an N-turn coil, the mmf F is expressed by NI (A-turn), i.e., $F = NI$. For a given mmf, the magnitude of the developed flux depends on the length of the closed-path through which the flux passes. The longer the length, the lower the flux density will be. Thus the magnetic field intensity H, which represents mmf per length, will be used instead of the mmf.

$$mmf = F = NI = \oint H dl \rightarrow H = \frac{mmf}{l} = \frac{NI}{l} \quad \text{(A/m)}$$

Also, for a given magnetic field intensity, the flux density will be different depending on the material through which the flux passes. For example, high flux density can be created in iron or steel but not in air gap as in the following figure.

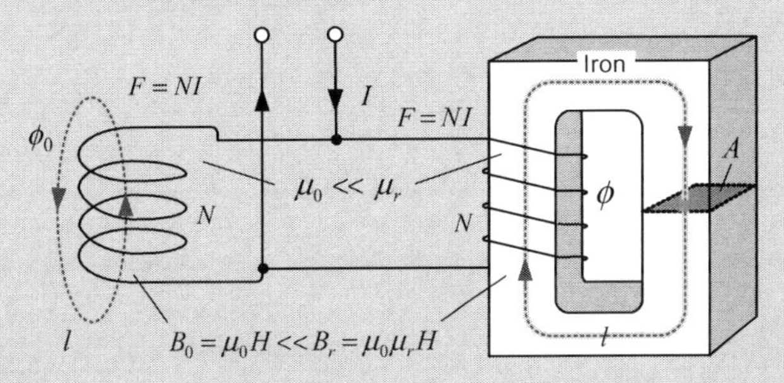

This magnetic characteristic can be described in terms of relative permeability μ_r in reference to permeability μ_0 of free space. Thus the magnetic flux density B produced by the current can be represented as

$$B = \mu_0 \mu_r H \quad \text{(Wb/m}^2\text{)}$$

(Continued)

MAGNETOMOTIVE FORCE (MMF) AND MAGNETIC FLUX (CONTINUED)

The total magnetic flux flowing through the cross section A of a magnetic circuit in an uniform magnetic flux density B can be expressed as

$$\phi = \int_A B \cdot ndA = BA\cos\theta \quad \text{(Wb)}$$

where θ is the angle between the magnetic field lines and the surface A. Similar to the relation between the current and the voltage by Ohm's law in the electric circuit, there is a following relationship between the mmf F and the flux ϕ in the magnetic circuit.

$$F = \Re\phi$$

where $\Re$ is called reluctance and is inversely proportional to permeability.

In the system of Fig. 1.13, the electric energy dW_e supplied during the differential time interval is expressed from input power $P(=vi)$ as

$$dW_e = Pdt = vidt \tag{1.5}$$

The supplied electric energy after subtracting the copper loss produced by current in the coil is stored as the magnetic field energy. i.e.

$$dW_f = (vi - Ri^2)dt = (vi - Ri)dt = eidt \tag{1.6}$$

where e is back-EMF voltage induced in the coil.

Since the back-EMF voltage e on the N-turn coil is proportional to the rate of change of flux linkage λ with time, i.e., $e = d\lambda/dt$, Eq. (1.5) can be given by

$$dW_f = eidt = id\lambda \tag{1.7}$$

When the flux linkage is increased from zero to λ, the total energy stored in the magnetic system is

$$dW_f = \int_0^{\lambda} id\lambda \tag{1.8}$$

This magnetic energy is the area to the left of the $\lambda - i$ curve as shown in Fig. 1.14. On the other hand, the area to the right of $\lambda - i$ curve is known as *coenergy* W'_f, which is a useful quantity for the calculation of force and can be expressed as

$$W'_f = \int_0^{\lambda} \lambda di \tag{1.9}$$

For the linear system in which the $\lambda - i$ curve is a straight line, the magnetic energy is equal to the coenergy. The magnetic system that has an air gap in the flux path becomes a linear system.

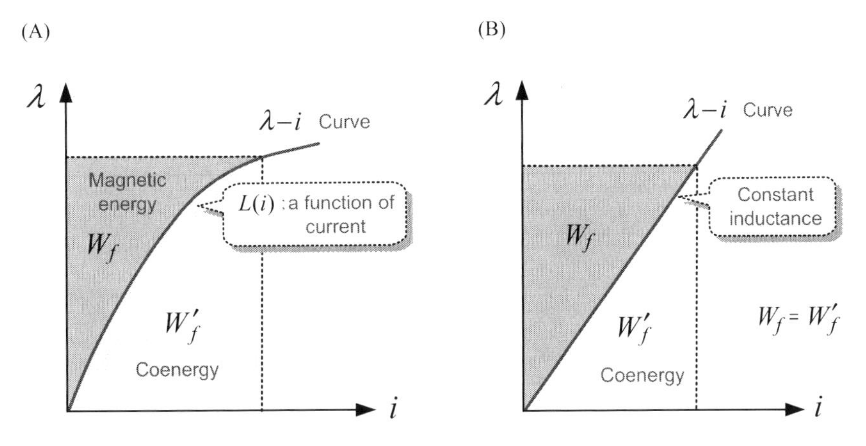

FIGURE 1.14

Magnetic energy in the flux–current characteristic. (A) Linear system and (B) nonlinear system.

BACK-ELECTROMOTIVE FORCE (BACK-EMF), OR INDUCED VOLTAGE *e*

Faraday's law states that the voltage e induced in the turn of a coil is directly proportional to the rate of change in the flux ϕ passing through that turn with respect to time. If there are N turns in the coil and the same flux passes through each turn of the coil, then the total voltage induced on the coil is given by

$$e = N\frac{d\phi}{dt} = \frac{d\lambda}{dt}$$

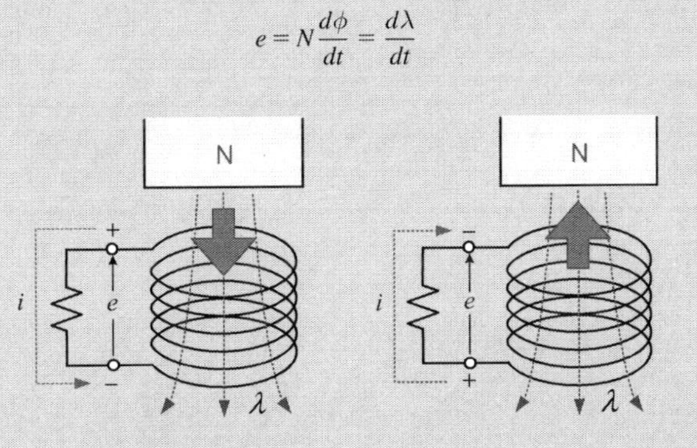

According to Lenz's law, the direction of the induced voltage opposes the change of the flux that causes it, and thus a minus sign is included in the equation above.

1.2.2 LINEAR MOTION DEVICE

Now, we will look at the force developed inside a movable energy conversion device. Before looking into rotating machines, we will first study the linear motion device, which consists of a fixed part (also called *stator*) and a moving part as shown in Fig. 1.15.

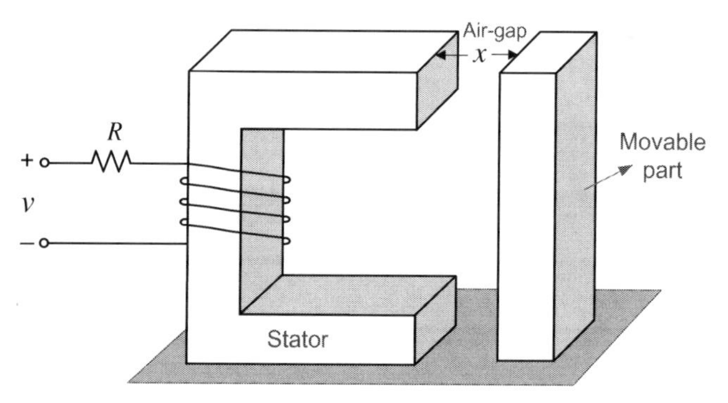

FIGURE 1.15

Linear motion device.

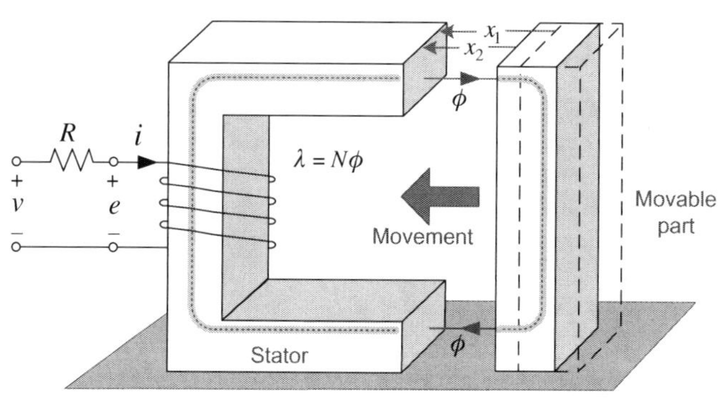

FIGURE 1.16

Movement of the movable part.

If a magnetic flux is produced by the current flowing through the coil of N turns in this device, then, as seen in everyday experiences, we can easily know that the movable part will move toward the stator as shown in Fig. 1.16. Now, we will discuss the force that causes this motion.

The current I of the coil is assumed to remain the same before and after the movable part moves. Suppose that the movable part moves from position x_1 to position x_2. The length of the air gap is changed as a result of the movement. This also changes the $\lambda - i$ characteristic of the system as shown in Fig. 1.17. The air gap decreases according to the movement, resulting in a decrease of reluctance so that the flux linkage increases from λ_1 to λ_2. The operating point will move from A to B. The trajectory of the movement depends on the moving speed. So, we will find the force acting on the movable part for the two extreme moving speeds.

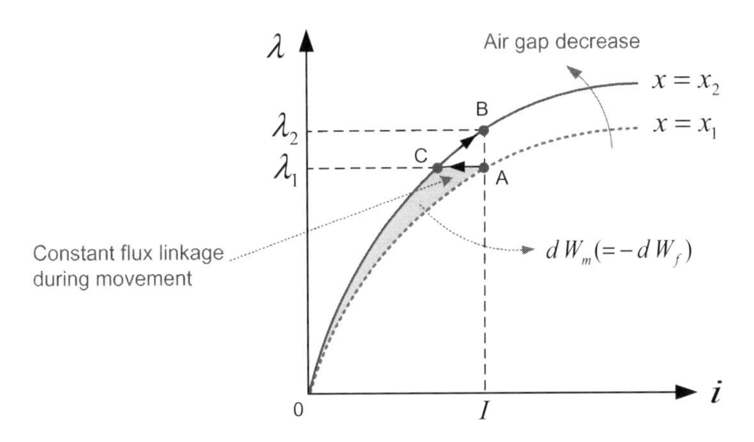

FIGURE 1.17

Flux linkage–current characteristic (rapid movement of the movable part).

First, suppose that the movement occurs very rapidly. In this case, the movement is so rapid that it may be completed before any significant change has occurred in flux linkage λ. Thus, during the movement, the flux linkage λ_1 remains constant so that the operating point moves from A to C as shown in Fig. 1.17.

After the completion of the movement, the flux linkage will be changed from λ_1 to λ_2 and the operating point will move from C to B. During the movement, the flux linkage does not change and this results in zero back-EMF, so the electrical energy input will be zero, i.e. $dW_e = eidt = id\lambda = 0$. Therefore Eq. (1.1) can be written as

$$dW_m = -\,dW_f \tag{1.10}$$

This equation indicates that the mechanical energy needed for the movement is supplied entirely by the magnetic energy. The amount of the reduced magnetic energy for the movement corresponds to the shaded area OAC of the $\lambda - i$ characteristic.

Considering the force to be defined as the mechanical work done dW_m per displacement dx, the developed force during this movement from Eq. (1.10) is

$$f_m = \frac{dW_m}{dx} = -\left.\frac{dW_f(i,x)}{dx}\right|_{\lambda=\text{constant}} \tag{1.11}$$

This implies that the force acts in the direction at which the magnetic energy of the system is decreasing.

Now, we will express the force of Eq. (1.11) in terms of the coil current. Assume for simplicity that the $\lambda - i$ characteristic is linear. The magnetic energy of Eq. (1.8) can be expressed as

$$W_f = \int i d\lambda = \int \frac{\lambda}{L(x)} d\lambda = \frac{\lambda^2}{2L(x)} = \frac{1}{2}L(x)i^2 \tag{1.12}$$

By substituting Eq. (1.12) into the force of Eq. (1.11), we obtain

$$\begin{aligned} f_m &= -\left.\frac{dW_f(i,x)}{dx}\right|_{\lambda=\text{constant}} = -\frac{d}{dx}\left(\frac{\lambda^2}{2L(x)}\right)\Bigg|_{\lambda=\text{constant}} \\ &= \frac{\lambda^2}{2L^2(x)}\frac{dL(x)}{dx} = \frac{1}{2}i^2\frac{dL(x)}{dx} \end{aligned} \tag{1.13}$$

Eq. (1.18) indicates that *the force acts in the direction that increases the inductance of the magnetic system.* Since the direction that increases the inductance is the direction that decreases the length of the air gap, the force is exerted on the movable part moving toward the stator.

From this, we can explain the attractive force between an electromagnet (or magnet) and magnetic materials. When an electromagnet and a magnetic material cling to each other without an air gap, the system has the greatest inductance. Thus an attractive force between the two objects is developed to move them toward each other.

Next, suppose that the movable part moves very slowly. In this case, the current $i=(v-e)/R$ remains constant during the movement. This is because the back-EMF $e(=d\lambda/dt)$ is negligibly small. Therefore the operating point on the $\lambda-i$ characteristic moves upward from A to B as shown in Fig. 1.18.

During this movement the change in the electric energy is given by

$$dW_e = eidt = id\lambda = i(\lambda_2-\lambda_1) \tag{1.14}$$

This corresponds to the area $ABEF$. In this case, the amount of change in the stored magnetic energy from Eq. (1.8) is given by

$$dW_f = \int_0^{\lambda_2} id\lambda - \int_0^{\lambda_1} id\lambda = \text{Area } OBE - \text{Area } OAF \tag{1.15}$$

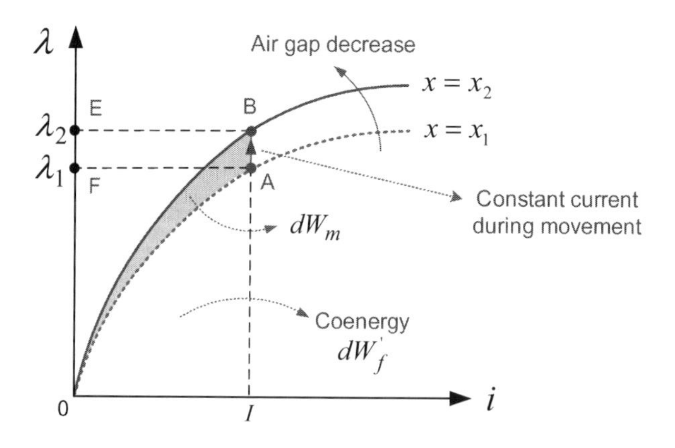

FIGURE 1.18

Flux linkage–current characteristic (very slow movement of the movable part).

Therefore, from Eq. (1.1), the mechanical energy required for the movement is

$$dW_m = dW_e - dW_f \tag{1.16}$$

By substituting Eqs. (1.14) and (1.15) into Eq. (1.16), it can be readily seen that the mechanical energy corresponds to the shaded area *OAB* of the $\lambda - i$ characteristic and is equal to the increments of coenergy dW'_f. Consequently, the used mechanical energy for the movement is equal to the increments of coenergy dW'_f during the movement as shown in the following.

$$dW_m = dW'_f \tag{1.17}$$

Therefore the developed force during this movement is

$$f_m = \frac{dW_m}{dx} = \left.\frac{dW'_f(i,x)}{dx}\right|_{i=\text{constant}} \tag{1.18}$$

This force acts in the direction of the increasing coenergy of the system.

The force for a slow movement is expressed in the same way as for a rapid movement. For a linear system $dW_f = dW'_f$, by substituting Eq. (1.12) into (1.18), we obtain

$$\begin{aligned} f_m &= \left.\frac{dW'_f(i,x)}{dx}\right|_{i=\text{constant}} = \left.\frac{dW_f(i,x)}{dx}\right|_{i=\text{constant}} \\ &= \left.\frac{d}{dx}\left(\frac{1}{2}L(x)i^2\right)\right|_{i=\text{constant}} = \frac{1}{2}i^2\frac{dL(x)}{dx} \end{aligned} \tag{1.19}$$

This is the same conclusion as was reached in Eq. (1.13), meaning that the force is the same regardless of the moving speed.

This force can also be expressed in terms of the *reluctance* $\mathfrak{R}$. The magnetic energy of Eq. (1.8) expressed as a function of reluctance $\mathfrak{R}$ is

$$W_f = \int_0^\lambda i d\lambda = \int_0^\phi F d\phi = \int_0^\phi \mathfrak{R}\phi d\phi = \frac{1}{2}\mathfrak{R}(x)\phi^2 \tag{1.20}$$

From Eq. (1.20), the force of Eq. (1.11) is

$$f_m = -\left.\frac{d}{dx}\left(\frac{1}{2}\mathfrak{R}(x)\phi^2\right)\right|_{\phi=\text{constant}} = -\frac{1}{2}\phi^2\frac{d\mathfrak{R}(x)}{dx} \tag{1.21}$$

This means that *the force acts in the direction that decreases the reluctance of the magnetic system.* Since the decrease in the air-gap length causes the reluctance to decrease, the force is exerted on the movable part moving toward the stator. Next, we will discuss the torque production in rotating machines.

1.2.3 ROTATING MACHINE

We will first start examining the machine in Fig. 1.19, which produces a rotational motion. The rotating machine consists of a fixed part (called *stator*) and a moving part (called *rotor*).

The rotor is mounted on a shaft and is free to rotate between the poles of the stator.

The force that causes a rotor to rotate can be expressed as torque, which is the mechanical work done per rotational distance or angle θ as follows:

$$T = \frac{dW_m}{d\theta} \tag{1.22}$$

In the machine shown in Fig. 1.18, when current i_s flows in the stator coil, a magnetic flux is produced, and the developed torque from Eqs. (1.12) and (1.22) acting on the rotor can be expressed as

$$T = \frac{1}{2} i_s^2 \frac{dL(\theta)}{d\theta} = -\frac{1}{2}\phi^2 \frac{d\Re(\theta)}{d\theta} \tag{1.23}$$

This torque results from the variation of reluctance (or inductance) with rotor position, and thus is called the *reluctance torque*. A machine using this torque is known as the *reluctance motor*. For motors of a cylindrical rotor configuration, since the reluctance does not vary with the rotor position, the reluctance torque cannot be produced.

Now, we will take a look at the necessary condition for ensuring that this machine continuously rotates to serve as a motor. First of all, it is not hard to consider that, if there is a DC current flowing in the coil, this machine never operates as a continuously rotating motor. Therefore we will consider the case in which the winding is excited from an AC current $i_s(=I_m \cos\omega_s t)$. In this case, the torque of Eq. (1.23) can be expressed in terms of the stator current i_s and the stator self-inductance L_{ss} as

$$T = \frac{1}{2} i_s^2 \frac{dL_{ss}(\theta)}{d\theta} \tag{1.24}$$

where the inductance $L_{ss}(=\lambda_s/i_s)$ is defined as the ratio of the total flux linkage λ_s in the stator coil to the stator current i_s generating the flux.

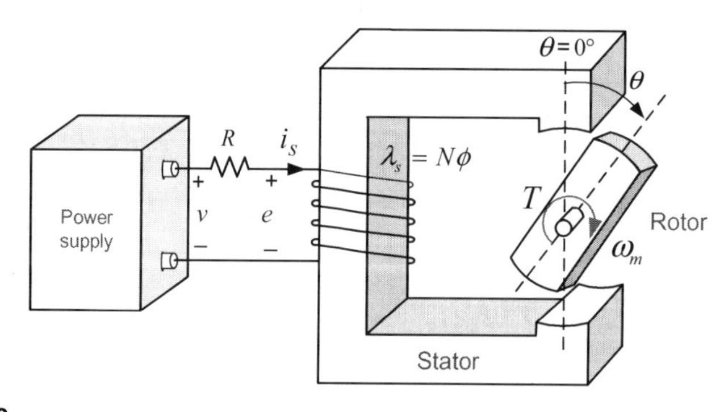

FIGURE 1.19

Rotating machine.

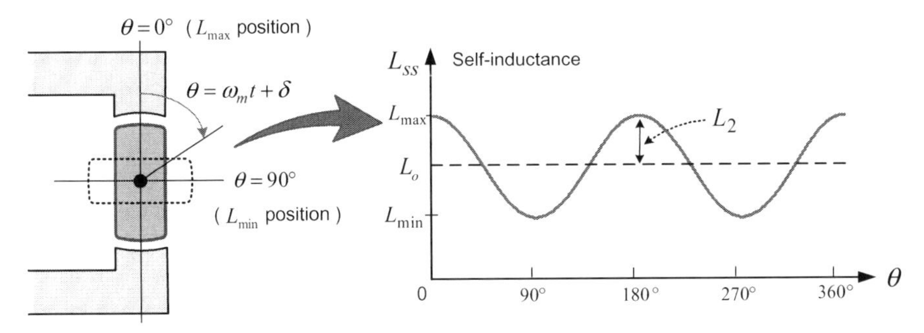

FIGURE 1.20

Self-inductance L_{ss} profile with respect to the rotor position.

The self-inductance L_{ss} in the machine of Fig. 1.19 varies with the angular position θ of the rotor as shown in Fig. 1.20.

There are two cycles of inductance during one revolution of a rotor. When $\theta = 0°$, the air gap is the smallest and the inductance L_{ss} becomes the maximum value $L_{\max}$. When $\theta = 90°$, the air gap is the largest and the inductance becomes the minimum value $L_{\min}$. In this way, the stator self-inductance L_{ss} varies with the rotor position θ in a sinusoidal manner and can be expressed as

$$L_{ss}(\theta) = L_0 + L_2 \cos 2\theta \tag{1.25}$$

where $L_0 = (L_{max} + L_{min})/2, \quad L_2 = (L_{max} - L_{min})/2.$

Substitution of Eq. (1.25) into Eq. (1.24) yields

$$T = \frac{1}{2} i_s^2 \frac{dL_{ss}(\theta)}{d\theta} = -I_m^2 L_2 \sin 2\theta \cos^2 \omega_s t \tag{1.26}$$

If the rotor is rotating at a constant angular velocity ω_m, then the angular position θ of the rotor can be given by

$$\theta = \omega_m t + \delta \tag{1.27}$$

where δ is the initial angular position of the rotor.

SELF-INDUCTANCE AND MUTUAL-INDUCTANCE

The inductance L of a coil is defined as the flux linkage per ampere of current in the coil.

- Self-inductance: Self-inductance is the flux linkage produced in the winding by the current in that same winding divided by that current.

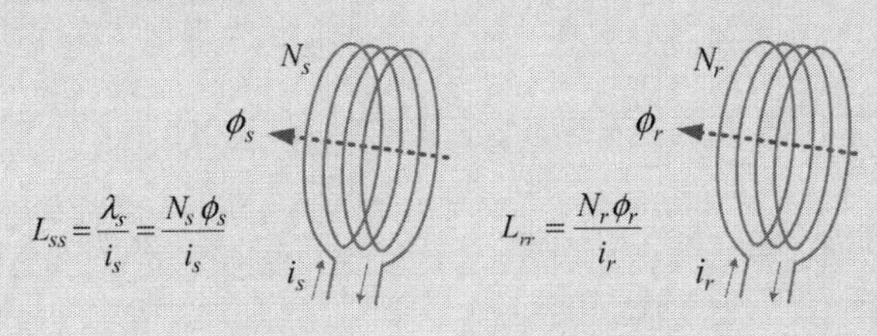

(Continued)

SELF-INDUCTANCE AND MUTUAL-INDUCTANCE (CONTINUED)

- Mutual-inductance: Mutual-inductance is the flux linkage produced in one winding by the current in the other winding divided by that current.

$$L_{sr} = \frac{N_s \phi_{sr}}{i_r}, \quad L_{rs} = \frac{N_r \phi_{rs}}{i_s}$$

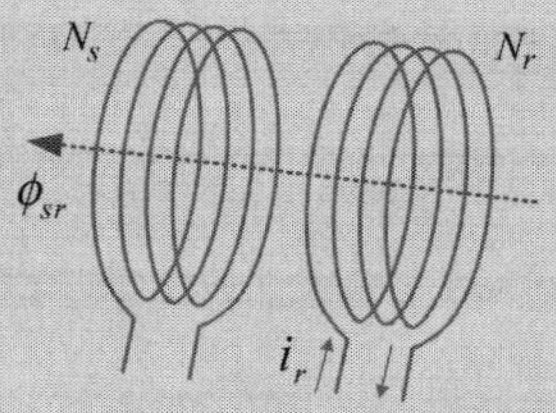

Also, we consider ω_m to be positive when the position θ is increasing (clockwise motion). Substitution of Eq. (1.27) into Eq. (1.26) yields

$$\begin{aligned} T &= -I_m^2 L_2 \sin 2(\omega_m t + \delta)\frac{1+\cos 2\omega_s t}{2} \\ &= -\frac{1}{2}I_m^2\left\{L_2 \sin 2(\omega_m t + \delta) + \frac{1}{2}\sin 2([\omega_m + \omega_s]t + \delta) + \frac{1}{2}\sin 2([\omega_m - \omega_s]t + \delta\right\} \end{aligned} \tag{1.28}$$

The three sin terms on the right side of Eq. (1.28) are functions of time t whose average value is zero. The average value of this torque is, therefore, zero.

This means that a constant torque cannot be developed in this machine for any direction. Therefore this machine cannot operate as a continuously rotating motor. However, if $\omega_m = 0$ or $\omega_m = \pm\omega_s$, then a nonzero average torque can be produced. The case of $\omega_m = 0$ will be excluded since this indicates nonrotating. Consequently the requirement for a continuous rotation is $\omega_m = \omega_s$. This requires that the rotor must rotate at the speed equal to the angular frequency ω_s of the exciting current, which is also called *synchronous speed.* In this case this machine can operate as a continuously rotating motor and its average torque becomes

$$T_{avg} = -\frac{1}{4}I_m^2 L_2 \sin 2\delta = -\frac{1}{8}I_m^2(L_{max} - L_{min})\sin 2\delta \tag{1.29}$$

The average developed torque depends on the inductance difference $(L_{max} - L_{min})$ and the initial rotor position δ. The maximum torque occurs when $\delta = -45°$. Since the torque in this motor caused by the variation of reluctance (or inductance) with rotor position can be developed only at the synchronous speed, this motor is called *synchronous reluctance motor.*

Next, we will examine the necessary condition for a continuous rotation of the *doubly fed machine* in which both the stator and the rotor have windings carrying currents as shown in Fig. 1.21. Typical electric motors such as a DC motor, an induction motor, and a synchronous motor belong to the doubly fed machine.

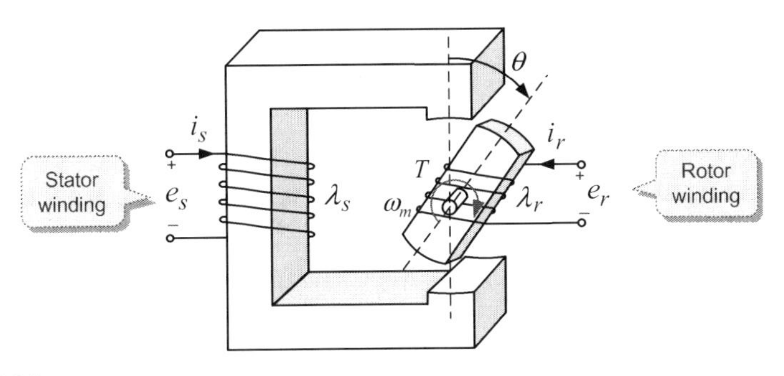

FIGURE 1.21

Doubly fed machine.

Modified from G.R. Slemon, A. Straughen, Electric Machines, Addison-Wesley, 1980 (Chapter 3); P.C. Sen, Principles of Electric Machines and Power Electronics, John Wiley & Sons, 1996.

Assume that this machine is a linear system, in which the magnetic energy and the coenergy are the same, i.e., $W_f = W'_f$. Similar to the force of Eq. (1.19), the torque of this machine can be obtained from

$$T = \left.\frac{dW'_f(i,\theta)}{d\theta}\right|_{i=\text{constant}} = \left.\frac{dW_f(i,\theta)}{d\theta}\right|_{i=\text{constant}} \tag{1.30}$$

To evaluate this torque, we need to obtain the magnetic energy W_f in this machine. When the magnetic field is established by winding currents i_s and i_r, the stored magnetic energy W_f can be easily derived by locking the rotor at an arbitrary position, which makes it produce no mechanical output. If there is no mechanical output, then the magnetic field energy increment during dt is

$$dW_f = dW_e = e_s i_s dt + e_r i_r dt = i_s d\lambda_s + i_r d\lambda_r \tag{1.31}$$

Here, the stator flux linkage λ_s consists of λ_{ss} produced by the stator current i_s and λ_{sr} produced by the rotor current i_r. This stator flux linkage can be expressed as functions of currents and inductances of windings as

$$\lambda_s = \lambda_{ss} + \lambda_{sr} = L_{ss} i_s + L_{sr} i_r \tag{1.32}$$

where $L_{ss}(=\lambda_{ss}/i_s)$ is the self-inductance of the stator winding, and $L_{sr}(=\lambda_{sr}/i_r)$ is the mutual-inductance between the stator and rotor windings.

The flux linkage λ_r in the rotor winding also consists of λ_{rr} produced by the rotor current i_r and λ_{rs} produced by the stator current i_s. This rotor flux linkage λ_r is expressed by

$$\lambda_r = \lambda_{rr} + \lambda_{rs} = L_{rr} i_r + L_{rs} i_s \tag{1.33}$$

where $L_{rr}(=\lambda_{rr}/i_r)$ is the self-inductance of the rotor winding, and $L_{rs}(=\lambda_{rs}/i_s)$ is the mutual-inductance between the stator and rotor windings. For a linear magnetic system, $L_{sr} = L_{rs}$.

In the doubly fed machine in Fig. 1.21, both self-inductances L_{ss}, L_{rr} and mutual-inductances L_{sr}, L_{rs} vary with the rotor position θ. For example, when $\theta = 0°$, at which the length of the air gap is the minimum, the reluctance of the system is the minimum and the inductance becomes the largest. Thus λ_s and λ_r are functions of rotor position θ.

Substitution of Eqs. (1.32) and (1.33) into Eq. (1.31) yields

$$\begin{aligned} dW_f &= i_s d(L_{ss} i_s + i_r dL_{sr} i_r) + i_r d(L_{sr} i_s + i_r dL_{rr} i_r) \\ &= L_{ss} i_s di_s + L_{rr} i_r di_r + L_{sr} d(i_s i_r) \end{aligned} \tag{1.34}$$

Assuming that the rotor current is increased from zero to i_r after the stator current is increased from zero to i_s, the total stored magnetic energy will be

$$\begin{aligned} W_f &= L_{ss} \int_o^{i_s} i_s di_s + L_{rr} \int_0^{i_r} i_r di_r + L_{sr} \int_0^{i_s, i_r} d(i_s i_r) \\ &= \frac{1}{2} L_{ss} i_s^2 + \frac{1}{2} L_{rr} i_r^2 + L_{sr} i_s i_r \end{aligned} \tag{1.35}$$

By substituting Eq. (1.35) into Eq. (1.30), the torque is expressed by

$$\begin{aligned} T &= -\left. \frac{dW_f(i, \theta)}{d\theta} \right|_{i=\text{constant}} \\ &= \frac{1}{2} i_s^2 \frac{dL_{ss}}{d\theta} + \frac{1}{2} i_r^2 \frac{dL_{rr}}{d\theta} + i_s i_r \frac{dL_{sr}}{d\theta} \end{aligned} \tag{1.36}$$

This torque expression is derived in the condition in which the rotor is locked to produce no mechanical output. Even if the rotor is rotating, we can obtain the same torque expression [3].

In Eq. (1.36), the first two terms on the right side represent reluctance torques involving a change of self-inductances according to the rotor position θ. The last term involving a change of mutual-inductance expresses the electromagnetic torque formed by the interaction of two fields produced by the stator and rotor currents i_s and i_r. Typical rotating motors such as a DC motor, an induction motor, and a synchronous motor are based on this torque.

In the previous section, we saw that the reluctance motor must run at the speed equal to the angular frequency of the exciting current for developing an average torque. In the doubly fed machine, if the stator and the rotor currents differ in their operating frequency, then the motor will need to be synchronized at two different speeds for producing an average torque. Only one of the two can be satisfied at a time, and the other torque term will be oscillatory. This will result in an unwanted speed oscillation. However, we can solve this problem by eliminating one or both of the reluctance torque terms in Eq. (1.36).

First, we will eliminate the reluctance torque term involving a change in the stator self-inductance. We can realize this by making the rotor configuration cylindrical as shown in Fig. 1.22.

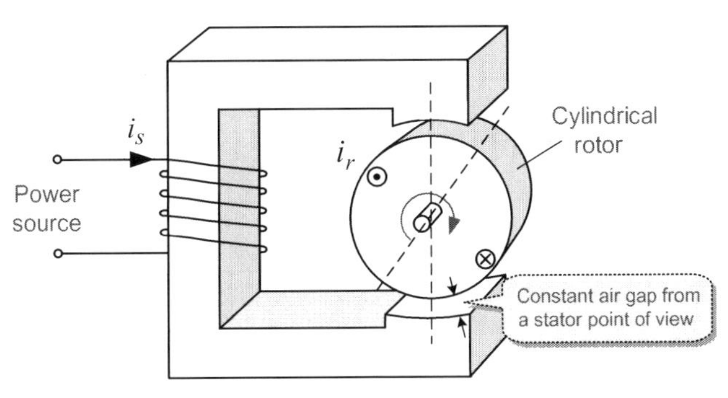

FIGURE 1.22

Cylindrical rotor configuration.

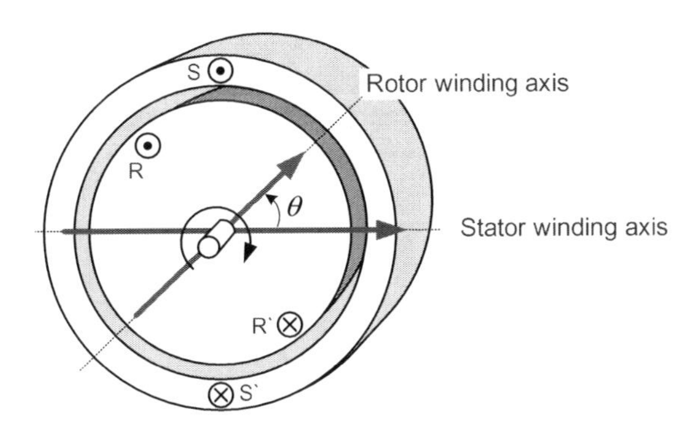

FIGURE 1.23

Cylindrical stator and rotor configurations.

In the cylindrical rotor configuration, the stator self-inductance L_{ss} becomes constant since the magnetic path becomes constant for the flux produced by the stator winding regardless of the rotor position. Thus

$$\frac{dL_{ss}}{d\theta} = 0 \tag{1.37}$$

Therefore the reluctance torque term due to the stator excitation will disappear from Eq. (1.36).

Furthermore, if we make the stator into a hollow cylinder coaxial with the rotor as shown in Fig. 1.23, then the rotor self-inductance L_{rr} becomes constant regardless of the rotor position. Thus

$$\frac{dL_{rr}}{d\theta} = 0 \tag{1.38}$$

This will lead to a disappearance of the second reluctance torque term due to rotor excitation.

We can see that no reluctance torque production results in a cylindrical machine that has both the stator and the rotor in the cylindrical configuration. This cylindrical machine produces only the torque involving a change of mutual-inductance between the stator and the rotor windings as given by

$$T = i_s i_r \frac{dL_{sr}}{d\theta} \tag{1.39}$$

Even in the cylindrical machine, this mutual-inductance can be varied by changing the relative positions of the stator and the rotor windings according to the rotor position.

Now, we will look at the necessary condition for the cylindrical machine to produce a torque continuously. Suppose that the currents in the stator and the rotor windings are

$$i_s = I_{sm}\cos \omega_s t \tag{1.40}$$

$$i_r = I_{rm}\cos(\omega_r t + \alpha) \tag{1.41}$$

If the axes of the two windings are aligned, i.e., when $\theta = 0°$ as in Fig. 1.23, then the linking flux of the two windings will become a maximum, which will make the mutual-inductance its maximum as well. Thus the mutual-inductance according to the position of the rotor can be expressed as

$$L_{sr} = L_m\cos\theta \tag{1.42}$$

where L_m is the maximum mutual-inductance and θ is the angle between the stator and the rotor winding axes. The position of the rotor at instant t with the angular velocity ω_m of the rotor is

$$\theta = \omega_m t + \delta \tag{1.43}$$

where δ is the initial rotor position at $t = 0$.

Substitution of Eqs. (1.40), (1.41) and (1.42) into Eq. (1.39) yields

$$\begin{aligned} T &= i_s i_r \frac{dL_{sr}}{d\theta} \\ &= -I_{sm} I_{rm} L_m \cos\omega_s t \cdot \cos(\omega_r t + \alpha)\cdot \sin(\omega_m t + \delta) \\ &= -\frac{I_{sm} I_{rm} L_m}{4}\left[\sin\{(\omega_m + (\omega_s + \omega_r))t + \alpha + \delta\}\right. \\ &\quad + \sin\{(\omega_m - (\omega_s + \omega_r))t + \alpha + \delta\} \\ &\quad + \sin\{(\omega_m + (\omega_s - \omega_r))t + \alpha + \delta\} \\ &\quad \left. + \sin\{(\omega_m - (\omega_s - \omega_r))t + \alpha + \delta\}\right] \end{aligned} \tag{1.44}$$

Since averaging over one period of the sinusoidal function is zero, the average torque of Eq. (1.44) will also be zero. However, an average torque can be developed if the coefficient of t in sinusoidal terms of Eq. (1.44) is zero. This needs to satisfy the following condition.

$$|\omega_m| = |\omega_s \pm \omega_r| \tag{1.45}$$

Typical electric motors such as a DC motor, an induction motor, and a synchronous motor meet this condition in different ways, and thus, can rotate continuously.

Now, we will take a look at how each of these three motors meets this requirement for the average torque production. First, we will neglect the case in which both the stator and the rotor windings are excited by the DC ($\omega_s = \omega_r = 0$), because Eq. (1.45) results in $\omega_m = 0$, which indicates the state of nonrotating.

1.2.3.1 Direct current motor

If the stator is excited with a DC power ($\omega_s = 0$) and the rotor is excited with an AC power of an angular frequency ω_r, then the necessary condition of Eq. (1.45) for the development of an average torque is

$$|\omega_m| = |\omega_r| \tag{1.46}$$

This implies that the rotor needs to rotate at the same frequency as the frequency ω_r of the rotor current. A DC motor as shown in Fig. 1.24A is the one that satisfies this condition. In the DC motor, by the action of the commutator and brush, a rotor current becomes an AC, whose angular frequency is naturally equal to the rotor speed. Thus, the DC motor has a special mechanical structure, which allows it to always meet the condition of Eq. (1.46). From Eqs. (1.44) and (1.46), the average torque in this motor is

$$T_{avg} = -\frac{I_{sm}I_{rm}}{2}L_m \sin\delta \tag{1.47}$$

In the DC motor, δ, which indicates the angle between the stator field flux and the rotor mmf, is always 90° electric angle structurally, so the maximum torque per ampere can be obtained as

$$T_{avg} = -\frac{L_m}{2}I_{sm}I_{rm} \tag{1.48}$$

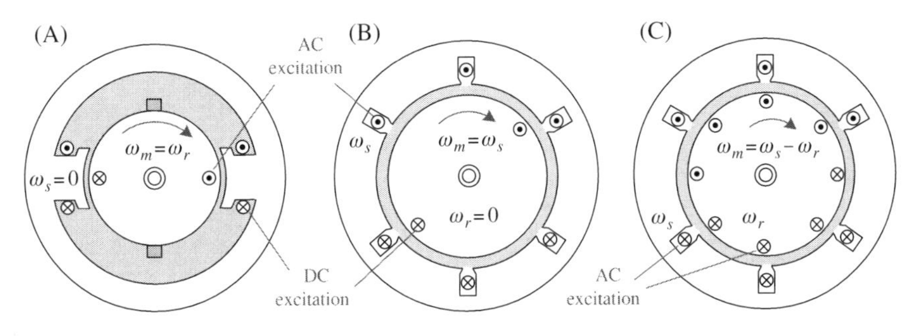

FIGURE 1.24

Doubly fed motors. (A) DC motor, (B) synchronous motor, and (C) induction motor.

1.2.3.2 Synchronous motor

If the stator is excited with an AC power source of an angular frequency ω_s and the rotor is excited with a DC power as shown in Fig. 1.24B, then Eq. (1.45) can only be satisfied if

$$|\omega_m| = |\omega_s| \tag{1.49}$$

To achieve this, the rotor must rotate at an angular frequency synchronized to the frequency ω_s of the stator current. A motor satisfying this condition becomes a synchronous motor. The name "*synchronous motor*" is due to this fact. From Eq. (1.44), the average torque in this motor can be given as

$$T_{avg} = -\frac{I_{sm}I_{rm}}{2}L_m \sin\delta \tag{1.47}$$

Unlike in the DC motor, the average torque in this motor varies according to the angle δ between the stator and the rotor field fluxes even though the same currents are flowing. We can easily see that a synchronous motor has no starting torque because it cannot meet the requirement of $\omega_m \neq \omega_s$ when starting.

1.2.3.3 Induction motor

When the stator and the rotor windings are excited with AC power of different frequencies ω_s and ω_r, respectively, a necessary condition should be

$$\omega_m = \omega_s - \omega_r \tag{1.50}$$

This means that the rotor has to rotate at the angular frequency of $\omega_s - \omega_r$. This can be achieved in an induction motor as shown in Fig. 1.24C. In an induction motor, the frequency ω_r of the currents induced in the rotor windings by electromagnetic induction becomes naturally $\omega_s - \omega_m$, which is the difference between the stator current and the rotor speed in the angular frequency. Thus, an induction motor always satisfies Eq. (1.51). From Eq. (1.44), the average torque in this motor can be given as

$$T_{avg} = -\frac{I_{sm}I_{rm}}{2}L_m \sin(\alpha + \delta) \tag{1.51}$$

Unlike in the DC motor, the average torque in the induction motor varies according to the angle δ between the stator and the rotor field fluxes even though the same currents are flowing.

As we saw previously, electric motors can be classified into cylindrical motors and reluctance motors. Cylindrical motors exploit the torque produced by varying the mutual-inductance between the windings as mentioned before. These cylindrical motors with windings on both the stator and the rotor can produce a larger torque even though they are more complex in construction. Therefore, most motors are of the cylindrical type. In comparison, reluctance motors of noncylindrical configuration utilize the torque produced by the variation of inductance (or reluctance) of the magnetic path. Reluctance motors are simple in construction, but the torque developed in these motors is small.

1.3 MECHANICAL LOAD SYSTEM

An electric motor is an electromechanical device that converts electrical energy into mechanical torque. When the torque developed by the motor is transferred to the load connected to it, the load variables such as speed, position, air-flow, pressure, and tension will be controlled. Now, we will discuss the characteristics of the load connected a motor and the mathematical representation that describes the motor-driven system.

1.3.1 DYNAMIC EQUATION OF MOTION

Consider that a driving force F_M acts on an object of mass M as shown in Fig. 1.25A, so that the object moves at a speed v[4]. From Newton's second law, which states that the rate of change of momentum of an object is directly proportional to the applied force, we have

$$F_M - F_L = \frac{d}{dt}(Mv) = M\frac{dv}{dt} + v\frac{dM}{dt} \tag{1.53}$$

where F_L is the load force, which acts in a direction opposite to F_M, and Mv is the object's momentum.

If the mass M of the load is constant, then the second term of the derivative in Eq. (1.53) is zero. Thus, Eq. (1.53) is given by a well-known equation derived from Newton's second law as

$$F_M - F_L = M\frac{dv}{dt} = Ma \tag{1.54}$$

where $a(=dv/dt)$ is the object's acceleration.

Next, for a rotational motion in Fig. 1.25B, the rotational analogue of Eq. (1.53) is given by

$$T_M - T_L = \frac{d}{dt}(J\omega) = J\frac{d\omega}{dt} + \omega\frac{dJ}{dt} \tag{1.55}$$

where T_M and T_L are driving and load torque, respectively, and ω is the angular speed of the rotating object. J denotes the moment of inertia of the rotating object and $J\omega$ is the object's angular momentum.

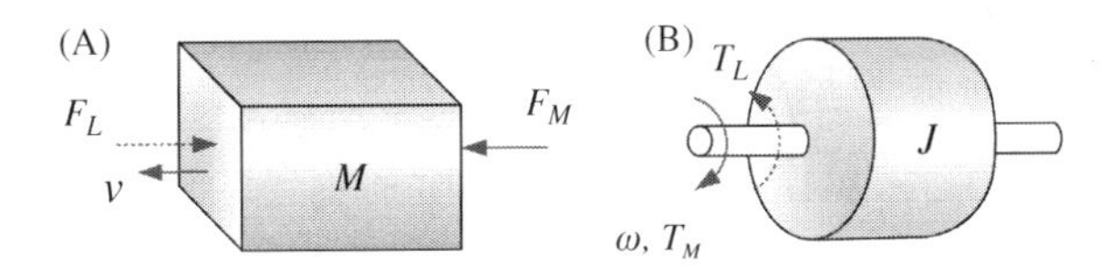

FIGURE 1.25

(A) Linear motion and (B) rotational motion objects.

When the moment of inertia J is assumed to be constant, we can rewrite Eq. (1.55) as

$$T_M - T_L = J\frac{d\omega}{dt} = J\alpha \tag{1.56}$$

where $\alpha(=d\omega/dt)$ is the object's angular acceleration.

We can see that if $T_M > T_L$, then the object will accelerate, and if $T_M < T_L$, then it will decelerate. In the case of $T_M = T_L$, the speed will not be changed. Thus, when driving a load, to keep the load speed constant, the motor torque must be equal to the load torque.

MOMENT OF INERTIA: J (kg m^2)

The tendency of a rotating object to continue in its original state of rotary motion is referred to as the *moment of inertia J*. The rotational kinetic energy of a rigid object with angular velocity ω is expressed as $\frac{1}{2}J\omega^2$ in terms of its moment of inertia. The moment of inertia of an object is also the measure of how fast the object can be accelerated or decelerated.

The moment of inertia in a rigid object is related to the distribution of mass as well as the mass m of the object. Therefore the moment of inertia of an object can depend on its shape. For example, if two objects have the same mass but are different in radius as shown in the following figure, then they may have different moments of inertia.

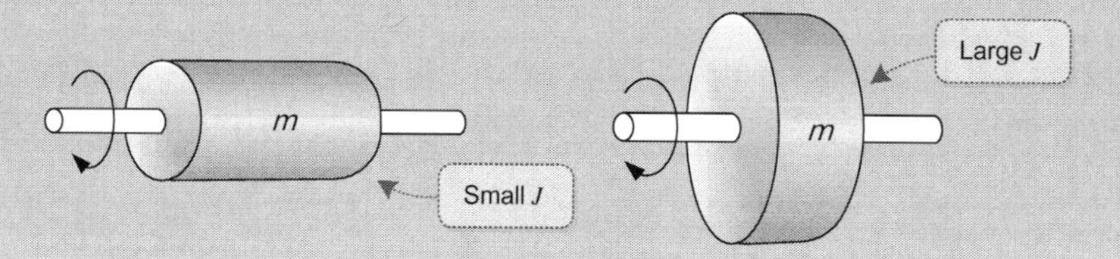

If the radius of an object is larger, then the moment of inertia is larger. The reason for this is as follows. A rotating object consists of many small particles at various distances from the axis of rotation. When the object is rotating, the particles which are far away from the axis of rotation will move faster than the ones which are located near the axis of rotation. Thus, the particles which are far away from the axis of rotation require more kinetic energy for a rotation. The moment of inertia I for each particle is defined as its mass multiplied by the square of the distance r from the axis of rotation to the particle, i.e., $I = mr^2$. Thus, for the same object mass, the further out from the axis of rotation its mass is distributed, the larger the moment of inertia of the object is. For solid or hollow cylinders, the moment of inertia can be calculated by the equations shown in the next figures.

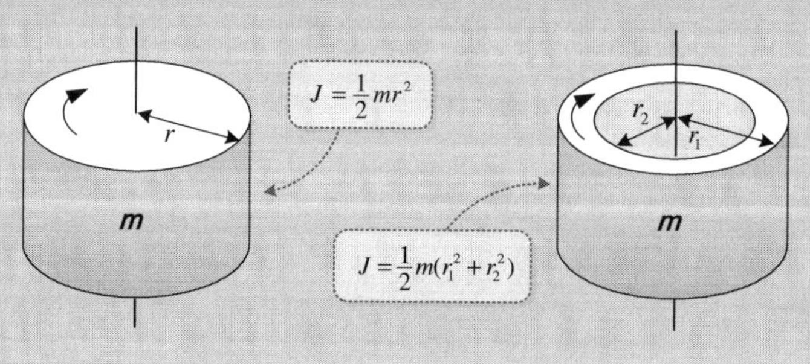

(Continued)

MOMENT OF INERTIA: J (kg m^2) (Continued)

Instead of J, $GD^2(=4J)$, also known as *flywheel effect*, is generally used in industrial applications. Here, G is the mass (kg) and D is the diameter (m).

There is maximum permissible load inertia to a motor. It is important to ensure that the inertia of the motor matches the inertia of the driven load. Ideally, it is desirable to have a 1:1 inertia ratio between the load and the motor. For traction motors, the motor inertia should be large enough to drive the load of a large inertia. To have a large inertia, the motor will be designed with a rotor configuration of a large diameter and a short axial length. On the other hand, for servo motors operated in frequent acceleration/deceleration, the inertia of the rotor should be small. Thus, the rotor will be designed to have a small diameter and a long axial length.

Fig. 1.26 shows a simplified configuration of a motor drive system consisting of a driving motor and a driven mechanical load, in which the torque of the motor is transferred to the load through the shaft and coupling.

From the motor's viewpoint, a mechanical load can be seen as a load torque T_L connected to its shaft. Thus, from Eq. (1.56), the equation of motion of a motor drive system can be expressed as

$$T_M = (J_M + J_L)\frac{d\omega}{dt} + T_L \tag{1.57}$$

where J_M and J_L are the motor inertia and the load inertia, respectively.

When the motor is driving a mechanical load, the torque required to run the load is the load torque T_L, which varies with the type of load.

In addition to Eq. (1.57), when a load begins to move and is in motion, a friction force occurs, resisting the motion. Therefore, the equation of motion needs to include the friction force T_F as

$$T_M = (J_M + J_L)\frac{d\omega}{dt} + T_F + T_L \tag{1.58}$$

The friction force is usually proportional to the speed, given by $T_F = B\omega$, where B is the coefficient of friction. Finally, we have an equation of motion, which expresses the dynamic behavior of a motor drive system as

$$T_M = J\frac{d\omega}{dt} + B\omega + T_L \tag{1.59}$$

where $J = J_M + J_L$.

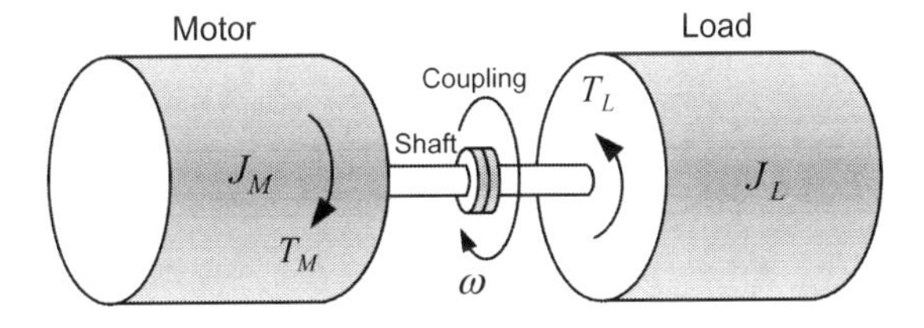

FIGURE 1.26

Simplified motor drive system.

A motor-driven system has mechanical transmission devices such as shaft, gearbox, or coupling to connect the motor and the load. In a real system, the stiffness of these components is limited, and the transmissions are flexible. Therefore, if the coupling or shaft between the motor and the load is long or elastic, then it may be subjected to torsion. In this case, we need to add the torque term $K_{sh}\theta$ produced by torsional deformation of the shaft to Eq. (1.59), where K_{sh} (Nm/rad) is the coefficient of stiffness.

LOAD TORQUE T_L

There are several types of loads driven by a motor. In most loads, the required torque is normally a function of speed. There are three typical loads whose speed-torque characteristics are shown in the following figure.

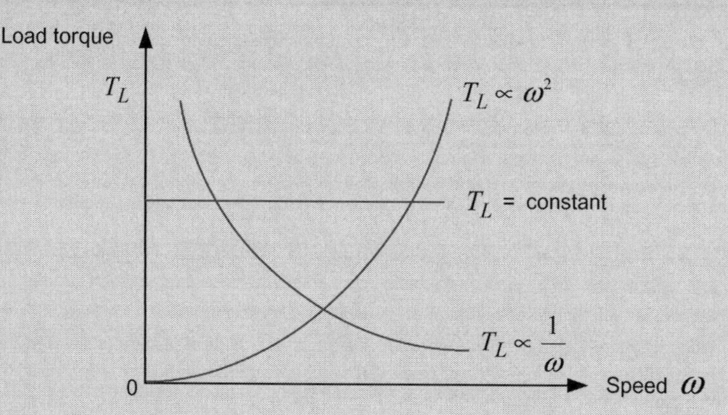

1. Constant torque load

   The torque required by this type of load remains constant regardless of the speed. Loads such as elevators, screw compressors, conveyors, and feeders have this type of characteristic.

2. Constant power load ($T_L \propto 1/\omega$)

   This load requires a torque, which is inversely proportional to the speed, so this load is considered a constant power load. This type is most often found in machine tool industry and drilling, milling industry.

3. Load proportional to the square of the speed ($T_L \propto \omega^2$)

   The torque of this load increases in proportion to the square of the speed. This is a typical characteristic of a fan, blower, and pump, which are the most commonly found in industrial drive applications.

The load in traction drives such as electric locomotives and electric vehicles shows the combined characteristic of a constant torque load in low speeds and a constant power load in high speeds.

1.3.1.1 Combination system of translational motion and rotational motion

Many motions in the industry such as an elevator, a conveyor, or electric vehicles are a combination of translational motion and rotational motion as shown in Fig. 1.27. In this case, we need to convert the load parameters such as load inertia to the motor shaft.

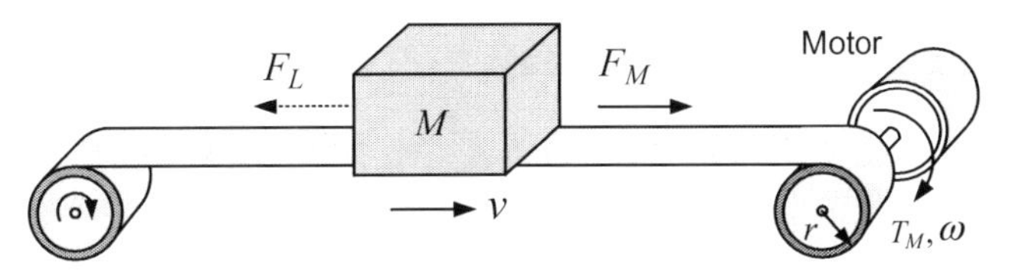

FIGURE 1.27

Combination system of translational motion and rotational motion.

The expression for the translational motion in Fig. 1.27 is given by

$$F_M - F_L = M\frac{dv}{dt} \tag{1.60}$$

The relationships between force and torque, and velocity and angular velocity are

$$T_M = rF_M \tag{1.61}$$

$$T_L = rF_L \tag{1.62}$$

$$\omega = \frac{v}{r} \tag{1.63}$$

By substituting Eqs. (1.61)–(1.63) into Eq. (1.60), we obtain

$$T_M - T_L = Mr^2\frac{d\omega}{dt} = J_e\frac{d\omega}{dt} \tag{1.64}$$

where $J_e = Mr^2$ represents the equivalent moment of inertia, which is reflected to the motor shaft side of the mass M in translational motion.

1.3.1.2 System with gears or pulleys

Often, the speed required by the load is too low compared to the nominal speed of the motor. Gears or pulleys between the motor and the load being driven are most often used to change the speed. In this case, we need to know how the load will be seen through the gears or pulleys at the motor side.

We will look at an example of two meshed gears as shown in Fig. 1.28.

The equation of motion at the load side is given by

$$T_L = J_L\frac{d\omega_L}{dt} + B\omega_L \tag{1.65}$$

Because the two gears will travel at an equal distance, we have

$$\omega_L = \frac{N_1}{N_2}\omega_M \tag{1.66}$$

where N_1 and N_2 are the number of teeth on the gears of the motor and load sides, respectively.

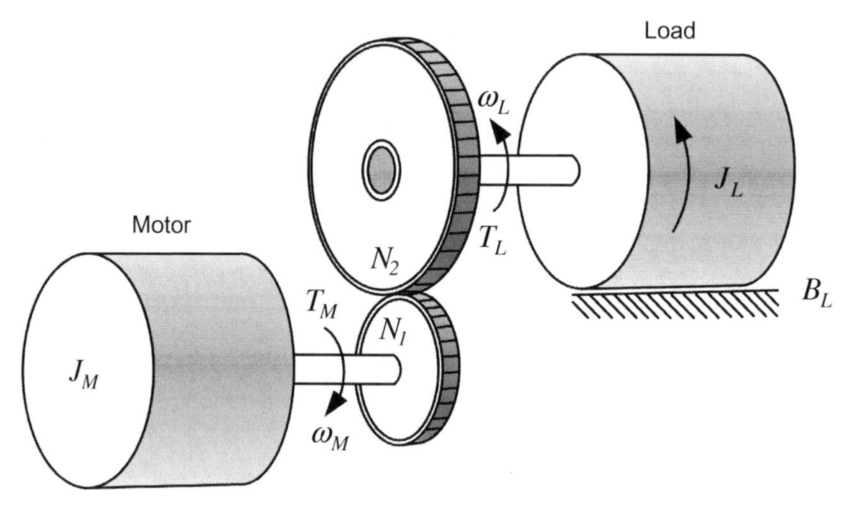

FIGURE 1.28

System with gears.

When neglecting friction and losses of the gears, the power is the same at the input and at the output of the gear, i.e.,

$$P = T_L\omega_L = T_M\omega_M \quad \rightarrow \quad T_L = \frac{N_2}{N_1}T_M \tag{1.67}$$

Substituting Eqs. (1.65) and (1.66) in Eq. (1.65) yields

$$T_L = J_L\left(\frac{N_1}{N_2}\right)\frac{d\omega_L}{dt} + B_L\left(\frac{N_1}{N_2}\right)\omega_L = \left(\frac{N_2}{N_1}\right)T_M \tag{1.68}$$

Finally, the equation of motion at the motor side is given by

$$\begin{aligned} T_M &= J_L\left(\frac{N_1}{N_2}\right)^2\frac{d\omega_M}{dt} + B_L\left(\frac{N_1}{N_2}\right)^2\omega_M \\ &= (J_M + J)\frac{d\omega_M}{dt} + B\omega_M \end{aligned} \tag{1.69}$$

where $J = J_L(N_1/N_2)^2$ and $B = B_L(N_1/N_2)^2$ are the load inertia J_L and the friction coefficient B_L converted to the motor side, respectively. It can be seen that the load parameters reflected back to the motor are a squared function of the gear ratio N_1/N_2.

1.3.2 OPERATING MODES OF AN ELECTRIC MOTOR

Operation of a motor can be divided into two modes according to the direction of the energy transfer.

When the motor rotates in the same direction as its developed torque in Fig. 1.29A, the motor will supply mechanical energy to the load. In this case, the motor operates as a typical *motoring mode*. On the other hand, when the

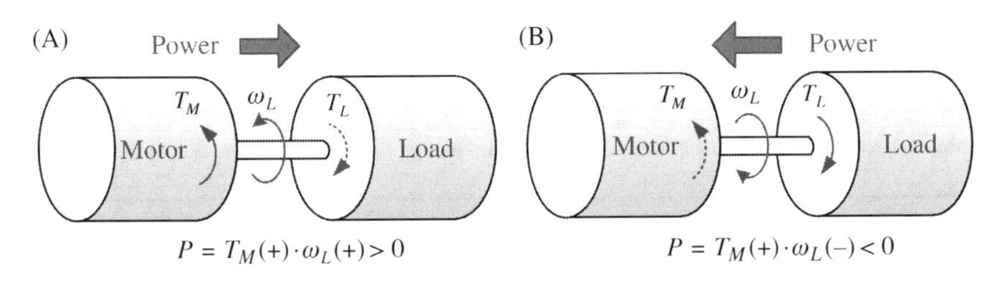

FIGURE 1.29

Operation mode of the motor. (A) Motoring mode and (B) generating (braking) mode.

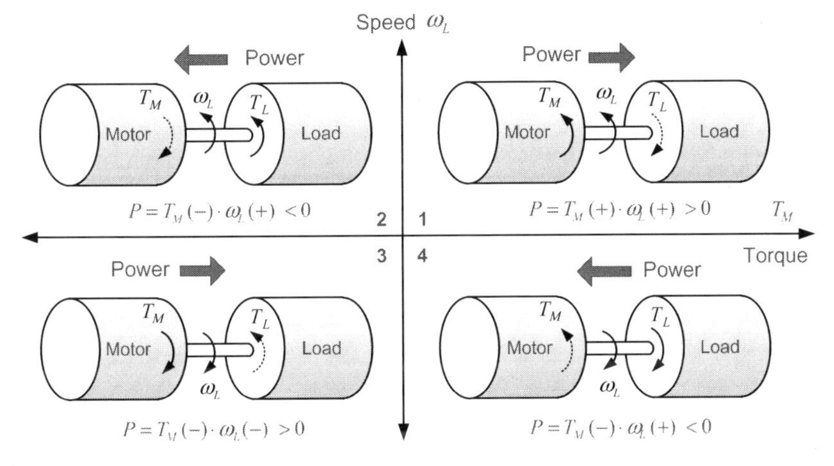

FIGURE 1.30

Four-quadrant operation modes.

motor is rotating in a direction opposite to its developed torque in Fig. 1.29B, the mechanical energy of the load is supplied to the motor. In this case, the motor is acting as a generator. It corresponds to the *generating mode*, in which the motor absorbs the load energy, providing a braking action.

When braking, the mechanical energy of the load is converted into electrical energy through the motor, and the generated electrical power can be returned to the power source. This method called *regenerative braking* is the most efficient braking method, but it requires a high-cost power converter. An alternative method for dealing with the generated electrical power is to dissipate it as heat in the resistors. This is called *dynamic braking*, which is inefficient but inexpensive.

Similarly, there are also operation modes of motoring and generating in the reverse driving. Consequently, four-quadrant operation modes are available in variable speed drive systems as shown in Fig. 1.30. Quadrants 1 and 2 are the motoring and generating operation modes, respectively, in the forward drive, while Quadrants 3 and 4 are the two operation modes in the reverse drive. Braking occurs in Quadrants 2 and 4. To achieve the four-quadrant operation modes in motor drive systems, the torque should be controlled in the positive and negative direction, and the speed should be controlled in the forward and reverse direction.

As an example of the four-quadrant operation, let us take a look at an elevator drive system as shown in Fig. 1.31. In an elevator system, the car carrying people or goods is connected to and is driven by a traction motor through ropes and gears. As a traction motor, induction motors are most widely used. In recent times, however, permanent magnet synchronous motors are used as well.

A counterweight attached to the opposite side of the ropes reduces the amount of power required to move the car. The weight of the counterweight is typically equal to the weight of the car plus 40–50% of the capacity of the elevator. When the car goes up or down, we can see the four-quadrant operation of the driving motor occurring due to the weight difference between the car and the counterweight as follows:

- Quadrant 1 operation: the forward motoring mode $(+T, +\omega)$

  When the car carrying people goes up and is heavier than the counterweight, the traction motor is needed to produce the forward direction torque, so it operates in the *forward motoring mode*.

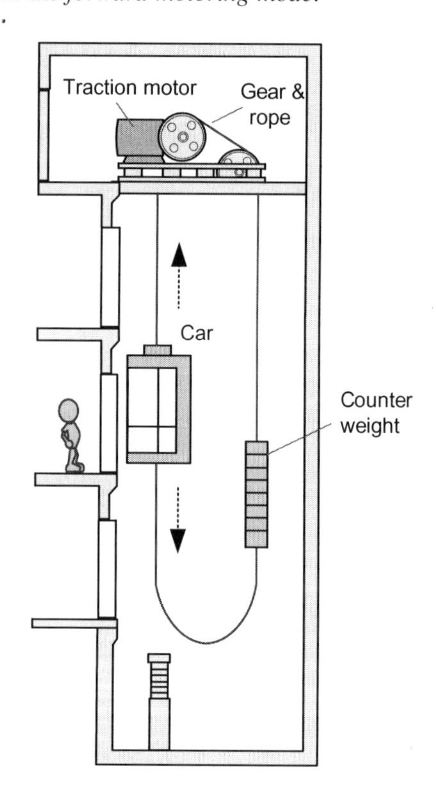

FIGURE 1.31

Four-quadrant operation in an elevator drive.

- Quadrant 2 operation: the forward braking mode ($-T$, $+\omega$)

  When an empty car goes up and is lighter than the counterweight, the traction motor is needed to produce the braking torque of the reverse direction, so it operates in the *forward braking (generating) mode*.
- Quadrant 3 operation: the reverse motoring mode ($-T$, $-\omega$)

  When an empty car goes down and is lighter than the counterweight, the traction motor is needed to produce the reverse direction torque, so it operates in the *reverse motoring mode*.
- Quadrant 4 operation: the reverse braking mode ($+T$, $-\omega$)

  When the car carrying people goes down and is heavier than the counterweight, the traction motor is needed to produce the braking torque of the forward direction, so it operates in the *reverse braking (generating) mode*.

Here, the going-up of the car is assumed as the forward direction.

1.4 COMPONENTS OF AN ELECTRIC DRIVE SYSTEM

For a given mechanical load, an adequate design of the motor drive system is essential to meet its required performance criteria such as speed-torque characteristic, operation speed range, speed regulation, efficiency, cost, duty cycle, etc.

A motor drive system consists normally of five parts as shown in Fig. 1.32. Within the available power supply, we have to choose these components appropriately to achieve the requirements of the drive performance. We will now briefly take a look at these five parts.

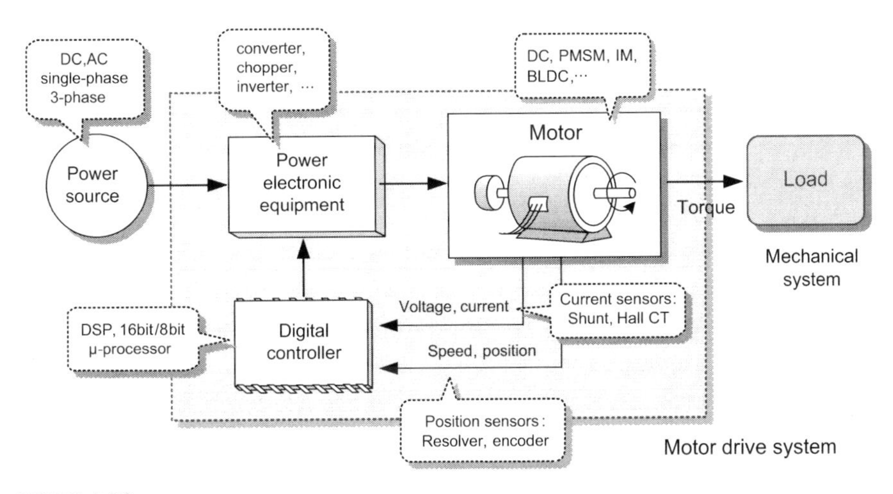

FIGURE 1.32

Configuration of a motor drive system.

1.4.1 POWER SUPPLY

Single-phase or three-phase AC voltage sources are most commonly used as a power supply that provides electric energy for motor drives. The three-phase AC voltage source is generally used for motors of several kilowatts or higher. In many cases, the AC voltage sources are converted into a DC source, which is converted into the form of power needed by a motor through using power electronic converters such as an inverter or a chopper. Recently, DC power has often been provided by renewable energy sources such as solar cells and fuel cells.

1.4.2 ELECTRIC MOTORS

For driving motors, we have to choose the one that meets the required speed-torque characteristic of the driven mechanical load. In addition, important performance criteria like efficiency, size, cost, duty cycle, and operating environment need to be considered. As mentioned earlier, there are two basic types of a driving motor: DC motor and AC motor. Some applications use special types such as a BLDC motor or a stepping motor, which cannot be classified into the basic types.

DC motors have an advantage of a simple speed/torque control for their excellent drive performance. However, due to the wear associated with their commutators and brushes, DC motors are less reliable and unsuitable for a maintenance-free operation. On the other hand, AC motors are smaller, lighter, and have more power density than DC motors. AC motors also require virtually no maintenance and thus are better suited for high speeds. However, AC drives with a high performance capability are more complex and expensive than DC drives.

1.4.3 POWER ELECTRONIC CONVERTERS

Power electronic converters play the role of taking electrical energy from the power system and turning it into a suitable form needed by a motor. The power electronic converter may be determined according to the given power source and the driving motor.

For DC drives, power electronic converters such as a controlled rectifier or a chopper can be used to adjust the DC power, which will be described in more detail in Chapter 2. In contrast, AC drives mostly use an inverter to adjust the voltage and frequency in the AC power, which will be also described in more detail in Chapter 7. In this case, a rectifier is often included to convert the AC power in the mains power system into the DC power.

These power electronic converters use power semiconductor devices such as gate turn-off (GTO) thyristor, integrated gate-commutated thyristor (IGCT), insulated gate bipolar transistor (IGBT), power metal oxide semiconductor field effect transistor (MOSFET), and power bipolar junction transistor (BJT). These switching devices are determined according to their power handling capability and their switching speed. The power handling capability can be ranked in an increasing order of

MOSFET, IGBT, and GTO thyristor, while the switching speed can be ranked in an increasing order of GTO thyristor, IGBT, and MOSFET. All these switching devices described above are based on the silicon (Si) semiconductor material. Recently, switching devices based on wide bandgap materials such as silicon carbide (SiC) or gallium nitride (GaN) are being recognized as a promising future device.

1.4.4 DIGITAL CONTROLLERS

The controller in charge of executing algorithms to control the motor is the most important part in the motor drive systems. This controller manipulates the operation of the power electronic converter to adjust the frequency, voltage, or current provided to the motor. Nowadays, digital controllers are usually used. A digital controller is based on microprocessor, microcontroller, or digital signal processor (DSP), which can be a 16-bit or 32-bit, and also either a fixed-point or floating-point.

There is a trade-off between the control performance and the cost. DC motor and BLDC motor drives requiring relatively simple controls may use 16-bit processors, while AC motor drives requiring complex controls such as a vector controlled induction or synchronous motors may need high-performance 32-bit processors or DSP.

1.4.5 SENSORS AND OTHER ANCILLARY CIRCUITS

To achieve a high-performance motor drive, a closed-loop control of position, speed or current is often adopted, and in this case, speed/position or current information is required.

Sensors are essential for obtaining this information. Shunt resistances, Hall effect sensors, or current transformers are mainly used as the current sensor. The shunt resistance is more cost effective, while the Hall effect sensors can usually give a high resolution and isolation between the controller and the system. Encoders and resolvers are widely used as speed/position sensors, which will also be discussed in Chapter 9.

In addition to these components, several circuits for protection, filtering, power factor correction, or harmonics reduction may be necessary to improve the reliability and quality of the drive system.

REFERENCES

[1] P.E. Scheihing, et al., United States Industrial Motor-Driven Systems Market Assessment: Charting a Road map to Energy Savings for Industry,' U.S. Department of Energy, April 2007.

[2] S.J. Chapman, Electric Machinery Fundamentals, third ed., McGrawHill, Boston, MA, 1999.

[3] G.R. Slemon, A. Straughen, (Chapter 3) Electric Machines, Addison-Wesley, 1980.

[4] W. Leonhard, Control of Electrical Drives, second ed., Springer-Verlag, Berlin Heidelberg New York, 1996.

CHAPTER

Control of direct current motors

2

Direct current (DC) motors have been widely used in adjustable speed drives or variable torque controls because their torque is easy to control and their range of speed control is wide. However, DC motors have a major drawback in which they need mechanical devices such as commutators and brushes for a continuous rotation. These mechanical parts require regular maintenance and hinder high-speed operations. Recently due to a major development in alternating current (AC) motor control technology, power electronics technology, and microprocessors, AC motors have been used instead of DC motors, but DC motors are still used in lower power ratings below several kilowatts.

In this chapter, we will study the current and the speed control of DC motors. Understanding the current and speed control techniques for DC motors is very important because these techniques are directly applicable to the AC motor control, which will be discussed in Chapter 6.

2.1 CONFIGURATION OF DIRECT CURRENT MOTORS

A DC motor has two windings as shown in Fig. 2.1. One is found in the stator and produces a field flux. This winding is called a *field winding*. The other one resides in the rotor and is called an *armature winding*. The armature current interacts with the stator field flux to produce torque on the rotor shaft. Small DC motors commonly have a permanent magnet instead of a field winding to produce the stator flux. A DC motor also has a mechanical commutation device consisting of brushes and commutators, which converts the DC current given from the DC power supply to AC current in the armature winding.

The DC motors can be mainly classified into two different categories: a *separately excited DC motor* and a *self-excited DC motor*. In a separately excited DC motor, the armature winding and field winding each have separate DC power supplies, while in a self-excited DC motor, the two windings share one DC power supply.

The self-excited DC motors can also be categorized into a *shunt motor*, *series motor*, and *compound motor* depending on the ways in which the two windings are connected to the DC power supply. Nowadays the DC motors mostly use a

Electric Motor Control. DOI: http://dx.doi.org/10.1016/B978-0-12-812138-2.00002-7

FIGURE 2.1

DC motor configuration.

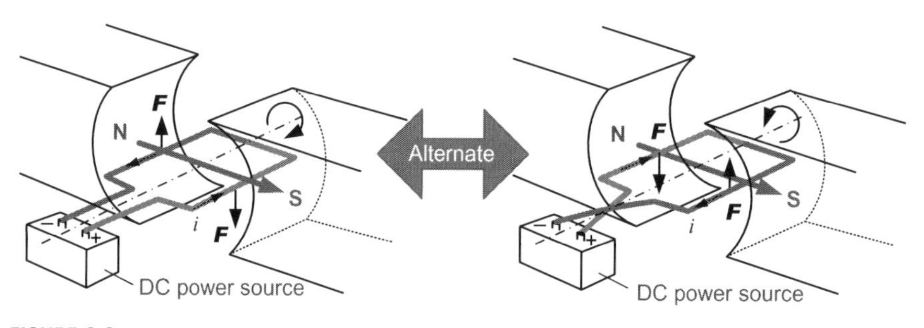

FIGURE 2.2

Force on a one-turn coil.

permanent magnet to produce the field flux and the DC power supply thus applies only to the armature winding.

Now we will learn about the roles of brushes and commutators, which are essential for the continuous rotation of DC motors. First, we will take a look at Fig. 2.2, which shows a one-turn coil in a rotor winding. When the coil is connected to the DC power supply, a current flows in the coil. We will assume that, in the initial stage, the flux distribution and the current flowing in the coil are formed like the figure on the left. In this condition, the force on the conductors on both sides makes the coil rotate in the clockwise direction according to the direction of the force given by *Fleming's left-hand rule*.

When the coil rotates and comes to the position as in the figure on the right, the force produced on the conductors on both sides will return the coil again to its initial position as shown in the left side of Fig. 2.2. Since the force produced on the coil is not produced continuously in one direction, the

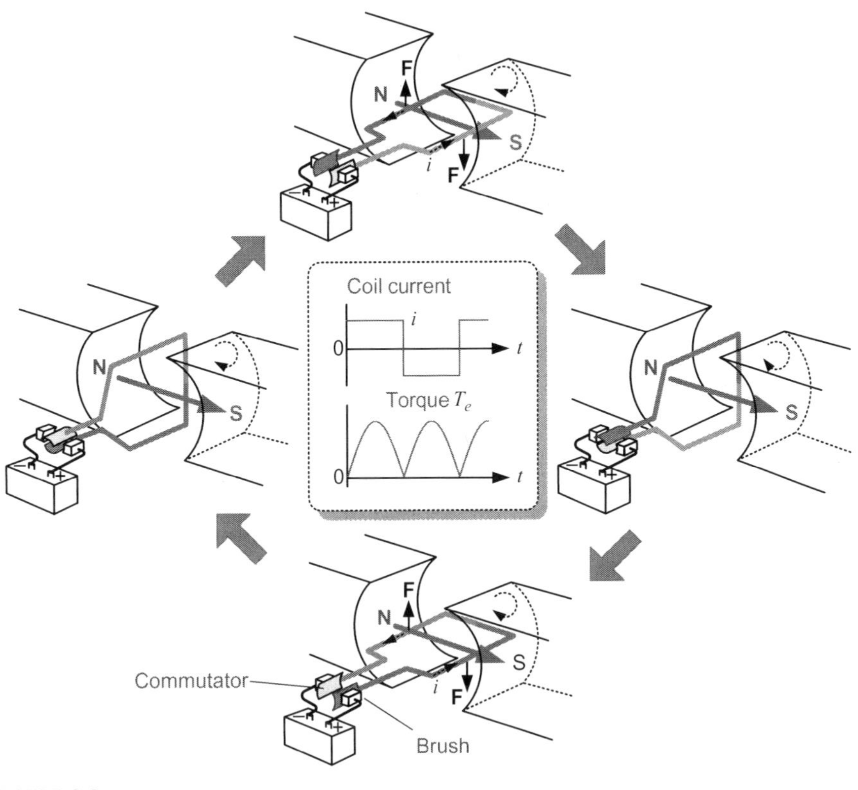

FIGURE 2.3

Continuous rotation by the use of brushes and commutators.

coil also cannot rotate in the same direction continuously. However, when the coil is at the position shown on the right, if the direction of its current is reversed, then the force produced on the conductors on both sides is reversed and thus, the force on the coil will remain in the same direction of clockwise. Thus the coil will rotate continuously in that direction. By the commutation actions of the brush and the commutator, the coil current can be reversed as shown in Fig. 2.3. It should be noted that the current that flows in the coil must be an alternating current in DC motors. The frequency of this alternating current is always the same as the angular velocity of the coil, which indicates the speed of the rotor. In Section 1.2.3, we have already seen that this meets the necessary condition for the continuous torque production in the doubly fed machine.

The armature winding of the DC motor has many coils to increase the average torque and reduce torque ripple as shown in Fig. 2.4. Each coil is connected to the DC power supply through a brush and its own isolated commutator.

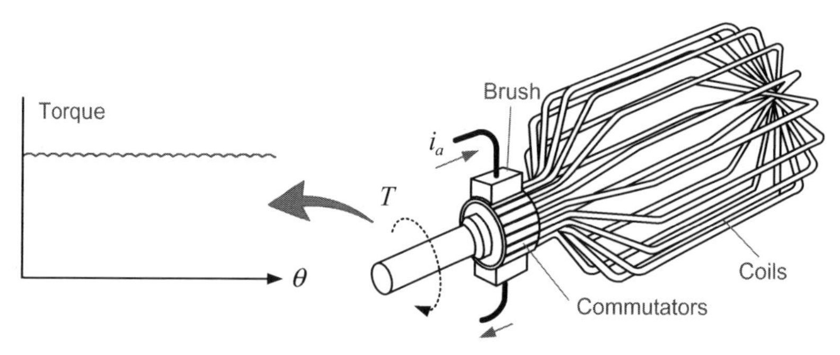

FIGURE 2.4

Armature winding.

As described earlier, in order for a DC motor to rotate continuously, it must have brushes and commutators in its structure. However, due to the wear and the arcing associated with their brushes and commutators, DC motors require much more maintenance and are limited in high-speed operation.

2.2 MODELING OF DIRECT CURRENT MOTORS

A complete dynamic model of the DC motor drive system can be expressed as the following four equations: *armature circuit*, *back-electromotive force (back-EMF)*, *torque*, and *mechanical load system.*

2.2.1 ARMATURE CIRCUIT

The voltage equation for the armature winding of a DC motor is given by

$$V_a = R_a i_a + L_a \frac{di_a}{dt} + e_a \tag{2.1}$$

where i_a is winding current, L_a is winding inductance, R_a is winding resistance, and e_a is back-EMF induced by the rotation of the armature winding in a magnetic field. The equivalent circuit of a DC motor derived from Eq. (2.1) is shown in Fig. 2.5.

We need a back-EMF e_a to obtain the current flowing through the armature winding for a voltage V_a applied to a DC motor from Eq. (2.1).

2.2.2 BACK-ELECTROMOTIVE FORCE

As shown in Fig. 2.6, when a conductor of length l meters is moving with a constant velocity v in the presence of a uniform magnetic field B, a voltage $e = Blv$ (called back-EMF) will be induced in the conductor. Similarly, in a DC motor, when the conductors of the armature winding with an angular velocity ω_m

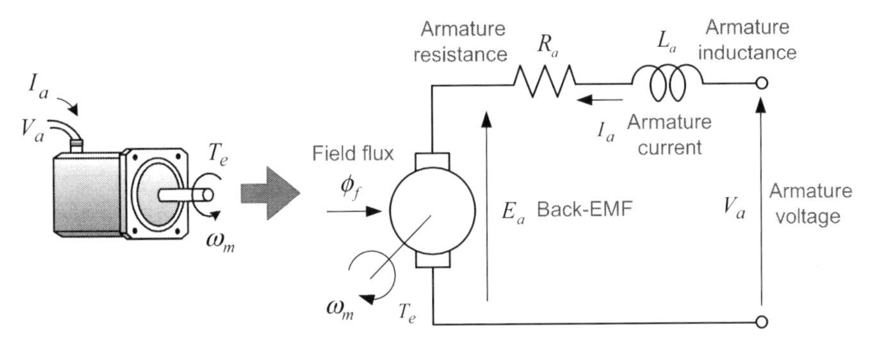

FIGURE 2.5

Equivalent circuit of a DC motor.

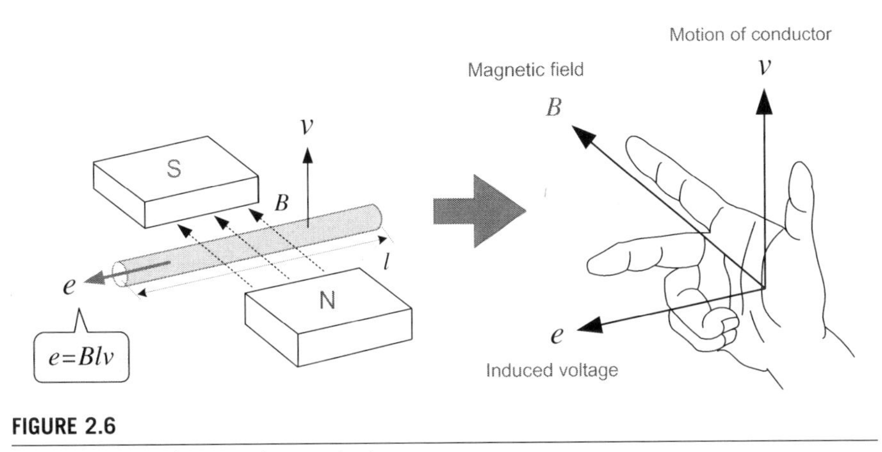

FIGURE 2.6

Induced voltage for a moving conductor.

are moving in the presence of the uniform stator magnetic flux ϕ_f, the back-EMF in the armature winding will be expressed as

$$e_a = k_e \phi_f \omega_m \quad (\leftarrow e = Blv) \tag{2.2}$$

where k_e is the back-EMF constant (Vs/rad/Wb). Commonly, since the flux remains constant, the back-EMF is proportional to the angular velocity.

With the armature current i_a calculated from the applied input voltage V_a and the back-EMF e_a, we can obtain the developed torque of a DC motor as the following.

2.2.3 TORQUE

As shown in Fig. 2.7, when a conductor of length l meters carrying a current of i amps is placed in a uniform magnetic field B, the force acting on the conductor

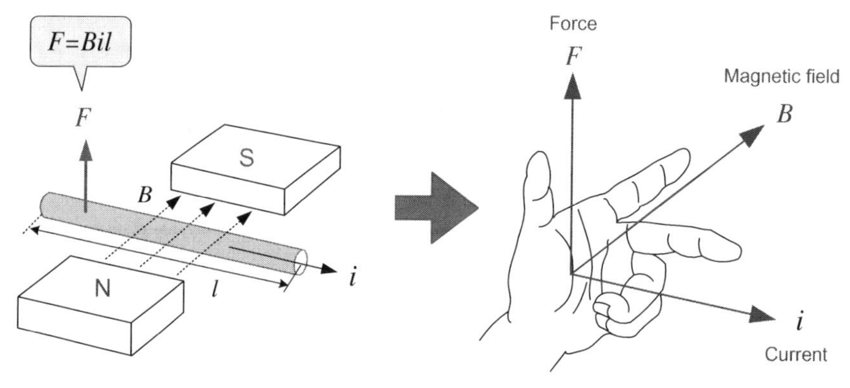

FIGURE 2.7

Force for a current carrying conductor.

will be $F = Bli$. Similarly, in a DC motor, the torque experienced by the conductors of the armature winding carrying a current i_a in the uniform stator magnetic flux ϕ_f is given by

$$T_e = k_T \cdot \phi_f \cdot i_a \quad (\leftarrow F = Bli) \tag{2.3}$$

where k_T is the torque constant (Nm/Wb/A). The numerical values of k_T and k_e constants are equal in SI units (the International System of Units), i.e., $k_T = k_e$.

It is interesting to note that the torque of a DC motor can be expressed as a simple arithmetical product of the stator magnetic flux ϕ_f and the armature current i_a. This is because the stator magnetic flux ϕ_f and the conductors carrying current i_a are always perpendicular to each other. Thus the DC motors inherently have a maximum torque per ampere characteristic.

The speed of a rotor can be determined by the developed torque as the following.

2.2.4 MECHANICAL LOAD SYSTEM

When a DC motor with the output torque in Eq. (2.3) is driving a mechanical load system, the rotor speed can be determined from the equation of motion in Eq. (2.4) as mentioned in Chapter 1.

$$T_e = J\frac{d\omega_m}{dt} + B\omega_m + T_L \tag{2.4}$$

where ω_m is the angular velocity of the rotor, T_L is the load torque, J is the moment inertia and B is the friction coefficient of the DC motor drive system, respectively.

From the above dynamic model represented by Eqs. (2.1)–(2.4), we can understand the steady-state as well as the transient characteristics of the current and the speed in the DC motor drive system. First, on the basis of this model, we will examine the speed–torque characteristic of the DC motor in the steady-state condition.

2.3 STEADY-STATE CHARACTERISTICS OF DIRECT CURRENT MOTORS

The term "steady-state" refers to a condition that exists after all initial transients or fluctuating conditions have damped out. The steady-state behavior of the DC motor is easily determined by making the time derivatives of the variables in Eqs. (2.1) and (2.4) zero, i.e., $di_a/dt = 0$, $d\omega_m/dt = 0$. From the steady-state equations, the speed–torque relation may be written as

$$\omega_m = \frac{V_a}{k\phi_f} - \frac{R_a}{(k\phi_f)^2}T_L \quad (k = k_T = k_e) \tag{2.5}$$

Here, the friction coefficient B is neglected.

We can see that, from this torque–speed relation, the speed of the DC motor decreases steadily with increasing load torque (or armature current) as shown in Fig. 2.8. Here, N is the speed expressed in terms of revolution per minute (r/min) for angular speed ω_m. From the curve, we can find the speed and the current at a certain load.

Eq. (2.5) implies that the steady-state speed of a DC motor can be adjusted by the armature voltage V_a or the stator field flux ϕ_f. Thus there are two speed control methods for the DC motor: *armature voltage control* and *field flux control*. In addition to the two methods, we can see from Eq. (2.5) that the speed can be controlled by varying the armature resistance R_a. However, this inefficient method is no longer in use.

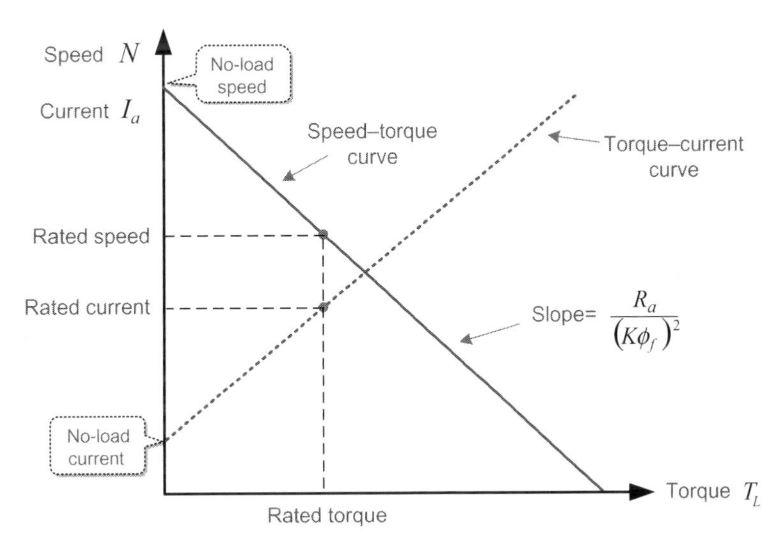

FIGURE 2.8

Speed–torque characteristic of a DC motor in the steady-state.

Now, we will take a look at the speed control characteristics of the DC motor by armature voltage control and field flux control.

2.3.1 ARMATURE VOLTAGE CONTROL

With a constant stator field flux, Eq. (2.5) expressing the torque–speed characteristic can be reduced to

$$\omega_m = \frac{V_a}{k\phi_f} - \frac{R_a}{(k\phi_f)^2}T_L = K_1V_a - K_2T_L \tag{2.6}$$

where $K_1 = V_a/k\phi_f$ and $K_2 = R_a/(k\phi_f)^2$

From this expression, we can see that the speed ω_m of the DC motor changes linearly with the armature voltage V_a. Thus, for a constant torque load such as elevators and cranes, the speed of the load can be controlled linearly by adjusting the armature voltage V_a of the DC motor as shown in Fig. 2.9.

To employ the armature voltage control, variable DC power supplies such as phase-controlled rectifiers or choppers are needed. Because a voltage higher than the rated voltage should not be applied to the motor, this armature voltage control method can provide a speed control range only up to the rated speed. To operate in a speed range above the rated speed, the DC motor has to adopt the field flux control, which will be described later. Permanent magnet DC motors, however, cannot use the field flux control method, so they are incapable of an operation above the rated speed.

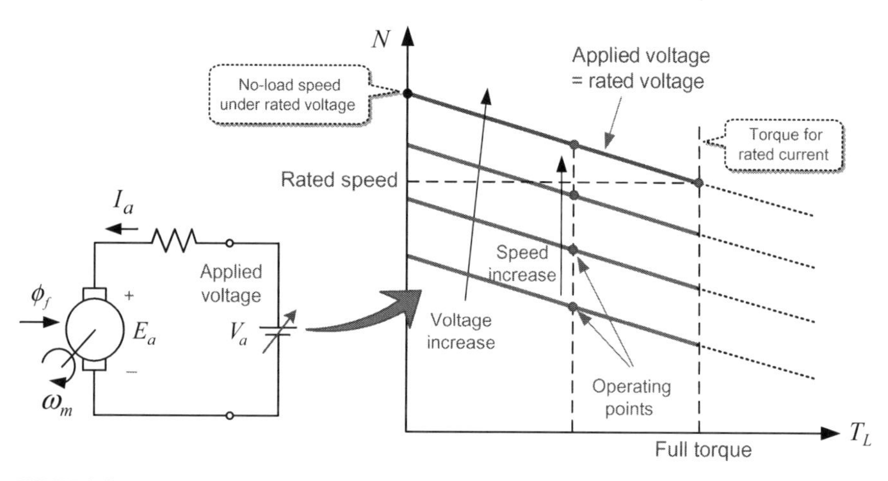

FIGURE 2.9

Armature voltage control.

2.3.2 FIELD FLUX CONTROL

Under a constant armature voltage, Eq. (2.5) can be rewritten as a function of field flux:

$$\omega_m = \frac{V_a}{k\phi_f} - \frac{R_a}{(K\phi_f)^2}T_L = \frac{K_1}{\phi_f} - \frac{K_2}{\phi_f^2}T_L \tag{2.7}$$

where $K_1 = V_a/k$ and $K_2 = R_a/k^2$.

From Eq. (2.7), we can see that the speed of the DC motor varies inversely with the field flux. Fig. 2.10 shows the torque–speed characteristic of the DC motor using the field flux control. When using this field flux control method, the field flux has to be controlled below the rated value to avoid magnetic saturation of iron core because the rated flux is commonly given as a value near the knee of the magnetization curve as shown in the left side of Fig. 2.10. If the field flux is reduced below the rated value, then the speed can increase above the rated speed.

When the speed of the DC motor can be controlled above the rated value by adjusting the field flux control, there are potential problems that can arise. When the field flux is reduced to increase the speed, the developed torque will also be reduced for the same armature current. If the armature current is increased to produce the same torque, then the efficiency will be lowered due to the increased copper losses. The armature current also may not be fully increased due to the thermal limit and the capacity of the DC power supply. Moreover the inductance of the field winding is normally designed to be large to produce the flux effectively. Thus the field flux cannot be changed rapidly, and this will result in a slow response of the speed regulation. In addition, in a high-speed range, the speed regulation will be poor for the applied load, and the resultant lower field

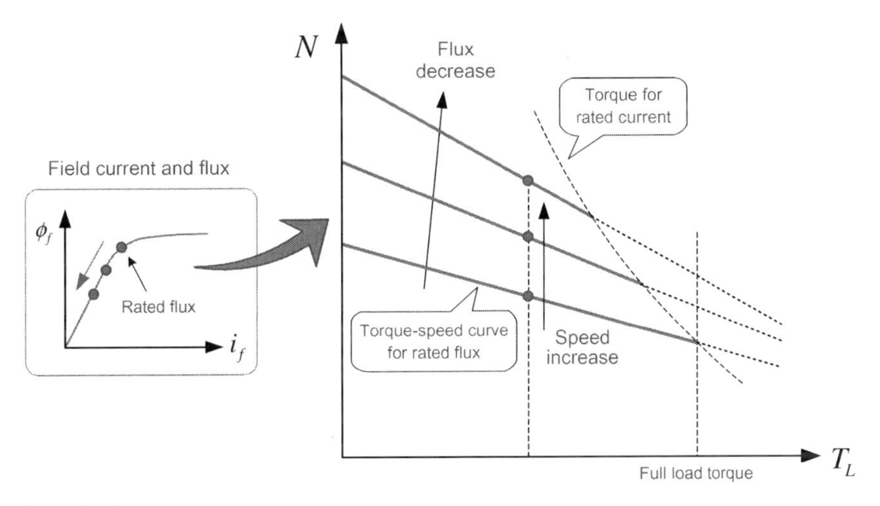

FIGURE 2.10

Field flux control.

flux will be easily influenced by the armature reaction. Besides these electrical problems, the DC motors also suffer from speed limitations due to mechanical commutation devices. On this account, the achievable speed range of this field flux control is generally limited up to three times the rated speed.

2.3.3 OPERATION REGIONS OF DIRECT CURRENT MOTORS

The operating speed range of the DC motor is divided into two regions as follows: *constant torque region* and *constant power region*. The operation in these two regions can be achieved through the armature voltage control and the field flux control.

2.3.3.1 Constant torque region ($0 \leq \omega_m \leq$ base speed ω_b)

In the operating region below the rated speed, the field flux is maintained at a constant and the torque production is controlled only by the armature current I_a. This region is called the *constant torque region* because the same torque can be developed by the same armature current at any speed within this region.

As the operating speed is increased, the back-EMF is increased. Thus the applied motor voltage should also be increased to maintain the same armature current, i.e., the same torque. Like this, in this region, the terminal voltage of the DC motor varies with its speed. Because the voltage applied to the motor should not exceed the rated value, the range of this region is normally limited to the speed at which the terminal voltage of the DC motor reaches the rated value. This speed may be different from the rated speed depending on the operating conditions. Thus it is usually referred to as the base speed ω_b.

2.3.3.2 Constant power region ($\omega_m \geq$ base speed): field-weakening region

Once the motor voltage reaches the rated value, the speed can no longer be controlled by the voltage. At speeds higher than the base speed, the applied voltage remains constant. Thus, instead of the voltage control, the speed can be increased by the reduction of the field flux (is called *field-weakening operation*). The reduction of the field flux can make the back-EMF constant regardless of the speed increase. This allows the armature current, which produces a torque, to flow. However, even though the same current is flowing, the output torque will be decreased with the operating speed due to the reduced field flux. Because the terminal voltage and the current remain constant in this region, the motor operates in a constant power, $P = (V_a \cdot I_a = T_e \cdot \omega_m)$. Thus this speed region is called the *constant power region*.

Fig. 2.11 shows the speed–torque characteristic of the DC motor, which is commonly called the *capability curve*. The capability curve demonstrates the achievable speed range and the output torque of a motor under the available voltage and current. For DC motor drives with the four-quadrant operation modes as described in Section 1.3.2, this speed–torque characteristic appears in all four quadrants. Considering the voltage drop in armature resistance and losses, the output torque in Quadrants 2 and 4 are somewhat larger than those in Quadrants 1

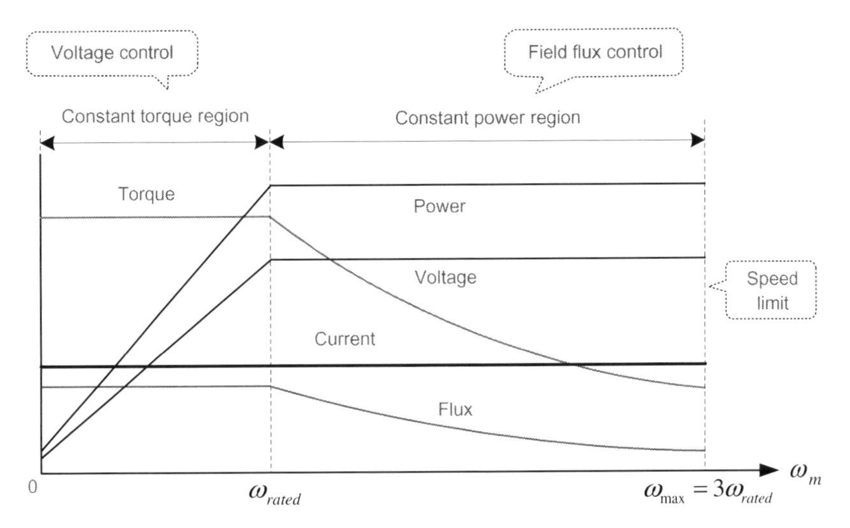

FIGURE 2.11

Capability curve of the DC motor.

and 3. For the four-quadrant operation, the direction of the output torque needs to be reversed to change the direction of the rotor rotation. In this case, changing the polarity of the armature current will be more effective. Even though the direction of the torque may be changed by reversing the polarity of field flux, its response is slow due to a large inductance of the field winding, and thus the stability of the drive system may be deteriorated.

EXAMPLE 1

The rated values for a separately excited DC motor are 5 kW, 140 V, 1800 r/min and the efficiency η is 91% (the losses such as mechanical losses and iron losses are ignored). The armature resistance is 0.25 Ω. Evaluate the following questions.

1. the armature current and the rated torque
2. the speed at full-load when the armature voltage is reduced by 10%
3. the speed at full-load current when the flux is reduced by 10% and the output torque available by armature current at this speed

Solution

1. From the relation of the input and output powers as

$$P_{out} = \frac{P_{in}}{\eta} = \frac{V_a \cdot I_a}{\eta}.$$

the armature current is

$$I_a = \frac{P_{out}}{V_a \cdot \eta} = \frac{5\ \text{kW}}{125\ \text{V} \cdot 0.91} = 44.96\ \text{A}.$$

From the output power $P_{out} = T_e \cdot \omega_m$, the rated torque is

$$T_e = \frac{P_{out}}{\omega_m} = \frac{3150\text{ W}}{\left(\frac{3000 \cdot 2\pi}{60}\right)} = 26.5\text{ N m}.$$

2. The armature voltage reduced by 10%, V_{new}, is 112.5V.
At this voltage, the speed is

$$\omega_m = \frac{E_a}{k\phi_f} = \frac{V_{new} - R_a I_a}{k\phi_f} = \frac{101.25}{0.60344}.$$
$$= 167.8\text{ rad/s } (= 1602.2\ r/\text{min})$$

Here, $k\phi_f = E_a/\omega_m = (V_{new} - R_a I_a)/\omega_m = 0.60344$.

3. The speed at the flux reduced by 10% is

$$\omega_m = \frac{E_a}{0.9k\phi_f} = \frac{(V_{new} - R_a I_a)}{0.9k\phi_f}$$
$$= \frac{113.75}{(0.9 \cdot 0.60344)} = 209.44\text{ rad/s}(= 2000\text{ r/min})$$

At this speed, the output torque available by armature current is

$$T_{e@2000} = 0.9k\phi_f I_a = 23.87\text{N m}.$$

2.4 TRANSIENT RESPONSE CHARACTERISTICS OF DIRECT CURRENT MOTORS

In Section 2.3, we studied the steady-state characteristics of DC motors, which demonstrate the speed control by the armature voltage or the field flux. The steady-state characteristic shows what the final speed is, whereas the transient response shows how to reach the final speed. In addition to the steady-state characteristic, the transient characteristic is very important for high-performance motor drives. This is because the transient response characterizes the dynamics of the drive system. As an example, transient responses to a step speed command are shown in Fig. 2.12.

Even though all three responses reach the same final speed, their response times and moving trajectories to reach the final speed are different. Although an oscillatory or slow response can ultimately achieve the speed command, it has no practical use. Therefore we need to take into account the transient characteristic as well as the steady-state characteristic for high-performance motor drives.

Now, let us take a look at the transient response of reaching the final speed by the applied armature voltage for a DC motor. An easy way to examine the transient response is to use the transfer function of the system. We can obtain the

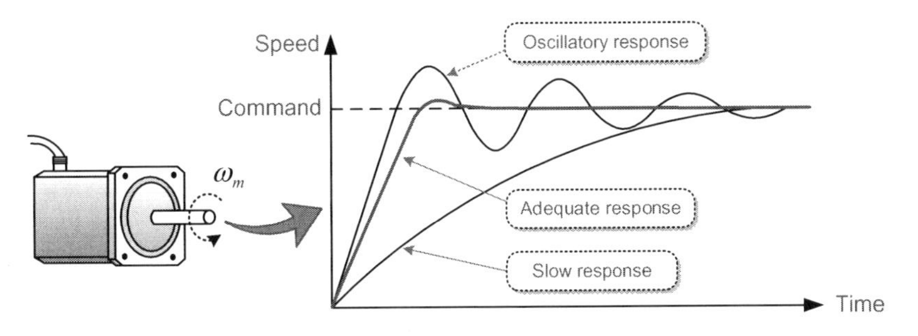

FIGURE 2.12

Transient response to a step speed command.

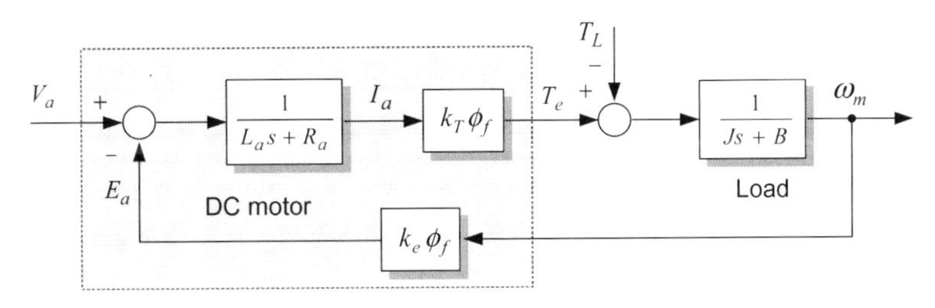

FIGURE 2.13

Block diagram for a DC motor.

transfer function for the DC motor drive system by taking the *Laplace* transform of the differential equations of Eq. (2.1) and (2.4) as follows.

$$\begin{aligned} v_a = R_a i_a + L_a \frac{di_a}{dt} + e_a \rightarrow V_a(s) &= (R_a + sL_a)I_a(s) + E_a(s) \\ &= (R_a + sL_a)I_a(s) + k_e\phi_f\omega_m(s) \end{aligned} \tag{2.8}$$

$$T_e = J\frac{d\omega_m}{dt} + B\omega_m + T_L \rightarrow T_e(s) = (Js + B)\omega_m(s) + T_L = k_T\phi_f I_a(s) \tag{2.9}$$

Here, s denotes the *Laplace* operator. Eqs. (2.8) and (2.9) together can be represented by the closed-loop block diagram as shown in Fig. 2.13.

From the block diagram in Fig. 2.13, the transfer function from the input armature voltage $V_a(s)$ to the output angular velocity $\omega_m(s)$ is given by

$$\begin{aligned} \frac{\omega_m(s)}{V_a(s)} &= \frac{\frac{1}{L_a s + R_a} \cdot k_e\phi_f \cdot \frac{1}{Js + B}}{1 + \left(\frac{1}{L_a s + R_a} \cdot k_e\phi_f \cdot \frac{1}{Js + B} k_T\phi_f\right)} \\ &= \frac{\frac{k}{JL_a}}{s^2 + \left(\frac{R_a}{L_a} + \frac{B}{J}\right)s + \left(\frac{R_a}{L_a}\frac{B}{J} + \frac{k^2}{JL_a}\right)} \quad (k = k_T\phi_f = k_e\phi_f) \end{aligned} \tag{2.10}$$

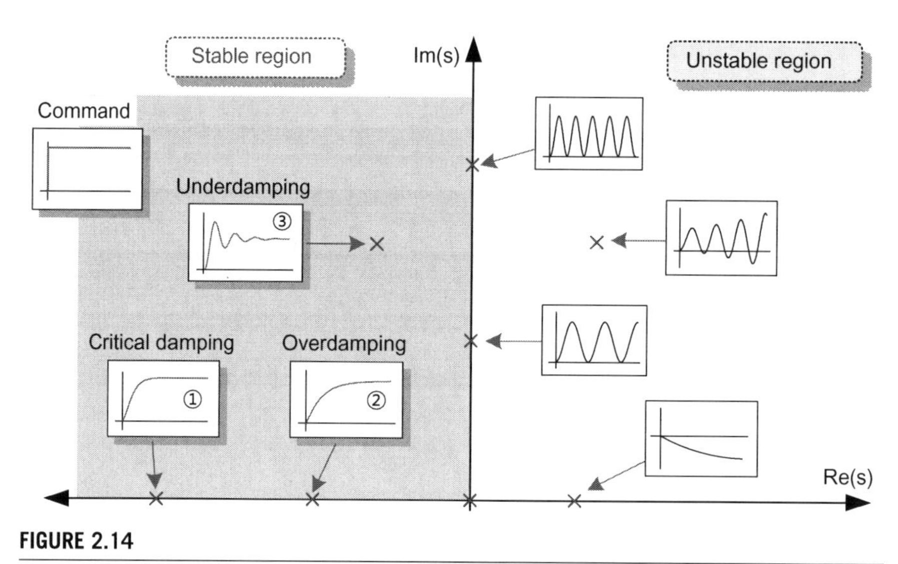

FIGURE 2.14

System pole locations and the corresponding transient responses.

This transfer function is a second-order system including two poles, which are important factors in determining the transient response of the system. Fig. 2.14 shows the pole locations of a system in the s-plane and the corresponding transient responses. The imaginary axis location of the pole determines the frequency of oscillations in the response. Thus, if the poles are complex roots, the response is oscillatory. The real axis location of the pole determines the speed of the exponential decay of the transient situation. Therefore, the farther the distance of the pole from the origin is to the left in the s-plane, the faster the transient situation disappears, i.e., the system reaches the final value faster.

CLOSED-LOOP TRANSFER FUNCTION

A system with a feedback loop has the following transfer function:

$$\frac{Y(s)}{R(s)} = \frac{G(s)}{1 + G(s)H(s)}$$

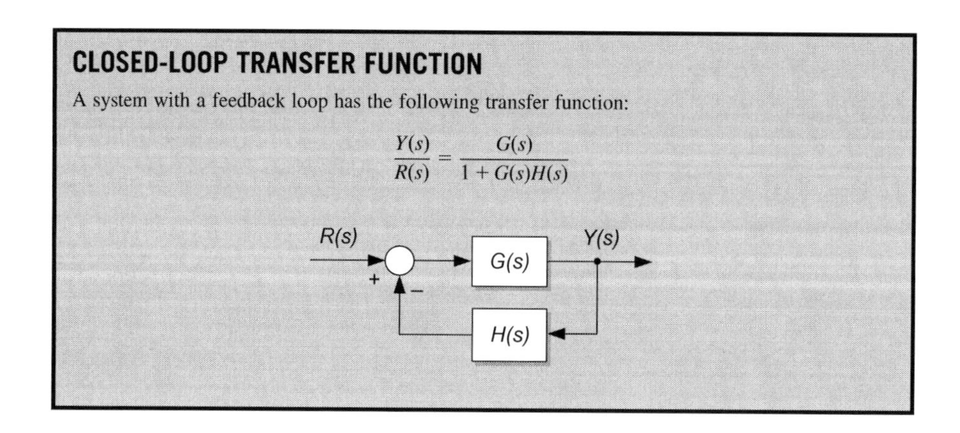

From the locations of the two poles in Eq. (2.10), we can predict the transient response of the speed to the applied armature voltage. Thus the transient response depends heavily on the system parameters such as motor parameters R_a and L_a and mechanical parameters J and B, as shown in the following.

Assuming that $B = 0$ in Eq. (2.10) to analyze the system easily, we can obtain a simplified transfer function as

$$\frac{\omega_m}{V_a} = \frac{\dfrac{k}{JL_a}}{s^2 + \left(\dfrac{R_a}{L_a}\right)s + \left(\dfrac{k^2}{JL_a}\right)} = \frac{\dfrac{1}{k}\left(\dfrac{R_a}{L_a} \cdot \dfrac{k^2}{JR_a}\right)}{s^2 + \left(\dfrac{R_a}{L_a}\right)s + \left(\dfrac{R_a}{L_a} \cdot \dfrac{k^2}{JR_a}\right)} \tag{2.11}$$

Letting the electromechanical time constant $T_m = JR_a/k^2$ and the electrical time constant $T_a = L_a/R_a$, Eq. (2.11) can be rewritten as

$$\frac{\omega_m}{V_a} = \frac{\dfrac{1}{k}\left(\dfrac{1}{T_m T_a}\right)}{s^2 + \dfrac{1}{T_a}s + \dfrac{1}{T_m T_a}} \tag{2.12}$$

The poles can be obtained from the roots of the denominator (also called the characteristic equation) of Eq. (2.12) as

$$s_{1,2} = -\frac{1}{2T_a} \pm \frac{1}{T_a}\sqrt{\frac{1}{4} - \frac{T_a}{T_m}} \tag{2.13}$$

If $T_m \geq 4T_a$, then the characteristic equation has two real roots. A DC motor drive system with a large J or R_a corresponds to this case, and its response is either ① or ② in Fig. 2.14. If $T_m < 4T_a$, then the characteristic equation has two complex roots. In this case, the system response is oscillatory such as ③ in Fig. 2.14.

We can also find out the response characteristic from the damping ratio for this second-order system. To do so, we compare this equation with the following prototype of the second-order system.

$$G(s) = \frac{\omega_n^2}{s^2 + 2\zeta\omega_n s + \omega_n^2} \tag{2.14}$$

Given Eq. (2.14), the undamped natural frequency ω_n and the damping ratio ζ for the DC motor drive system are given by

$$\omega_n = \frac{1}{\sqrt{T_a T_m}} = \frac{k}{\sqrt{JL_a}} \tag{2.15}$$

$$\zeta = \frac{1}{2}\sqrt{\frac{T_m}{T_a}} = \frac{1}{2}\frac{R_a}{k}\sqrt{\frac{J}{L_a}} \tag{2.16}$$

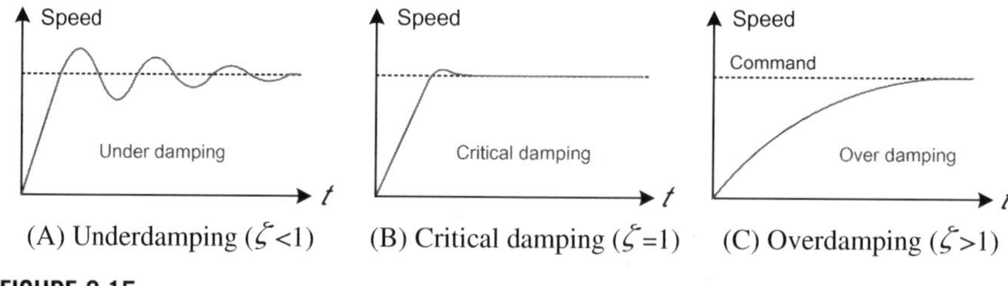

FIGURE 2.15

System responses according to the damping ratio. (A) Underdamping ($\zeta < 1$), (B) critical damping ($\zeta = 1$), and (C) overdamping ($\zeta > 1$).

There are three typical responses of the system according to the value of the damping ratio ζ: *underdamping* ($0<\zeta<1$), *critical damping* ($\zeta = 1$), or *overdamping* ($\zeta>1$) [1]. As shown in Fig. 2.15, in the case of underdamping ($0<\zeta<1$), the system comes rapidly to equilibrium but has a slight oscillatory response. As ζ is increased, the oscillation is reduced. In the case of overdamping ($\zeta>1$), the system is not oscillatory but has a very slow response. A critically damped system response ($\zeta=1$), which comes to equilibrium as fast as possible without oscillating, is the most desirable.

Eq. (2.16) implies that the damping ratio ζ, which implies the dynamic performance of the speed to the applied armature voltage, depends on the electromechanical time constant value T_m and the electrical time constant value T_a. If $T_m<4T_a$, then $\zeta<1$ and its speed response will be oscillatory. By contrast, in the case of $T_m \geq 4T_a$, the system will have a stable speed response without oscillation.

Normally, small motors have $T_m>4T_a$, while servo motors or high power motors have $T_m<4T_a$, resulting in an oscillatory response. Even though the system response is inherently oscillatory, we can achieve the desired response by adopting an adequate controller, which will be described later.

In most motor drive systems, the time response of an electric system is faster than that of a mechanical system. In that case, we can ignore the transient of the electric circuit so that $L_a = 0$. Thus Eq. (2.11) can be written as

$$\frac{\omega_m}{V_a} = \frac{k}{R_a J s + K^2} = \frac{1}{k}\frac{\omega_c}{s + \omega_c} \quad \left(\omega_c = \frac{k^2}{R_a J}\right) \tag{2.17}$$

This equation implies that, in a DC motor, the relationship between the speed ω_m and the armature voltage V_a can be considered as a first-order, low-pass filter with a cut-off frequency ω_c. This means that the speed ω_m is directly proportional to the armature voltage V_a with its changing rate below the ω_c, though there is a time delay between the two. Generally it can be assumed that the rotational speed of a DC motor is proportional to the voltage applied to it.

EXAMPLE 2

Consider a separately excited DC motor with the following parameters. The armature resistance is 0.28 Ω, the armature inductance is 1.7 mH, the inertia of moment J of this motor is 0.00252, and $k\phi_f = 0.4078$.

1. Calculate the locations of the two poles and predict the transient response of the speed to an armature voltage at no-load.
2. Predict the transient response when this motor is driving the load with the inertia of moment of $6J$.

Solution

1. From Eq. (2.13), the two poles are

$$s_{1,2} = -\frac{1}{2T_a} \pm \frac{1}{T_a}\sqrt{\frac{1}{4} - \frac{T_a}{T_m}} = -76.47 \pm j181.57.$$

Since the system has two complex poles, we can expect an oscillatory system response. Also, from Eq. (2.16), the damping ratio is

$$\zeta = \frac{1}{2}\sqrt{\frac{T_m}{T_a}} = \frac{1}{2}\frac{R_a}{k}\sqrt{\frac{J}{L_a}} = 0.3881.$$

Because $\zeta < 1$, we can expect the system to have an under-damped response and an overshoot. From Figure A, we can see the oscillatory speed response of this system to the step applied voltages.

2. When this motor is driving the load with the inertia of moment of $6J$, the system has two real poles of -152 and -0.0002. Thus the system response is not oscillatory. Also, since $\zeta = 1.0269$, we can expect the critically damped system response to be as Figure B.

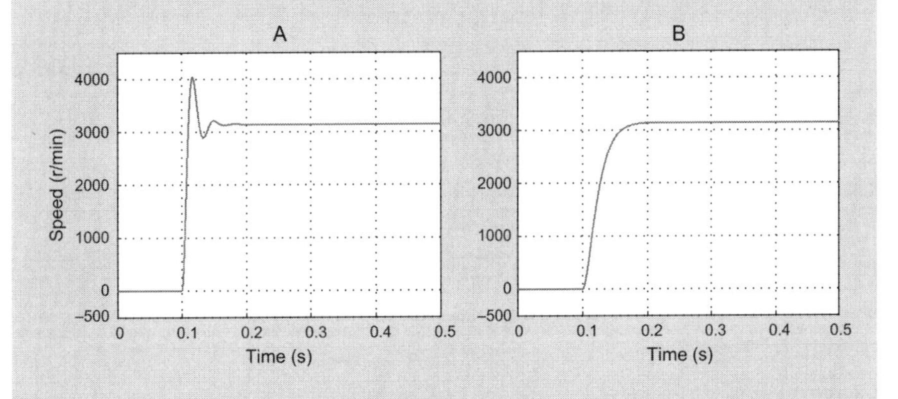

2.5 MOTOR CONTROL SYSTEM

An electric motor is an electromechanical device that converts electrical energy into mechanical torque. As a torque-producing device, the electric motor needs to have its torque controlled as a priority. When the torque produced by the motor is applied to a mechanical load, the torque will change the speed ω_m of the driven load and, in turn, change its position θ_m as shown in Fig. 2.16.

Therefore, in motor drives, the most effective way for controlling the position or speed of the load connected to the motor is to control the torque of the motor directly. However, the torque control is expensive due to the high cost of the torque sensor for motor torque feedback. We can see from Eq. (2.3) that the torque is directly proportional to the current under a constant field flux. This tells us that we can control the torque by using the current instead of the torque itself. Thus the current control is important for the speed or position control of a motor.

2.5.1 CONFIGURATION OF CONTROL SYSTEM

Fig. 2.17 illustrates the typical configuration of controllers in the motor drive systems, in which the controllers required for position, speed and current control are usually connected in a cascade.

Among these, the current controller must be placed at the most inner loop and has the widest bandwidth. As mentioned above, since the speed and position can be controlled through the current control, the response time for the current control must be the fastest among the three. Likewise, the response for the control of the inner loop should be sufficiently faster than that of the outer loop to improve the response performance and the stability of the outer control loop. For example, the bandwidth of the current loop needed to be at least five times wider than that of the speed loop to ensure that the current control will not have any influence on the speed control performance.

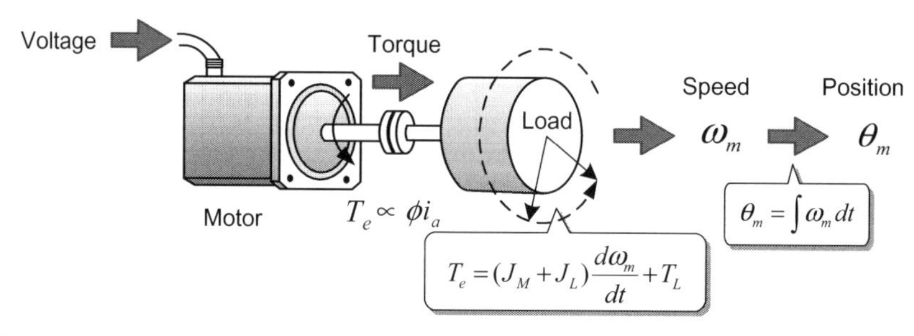

FIGURE 2.16

The motor drive system.

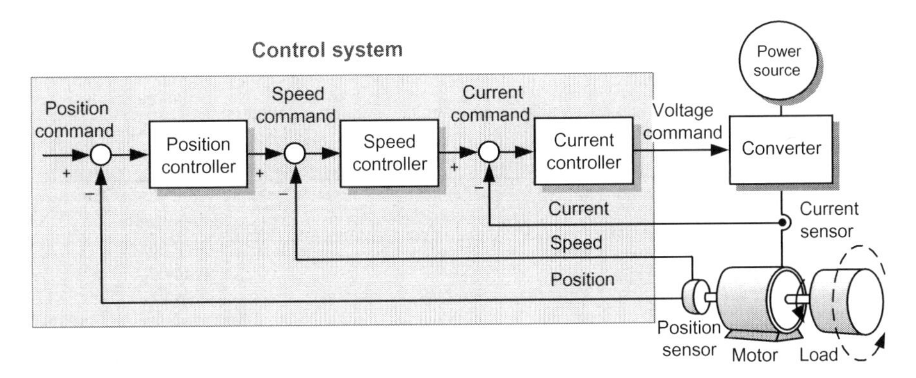

FIGURE 2.17

Configuration of the motor control system.

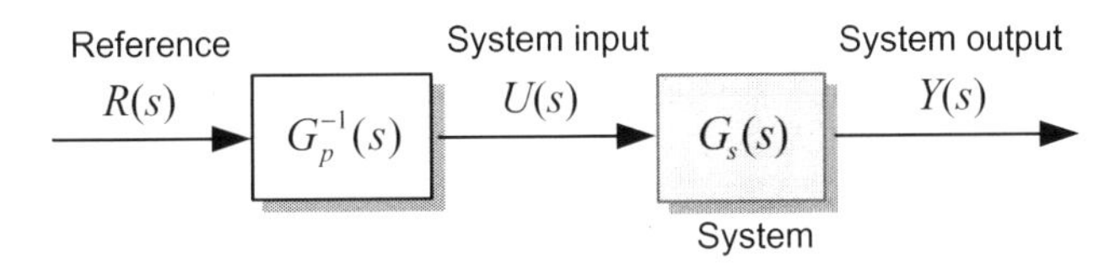

FIGURE 2.18

Open-loop control.

2.5.2 TYPES OF CONTROL

The aim of control is to adjust the control input so that the state or output of the system reaches the required goal. In the motor drive systems, the state or output may be the current, flux, speed, position, or torque, while the control input is only the input voltage. For control performance, the ability of reference tracking, i.e., the system output follows the reference input as accurately as possible, is the most important factor. The control system is also needed to have the disturbance rejection capability while tracking the reference input. This means that no disturbance can affect the system tracking the reference input as much.

There are two basic types of control: *open-loop control* and *closed-loop control*, also known as *nonfeedback* and *feedback*, respectively.

2.5.2.1 Open-loop control (or nonfeedback control)

Open-loop control is a type of control that determines the control input $U(s)$ of a system by using the system model $G_s(s)$ as shown in Fig. 2.18.

A drawback of the open-loop control is that it requires perfect knowledge of the system. Since the open-loop control does not use the feedback of the system output $Y(s)$, the error between the reference $R(s)$ and the output $Y(s)$ cannot be reflected in determining its control input $U(s)$. Therefore its output may be different from the desired goal due to the change in system parameters or disturbances.

Because of its simplicity and low cost, the open-loop control is often used in systems where it has the well-defined relationship between input and output and are not influenced by disturbances. However, this open-loop control is inadequate for the high-performance motor drive systems, which are easily affected by parameter variations and disturbances such as the back-EMF and load torque.

2.5.2.2 Closed-loop control (or feedback control)

Closed-loop control is a type of control that adjusts the control input $U(s)$ by the feedback of the output $Y(s)$, as shown in Fig. 2.19. The closed-loop control is generally used to control the position, speed, current, or flux in the motor drive systems.

The controller $G_c(s)$ adjusts its control input $U(s)$ to reduce the error between the output and the desired goal. Thus the role of the controller $G_c(s)$ in the closed-loop control is very critical to the system's performance. There are several types of widely used types of the feedback controller in the motor drive systems as follows.

2.5.2.2.1 Feedback control

Proportional controller (P controller)

In this type of controller, which is illustrated in Fig. 2.20, the controller output is proportional to the present error $e(t)(=r(t)-y(t))$ as given by

$$U(t) = K_p e(t) \tag{2.18}$$

where K_p represents the proportional gain.

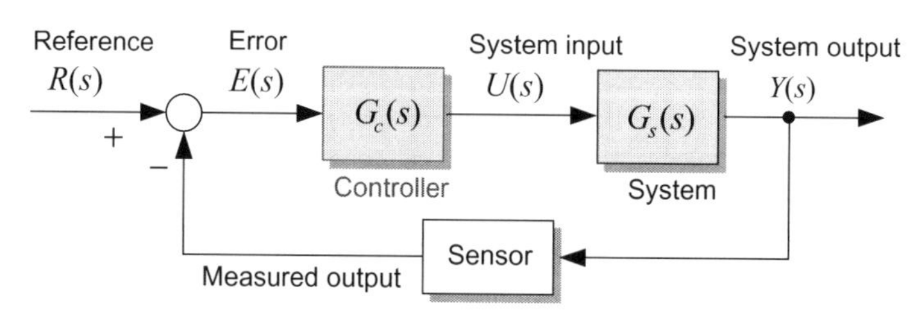

FIGURE 2.19

Closed-loop control.

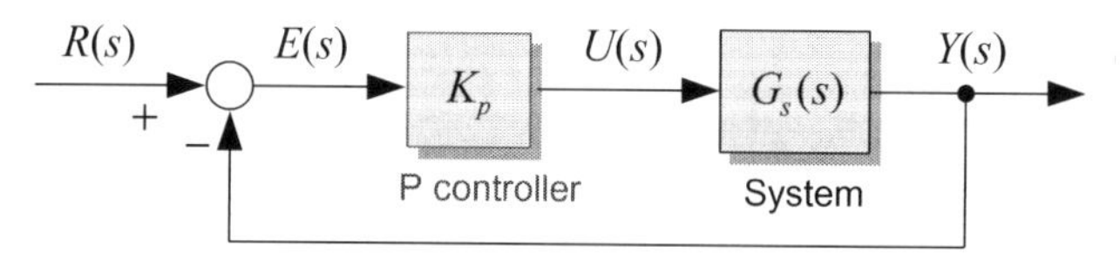

FIGURE 2.20

Proportional control.

An advantage of the proportional control is its immediate response to errors. The larger the proportional gain is, the faster the response is. Thus it should be noted that the proportional gain determines the control bandwidth. However, there is a limit to increasing the proportional gain because a large gain can lead to system instability or may require unrealizable control input. A drawback of the proportional control is that the error can never be zero. In other words, there exists a steady-state error in the P control. This is because when the error comes to zero, the output of the P controller becomes zero. Thus the system will have an error again. We can evaluate the steady-state error of the P controller from its close-loop transfer function of the block diagram in Fig. 2.20 as,

$$\frac{R(s)}{Y(s)} = \frac{G^o(s)}{1+G^o(s)} = \frac{K_pG_p(s)}{1+K_pG_p(s)} \tag{2.19}$$

where $G^o(s)(=K_pG_p(s))$ is the open-loop transfer function of the system including the proportional controller.

From Eq. (2.19), if the proportional gain K_p is very large, then $R(s)/Y(s)$ will be close to one and the error will be very small, but it will never reach zero.

Integral controller (I controller)

In this type of controller, the controller output is the integral of the error as shown in Fig. 2.21, so the past errors as well as the present error have an effect on the control input as given by

$$U(t) = K_i\int e(t)dt = \frac{K_p}{T_{pi}}\int e(t)dt \tag{2.20}$$

where K_i represents the integral gain and $T_{pi}(=K_p/K_i)$ represents the integral time constant.

The integral controller can remove the steady-state error completely for the non-time varying reference $R(s)$. This is because it continuously produces nonzero control input from the accumulated past errors even though the present error is zero. However, due to the accumulated errors in the integral term, it responds slowly to change in errors. For this reason, the integral controller is not normally used alone but is combined with the P controller.

Proportional–integral controller

A widely used controller in motor drive systems is the proportional–integral controller (PI controller), which is a combination of a proportional controller and an integral controller to achieve both a fast response and a zero steady-state error. The block diagram of a proportional–integral controller is shown in Fig. 2.22.

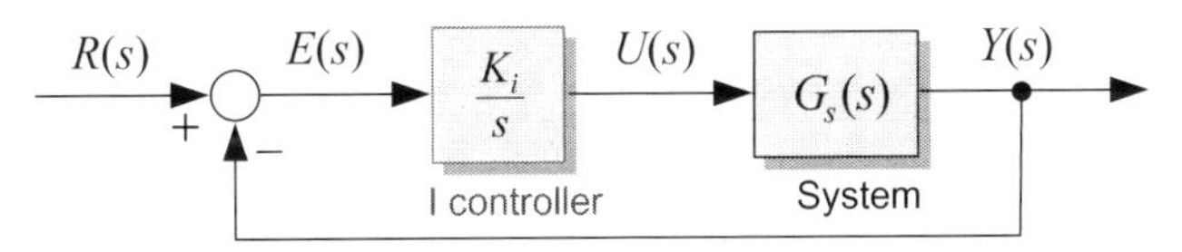

FIGURE 2.21

Integral control.

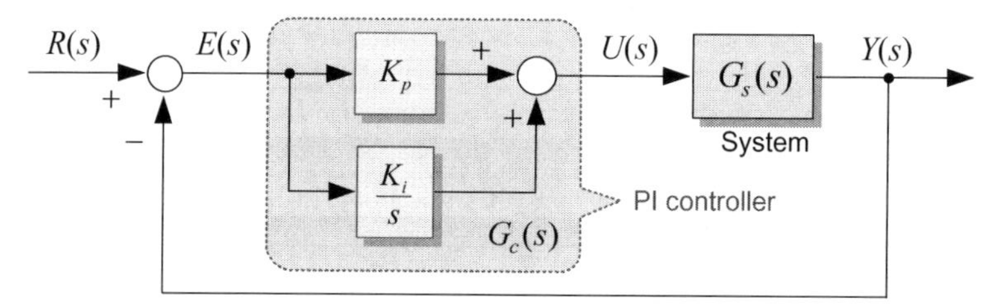

FIGURE 2.22

Proportional–integral control.

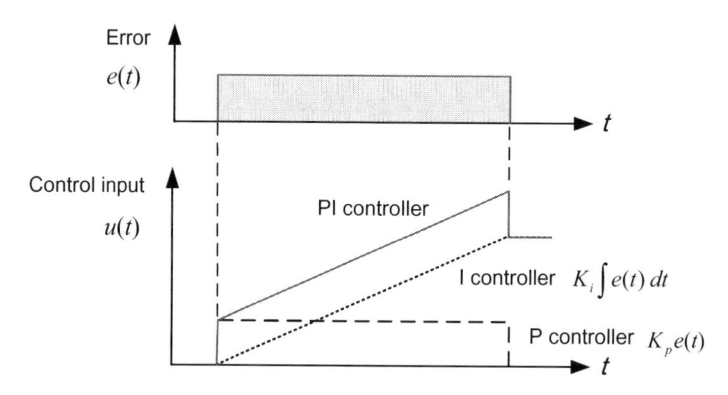

FIGURE 2.23

Control inputs for an error.

Fig. 2.23 compares the control inputs for an error in the three controllers described earlier.

2.5.2.3 Feedforward control

Disturbances are always present in the control process. In motor drive systems, e.g., the back-EMF acts as a disturbance on the current control, and the load torque acts as a disturbance on the speed and position control. We can see from the Section 2.5.2.2.1 that the feedback control operates to reduce the error between the output and the desired goal. Thus the feedback control can react to the disturbance only after the disturbance has an effect on the output. Accordingly, the feedback control cannot achieve a fast response to the disturbance.

Feedforward control is an effective way to rapidly eliminate the effect of the disturbance on the output. The feedforward control predicts and compensates for the disturbance by adding it to the control input in advance.

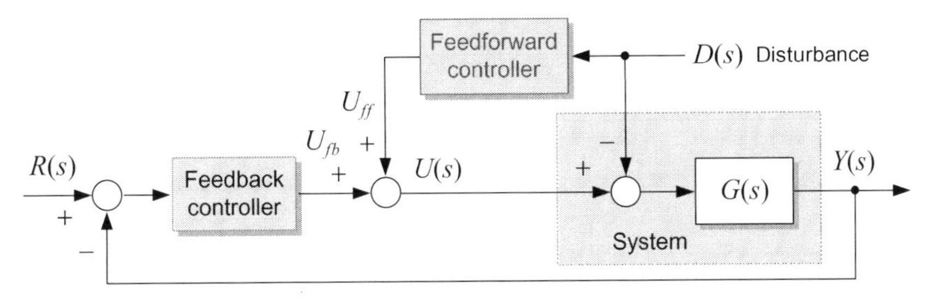

FIGURE 2.24

Feedforward control.

The feedforward control is always used along with the feedback control. Fig. 2.24 shows the block diagram of a typical system with the feedback control and the feedforward control. This control system can reduce the effect of the disturbance $D(s)$ on the output $Y(s)$ more than the system with the feedback control alone. For the feedforward control, accurate system parameters are required to predict the disturbance correctly.

In motor current control systems, the performance of the current control could be significantly improved by the feedforward control, which compensates for the back-EMF that acts as a disturbance on the current control. This feedforward control for current control of DC motors and AC motors will be described in detail later. For motor speed control, the performance of the speed control could also be significantly improved by feedforward control for the load torque.

2.5.3 DESIGN CONSIDERATION OF CONTROL SYSTEM

The PI controller has been widely used in motor drive systems as the controller for the current, speed, or flux linkage control because it could give a satisfactory performance. The performance of the PI controller depends heavily on its selected gains. Thus it is very important to select appropriate gains of the PI controller for obtaining a satisfactory performance in a given system.

Now, we will discuss the gain selection method for the current and speed PI controllers. Through the frequency response of the whole system including the PI controller, we will determine the P and I gains for achieving the desired transient and steady-state characteristics.

WHAT IS FREQUENCY RESPONSE?

The output of a linear system to a sinusoidal input is a sinusoid of the same frequency but with a different amplitude and phase. The *frequency response* refers to a system's open-loop responses at the steady state to sinusoidal inputs at varying frequencies as shown in the following figure.

(*Continued*)

WHAT IS FREQUENCY RESPONSE? (CONTINUED)

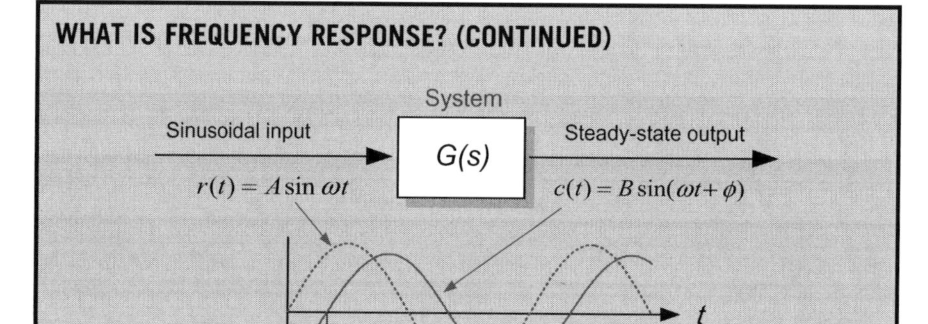

The frequency response is used to characterize the dynamics of a system and measure the ability of the system following an input reference. The frequency response is characterized by the magnitude and phase of the system's response as a function of frequency, which are calculated from the transfer function $G(s)$ of the system as

$$\left.\frac{C(s)}{R(s)}\right|_{s=j\omega} = G(s)|_{s=j\omega} = |G(j\omega)| \angle \phi(j\omega)$$

where $|G(j\omega)|$ and $\angle \phi(j\omega)$ represent the amplitude and phase angle between the input and output signals, respectively.

We can easily see the frequency response of a dynamic system through a *Bode plot*, which displays the magnitude (in dB) and the phase (in degrees) of the system response as a function of frequency.

When designing a control system, we should consider the three specifications: *stability, response time*, and *steady-state error*. We will begin with briefly reviewing these specifications.

2.5.3.1 Stability

First of all, the whole system, including the controller, must be stable. Therefore the stability is the most important requirement for the design of a control system. The response of an unstable system to a bounded input signal will continue to oscillate, but the magnitude of this oscillation will never diminish or get larger.

The stability of linear systems can be assessed by checking the poles of the transfer function of the system, i.e., the roots of the characteristic equation in the s-plane as we already saw in Fig. 2.14. A linear system is stable if *all* the poles lie in the left-half s-plane.

However, merely knowing whether a system is stable or unstable is not sufficient. We need to find out how stable the system is. We can measure the relative stability of a system by examining the *gain margin* and the *phase margin* obtained from the Bode plot of the frequency response. A brief description of the gain margin and the phase margin is presented below.

2.5.3.1.1 Gain margin and phase margin

Phase margin and gain margin are the measures of closed-loop control systems stability. They are illustrated in Fig. 2.25.

1. Gain margin
   Gain margin is defined as the amount of change in open-loop gain needed to make a closed-loop system unstable. The gain margin is the difference between 0 dB and the gain at the phase cross-over frequency that gives a phase of $-180°$. If the gain $|GH(j\omega)|$ at the frequency of $\angle GH(j\omega) = -180°$ is greater than 0 dB as shown in the left of Fig. 2.25, meaning a positive gain margin, then the closed-loop system is stable.
2. Phase margin
   Phase margin is defined as the amount of change in open-loop phase needed to make a closed-loop system unstable. The phase margin is the difference in phase between $-180°$ and the phase at the gain cross-over frequency that gives a gain of 0 dB. If the phase $\angle GH(j\omega)$ at the frequency of $|GH(j\omega)| = 1$ is greater than $-180°$ as shown in the left of Fig. 2.25, meaning a positive phase margin, the closed-loop system is stable.

In general, as the gain of a system increases, the system becomes less stable. The gain margin and the phase margin indicate how much the gain increases until the system becomes unstable.

Let us take a look at the relationship between the gain and phase margins and the stability of a system. For example, assume that the phase margin of a system is zero degree, i.e., $\angle GH(j\omega) = 180°$ at a frequency ω of $|G(j\omega)| = 1$ dB. In this case, as shown in Fig. 2.26, the phase difference between $Y(s)$ and $R(s)$ is 180°,

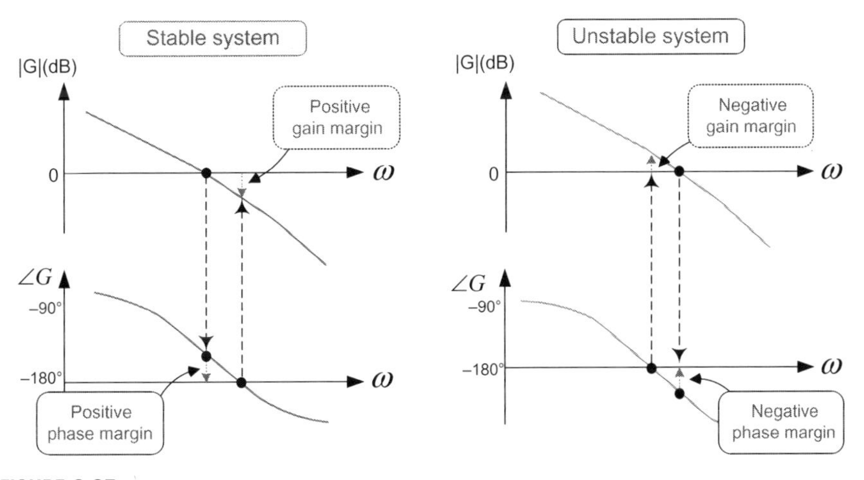

FIGURE 2.25

Gain margin and phase margin.

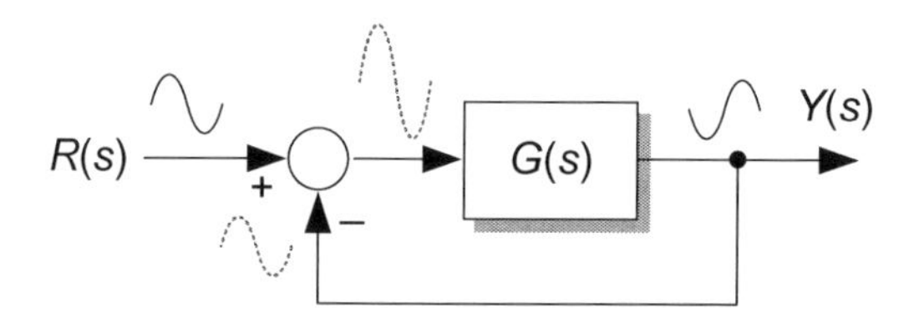

FIGURE 2.26

Example of an unstable system.

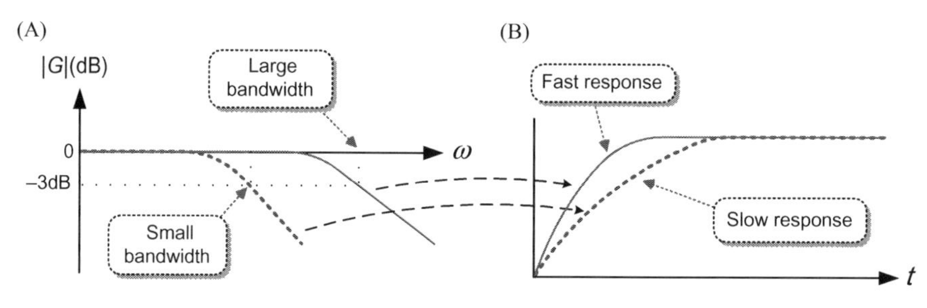

FIGURE 2.27

Speed of response versus system bandwidths. (A) Frequency responses and (B) step responses.

and this closed-loop system thus becomes a positive feedback system. This system will continue to amplify its output and become unstable.

If a closed-loop system is stable, both the gain margin and the phase margin need to be positive. In general, the phase margin of 30–60 degrees and the gain margin of 2–10 dB are desirable in the closed-loop system design. A system with a large gain margin and phase margin is stable but has a sluggish response, while the one with a small gain margin and phase margin has a less sluggish response but is oscillatory.

2.5.3.2 Response time/speed of response

How quickly a system responds to the change in input is an important measure of the system's dynamic performance. The speed of response of a closed-loop system may be proportional to its control bandwidth. Thus the control bandwidth is a measure of the speed of response.

We can see the system responds according to its control bandwidth from Fig. 2.27, which is showing the responses to a step command versus system bandwidths. Generally, the wider the system's control bandwidth, the faster the speed of response is. However, if the control bandwidth of a system is too large, the system may be sensitive to noise and become unstable.

This control bandwidth is mainly dependent on the proportional gain of the controller. Thus we need to adjust the P gain to obtain the desired response speed.

WHAT IS BANDWIDTH?

Bandwidth describes how well a system follows a sinusoidal input. We can easily find the bandwidth from the system's closed-loop frequency response. As in the following figure, the bandwidth is the range of frequencies for which the magnitude $|G|$ of the frequency response is greater than 3 dB, i.e., $0 \leq \omega \leq \omega_b$.

The bandwidth is a measure of the feedback system's ability to follow the input signal. Sinusoidal inputs with a frequency less than the bandwidth frequency ω_b are followed reasonably well by the system, whereas sinusoidal inputs with a frequency greater than the bandwidth frequency are attenuated in magnitude by a factor of 1/10 or greater (and are also shifted in phase). The concept of the bandwidth is very important in control systems because it describes how quickly a system follows its command.

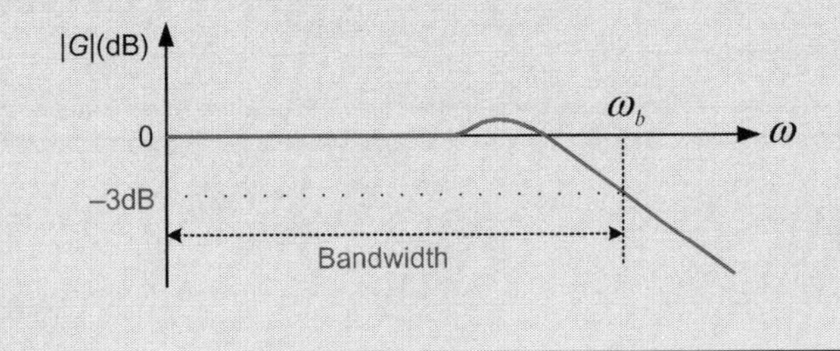

2.5.3.3 Steady-state error

A steady-state error is defined as the difference between the desired value and the actual value of a system when the response has reached the steady state. We can calculate the steady-state error of the system using the *final value theorem*. This theorem is given for the unit feedback system in Fig. 2.28 as

$$\text{Final value theorem: } e_{\infty} = \lim_{t \to \infty} e(t) = \lim_{s \to 0} sE(s) = \lim_{s \to 0} \frac{s}{1 + G(s)} R(s) \tag{2.21}$$

$$\left(\text{Error: } E(s) = R(s) - Y(s) = \frac{R(s)}{1 + G(s)} \right)$$

As an example, let us evaluate the steady-state error for a step input command in the control system for a DC motor, where the speed or current command is

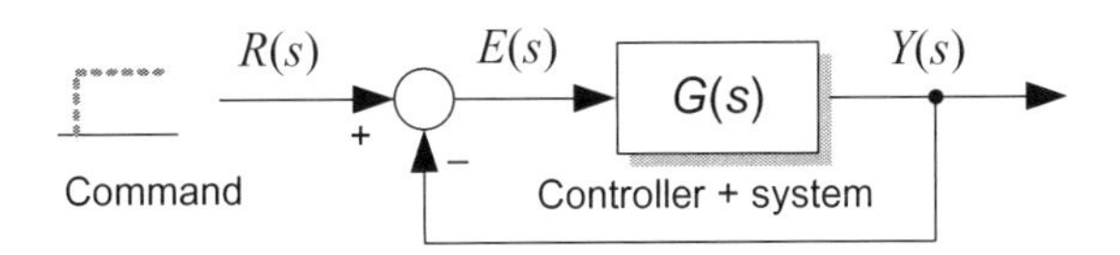

FIGURE 2.28

Unit feedback system.

generally given as a step signal. Assume that the control system for the DC motor is a unit feedback system as in Fig. 2.28, and the transfer function of the current control system is $G(s)$. For the unit step input ($r(t) = 1 \rightarrow R(s) = 1/s$) to this system, the steady-state error is given from the final value theorem by

$$e_{\infty} = \lim_{s \to 0} \frac{s}{1 + G(s)} R(s) = \lim_{s \to 0} \frac{s}{1 + G(s)} \frac{1}{s} = \lim_{s \to 0} \frac{1}{1 + G(s)} \tag{2.22}$$

The steady-state error depends on the system type (0, I, or II).

First, suppose a control system of type 0 expressed as $G(s) = K/(Ts + 1)$. In this system the steady-state error is given as

$$e_{\infty} = \frac{1}{1 + G(0)} = \frac{1}{1 + K} \tag{2.23}$$

The magnitude of the error depends on the system gain K and will be never zero.

Next, suppose a type I system of $G(s) = K/s(Ts + 1)$. This control system has no steady-state error as

$$e_{\infty} = \frac{1}{1 + G(0)} = \frac{1}{1 + \infty} = 0 \tag{2.24}$$

From this, it can be seen that if a control system includes an integrator $1/s$, then the steady-state error to a step reference will be zero.

2.6 CURRENT CONTROLLER DESIGN

The PI controller is the most commonly used in the DC motor drive systems for current control. In this section we will introduce how to design a current controller, i.e., how to select the P and I gains of a PI current controller for the desired performance.

The PI controller consists of a proportional term, which depends on the present error, and an integral term, which depends on the accumulation of past and present errors as follows:

$$G_{pi}(s) = K_p + \frac{K_i}{s} \tag{2.25}$$

where K_p and K_i are the proportional and integral gain, respectively.

Fig. 2.29 shows the frequency response of the PI controller. The PI controller has an infinite DC gain (i.e., gain when $\omega = 0$) due to an integrator.

The transfer function of Eq. (2.25) can be rewritten as

$$G_{pi}(s) = K_p \left(1 + \frac{1}{T_{pi} s} \right) = K_p \frac{\left(s + \frac{1}{T_{pi}} \right)}{s} \tag{2.26}$$

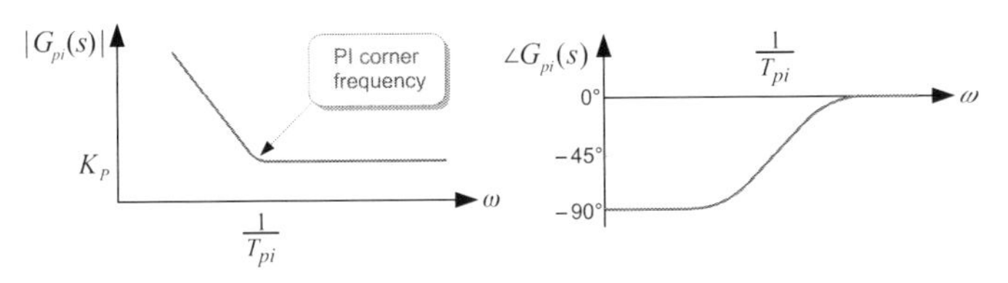

FIGURE 2.29

Frequency response of the PI controller.

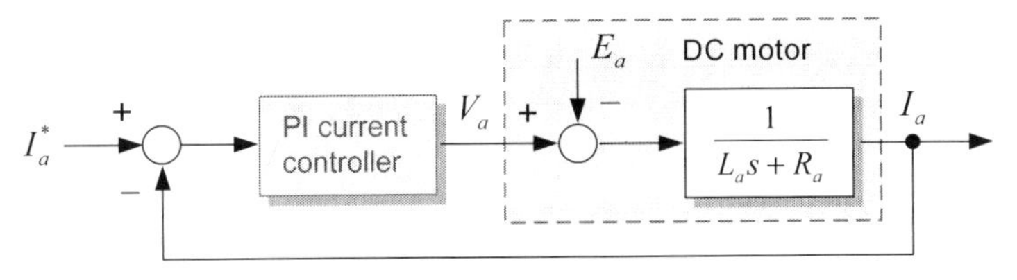

FIGURE 2.30

Block diagram of the DC motor including a current controller.

where $T_{pi} = K_p/K_i$ is the integral time constant of the PI controller and $1/T_{pi}$ is called the corner frequency of the PI controller.

From Eq. (2.26), we can see that the PI controller can add one pole at $s = 0$ and one zero at $s = -K_i/K_p$ to the transfer function of the system. The location of this zero has a strong influence on the system performance, which will be described in detail later. If the zero is correctly placed, then the damping of the system can be improved. As mentioned earlier, the pole at $s = 0$ of the PI controller causes the system type to be increased by adding an integrator $1/s$ to the transfer function, improving the steady-state error. However, the increased system type may make the system unstable when gains are large. In general, the PI controller can improve the steady-state error but at the expense of stability. In addition, the PI controller can reduce the maximum overshoot and improve the gain and phase margins as well as the resonant peak, but it reduces the bandwidth and extends the rise time. The PI controller also has a characteristic of a low-pass filter that reduces the noise.

Now, let us evaluate the performance of the PI controller for current control of a DC motor. The transfer function of a DC motor system that includes a PI current controller as shown in Fig. 2.30 is given as

$$I(s) = \frac{K_{pc}s + K_{ic}}{L_a s^2 + (R_a + K_{pc})s + K_{ic}} I^*(s) - \frac{s}{L_a s^2 + (R_a + K_{pc})s + K_{ic}} E(s) \tag{2.27}$$

where K_{pc} and K_{ic} represent the proportional and integral gains of the current controller, respectively.

We can see from Eq. (2.27) that the back-EMF $E(s)$ of the DC motor will act as a disturbance to the current control system. If the current reference of the

DC motor remains a constant value (i.e., $s = 0$), then $E(s)$ will have no influence on the current control performance. However, when the current reference $I^*(s)$ is changed, $E(s)$ may have an influence on the dynamic performance of the actual current $I(s)$ following its reference.

For a large system inertia, the effect of $E(s)$ on the current control can be ignored since the variation of $E(s)$ due to the speed variation is very small. However, for servo motors designed with a small inertia for a fast speed response, the effect of $E(s)$ cannot be ignored. To obtain a good performance of the current control by eliminating this undesirable effect of the back-EMF, we can adopt the feedforward control as described in the Section 2.5.2.3. Since the back-EMF of a DC motor is $e_a = k_e\phi_f\omega_m$, we can easily estimate it from the speed information. When the disturbance $E(s)$ is compensated by the feedforward control as in Fig. 2.31A, the DC motor will be simplified as a R-L circuit shown in Fig. 2.31B.

For the case of Fig. 2.31B, the current control will be easier and its performance will be improved. In this case, Eq. (2.27) becomes

$$\frac{I(s)}{I^*(s)} = \frac{K_{pc}s + K_{ic}}{L_a s^2 + (R_a + K_{pc})s + K_{ic}} \tag{2.28}$$

From Eq. (2.28), it can be seen that the current control characteristic depends on the performance of the current controller, which will be determined by the P and I gains selected for the PI controller.

Now, we will look at how the proper gains are selected to achieve the desired performance of the current control. Here, we will introduce a gain selection method based on the frequency response of the system.

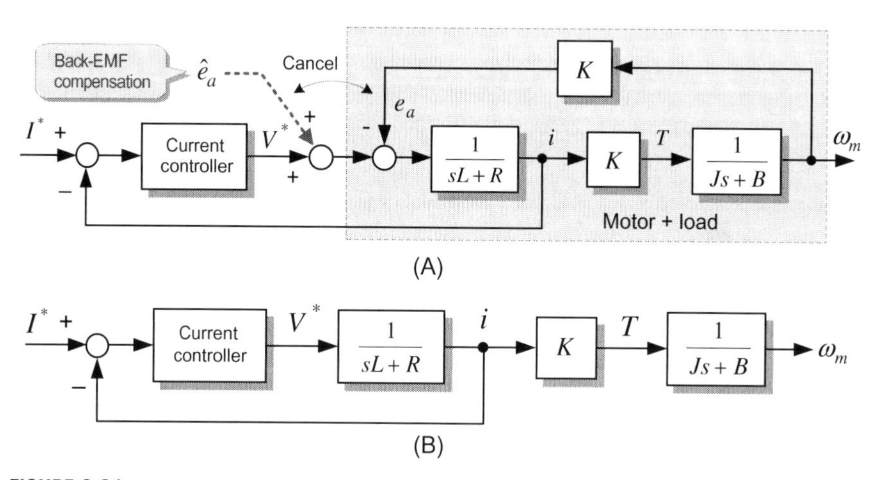

FIGURE 2.31

Current control system. (A) Feedforward control of the back-EMF and (B) system after compensating the back-EMF.

2.6.1 PROPORTIONAL–INTEGRAL CURRENT CONTROLLER

A current control system containing a PI controller is shown in Fig. 2.32. Here, assume that the feedforward control compensates the back-EMF of the DC motor perfectly.

In this PI current controller, the relationship between the current error and its output voltage is given as

$$V_a = K_{pc}\left(1 + \frac{1}{T_{pi}s}\right)(I_a^* - I_a) \tag{2.29}$$

where $T_{pi}(=K_{pc}/K_{ic})$ is the integral time constant of the PI current controller and K_{ic} is the integral gain.

There are several PI gain tuning rules such as the famous Ziegler–Nichols method. Here, we will adopt the technique called the *pole-zero cancellation method.* By using the pole-zero cancellation, we can remove the current control characteristic of the DC motor itself so that the PI controller may determine the performance of the current control.

The open-loop transfer function $G_c^o(s)$ of this system is given by

$$G_c^o(s) = K_{pc}\left(\frac{s + \frac{1}{T_{pi}}}{s}\right)\cdot\frac{1}{L_a s + R_a} = K_{pc}\frac{\left(s + \frac{K_{ic}}{K_{pc}}\right)}{s}\cdot\frac{\frac{1}{L_a}}{\left(s + \frac{R_a}{L_a}\right)} \tag{2.30}$$

If the zero $(-K_{ic}/K_{pc})$ of the PI controller is designed to cancel the pole $(-R_a/L_a)$ of the DC motor by the pole-zero cancellation method, i.e.

$$\frac{1}{T_{pi}} = \frac{K_{ic}}{K_{pc}} = \frac{R_a}{L_a} \tag{2.31}$$

then, Eq. (2.30) becomes

$$G_c^o(s) = \frac{1}{\left(\frac{L_a}{K_{pc}}\right)s} \tag{2.32}$$

Fig. 2.33 shows the bode plot of the open-loop frequency response from $G_c^o(s)$. The phase is -90° at the gain cross-over frequency ω_{cc} where the magnitude $|G_c^o(j\omega)| = 0$ dB. Thus the gain margin is positive and the system will be

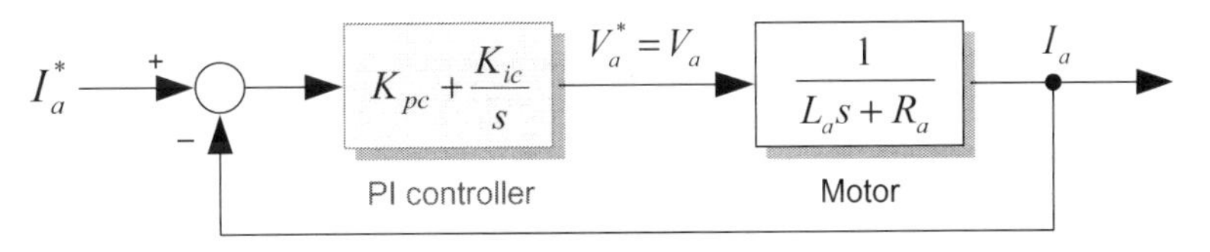

FIGURE 2.32

PI current control system.

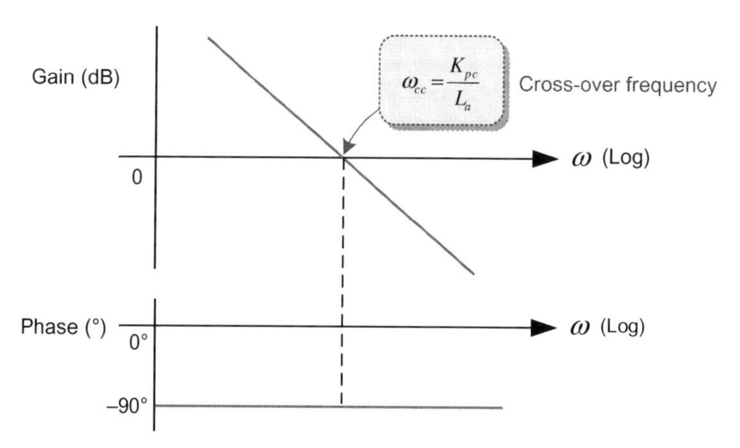

FIGURE 2.33

Open-loop frequency response.

stable. Also, in this case, the transfer function of Eq. (2.32) becomes the system of type = 1 and, thus, we can expect the steady-state error to be zero as

$$e_{\infty} = \lim_{s \to 0} \frac{1}{1 + G_c^o(s)} = \frac{1}{1 + \infty} = 0 \tag{2.33}$$

From Eq. (2.32), the gain cross-over frequency ω_{cc} can be determined by

$$\left| G_c^o(j\omega_{cc}) \right| = \frac{1}{\left| \frac{L_a}{K_{pc}} j\omega_{cc} \right|} = 1 \rightarrow \omega_{cc} = \frac{K_{pc}}{L_a} \tag{2.34}$$

The gain cross-over frequency ω_{cc} of the open-loop frequency response is equal to the gain cut-off frequency of the close-loop frequency response, which indicates the bandwidth of the current control system. Let us check the bandwidth of this current control system.

The closed-loop transfer function of this system is given by

$$\frac{I(s)}{I^*(s)} = G_c^c(s) = \frac{G_c^o(s)}{1 + G_c^o(s)} = \frac{1}{\left(\frac{L_a}{K_{pc}} \right) s + 1} = \frac{\omega_{cc}}{s + \omega_{cc}} \tag{2.35}$$

Its frequency response is shown in Fig. 2.34. We can see that the bandwidth of this system is given as ω_{cc} by letting $\left| G_c^c(j\omega) \right|$ equal to $1/\sqrt{2}(=-3\ \text{dB})$ as

$$\left| G_c^c(j\omega) \right| = \left| \frac{\omega_{cc}}{j\omega + \omega_{cc}} \right| = \frac{1}{\sqrt{2}} \tag{2.36}$$

Thus the bandwidth is equal to the gain cross-over frequency of the open-loop frequency response.

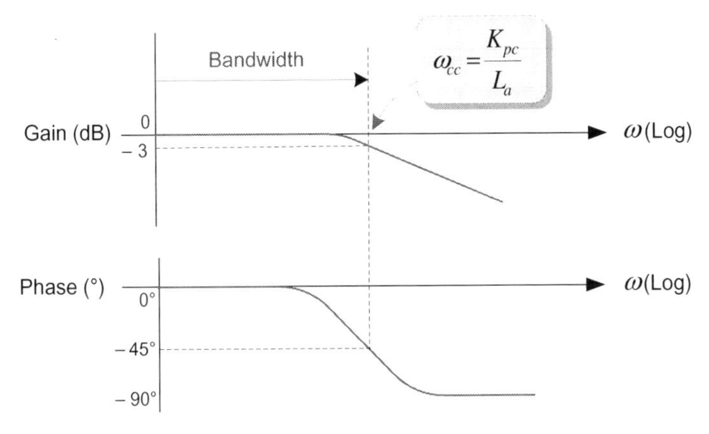

FIGURE 2.34

Closed-loop frequency response.

Like this, if the zero of the PI controller is designed to cancel the pole of the system, the current control system is expressed simply as a first-order system with a cut-off frequency ω_{cc}, which is stable. In this case, there will be no overshoot in the response and the time required for the response to reach the final value will be about four times the time constant of the system.

Based on the explanation given above, the proportional and integral gains of the current controller needed to achieve the required bandwidth can be determined as follows.

If the required control bandwidth is ω_{cc}, then the proportional gain K_{pc} can be obtained from Eq. (2.34) and the integral gain K_{ic} can be obtained from Eq. (2.31) as

$$\text{Proportional gain: } K_{pc} = L_a \cdot \omega_{cc} \tag{2.37}$$

$$\text{Integral gain: } K_{ic} = \frac{R_a}{L_a} K_{pc} = R_a \cdot \omega_{cc} \tag{2.38}$$

It is important to note that the gains of the controller depend on the motor parameters L_a and R_a. Therefore the gain values to achieve the same current control performance may differ from motor to motor. For example, a motor with a large value of winding inductance, whose current is difficult to change, needs a larger proportional gain to obtain an equal bandwidth of the current control than that with a small value of winding inductance, whose current is easy to change. Therefore, accurate information about the motor parameters is essential to achieve the required current control performance.

2.6.1.1 Selection of the bandwidth for current control

The value of the proportional gain K_{pc} determines how fast the system responds, whereas the value of the integral gain K_{ic} determines how fast the steady-state error is eliminated. When the value of these gains is larger, the control performance is better. However, large gains may lead to an oscillatory response and result in an unstable system. As can be seen in Eq. (2.37) and (2.38), since the gains of a current controller are determined by its bandwidth ω_{cc}, we first need to select an appropriate bandwidth ω_{cc} for the target current control system. A wider bandwidth will lead to a faster response but an unstable system.

The bandwidth ω_{cc} of a current controller is limited by the following two factors: the switching frequency of the power electronic converter realizing the output of the current controller and the sampling period for detecting an actual current for feedback in the digital controller. Since a motor current cannot change faster than the switching frequency of the power electronic converter, the switching frequency limits the bandwidth of the current control. If the current is sampled twice every switching period, as a rule of thumb, the maximum available bandwidth can be up to 1/10 of the switching frequency. On the other hand, if it is sampled once every switching period, the maximum available bandwidth can be up to 1/20 of the switching frequency. For this case, it is desirable to restrict the bandwidth to 1/25 of the sampling frequency [1,2].

As an example, suppose that the switching frequency of a chopper, which provides the voltage applied to a DC motor, is 5 kHz. In this case, the maximum available bandwidth of the current controller can be up to at most 1/10 of the switching frequency, i.e., 5 kHz/10 = 500 Hz($\approx$3100 rad/s). However, for a stable current control or if the current is sampled once every switching period, it is better to restrict to 1/20 of the switching frequency, i.e., 5 kHz/20 = 250 Hz ($\approx$1550 rad/s).

Fig. 2.35 compares the current control performances according to the gains, which are determined by the gain selection method as described earlier. We can identify the different responses according to the control bandwidth.

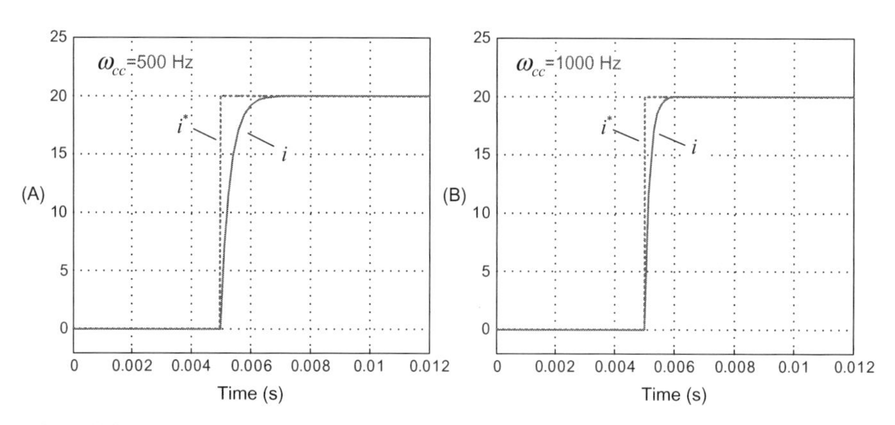

FIGURE 2.35

Current control performance according to the control bandwidth.

2.6.2 ANTI-WINDUP CONTROLLER

As stated before, we can see that the integral controller can effectively eliminate the steady-state error. This is because of the nature of the integral controller producing its output from the accumulated past errors. However, this nature is often the cause of control performance degradation in cases where the output of the controller is limited.

The output of the PI current controller, which indicates the voltage reference applied to a DC motor, should be limited to a feasible value by the following reasons. First, a voltage exceeding its rated value should not be applied to a motor. Furthermore, power electronic converters, which produce the voltage applied to the motor, normally have a restricted output voltage due to the limited availability of the input voltage and the voltage rating of switching devices.

Once the output of the PI controller exceeds its limit due to sustained error signals for a significant period of time, the output will be saturated, but the integrator in the controller may have a large value by its continual integral action. This phenomenon is known as *integral windup*. When the windup occurs, the controller is unable to immediately respond to the changes in the error due to a large accumulated value inside the integrator. It is required that the error have an opposite sign for a long time until the controller returns to the normal state. This system becomes an open-loop system because the output remains at its limit irrespective of the error. As a result, the system will exhibit a large overshoot and a long setting time. To avoid such integrator windup, the magnitude of the integral term should be kept at a proper value when the saturation occurs, so that the controller can resume the action as soon as the control error changes.

There are several methods to avoid the integrator windup such as *back calculation, conditional integration*, and *limited integration* [3,4]. The anti-windup scheme of back calculation as shown in Fig. 2.36 is widely used because of its satisfactory dynamic characteristic.

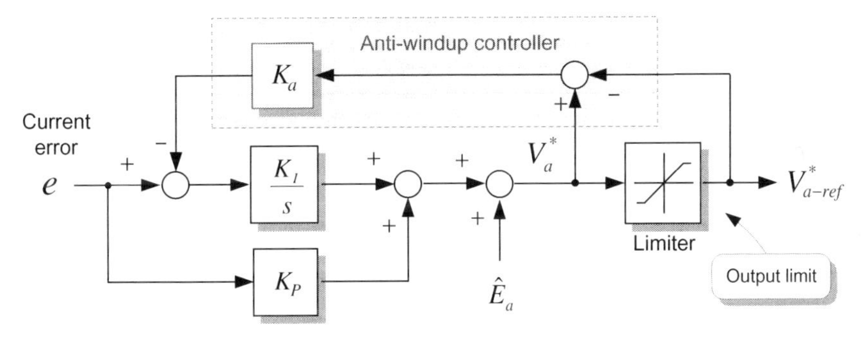

FIGURE 2.36

Anti-windup control by the back calculation method.

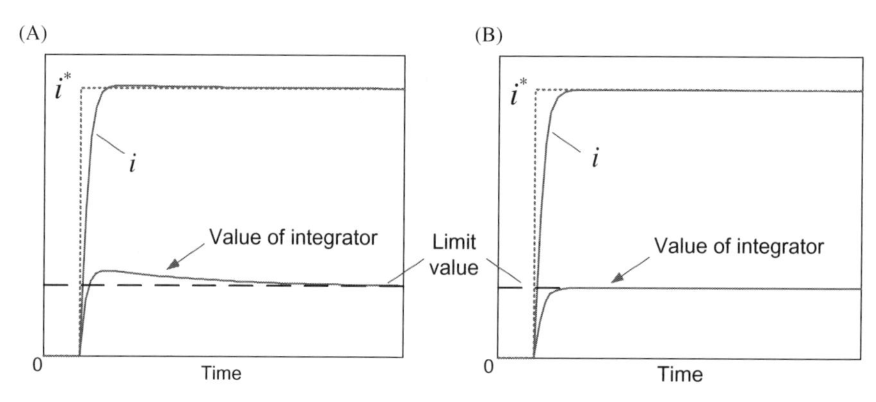

FIGURE 2.37

Anti-windup control. (A) Without anti-windup control and (B) with anti-windup control.

In this scheme, when the output is saturated, the difference between the controller output and the actual output is fed back to the input of the integrator with a gain of K_a so that the accumulated value of the integrator can be kept at a proper value. The gain of an anti-windup controller is normally selected as $K_a = 1/K_p$ to eliminate the dynamics of the limited voltage.

Fig. 2.37 shows the phenomenon of integrator windup for a PI current controller, which is generated by a large change in the reference value. Fig. 2.37A shows the performance of a current controller without an anti-windup control. Due to its saturated output voltage, the actual current exhibits a large overshoot and a long setting time. On the other hand, Fig. 2.37B shows a current controller with an anti-windup control. When the output is saturated, the accumulated value of the integrator can be kept at a proper value, resulting in an improved performance.

2.6.2.1 Gains selection procedure of the proportional–integral current controller

1. Find the switching frequency f_{sw} of the system.
2. Select the control bandwidth ω_{cc} of the current controller to be within 1/10–1/20 of the switching frequency f_{sw} and below 1/25 of the sampling frequency.
3. Calculate the proportional and integral gains from the selected bandwidth ω_{cc} as

$$\text{Proportional gain: } K_{pc} = L_a \cdot \omega_{cc}$$

$$\text{Integral gain: } K_{ic} = R_a \cdot \omega_{cc}$$

4. Select the gain of the anti-windup controller as $K_a = 1/K_{pc}$.

The procedures **1** and **2** are interchangeable with each other, i.e., the switching frequency can be determined by the required bandwidth ω_{cc} for current control.

2.7 SPEED CONTROLLER DESIGN

When controlling the speed of a motor, the speed controller must be placed in the outer loop of the current controller as shown in Fig. 2.38. In this case, if the bandwidth of current control is chosen to be sufficiently wider than that of speed control, then the current control will not have any influence on the speed control performance, and the stability of the speed control will be improved.

Like the current controller, a PI controller is widely used for speed control. Now, we will explain how the PI speed controller is designed, i.e., how to select the proper gains to achieve the desired speed control performance.

2.7.1 PROPORTIONAL–INTEGRAL SPEED CONTROLLER [5]

The transfer function of a PI speed controller is given by

$$G_{pi}(s) = K_{ps} + \frac{K_{is}}{s} \tag{2.39}$$

where K_{ps} and K_{is} are the proportional and integral gains of a PI speed controller, respectively.

From the block diagram in Fig. 2.38, the open-loop transfer function of the speed control system is given by

$$G_s^o(s) = G_{pi}(s)G_c^c(s)\cdot\frac{K_T}{Js} = \left(K_{ps} + \frac{K_{is}}{s}\right)\cdot\frac{\omega_{cc}}{s+\omega_{cc}}\cdot\frac{K_T}{Js} \tag{2.40}$$

This open-loop characteristic is the sum of the frequency responses of three parts: a PI speed controller, a PI current controller, and a mechanical system of a motor and a load. Here, we assume that the current controller is a first-order system with a bandwidth of ω_{cc}. Also, the load torque T_L is neglected and $K_T = k_T\phi_f$.

The bode plot of the open-loop frequency responses for a speed control system is shown in Fig. 2.39. Here, ω_{cs} is the gain cross-over frequency, which is equal to the bandwidth of the speed controller, and $\omega_{pi}(=K_{is}/K_{ps})$ is the PI corner frequency. As a rule of thumb, ω_{pi} is selected as one-fifth of ω_{cs}. If ω_{pi} is too small, the

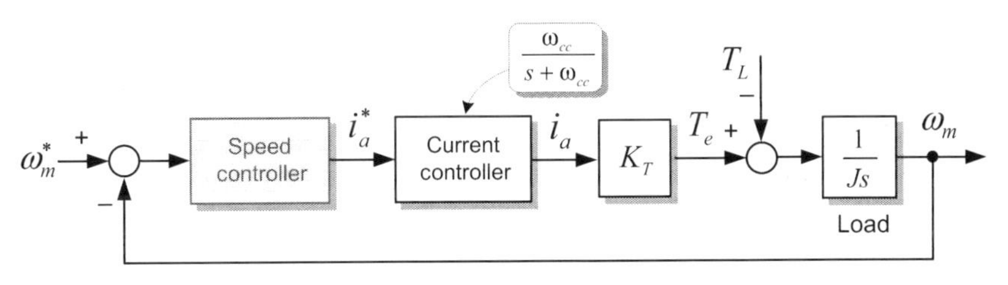

FIGURE 2.38

The speed control system.

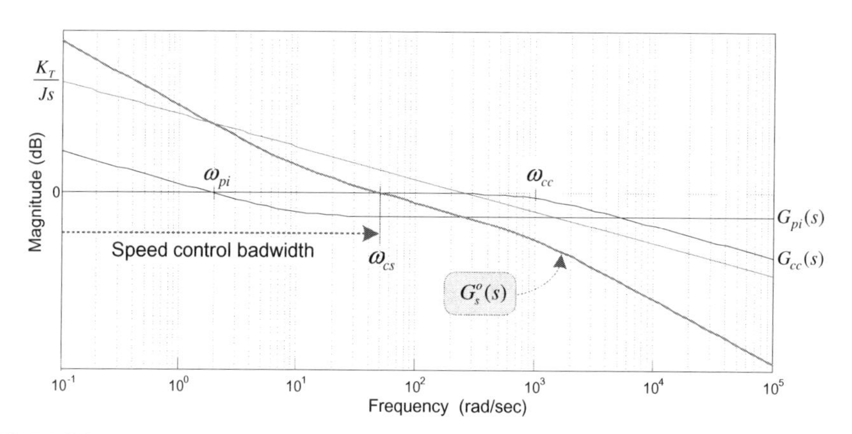

FIGURE 2.39

Bode plot of the open-loop frequency responses for a speed control system.

steady-state performance will be degraded, but if it is too large, the system will exhibit an overshoot.

As stated before, in order for current control to not exert any influence on speed control, the current control bandwidth ω_{cc} should be at least five times wider than the speed control bandwidth ω_{cs}. Under this condition, the gain of the current controller around the cross-over frequency ω_{cs} becomes almost one, i.e.,

$$G_c^c(s) = \frac{\omega_{cc}}{s + \omega_{cc}} \approx 1 \tag{2.41}$$

Also, if the PI corner frequency ω_{pi} is much smaller than the speed control bandwidth ω_{cs}, then the transfer function of the PI current controller around ω_{cs} can be reduced as

$$G_{pi}(s) = K_{ps} + \frac{K_{is}}{s} \approx K_{ps} \tag{2.42}$$

Consequently the open-loop transfer function of Eq. (2.40) around the cross-over frequency ω_{cs} can be simplified as

$$G_s^o(s) \approx K_{ps} \cdot \frac{K_T}{Js} \tag{2.43}$$

The gain of this open-loop transfer function at frequency ω_{cs} is 0 dB, i.e.,

$$\left|G_s^o(j\omega_{cs})\right| = 0 \text{ dB} \tag{2.44}$$

From this, we can obtain the proportional gain K_{ps} and the integral gain K_{is} as

$$\text{Proportional gain: } K_{ps} = \frac{J\omega_{cs}}{K_T} \tag{2.45}$$

$$\text{Integral gain: } K_{is} = K_{ps} \cdot \omega_{pi} = K_{ps} \cdot \frac{\omega_{cs}}{5} = \frac{J\omega_{cs}^2}{5K_T} \tag{2.46}$$

It is important to note that the gains of a speed controller depend on the system inertia J and the torque constant K_T of the target motor. Since J and K_T can be different from each system, the gains of a speed controller may vary with the applied system. From Eq. (2.45), it can be seen that, to achieve an equal bandwidth of speed control, the value of the proportional gain needs to be large for a large J and small for a large K_T.

Like the current controller, the PI speed controller needs to include an anti-windup controller to prevent integrator saturation.

2.7.1.1 Selection of the bandwidth of speed control

The speed control bandwidth ω_{cs} that determines the dynamic performance of the speed controller is limited by the bandwidth of the current controller and the sampling frequency of speed (i.e., the period of the speed control). As mentioned before, the bandwidth of speed control should be five times less than that of current control, and the sampling frequency of speed should also be considered. It is natural that the bandwidth of a speed controller cannot be larger than the frequency measuring the speed. We can select the speed control bandwidth within the limit of 1/10 or 1/20 of the sampling frequency.

As an example, suppose that the bandwidth of a current controller is 500 Hz and the sampling period to measure the speed is 2 ms. In this case, the available bandwidth of the speed controller can be up to one-fifth of the current controller, i.e., 500 Hz/5 = 100 Hz($\approx$628 rad/s). However, since the sampling frequency of speed is 500 Hz(=1/2 ms), it is desirable that the bandwidth of the speed controller be less than 50 Hz (=500/10).

Gains selection procedure of the proportional–integral speed controller

1. Identify the bandwidth of the current controller.
2. Select the bandwidth ω_{cs} of speed control to be five times less than the bandwidth of current control and ten times less than the sampling frequency of speed.
3. Calculate the proportional and integral gains from the selected bandwidth ω_{cs} as

   Proportional gain: $K_{ps} = \dfrac{J\omega_{cs}}{K_T}$

   Integral gain: $K_{is} = K_{ps} \cdot \omega_{pi} = K_{ps} \cdot \dfrac{\omega_{cs}}{5} = \dfrac{J\omega_{cs}^2}{5K_T}$
4. Select the gain of an anti-windup controller as $K_a = 1/K_{ps}$

We can normally obtain appropriate gains of the speed controller using the design method as stated just above. However, inaccurate information of the system parameters, J and K_T, may lead to incorrect gains for achieving the required bandwidth, so we may need to adjust additionally the gains by monitoring the transient response as shown in Fig. 2.40. If the system has an oscillatory response, we need to decrease the gains. On the other hand, if the system has a very slow response, we need to increase the gains.

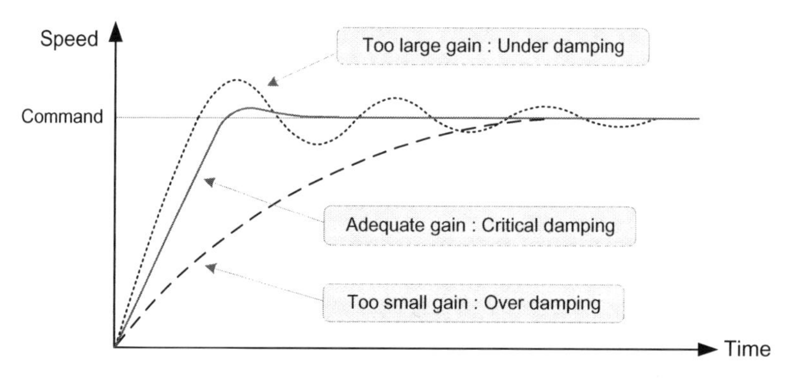

FIGURE 2.40

Transient responses according to different controller gains.

2.7.1.3 Drawback of a proportional–integral speed controller

From the open-loop transfer function $G_s^o(s)$of Eq. (2.40), the closed-loop transfer function $G_s^c(s)$ of the speed control system is given by

$$G_s^c(s) = \frac{G_c^o(s)}{1+G_c^o(s)} = \frac{\left(\frac{K_T K_{ps}}{J}\right)s + \left(\frac{K_T K_{is}}{J}\right)}{s^2 + \left(\frac{K_T K_{ps}}{J}\right)s + \left(\frac{K_T K_{is}}{J}\right)} \tag{2.47}$$

Here, the gain of the current controller is assumed to be 1.

By comparing this transfer function with the prototype form, $s^2 + 2\zeta\omega_n s + \omega_n^2$, of the second-order system, the damping ratio ζ can be obtained as

$$\zeta = \frac{1}{2}\sqrt{\frac{K_T K_{is}}{J}\cdot\frac{K_{ps}}{K_{is}}} = \frac{1}{2}\sqrt{\frac{\omega_{cs}}{\omega_{pi}}} \tag{2.48}$$

We can see that the damping ratio ζ depends on the ratio of ω_{cs} to ω_{pi}. The condition of $\omega_{pi} = \omega_{sc}/5$ given in the Section 2.7 results in $\zeta = \sqrt{5}/2$. Because ζ is greater than 1, we expect the system response by the PI controller to be overdamped and have no overshoot. However, the zero of $s = -K_{is}/K_{ps}$ in the PI controller causes an overshoot in the system response. The closer the zero is located to the imaginary axis in the s-plane, the greater its effects will be [1]. We can easily see the effect of this zero by dividing Eq. (2.47) into two parts as:

$$\frac{\frac{K_T K_{is}}{J}}{s^2 + \left(\frac{K_T K_{ps}}{J}\right)s + \left(\frac{K_T K_{is}}{J}\right)} + \frac{\left(\frac{K_T K_{ps}}{J}\right)s}{s^2 + \left(\frac{K_T K_{ps}}{J}\right)s + \left(\frac{K_T K_{is}}{J}\right)}$$
$$\rightarrow \frac{\omega_n^2}{s^2 + 2\zeta\omega_n s + \omega_n^2} + \frac{2\zeta\omega_n s}{s^2 + 2\zeta\omega_n s + \omega_n^2} \tag{2.49}$$

The responses of these two parts to a unit step command are shown in Fig. 2.41. It can be seen that the second term of Eq. (2.49) gives an overshoot to the system response. It is interesting to note that, in current control, this zero has no effect on the system response because the zero is canceled out by the system pole.

The overshoot problem by the zero in the PI controller can be solved by an *integral and proportional (IP) controller.*

2.7.2 INTEGRAL-PROPORTIONAL CONTROLLER

We can remove the zero from Eq. (2.47) by adjusting the position of the P control part in the PI controller as shown in Fig. 2.42. This controller is known as the *IP controller.*

The transfer function of the IP control system is given as

$$\frac{\omega_m}{\omega_m^*} = \frac{\dfrac{K_T K_{is}}{Js}}{s^2 + \left(\dfrac{K_T K_{ps}}{J}\right)s + \dfrac{K_T K_{is}}{J}} \tag{2.50}$$

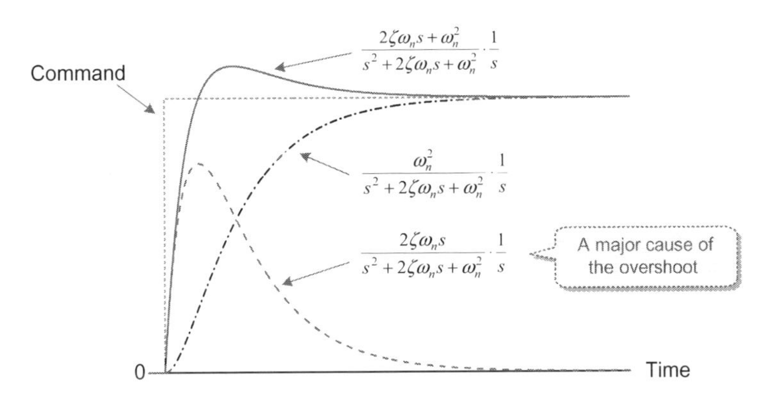

FIGURE 2.41

Response to a unit step command.

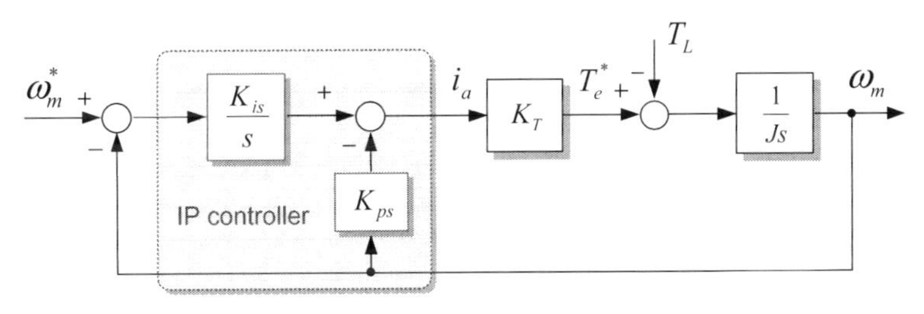

FIGURE 2.42

Block diagram of the IP controller.

The poles of this transfer function are the same as those of the PI controller but this transfer function has no zero. Thus the IP controller has no problem with the overshoot caused by the zero.

In the IP controller, the system response to the speed reference depends only on the damping ratio in Eq. (2.50). The critical damped response (i.e., $\zeta = 1$) is obtained by allowing $\omega_{pi} = \omega_{sc}/4$.

From reconstructing Fig. 2.42 as Fig. 2.43, we can see that an IP controller is equivalent to a PI controller except for the first-order, low-pass filter with the time constant of K_{ps}/K_{is} on the input side. Therefore, an IP controller exhibits no overshoot due to low-pass filtering of its input command, but its response will be slower than that of a PI controller.

Although the speed responses to speed reference of PI and IP controllers are different from each other, the speed responses to disturbance torque are the same. The transfer function of the speed response to a disturbance torque T_L for an IP controller can be obtained from Fig. 2.42 as

$$\frac{\omega_m}{T_L} = -\frac{s}{Js^2 + K_T K_{ps} s + K_T K_{ps}} \tag{2.51}$$

It is not hard to ascertain that the transfer function of the PI controller is the same as Eq. (2.51). Therefore the speed responses to a disturbance torque for the two controllers are identical. The speed responses of the PI and IP controllers under an equal control bandwidth are shown in Fig. 2.44A. It can be seen that the IP controller exhibits no overshoot, but its response is slower than that of the PI controller. On the other hand, the speed responses to a disturbance torque for the two controllers are the same, as we would expect. If the bandwidths are adjusted for the two controllers to have the same overshoot in their response to speed reference, the IP controller can achieve a larger bandwidth than the PI controller. For the same overshoot, the bandwidth of the IP controller can be set as 5.44 times larger than that of the PI controller. In this condition, the IP controller will exhibit an improved speed response to the disturbance torque as shown in Fig. 2.44B.

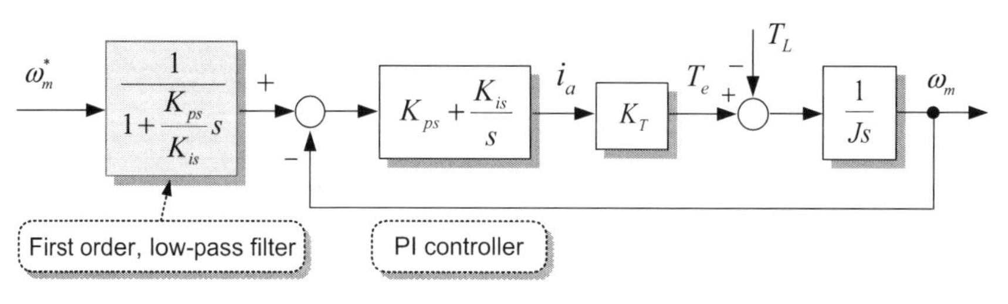

FIGURE 2.43

Equivalent block diagram of the IP controller.

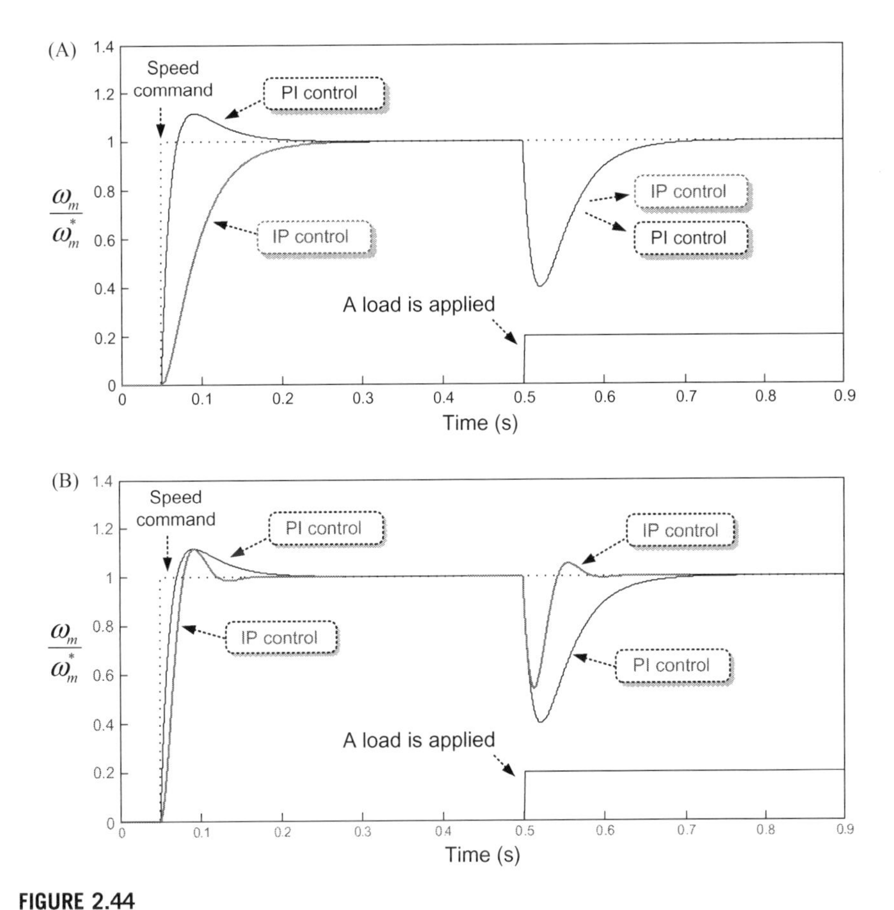

FIGURE 2.44

Speed responses of PI and IP controllers. (A) Same bandwidth and (B) same overshoot.

By combining these two controllers, we can make a compromise between the response to speed reference and the response to disturbance torque. This combined controller is shown in Fig. 2.45. The characteristic of the combined controller depends on the control variable α. When $\alpha = 0$, it becomes an IP controller, whereas if $\alpha = 1$, it becomes a PI controller. In case of $0 < \alpha < 1$, a combined characteristic of the two controllers can be obtained.

A PI speed controller will exhibit a satisfactory performance if its bandwidth is large enough. In many applications such as elevators that are susceptible to mechanical resonance and flexible joint robots, a speed controller often cannot have a large enough bandwidth, so we cannot expect a satisfactory response. In this case, we need to improve the control performance by combining a PI controller with other control schemes such as acceleration feedforward compensation or Fuzzy algorithm [6].

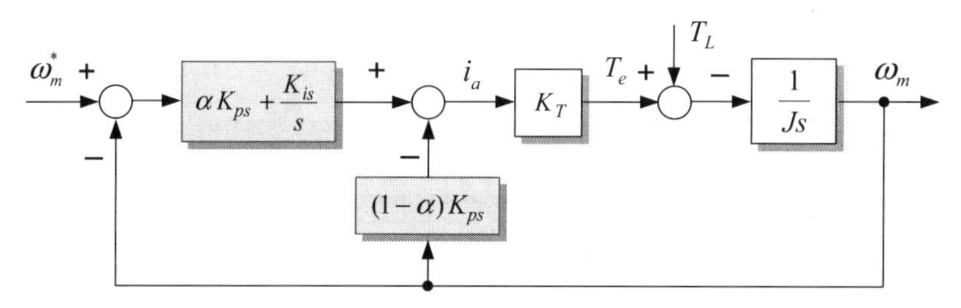

FIGURE 2.45

Combined controller.

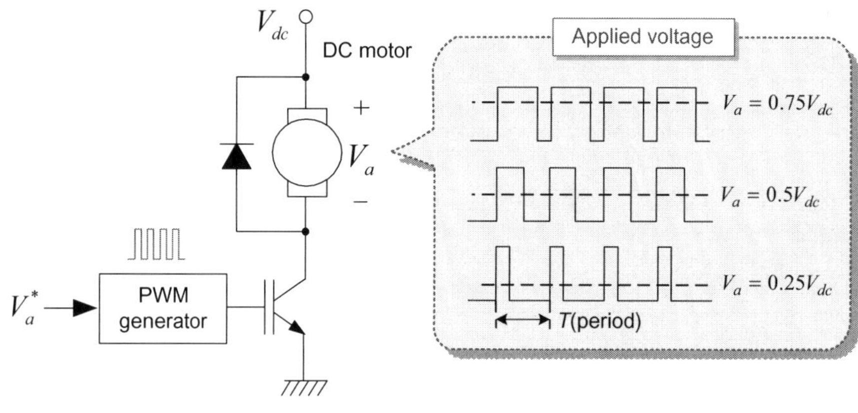

FIGURE 2.46

Simple driving circuit for a DC motor.

2.8 POWER ELECTRONIC CONVERTER FOR DIRECT CURRENT MOTORS

The voltage applied to a DC motor can be controlled by using a power electronic converter. Fig. 2.46 shows a simple driving circuit for the forward motoring mode operation. Commonly, in a driving circuit, the switching states of the power devices are determined by the pulse width modulation (PWM) technique to achieve the required voltage command. The voltage command is given by the current controller as described earlier.

A four-quadrant chopper (or H-bridge circuit) is often used to operate a DC motor in both motoring and braking modes for both forward and reverse directions. The H-bridge circuit is implemented with four power switching devices as shown in Fig. 2.47.

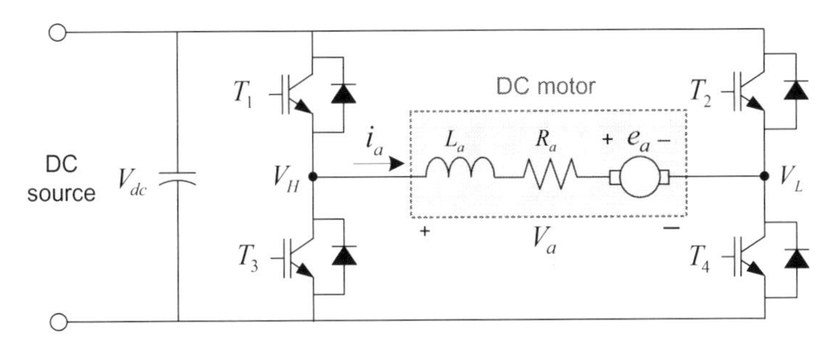

FIGURE 2.47

H-bridge circuit.

This circuit is widely used in DC servo drive systems because it can provide high-performance current control with a bandwidth up to several thousand (rad/s). In this circuit, there are two PWM techniques for adjusting the output voltage: *bipolar* and *unipolar switching schemes*.

2.8.1 SWITCHING SCHEMES

2.8.1.1 Bipolar switching scheme

The switching states of devices are determined by comparing the voltage reference V_a^* with the triangular carrier wave as shown in Fig. 2.48.

In this scheme, if the voltage reference V_a^* is larger than the triangular carrier wave, then both T_1 and T_4 will be turned on, and the output V_a will become $+V_{dc}$. On the other hand, if the reference voltage V_a^* is less than the triangular carrier wave, both T_2 and T_3 will be turned on, and the output V_a will become $-V_{dc}$. In this manner, two switching devices (T_1 and T_4, T_2 and T_3) placed diagonally will be turned on or off simultaneously. Since these switching actions give an output voltage of either positive or negative DC voltage (i.e., $+V_{dc}$ and $-V_{dc}$), this scheme is called the bipolar switching scheme.

In this scheme, the switching frequency equals to the frequency $(1/T_s)$ of the triangular carrier wave. This scheme has an advantage of smoothly changing the average output voltage from $+V_{dc}$ to $-V_{dc}$ as shown in Fig. 2.49. However, the instantaneous output voltage becomes as wide as $2V_{dc}$, resulting in a large current ripple of the motor.

2.8.1.2 Unipolar switching scheme

In this scheme, the switching of the devices is also determined by comparing the voltage reference with the triangular carrier wave as in the bipolar switching scheme, but the switching in the H-pole and the L-pole is performed independently according to its own reference as shown in Fig. 2.50. The reference voltage V_H of the H-pole is given as V_a^*. Meanwhile, the reference voltage V_L of the L-pole is given as $-V_a^*$. The output voltage (i.e., the motor voltage) V_a is given by the voltage difference between the H-pole and the L-pole, i.e., $V_H - V_L$. Since

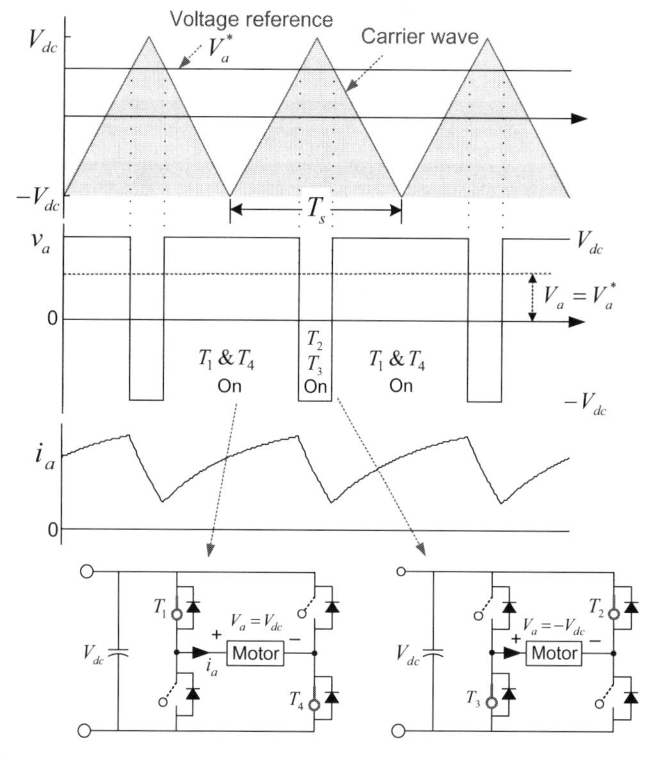

FIGURE 2.48

Bipolar switching scheme.

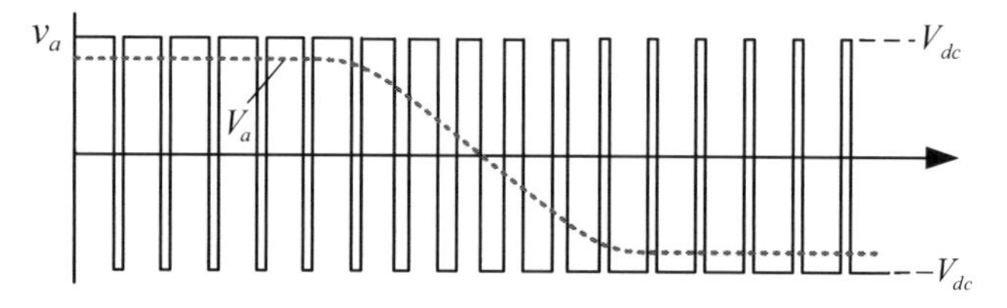

FIGURE 2.49

Output voltage of bipolar switching.

this switching action gives an output voltage of 0 and $+V_{dc}$ or 0 and $-V_{dc}$, this scheme is called the unipolar switching scheme.

In this scheme, the output voltage fluctuates by V_{dc}, so the current ripple of the motor will be two times smaller than that of the bipolar switching scheme and the torque ripple will also be smaller as shown in Fig. 2.51. Furthermore, the output voltage changes effectively with the doubling of the switching frequency. However, since this scheme is difficult to provide a small average voltage, the sign transition of the average voltage may be discontinuous as described in Fig. 2.52. As a result, it is difficult to change the rotational direction of the motor smoothly.

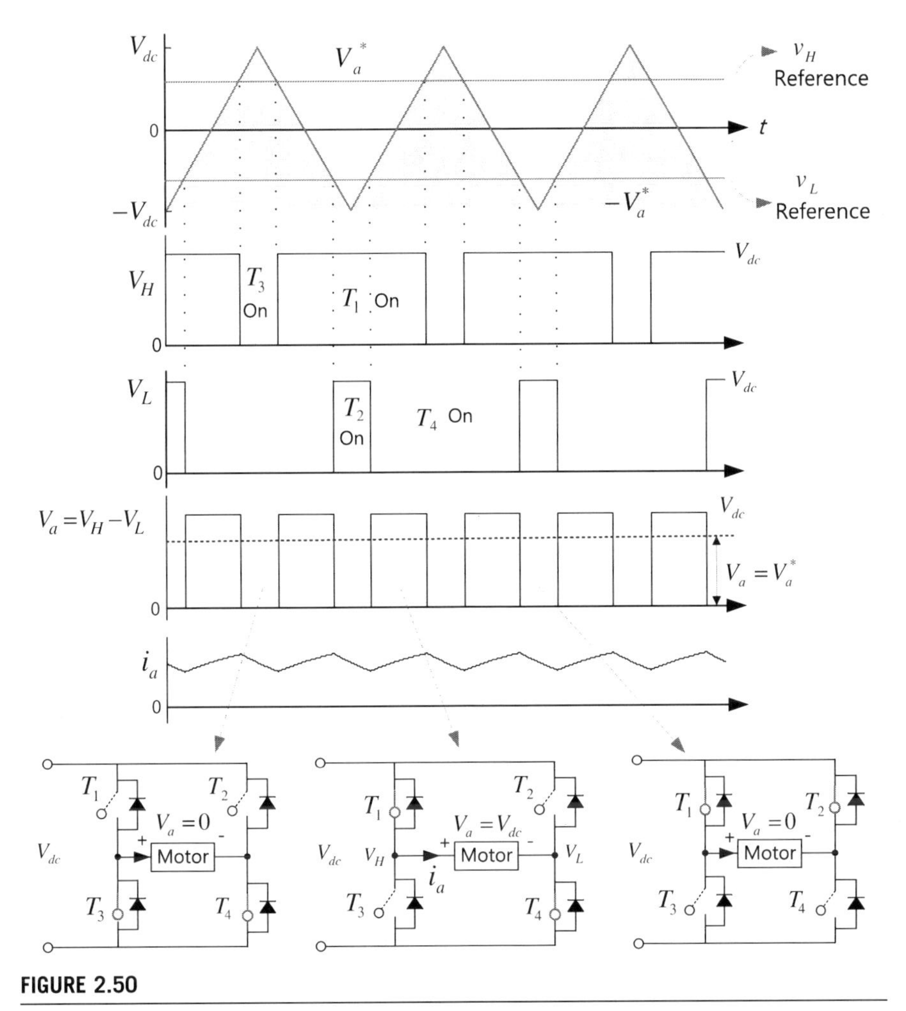

FIGURE 2.50

Unipolar switching scheme.

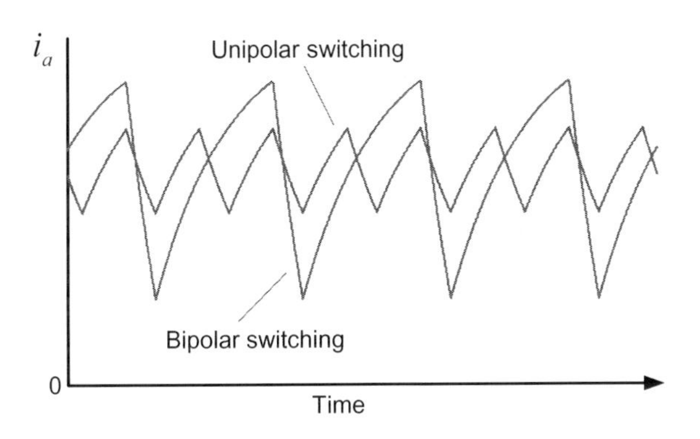

FIGURE 2.51

Comparison of the current ripple for the two switching schemes.

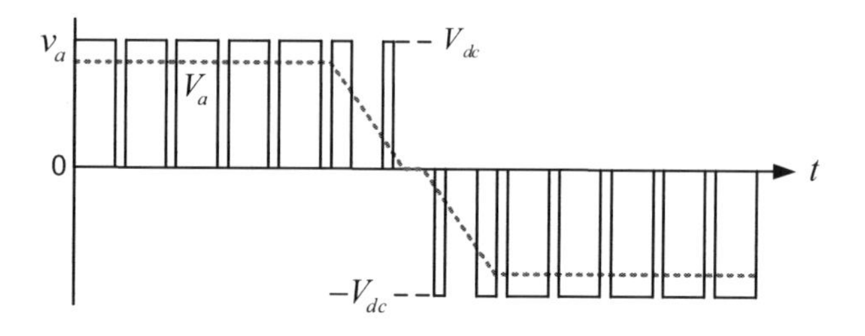

FIGURE 2.52

Output voltage of unipolar switching.

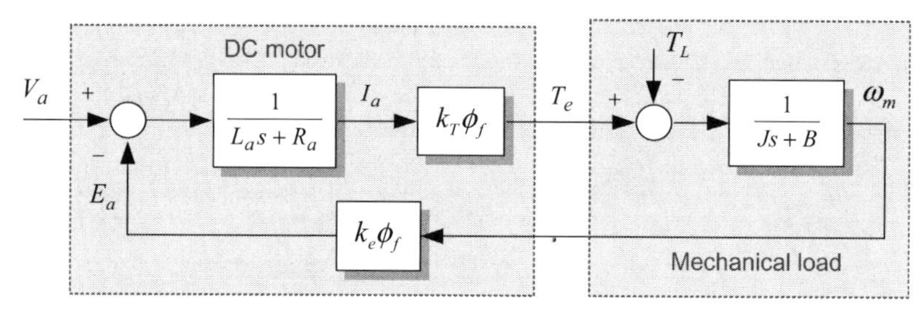

FIGURE 2.53

Block diagram of a permanent magnet DC motor drive system.

2.9 SIMULATION OF DIRECT CURRENT MOTOR DRIVE SYSTEM: MATLAB/SIMULINK

Now, we will examine the performance of the current and speed controllers as mentioned in the Sections 2.6 and 2.7 through a simulation of a DC motor drive system using MATLAB/Simulink.

A block diagram of a DC motor drive system is shown in Fig. 2.53. The system consists of a DC motor part and a mechanical load part.

The modeling of these parts for MATLAB/Simulink simulation is as follows:

2.9.1 DIRECT CURRENT MOTOR MODELING

In the electric circuit of a DC motor, the input is the applied armature voltage, and the output is the armature current. Thus, to obtain the armature current, the voltage equation of Eq. (2.1) can be rewritten as

$$V_a = R_a i_a + L_a \frac{di_a}{dt} + e_a$$
$$\rightarrow i_a(t) = \frac{1}{L_a} \int [V_a(\tau) - R_a i_a(\tau) - e_a(\tau)] d\tau \tag{2.52}$$

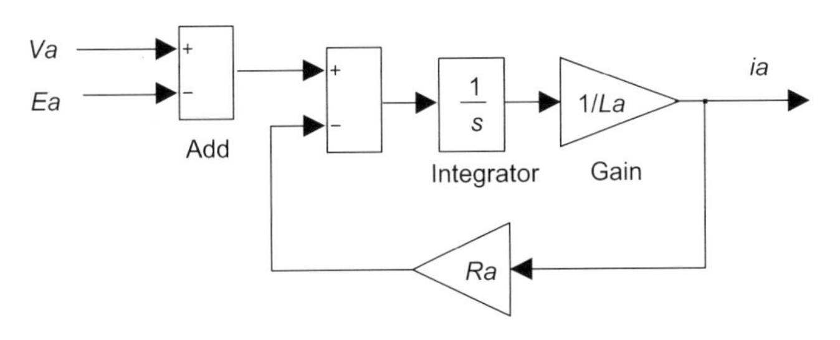

FIGURE 2.54

Simulink block diagram of the electric circuit in a DC motor.

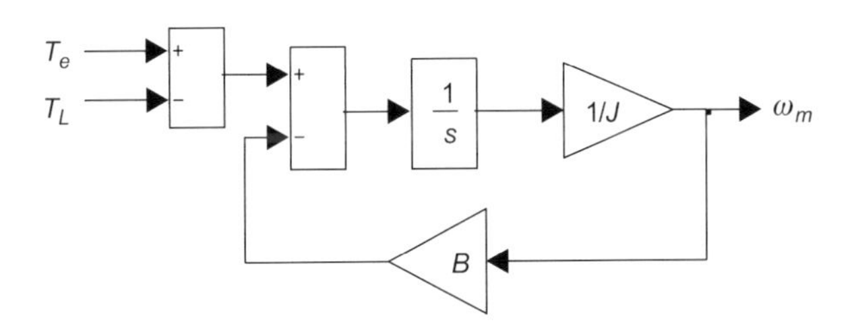

FIGURE 2.55

Simulink block diagram of the mechanical load system.

From this equation, the simulation model of the electric circuit of a DC motor is given as Fig. 2.54.

2.9.2 MECHANICAL SYSTEM MODELING

In the mechanical system, the input is the developed torque of the DC motor, and the output is the speed of the load. To obtain the speed, the motion equation of Eq. (2.4) can be rewritten as

$$T_e = k_T\phi i_a = J\frac{d\omega_m}{dt} + B\omega_m + T_L$$
$$\rightarrow \omega_m(t) = \frac{1}{J}\int[T_e(\tau) - B\omega_m(\tau) - T_L(\tau)]d\tau \tag{2.53}$$

From this equation, the simulation model of the mechanical load system is given as Fig. 2.55.

Fig. 2.56 shows a complete model for a DC motor drive system, which combines the electric circuit, mechanical system, torque, and back-EMF equations.

Here, we assume $K = k_e\phi = k_T\phi$ and $T_L = 0$. To apply a step input voltage to a DC motor, the step block (V_a) is added to the simulation model.

Now, we will carry out a simulation by using this complete model. The parameters and the rating of the DC motor used for simulations are listed as M-file in Fig. 2.57. We can set the simulation time and the solver option by using the *simulation/configuration parameters* menu.

First, we will examine the speed response to a step input voltage. Fig. 2.58 compares the speed responses of this system to two different step input voltages. The final speed depends on the voltage applied to the motor. The rotational speed

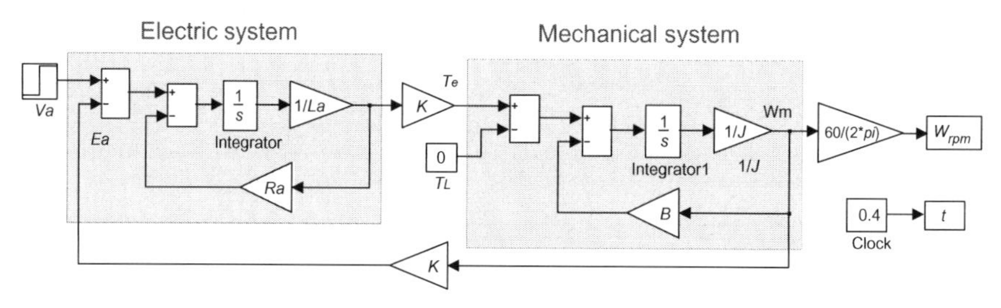

FIGURE 2.56

Complete Simulink model for a DC motor drive system.

```
1     %-- DC Motor Parameters -----------------------------------------------
2 -   Power=3336;                   % Power[W]
3 -   Va_rated=140;                 % Rated Voltage [V]
4 -   Ia_rated=25;                  % Rated Current [A]
5 -   Wm_rated=3000*2*pi/60;        % Rated Angular Velocity[rad/s]
6 -   Te_rated=Power/Wm_rated;      % Rated Torque [Nm]
7
8 -   Ra=0.26;                      % Amature Resistance [Ohm]
9 -   La=1.7e-3;                    % Amature Inductance [H]
10 -  J=.00252;                     % Moment of Inertia [kg-m^2]
11 -  B=.0;                         % Coefficient of Viscous Friction [kgm^2/sec]
12
13 -  Kt=Te_rated/Ia_rated;
```

FIGURE 2.57

M-file for simulation.

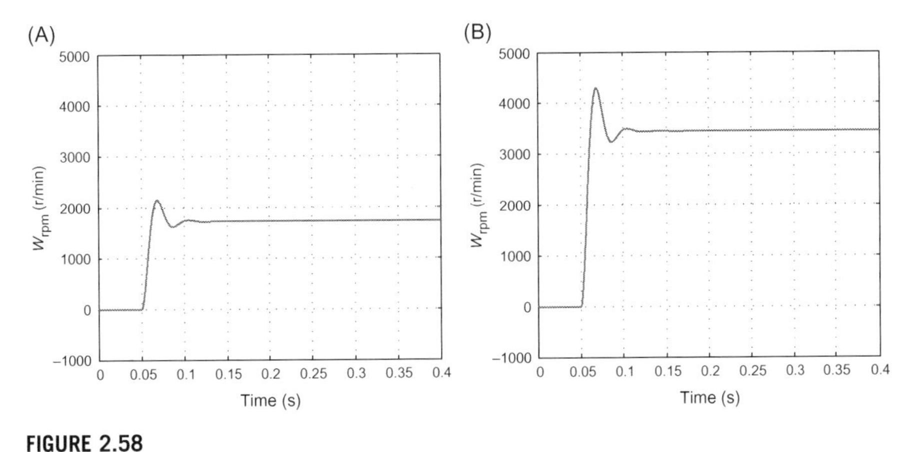

FIGURE 2.58

Speed response to a step input voltage. (A) $V_a = 70$ V and (B) $V_a = 140$ V.

W_{rpm} of the DC motor is proportional to the armature voltage V_a, as can be expected. The parameters of the motor give a damping ratio ζ of 0.37. Because $\zeta < 1$, we can expect the system to have an under-damped response and an overshoot. This is verified by this simulation results.

Next, we will model a PI current controller and then connect it with the DC motor drive system.

2.9.3 PROPORTIONAL–INTEGRAL CURRENT CONTROLLER MODELING

The entire system containing a PI current controller is shown in Fig. 2.59.

The subsystem "PMDC" represents all the blocks in Fig. 2.56. This system also includes an anti-windup controller, back-EMF feedforward compensation, and a limiter. The limiter restricts the output of the current controller to the rated voltage of the motor. The proportional and integral gains can be obtained from Eqs. (2.37) and (2.38), and the anti-windup gain is $K_a = 1/K_p$.

Fig. 2.60 compares the responses to the step current command ($i_a^* = 20$A) for gains according to the control bandwidths of 500 and 1000 Hz. As we have expected, the PI current controller does not give a steady-state error, and a larger bandwidth can improve the response time.

The two responses show that the actual currents reach their references at 2 and 1 ms, respectively. This indicates that the current control system has bandwidths of 500 and 1000 Hz, respectively, as designed.

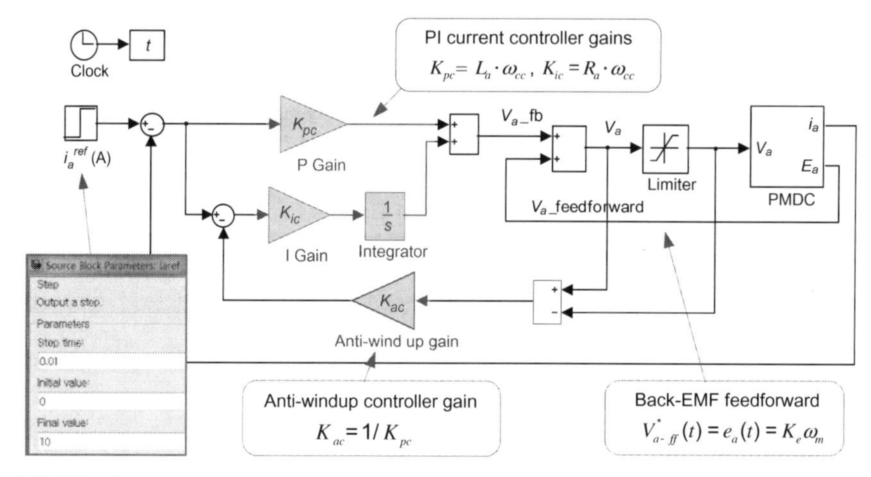

FIGURE 2.59

System containing a PI current controller.

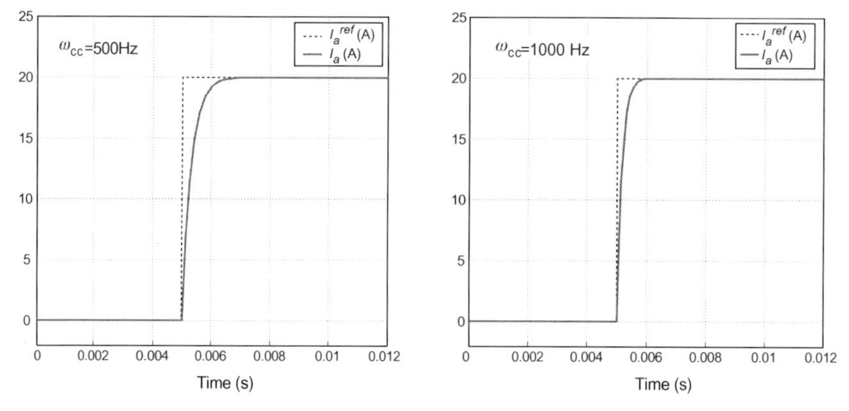

FIGURE 2.60

Step response according to control bandwidth.

2.9.4 PROPORTIONAL–INTEGRAL SPEED CONTROLLER MODELING

The entire DC motor drive system containing a PI speed controller is shown in Fig. 2.61. The subsystem "PMDC with CC" represents all the blocks in Fig. 2.59. This controller also has an anti-windup controller.

The system responses to the step speed command of 2500 r/min according to two different bandwidths are shown in Fig. 2.62. Again, we can make sure that a larger bandwidth can lead to a faster response.

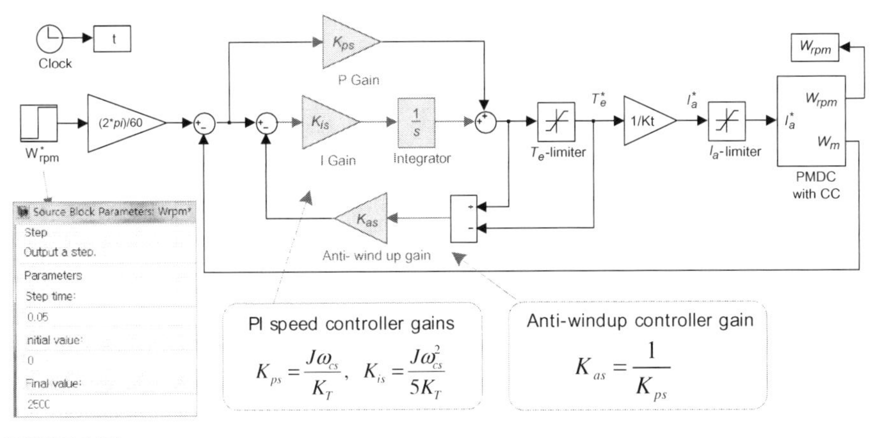

FIGURE 2.61

DC motor drive system block containing a PI speed controller.

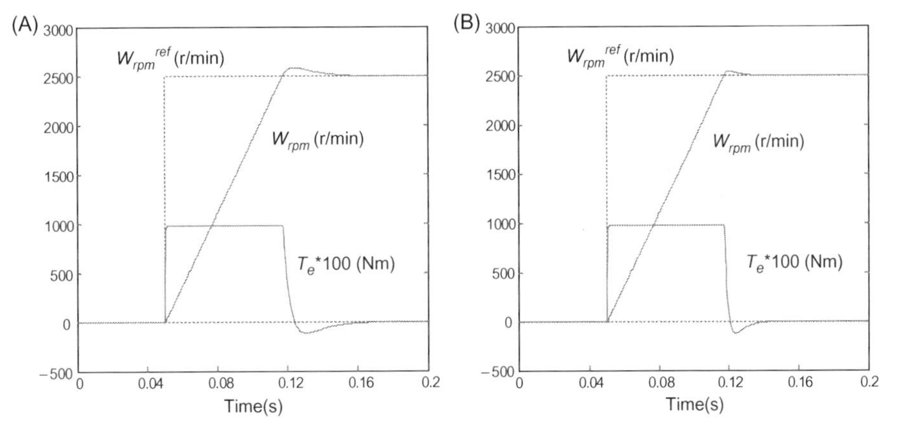

FIGURE 2.62

System responses according to two different bandwidths. (A) $\omega_{cs} = 50$ Hz and (B) $\omega_{cs} = 100$ Hz.

2.9.5 FOUR-QUADRANT CHOPPER MODELING

The simulation block containing a four-quadrant chopper, which generates the voltage applied to the DC motor, is shown in Fig. 2.63. We assume that this chopper adopts a unipolar switching scheme with a switching frequency of 5 kHz.

The switching signals produced by the unipolar switching scheme are shown in Fig. 2.64A. The PWM output voltage of the chopper and the current of the motor are shown in Fig. 2.64B.

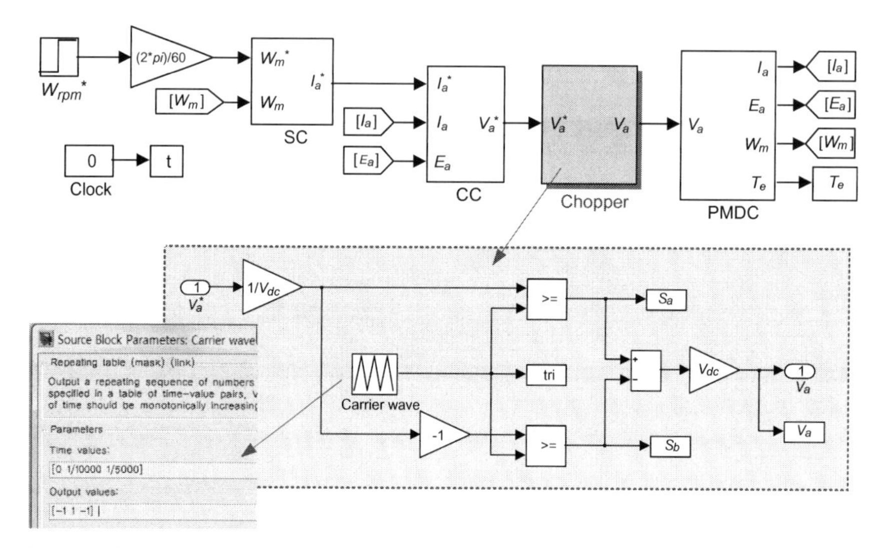

FIGURE 2.63

Simulation block containing a four-quadrant chopper.

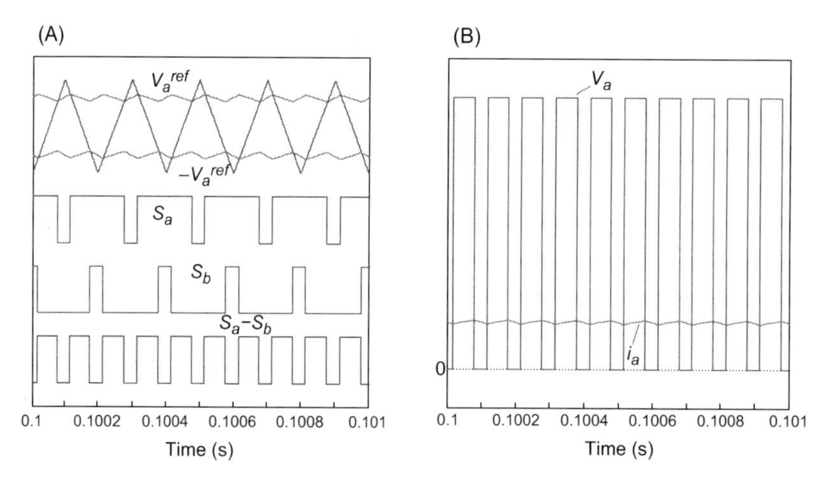

FIGURE 2.64

Unipolar switching scheme. (A) Switching signals and (B) PWM output voltage and current.

For the step speed reference of 2500 r/min, the speed and torque of the motor are shown in Fig. 2.65. When compared with Fig. 2.61, the current and torque waveforms powered by the chopper exhibit ripple components due to PWM switching.

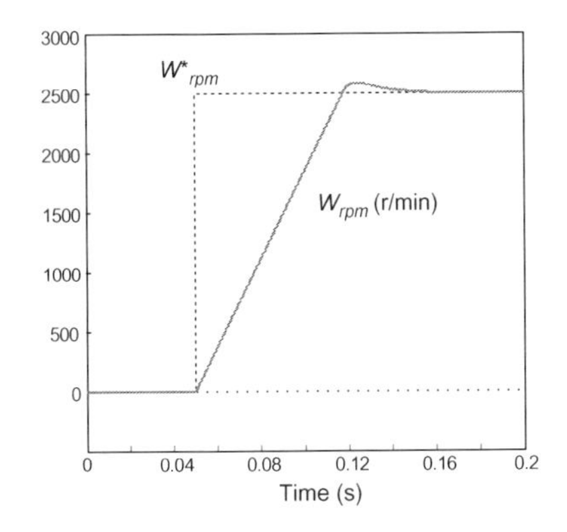

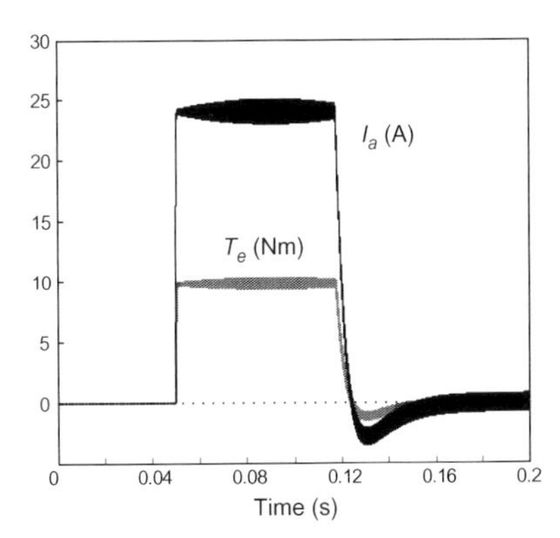

FIGURE 2.65

Speed, current, and output torque.

REFERENCES

[1] G. Franklin, J.D. Powell, A. Emami-Naeini, Feedback Control of Dynamic Systems, fifth ed., Prentice Hall, Upper Saddle River, NJ, 2006.

[2] Y. Peng, D. Vrancic, R. Hanus, Design and analysis of pulsewidth-modulated amplifiers for DC servo systems, IEEE Trans. Ind. Electron. Control Instr. IECI-23 (1) (1976) 47–55.

[3] Y. Peng, D. Vrancic, R. Hanus, Anti-windup, bumpless, and conditioned transfer techniques for PID controllers, IEEE Control Syst. Mag. 16 (2) (1996) 48–57.

[4] H.-B. Shin, New antiwindup PI controller for variable-speed motor drives, IEEE Trans. Ind. Electron. 45 (3) (1998) 445–450.

[5] H. Sugimoto, M. Koyama, S. Tamai, Practical Theory and Design of an AC Servo System, ISBN-10: 4915449580 (in Japanese) (Chapter 7), Tokyo, 1990.

[6] J.S. Yu, S.H. Kim, B.K. Lee, C.Y. Won, J. Hur, Fuzzy-logic-based vector control scheme for permanent-magnet synchronous motors in elevator drive applications, IEEE Trans. Ind. Electron. 54 (4) (2007) 2190–2200.

CHAPTER

Alternating current motors: synchronous motor and induction motor

3

The rotation of alternating current (AC) motors exploits the force developed by the interaction between two magnetic fields rotating at the same speed as stated in Chapter 1. There are two classifications of an AC motor: *synchronous motor* and *asynchronous motor*. An asynchronous motor is more commonly known as an *induction motor*.

Synchronous and induction motors have the same stator configuration for producing the rotating magnetic field. However, they have different rotor constructions due to the difference in their ways of generating the rotor magnetic field. In synchronous motors, typically, the rotor magnetic field is generated from the field winding excited by a direct current (DC) power source. For small- to medium-sized synchronous motors, permanent magnets are commonly used for generating the rotor magnetic field. On the other hand, in induction motors, the rotor magnetic field is produced by an AC power. It is interesting to note that this AC power is transferred from a stator by electromagnetic induction.

Induction motors have several advantages such as simple construction, reliability, ruggedness, low maintenance, low cost, high-speed operation capability, and the ability to be operated by a direct connection to an AC power source. On the other hand, synchronous motors are superior to induction motors in efficiency, power density, dynamic response, and power factor.

In this chapter, we will study the basic operation principle and characteristics of AC motors in more detail. We will begin by examining the more general type of motor, the induction motor.

3.1 INDUCTION MOTORS

The fundamental principle of induction motors was first demonstrated by the experiment known as Arago's Disk in 1824. From this experiment, it was found that, when a magnet rotates along the rim of a copper disk (a nonmagnetic substance), the disk rotates in the direction of the magnet at a smaller speed (Fig. 3.1). In 1832, Faraday demonstrated that this phenomenon is due to the currents induced in the copper disk. The discovery of this phenomenon ultimately led to the development of the fundamental concept of an induction motor. When a magnet passes along the rim of a copper disk, an electromotive force (EMF) is

Electric Motor Control. DOI: http://dx.doi.org/10.1016/B978-0-12-812138-2.00003-9

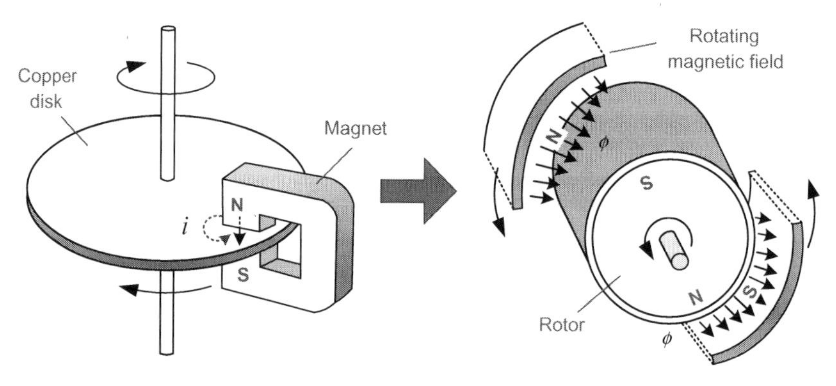

FIGURE 3.1

Arago's disk and the fundamental concept of an induction motor.

induced at the part in the disk that experiences a changing magnetic field. The induced voltage causes an eddy current to flow in the disk. As a result, since the current-carrying part is built under the magnetic field of the magnet, a force (known as *Lorentz force*) is created on the current-carrying part. The direction of the force is the same as that of the magnet's movement, and this force makes the copper disk to rotate along with the magnet.

In an induction motor based on this fundamental concept, such rotating magnet is realized by a rotating magnetic field generated from three-phase stator windings connected to a three-phase AC power source. The rotor, then, follows this rotating magnetic field.

Induction motors are most widely used as general-purpose motors in many industrial applications because of their low cost and rugged construction. Traditionally, induction motors had been connected directly to the line voltage of 60 or 50 Hz and operated at a nearly constant speed. However, recently, variable speed drives have been made possible by power electronic converters such as inverter, so induction motors are widely used in many applications requiring speed control.

3.1.1 STRUCTURE OF INDUCTION MOTORS

The structure of a typical induction motor is shown in Fig. 3.2. An induction motor has cylindrical stator and rotor configurations, which are separated by a uniform radial air gap. The stator and the rotor are made up of an iron core, which has windings inserted inside. The iron core is not a single solid lump but consists of a stack of insulated laminations of silicon steel, usually with a thickness of about 0.3–0.5 mm to reduce eddy current losses. The iron core is made from a ferromagnetic material such as steel, soft iron, or various nickel alloys to produce magnetic flux efficiently and reduce hysteresis losses.

FIGURE 3.2

Structure of a typical induction motor.

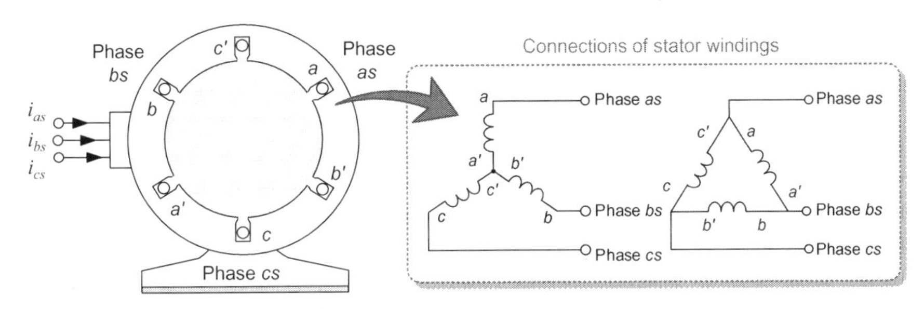

FIGURE 3.3

Stator windings.

3.1.1.1 Stator

In an induction motor, only the stator windings are fed by the three-phase AC power supply. It should be noted that the three-phase stator windings perform the roles of both the armature and field windings of a DC motor. These three-phase windings are placed in the slots that are axially cut along the inner periphery of the iron core as shown in Fig. 3.3. They are displaced from each other by 120 electrical degrees along the periphery and are typically connected in delta for low-supply voltage or in wye for high-supply voltage.

All turns in each winding are continuously distributed in the numerous slots spread around the periphery, such that the winding density can be sinusoidal as shown in Fig. 3.4. The purpose of this arrangement is to establish a sinusoidal flux distribution in the air gap when currents flow through them. This winding

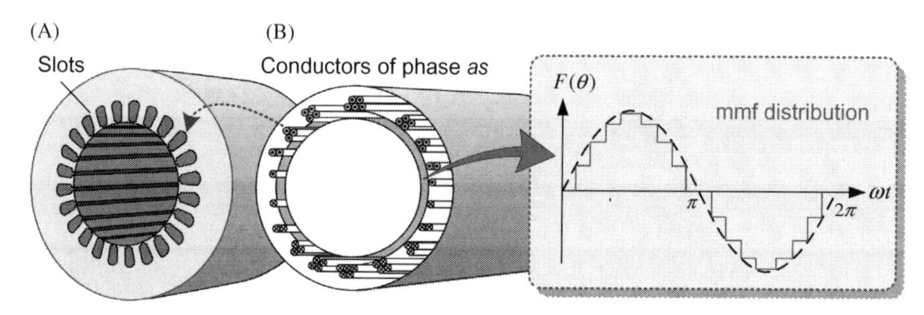

FIGURE 3.4

Stator phase winding. (A) Stator core and (B) sinusoidally distributed winding and mmf.

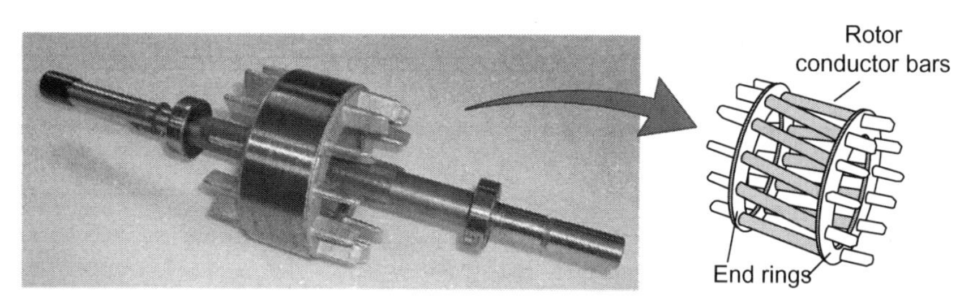

FIGURE 3.5

Squirrel-cage rotor.

type is called *distributed winding*. The distributed winding configuration increases the utilization of the iron core and reduces magnetomotive force (mmf) space harmonics, resulting in a lower torque ripple compared to that of the *concentrated winding*, in which all coils of the phase winding are placed in one slot under a pole.

3.1.1.2 Rotor

Similar to the stator, a rotor has a cylindrical iron core that consists of silicon steel laminations. It should be noted that the electric circuit (windings or conductors) in the iron core of the rotor is not fed by an external power source but its current flows by the induced EMF.

There are two types of a rotor used in induction motors: *squirrel-cage rotor* and *wound rotor*.

3.1.1.2.1 Squirrel-cage type

A squirrel-cage rotor has a laminated iron core with slots for placing skewed conductors, which may be a copper, aluminum, or alloy bar. These rotor bars are short-circuited at both ends through end rings as shown in Fig. 3.5.

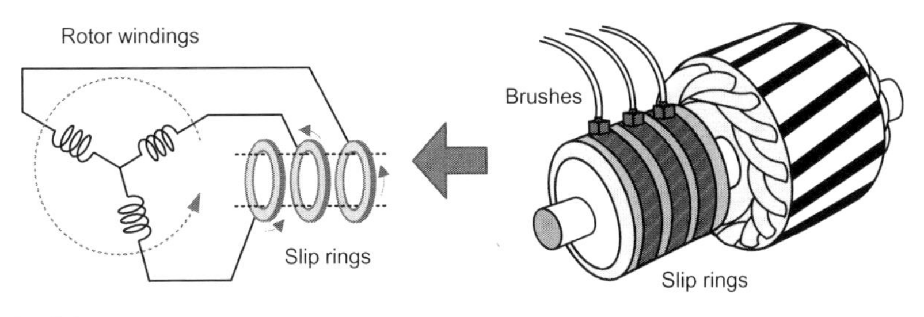

FIGURE 3.6

Wound-rotor.

This rotor gets its name from its structure's resemblance to a squirrel cage. The rotor conductors are not placed exactly parallel to the shaft but are skewed by one slot-pitch to reduce cogging torque, and this allows the motor to run quietly. Because of its simple and rugged construction, about 95% of the induction motors use the squirrel-cage rotor.

3.1.1.2.2 Wound-rotor type

Similar to the windings on a stator, a wound-rotor as shown in Fig. 3.6 has a set of three-phase windings, which are usually Y-connected. The rotor windings are tied to the slip rings on the rotor's shaft and thus can be accessible through the brushes. Due to this configuration, in a wound-rotor type induction motor, the rotor resistance can be varied by connecting external resistors to the rotor windings via the brushes. This allows the torque–speed characteristics of the induction motor to be varied as needed.

Now, we will discuss in detail the squirrel-cage induction motor, which is the most common type of an induction motor.

3.1.2 FUNDAMENTALS OF INDUCTION MOTORS

In an induction motor, when a three-phase AC voltage source is applied to the stator windings, which are displaced by 120 electrical degrees in space with respect to each other, the three-phase currents flowing in these windings will produce a rotating magnetic field in the air gap as shown in Fig. 3.7. The speed of this magnetic field is directly proportional to the frequency of the applied AC source voltage.

When the rotating magnetic field is linking to the conductors in the rotor, EMFs based on Faraday's law are induced in the conductors. The induced voltages cause currents in the shorted conductors. Torque is produced on the current-carrying conductors in the rotating magnetic field. Through this developed torque, the rotor will then start rotating in the same direction as the rotating magnetic field.

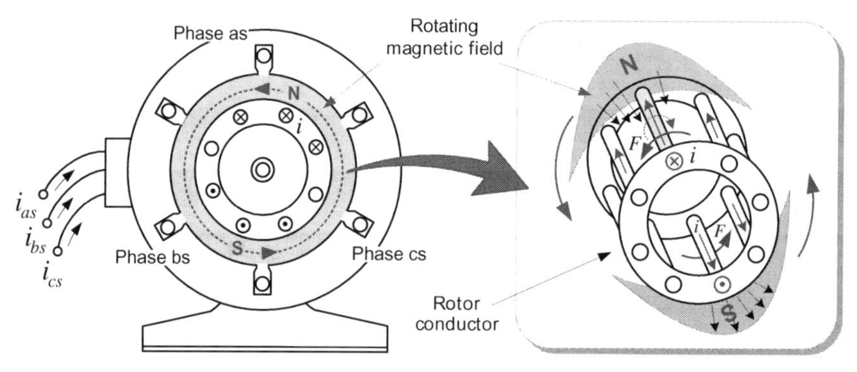

FIGURE 3.7

Rotation of an induction motor [1].

C.I. Hubert, Electric Machines, Theory, Operation, Applications, Adjustment, and Control, second ed., Prentice Hall, 2002, p. 135.

Supposing that the rotating magnetic field is moving counterclockwise, the induced voltages make the currents in the shorted conductors to flow in the direction as depicted in Fig. 3.7 [1]. This is because the induced currents will establish a flux that opposes the change imposed by the rotating magnetic field, in accordance with Lenz's law. When applying Fleming's left-hand rule to the induced currents and the rotating magnetic field, the force on the conductor is developed in the same direction as the rotating magnetic field.

As explained above, an induction motor has no external source provided to its rotor winding but exploits the rotor currents produced by electromagnetic induction. This is why it is called *induction motor*. When an induction motor is running, the rotor speed is always slightly less than the speed of the rotating magnetic field, i.e., synchronous speed. If the rotor rotates in the same speed as the rotating magnetic field, there will be no induced voltage and current in the short-circuited rotor windings and thus no torque. In induction motors, the rotor speed is just slow enough to cause the proper amount of the rotor current to flow so that the developed torque may be sufficient to drive the load. Thus, an increase in load will cause the rotor to slow down.

Now, we will take a look at the rotating magnetic field, which is the basis for the rotation of AC motors.

3.1.2.1 Rotating magnetic field

The rotating mmf is created by three-phase currents flowing through three-phase stator windings. To begin with, consider an mmf resulting from a current flowing in the winding of phase *as* as shown in Fig. 3.8. We will consider only the centered coil on the magnetic axis of the phase *as* distributed winding for simplicity.

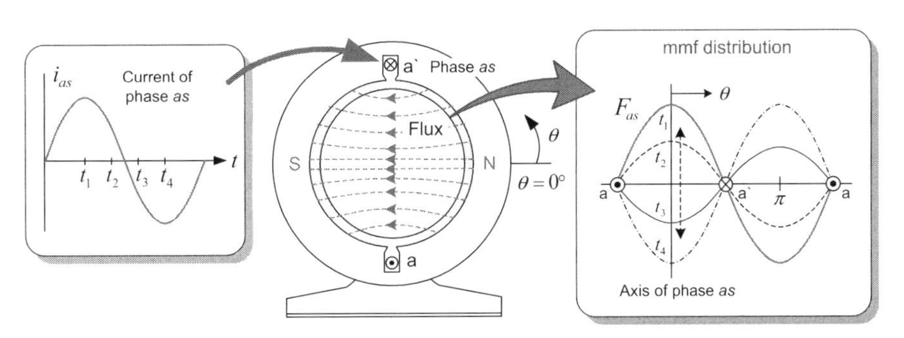

FIGURE 3.8

Phase *as* current and mmf.

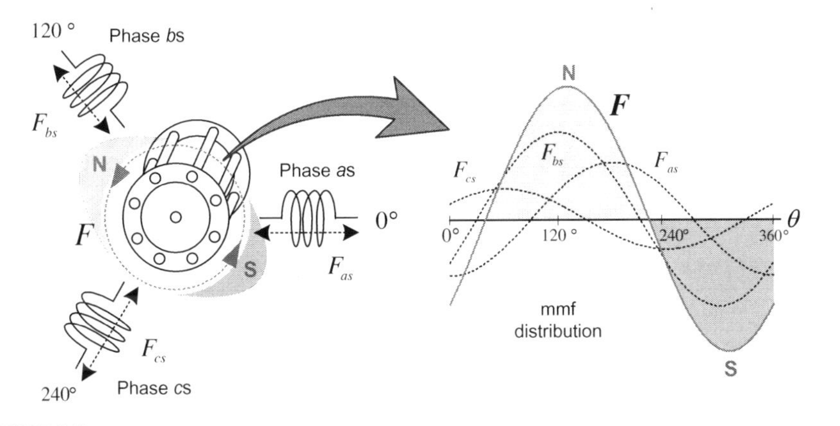

FIGURE 3.9

Resultant air-gap mmf.

Here, a sign of ⊗ indicates that the direction of the current is directed into the page, while a sign of ⊙ indicates that the direction of the current is directed out of the page. Also, the current flowing into the page is assumed to be positive. When a current flows in the phase *as* winding, it will produce a sinusoidally distributed mmf F_{as} centered on its axis. The amplitude of the mmf F_{as} varies with the instantaneous value of the current as shown in Fig. 3.8.

Similarly, the windings of the phase *bs* and phase *cs* produce their own mmfs, pulsating along their axes according to their currents as shown in Fig. 3.9. If these windings are displaced from each other by 120 electrical degrees in space and are supplied by symmetric three-phase AC sine currents, the vector sum of the three mmfs of these windings becomes a single rotating mmf vector F in the air gap. This is called the *rotating magnetic field*, and its moving speed is directly proportional to the frequency of the currents.

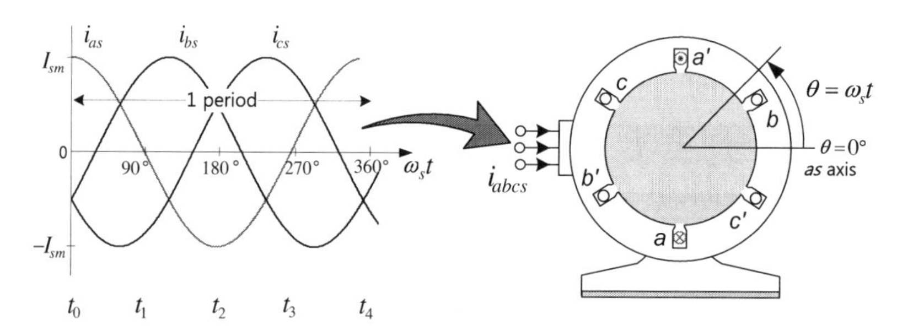

FIGURE 3.10

Three-phase windings and currents.

Let us describe mathematically this resultant air-gap mmf generated by the three-phase currents. The three-phase winding currents as shown in Fig. 3.10 are assumed as Eq. (3.1).

$$\begin{aligned} i_{as} &= I_{sm}\cos\,\omega_s t \\ i_{bs} &= I_{sm}\cos(\omega_s t - 120°) \\ i_{cs} &= I_{sm}\cos(\omega_s t + 120°) \end{aligned} \tag{3.1}$$

where $\omega_s(=2\pi f_s)$ is the angular frequency of the currents and f_s is the frequency of the currents.

The mmfs resulting from the three-phase currents flowing in windings with a total number of turns N_s are given by

$$\begin{aligned} F_{as}(\theta) &= N_s i_{as}\cos\,\theta \\ F_{bs}(\theta) &= N_s i_{bs}\cos\,(\theta - 120°) \\ F_{cs}(\theta) &= N_s i_{cs}\cos\,(\theta + 120°) \end{aligned} \tag{3.2}$$

Here, the axis of phase *as* is chosen as $\theta = 0°$.

From Eqs. (3.1) and (3.2), the resultant mmf is given as

$$\begin{aligned} F(\theta) &= F_{as}(\theta) + F_{bs}(\theta) + F_{cs}(\theta) \\ &= N_s I_{sm}\left[\cos\,\omega_s t\,\cos\,\theta + \cos(\omega_s t - 120°)\cos(\theta - 120°)\right] \\ &\quad + \cos(\omega_s t + 120°)\cos(\theta + 120°) \\ &= \frac{3}{2} N_s I_{sm}\,\cos(\omega_s t - \theta) \end{aligned} \tag{3.3}$$

Eq. (3.3) describes that the resultant mmf is rotating at the angular velocity ω_s over time with a constant amplitude ($\frac{3}{2}N_s I_{sm}$) and is sinusoidally distributed in the air gap as shown in Fig. 3.11.

Let us observe the movement of this resultant air-gap mmf according to the three-phase currents. Fig. 3.12 shows the locations of the resultant mmf produced

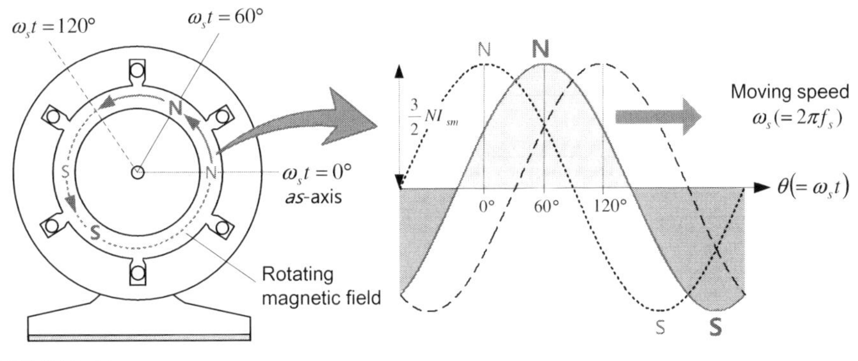

FIGURE 3.11

Resultant mmf.

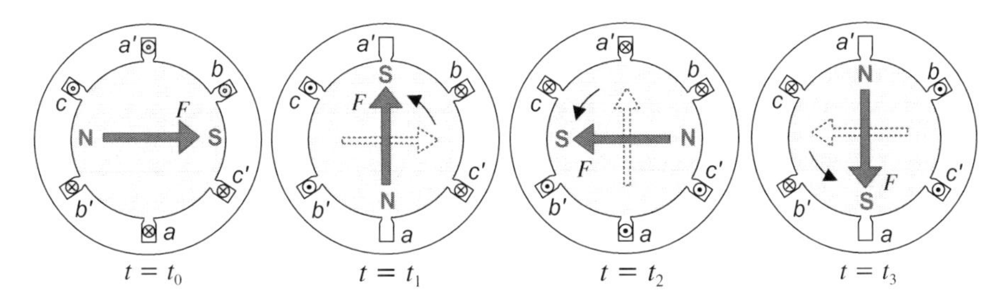

FIGURE 3.12

Movement of the resultant air-gap mmf per cycle of the current.

by the spatial distribution of the three-phase currents at four instants of time, which is shown in Fig. 3.10. Here, we will consider only the centered coils on the axes of the three-phase distributed windings for simplicity. We can see that the resultant air-gap mmf makes one revolution per cycle of the current. The direction of the rotation of the mmf can be reversed by reversing the phase sequence of the currents.

The resultant mmf produces one north magnetic pole and one south magnetic pole in the air gap, i.e., the number of magnetic poles is two. In this two-pole motor, the mmf makes one mechanical revolution per complete cycle of the current. In the four-pole motor that has two sets of three-phase windings, the mmf produces two north and south poles in the air gap as shown in Fig. 3.13. In this case, one cycle of the current corresponds to a half mechanical revolution of the mmf. Thus one revolution of the mmf needs two cycles of the current. The number of electrical cycles required to complete a mechanical rotation is

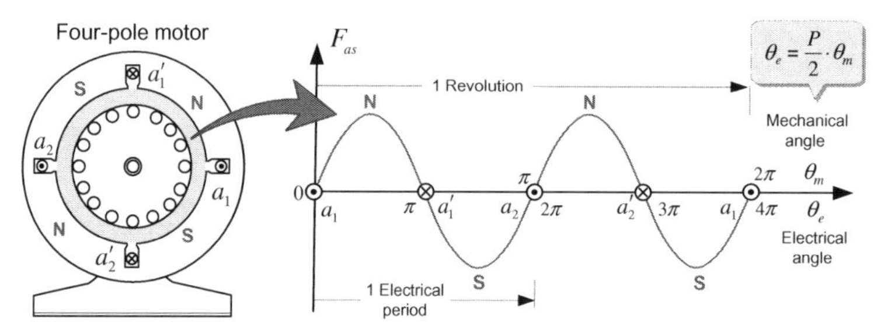

FIGURE 3.13

mmf Distribution by phase *as* current in the four-pole motor.

equal to the number of pole pairs. Thus, in a P-pole motor, there is a relationship between the mechanical angle θ_m and the electrical angle θ_e as

$$\theta_e = \frac{P}{2} \cdot \theta_m \tag{3.4}$$

When dealing with the performance of the motor, all calculations for electric quantities are carried out based on the electrical angle in radian. In a P-pole motor, the rotating speed n_s of the mmf by the three-phase currents with a frequency f is expressed in terms of revolution per minute (r/min) as

$$n_s = \frac{2}{P} \cdot f \cdot 60 = \frac{120f}{P} (r/\text{min}) \tag{3.5}$$

This speed of the rotating mmf is called the *synchronous speed*. For example, for a two-pole induction motor fed by currents of 60 Hz, the synchronous speed is 3600 r/min.

3.1.3 EQUIVALENT CIRCUIT OF INDUCTION MOTORS

To analyze the operating characteristic and performance of an induction motor, we usually use an equivalent circuit based on voltage equations that describe the behavior of the motor. In this section we will start with the steady-state equivalent circuit, and in Chapter 4, we will introduce the equivalent circuit that is useful for characteristics in the transient state as well as in the steady state.

Consider a two-pole, three-phase, wye-connected symmetrical induction motor as shown in Fig. 3.14.

It is assumed that the rotor windings, which may be a wound or a squirrel-cage type, are approximated as three-phase windings. Both the stator and rotor windings are distributed windings, but Fig. 3.14 shows only the centered coils on the axis of each phase. We will initially explore the voltage equations for the stator windings.

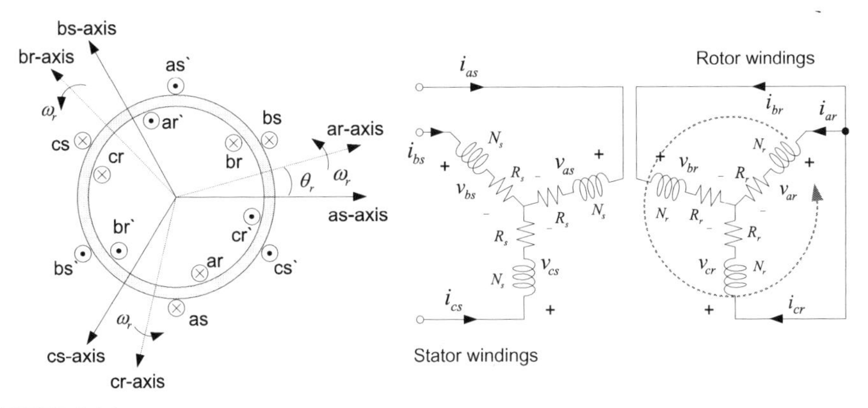

FIGURE 3.14

Two-pole, three-phase, wye-connected symmetrical induction motor.

3.1.3.1 Stator circuit

The voltage equation for the stator windings consists of the voltage drop of the winding resistance R_s and the induced voltage proportional to the rate of change over time of the stator flux linkage λ_s. Thus the voltage equation for the stator windings is written by

$$v_s = R_s i_s + \frac{d\lambda_s}{dt} \tag{3.6}$$

where R_s is the resistance of the stator winding and λ_s is the flux linkage of the stator winding.

When the stator voltage v_s is applied to the stator winding with the effective number of turns N_s, a stator current i_s flows in the winding. The current i_s produces a stator mmf ($N_s i_s$) and in turn produces a stator flux ϕ_s in the air gap. Most of the stator flux ϕ_s crosses the air gap and links to the rotor winding. This air-gap flux is termed mutual flux, which is often called *magnetizing flux*. Only the mutual flux is contributed to energy conversion, i.e., torque production. However, a small portion of the stator flux cannot cross the air gap but links to only the stator winding itself. This is termed *leakage flux*. Thus the stator flux linkage λ_s will be expressed as

$$\lambda_s = N_s(\phi_{ls} + \phi) = N_s(\phi_{ls} + \phi) \tag{3.7}$$

where ϕ is the magnetizing flux and ϕ_{ls} is the leakage flux.
Here, the flux linkage may be written as

$$\lambda_s = L_{ls} i_s + L_m i_m \tag{3.8}$$

where L_{ls} is leakage inductance, L_m is magnetizing inductance, and i_m is magnetizing current.

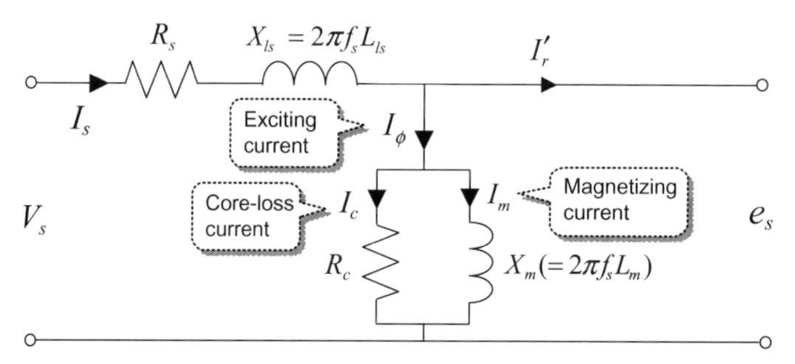

FIGURE 3.15

Equivalent circuit of the stator winding.

Substituting Eq. (3.8) into the voltage equation of Eq. (3.6) yields

$$v_s = R_s i_s + L_{ls}\frac{di_s}{dt} + L_m\frac{di_m}{dt} = R_s i_s + L_{ls}\frac{di_s}{dt} + e_s \tag{3.9}$$

Here, $e_s(=L_m(di_m/dt))$ is the induced voltage (called back-EMF) in the stator winding.

Based on Eq. (3.9), the equivalent circuit of the stator winding can be represented as shown in Fig. 3.15. Here, f_s is the frequency of the stator voltage.

To take into account the core loss such as eddy current loss and hysteresis loss, this circuit includes the equivalent resistance R_c representing the core loss. This resistance is connected in parallel with the magnetizing inductance. Accordingly, the exciting current i_ϕ to produce the flux is comprised of a magnetizing current i_m and a core loss current i_c.

Though an induction motor and a transformer have similar equivalent circuits as shown in Fig. 3.15, they differ in their value of parameters. Because there must be an air gap in induction motors, the magnetizing inductance L_m in induction motors will be a much smaller than that of a transformer. This means that a higher magnetizing current is required to achieve a given flux level. For transformers, the magnetizing current accounts for 1–5% of the rated current, while for induction motors it accounts for 30–50% of the rated current.

We need to know the induced voltage e_s to complete the voltage equation of Eq. (3.9). The induced voltage is proportional to the rate of change over time of the stator flux linkage. To obtain the induced voltage, we need to find out the stator flux, which also requires finding of the flux density.

The rotating magnetic field discussed in the Section 3.1.2.1 may be considered simply as a magnet rotating in the air gap as shown in Fig. 3.16.

The rotating magnetic field will induce voltages in both the stator and rotor windings. When the rotating magnetic field is aligned with the magnetic axis of phase *as* (i.e., $\theta = 0°$) as shown in Fig. 3.17, the flux density distribution in the air gap can be expressed as

$$B(\theta) = B_{max}\cos\theta \tag{3.10}$$

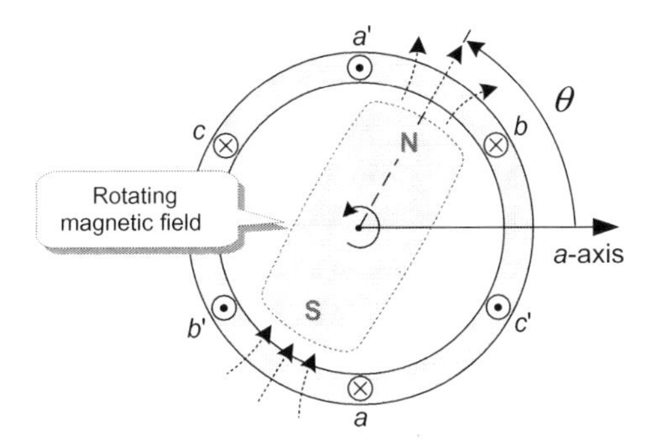

FIGURE 3.16

Rotating magnetic field.

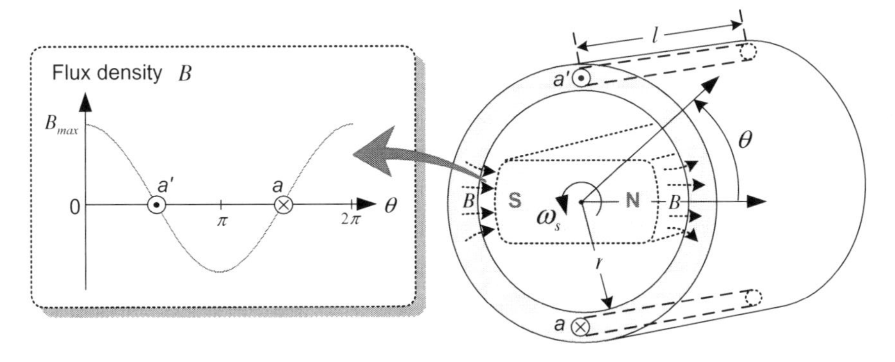

FIGURE 3.17

Flux density distribution in the air gap.

The air-gap flux per pole in the air gap is

$$\phi = \int_{-\pi/2}^{\pi/2} B(\theta) lr d\theta = 2B_{max} lr \tag{3.11}$$

where l is the axial length of the stator and r is the radius to the mean of the air gap.

As the rotating magnetic field moves along the air gap, the flux linking the windings will vary with its position as shown in Fig. 3.18.

The flux linkage λ_a for N_s-turn coils of phase *as* will be the maximum $N_s\phi$ at $\theta = 0°$ and zero at $\theta = 90°$. Thus λ_a will vary as the cosine of the position angle $\theta (= \omega_s t)$ as

$$\lambda_a = N_s \phi \cos \omega_s t \tag{3.12}$$

The voltage induced in the phase *as* winding is obtained from Faraday's law as

$$e_a = -\frac{d\lambda_a}{dt} = \omega_s N_s \phi \sin \omega_s t = E_{\max} \sin \omega_s t \tag{3.13}$$

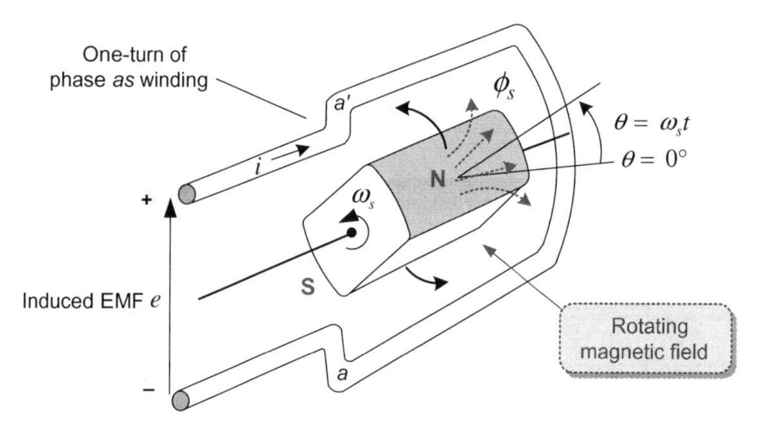

FIGURE 3.18

Flux linking the one-turn coil.

Similarly the voltages induced in the other windings are given by

$$e_b = E_{max}\sin(\omega_s t - 120°) \tag{3.14}$$

$$e_c = E_{max}\sin(\omega_s t + 120°) \tag{3.15}$$

The root mean square (rms) value of the induced voltage is

$$E_{rms} = \frac{E_{max}}{\sqrt{2}} = \frac{\omega_s N_s \phi}{\sqrt{2}} = 4.44 f_s N_s \phi \tag{3.16}$$

where N_s is the total number of the series turns per phase forming a concentrated full-pitch winding. A *full-pitch winding* means that the coil-pitch (i.e., the distance between two sides of a coil) is equal to the pole-pitch (i.e., the distance between two adjacent poles). If the coil-pitch is less than the pole-pitch, then it is called *fractional-pitch* or *short-pitch winding*. In an actual distributed winding, the voltages induced in coils placed in different slots are not in time phase, and therefore the phasor sum of the voltages is less than their numerical sum when all coils are connected in series for the phase winding. Considering this, a reduction factor, called the *winding factor* $K_{\omega s}$, is used. The winding factor $K_{\omega s}$ for most motors is about 0.85–0.95. Consequently, for a distributed winding, the rms-induced voltage per phase is

$$E_s = 4.44 f_s N_s \phi K_{\omega s} \tag{3.17}$$

It is important to note that the stator flux is directly proportional to the ratio of the induced voltage to the stator frequency, i.e., $\phi \propto E_s/f_s$. This fact gives a useful basis for the speed control of induction motors that will be described later.

Next, let us acquire the equivalent circuit of the rotor winding.

3.1.3.2 Rotor circuit

The rotating magnetic field in the air gap induces voltages in the rotor circuit as well as the stator circuit. This induced voltage becomes the source voltage applied to the rotor circuit. Similar to Eq. (3.17), the voltage induced in the rotor circuit can be expressed as

$$E_r = 4.44 N_r \phi K_{\omega r} f_r \tag{3.18}$$

where N_r is the total number of turns per phase, $K_{\omega r}$ is the winding factor for the rotor windings, and f_r is the frequency of the voltage induced in the rotor winding.

Since the rotor winding experiences a flux variation due to the difference in speed between the rotating magnetic field and the rotor, the frequency f_r of the induced voltage in the rotor circuit becomes the difference $f_s - f$ between the frequency f_s of the rotating magnetic field and the rotating frequency f of the rotor. Thus the induced voltage E_r depends on the rotor speed as

$$E_r = 4.44 N_r \phi K_{\omega r} \times f_r = 4.44 N_r \phi K_{\omega r} \times (f_s - f) \tag{3.19}$$

At standstill ($f = 0$), the frequency f_r is the same as the stator frequency f_s. Thus the induced voltage in the rotor winding will be the maximum as

$$E_{r0} = 4.44 N_r \phi K_{\omega r} f_s \tag{3.20}$$

In this case, similar to a transformer, the ratio of the voltages induced in the two windings is equal to the ratio of their number of turns.

The induced voltage in the shorted rotor winding produces a rotor current. Torque in the rotor winding is developed by the interaction between this rotor current and the stator rotating magnetic field. The rotor will then start rotating. Assume that the rotor eventually reaches a steady-state speed n (r/min). This rotor speed is always less than the speed of the stator rotating magnetic field (i.e., the synchronous speed). The difference between the synchronous speed n_s and the rotor speed n is called the *slip speed*. The slip speed expressed as a fraction of the synchronous speed is called the *slip* s and can be defined as

$$\text{Slip:} \quad s = \frac{n_s - n}{n_s} \tag{3.21}$$

It should be noted that the slip is a very important factor in the induction motor because most of its performance characteristics of an induction motor, such as the developed torque, current, efficiency and power factor, depends on the operating slip. The operating slip depends on the load. An increase in the load will cause the rotor to slow down and increase the slip. A decrease in the load will cause the rotor to speed up and decrease the slip. A typical induction motor operates in the slip range of 0.01–0.05, i.e., 1–5%. For example, a four-pole motor operated at 60 Hz has a synchronous speed of 1800 r/min. If the rotor speed at full load is 1765 r/min, then the slip is 1.9%.

Now, let us examine the induced voltage in the rotor winding when the rotor is rotating at a speed n (slip s). The frequency f_r in the rotor circuit at the slip s is called the *slip frequency* and is given as

$$f_r = f_s - f = sf_s \tag{3.22}$$

From Eq. (3.20), the induced voltages in the rotor circuit at the slip s becomes

$$E_r = 4.44 N_r \phi K_{\omega r} \times f_r = 4.44 N_r \phi K_{\omega r} \times sf_s = sE_{r0} \tag{3.23}$$

We can see that the magnitude of the voltage induced in the rotor winding is directly proportional to the slip. When the rotor is stationary (i.e., $s = 1$), the largest voltage is induced in the rotor circuit. As the rotor speed is increased, the induced voltage is decreased. If the rotor speed is equal to the synchronous speed, then the induced voltage becomes zero.

Now, we are ready to discuss the equivalent circuit of the rotor winding. From Eq. (3.23), we can see that the source voltage in the rotor circuit is sE_{0r}. With the resistance R_r and the leakage inductance L_{lr} of the rotor circuit, the rotor equivalent circuit is depicted as Fig. 3.19A. The rotor equivalent circuit of Fig. 3.19A is at the rotor frequency f_r. Thus we cannot combine this rotor circuit with the stator equivalent circuit of Fig. 3.16 together in a single circuit because the two circuits differ in the operating frequency. Therefore, we need to adjust the rotor frequency to combine the two circuits. By dividing the voltage and the impedance of Fig. 3.19A by the slip s, we will obtain the rotor equivalent circuit of Fig. 3.19B. In the circuit, the rotor current is the same as in Fig. 3.19A, but its operating frequency is equal to the stator frequency f_s. Thus, this becomes the rotor equivalent circuit as seen from the stator. These two circuits can now be joined together by considering the turns ratio $a(= N_s/N_r)$ of the stator winding and the rotor winding.

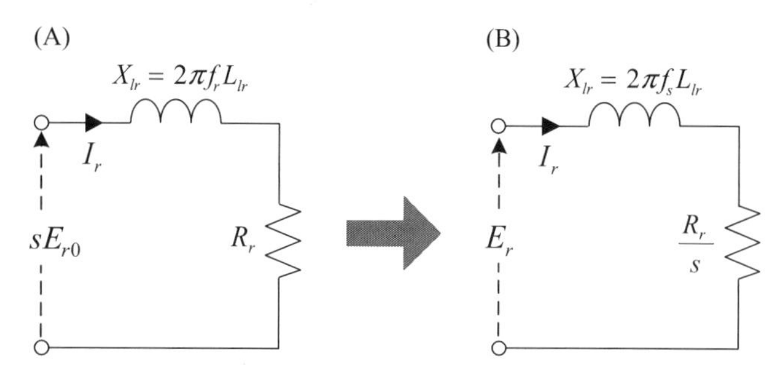

FIGURE 3.19

Rotor equivalent circuit. (A) Referred to rotor side and (B) referred to stator side.

The complete equivalent circuit per phase of a three-phase induction motor is shown in Fig. 3.20.

This equivalent circuit is similar to that of a transformer, except for the effects of varying speed. The quantities that are reflected from the rotor to the stator are denoted by the prime symbol ('). From this point on, we will skip the prime symbol.

Similar to the three-phase currents of the stator winding, the three-phase currents induced in the rotor winding also produce a rotating magnetic field F_r in the air gap as shown in Fig. 3.21A. From the rotor structure's point of view, this rotor rotating magnetic field F_r rotates at the speed $n_r(=n_s-n)$. Since the rotor itself is rotating at n, from the stator structure's point of view, the rotor rotating magnetic field rotates at the synchronous speed $n_s(=n_r+n)$. Thus, both the stator magnetic field and the rotor magnetic field always rotate in the air gap at

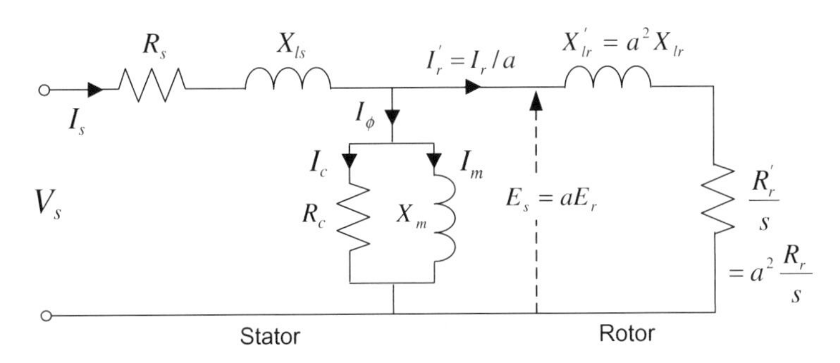

FIGURE 3.20

Complete per phase equivalent circuit of a three-phase induction motor.

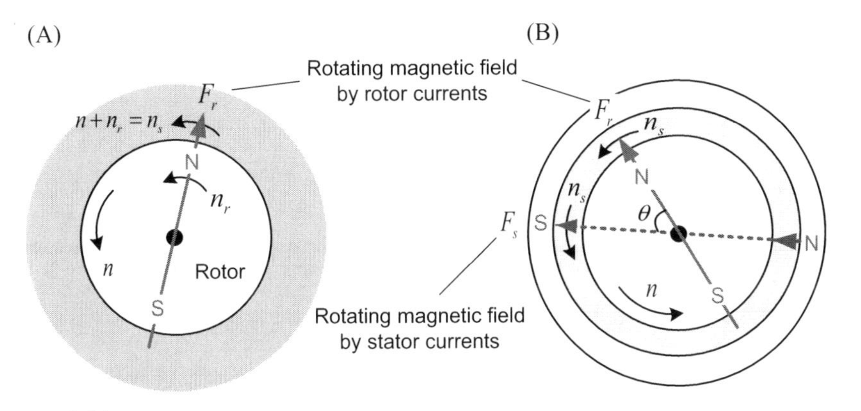

FIGURE 3.21

Rotating magnetic fields. (A) Magnetic field by rotor current and (B) magnetic fields by stator and rotor currents.

the same synchronous speed. These stator and rotor magnetic fields, therefore, remain stationary with respect to each other as shown in Fig. 3.21B. The interaction between these two fields is considered to produce the torque.

3.1.4 CHARACTERISTICS OF INDUCTION MOTORS

In this section, we will discuss the characteristics of an induction motor such as the current, input power factor, output torque, and efficiency from the equivalent circuit of Fig. 3.20. We will see that these characteristics depend on the slip, i.e., the operating speed.

3.1.4.1 Stator current

The stator current is the input current of an induction motor. From Fig. 3.20, neglecting the core losses, the input impedance as a function of slip is given as

$$Z_s = R_s + jX_{ls} + \left[X_m / \left(\frac{R_s}{s} + jX_{lr} \right) \right] = |Z_s| \angle \theta_s \tag{3.24}$$

where θ_s is input impedance angle.

From the applied stator voltage and the input impedance of Eq. (3.24), the stator current is given as

$$I_s = \frac{V_s}{Z_s} = I_\phi + I_r \tag{3.25}$$

The stator current is the sum of the exciting current I_ϕ and the rotor current I_r. With a constant stator voltage V_s, the stator current I_s as a function of speed (i.e., slip s) is shown in Fig. 3.22.

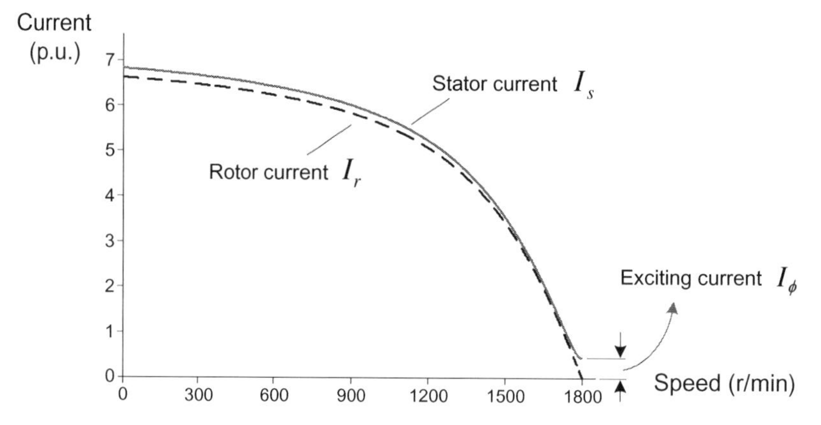

FIGURE 3.22

Stator current as a function of speed (5-hp, four poles).

At starting ($s = 1$), since the impedance is minimum, the stator current will be the maximum value, which may be 5–8 times larger than the rated current. As the speed increases, the slip s decreases. As the slip s decreases, the rotor current I_r decreases, and thus the stator current also decreases. At the synchronous speed, i.e., $s = 0$, $R_s/s \approx \infty$, so there is no rotor current, i.e., $I_r = 0$. Thus the stator current I_s will be equal to the exciting current I_ϕ.

3.1.4.2 Input power factor

From the input impedance angle θ_s of Eq. (3.24), the input power factor of an induction motor is written as

$$PF = \cos\theta_s \tag{3.26}$$

The power factor varies with speed as shown in Fig. 3.23. Since induction motors inherently draw a lagging current to produce the magnetic flux, the power factor is always smaller than one. The power factor also varies with the load applied to the motor. The power factor is minimum at no load, and it increases with the load. It usually reaches the peak at or near full load.

3.1.4.3 Output torque

We can derive a general expression for the output torque of an induction motor from the power-flow diagram as shown in Fig. 3.24.

The input power $P_{in}(=3V_sI_s\cos\theta_s)$ to an induction motor, excluding the stator losses such as the stator copper loss and the core loss, will be transferred to the rotor across the air gap and then transformed to mechanical power P_{mech}. We can derive the output torque using the mechanical power P_{mech}.

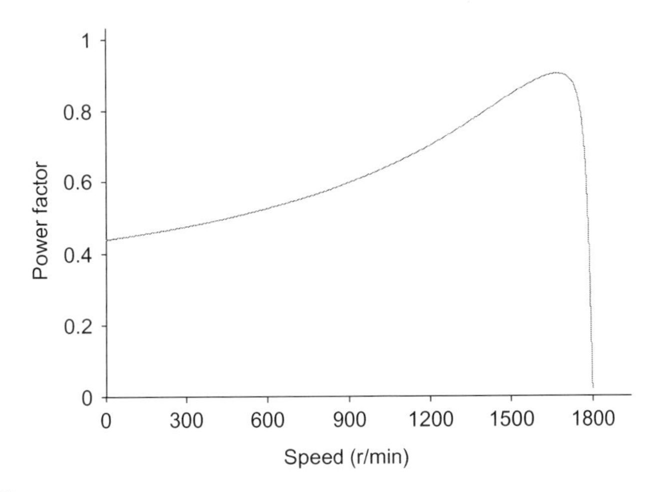

FIGURE 3.23

Input power factor (5-hp, four-pole motor).

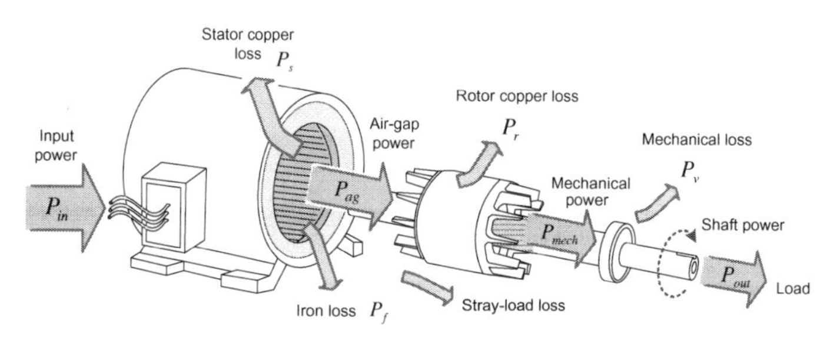

FIGURE 3.24

Power-flow diagram in an induction motor.

Redrawn from http://images.slideplayer.com/14/4186646/slides/slide_32.jpg.

The power delivered through the air gap from the stator to the rotor is called the *air-gap power* P_{ag}. Referring to the equivalent circuit given in Fig. 3.20, the air-gap power is written by

$$P_{ag} = I_r^2 \frac{R_r}{s} \tag{3.27}$$

Because the induction motors are operated typically at low slips of 0.01–0.05, the air-gap power is considerably large. This air-gap power P_{ag}, i.e., input power to the rotor, will be dissipated in the resistance R_r of the rotor circuit (i.e., rotor copper loss) and converted into a mechanical output. Thus, Eq. (3.27) can be divided into two components as

$$P_{ag} = \underbrace{I_r^2 R_r}_{\substack{\text{Rotor copper loss} \\ P_r}} + \underbrace{I_r^2 \frac{R_r}{s}(1-s)}_{\substack{\text{Mechanical output} \\ P_{mech}}} \tag{3.28}$$

$$P_r = I_r^2 R_r = s\,P_{ag} \tag{3.29}$$

$$P_{mech} = I_r^2 \frac{R_r}{s}(1-s) = (1-s)P_{ag} \tag{3.30}$$

The first term on the right-hand side of Eq. (3.28) represents the rotor copper loss P_r. The second term represents the mechanical output power P_{mech} that develops the output toque. Fig. 3.25 shows this power distribution in the rotor circuit based on Eq. (3.28).

We can see that the power distribution depends on the slip s. It is clear that the lower the slip of the motor, the lower the rotor copper loss and the higher the mechanical output power.

The mechanical power is equal to the output torque T_{mech} times the angular frequency ω_{mech} as

$$P_{mech} = I_r^2 \frac{R_r}{s}(1-s) = T_{mech} \cdot \omega_{mech} \tag{3.31}$$

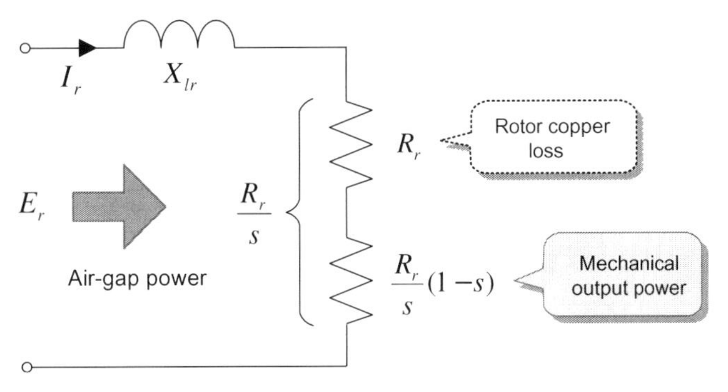

FIGURE 3.25

Distribution of the air-gap power.

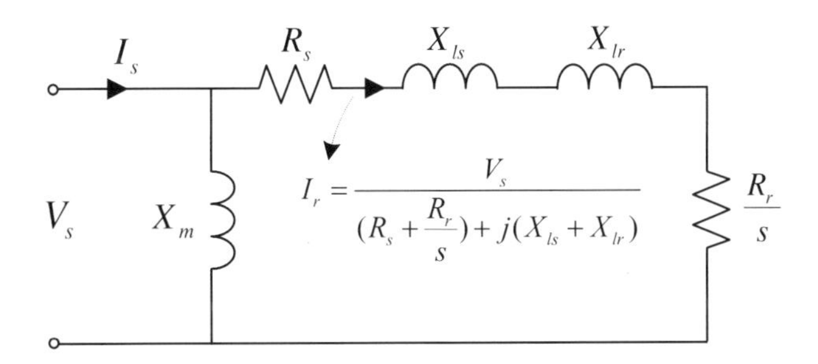

FIGURE 3.26

Approximated per phase equivalent circuit of an induction motor.

Thus the output torque of an induction motor is given by

$$T_{mech} = \frac{P_{mech}}{\omega_{mech}} = \frac{1}{\omega_{mech}} I_r^2 \frac{R_r}{s}(1-s) = \frac{1}{\omega_s} I_r^2 \frac{R_r}{s} \tag{3.32}$$

where $\omega_{mech} = (1-s)\omega_s$ and ω_s represents the synchronous angular frequency.

We need to express this torque equation in terms of the input voltage instead of a nonmeasurable rotor current. The rotor current can be easily obtained from the approximate equivalent circuit given in Fig. 3.26, in which the core loss resistance is ignored and the magnetizing inductance is shifted to the input side assuming $R_s + jX_{ls} \ll \omega_s L_m$.

From this circuit, the rotor current is given by

$$I_r = \frac{V_s}{\left(R_s + \frac{R_r}{s}\right) + j(X_{ls} + X_{lr})} \tag{3.33}$$

By substituting this rotor current into Eq. (3.32), the output torque (per phase) is expressed as

$$\text{Torque:} \quad T_{mech} = \frac{1}{\omega_s}\frac{V_s^2}{\left(R_s+\dfrac{R_r}{s}\right)^2+j(X_{ls}+X_{lr})^2}\frac{R_r}{s} \tag{3.34}$$

This is the steady-state average torque per phase for a given input voltage V_s, which is a function of the slip. The torque for a given slip is proportional to the square of the stator input voltage. The torque calculated by using this approximate circuit varies within 5% from that of the full circuit [2]. The shaft torque, which is available to the load at the shaft of the motor, can be obtained by subtracting the friction and windage torques from this output torque.

Let us examine the speed–torque characteristics of an induction motor based on Eq. (3.34). In the low-slip region, which is the normal operating range of an induction motor, the impedance of equivalent circuit parameters shows the following relations as

$$R_s + \frac{R_r}{s} \gg X_{ls} + X_{lr} \quad \text{and} \quad \frac{R_r}{s} \gg R_s \tag{3.35}$$

Thus the torque of Eq. (3.34) can be simplified as

$$T_{mech} \approx \frac{1}{\omega_s}\frac{V_s^2}{R_r}s \;\rightarrow\; T_{mech} \propto s \tag{3.36}$$

In this case, it is noteworthy that the output torque increases linearly with the slip s. On the other hand, in the low-speed range with larger values of the slip,

$$R_s + \frac{R_r}{s} \ll X_{ls} + X_{lr} \tag{3.37}$$

Thus the torque varies almost inversely with the slip s as

$$T_{mech} = \frac{1}{\omega_s}\frac{V_s^2}{(X_{ls}+X_{lr})^2}\frac{R_r}{s} \;\rightarrow\; T_{mech} \propto \frac{1}{s} \tag{3.38}$$

There is a speed at which the maximum torque, often referred to as *pull-out torque* or *breakdown torque*, is developed. The slip at the maximum torque can be obtained by solving $dT_{mech}/ds = 0$ as

$$s_{max} = \frac{R_r}{\sqrt{R_r^2 + (X_{ls}+X_{lr})^2}} \tag{3.39}$$

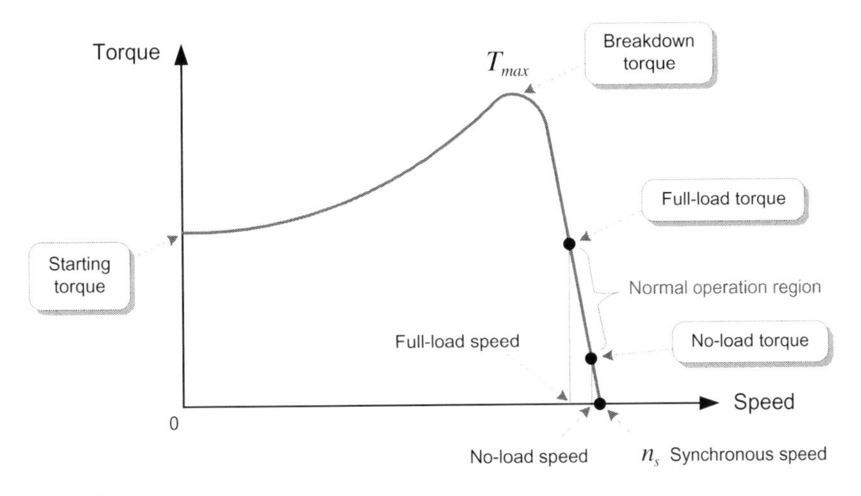

FIGURE 3.27

Speed versus torque curve for an induction motor.

It is important to note that the rotor resistance R_r is an important parameter in determining the slip at which the maximum torque occurs. From s_{max}, the maximum torque can be given as

$$T_{max} = \frac{1}{2\omega_s} \frac{V_s^2}{R_s + \sqrt{R_s^2 + (X_{ls} + X_{lr})^2}} \tag{3.40}$$

The value of the maximum torque is independent of the rotor resistance, which determines the maximum slip value. The speed–torque curve of a typical three-phase induction motor is shown in Fig. 3.27. This curve shows the speed versus the output torque when an induction motor is started with full voltage.

The normal operating range of an induction motor is near the synchronous speed, confined to less than 5% slip. In this low-slip region, the output torque increases linearly with the increasing slip. The slip increases approximately linearly with the increased load, and thus the rotor speed decreases approximately linearly with the load.

3.1.4.4 Stable operating point

When a motor is driving a mechanical load, the motor will operate in a steady state at a speed at which the torque developed by the motor is equal to the torque required by the load. As it can be seen from the motion equation in Eq. (3.41), when the motor torque T_{motor} exceeds the load torque T_{load}, the motor speed will increase. When the motor torque is less than the load torque, the motor speed will decrease. Thus the equilibrium point will be the speed at which the motor torque equals to the load torque.

$$T_{motor} - T_{load} = (J_{motor} + J_{load})\frac{d\omega}{dt} \tag{3.41}$$

There are two kinds of an equilibrium operating point: *stable point* and *unstable point*. At the stable operating point, the speed will be restored after a small departure from the original equilibrium point due to a disturbance in the input voltage or load. However, at the unstable operating point, the speed will not be restored.

In the induction motor drives, a stable operating point is in the range above the speed at which the maximum torque occurs in the speed–torque curve. As an example, consider Fig. 3.28 that shows the speed–torque curve of the induction motor along with two different load curves.

There exist three equilibrium operating points. Among these, point A is an unstable operating point, whereas points B and C are stable operating points. At point A, the system will not be able to maintain its original equilibrium state when there is a disturbance in the drive system. For example, assume that, at point A, there is a reduction in speed caused by a disturbance. In this case, a decrease in the speed causes the motor torque to be less than the load torque. As a result, the system will decelerate and eventually come to a stop. Similarly, an increase in the speed caused by a disturbance will make the motor torque larger than the load torque. Thus the operating point will move from point A to point B. Therefore point A is an unstable operating point. On the contrary, points B and C are stable operating points. At these points, the speed will return to their original points after the deviation in speed by a disturbance. Criterion for the stability is given by

$$\frac{d(T_{load} - T_{motor})}{d\omega} > 0 \tag{3.42}$$

Hence a stable operation needs the following conditions: when the speed is decreased, the motor torque developed needs to be greater than the load torque,

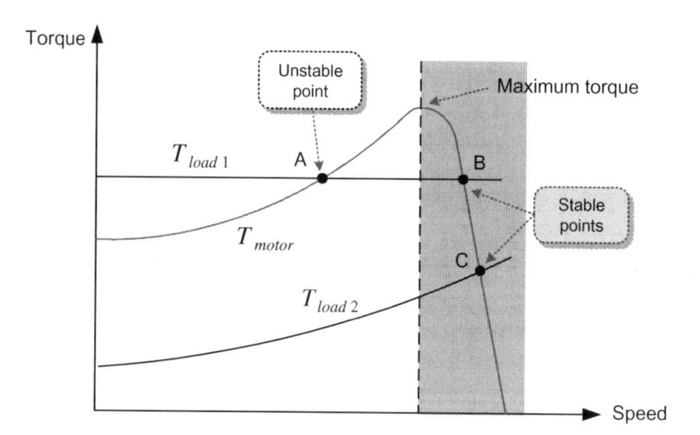

FIGURE 3.28

Operating points in motor drives.

whereas when the speed is increased, the motor torque needs to be less than the load torque.

3.1.4.5 Efficiency

Next, we will discuss the efficiency of an induction motor. Efficiency is defined as the ratio of the output power to the input power of the motor expressed as

$$\eta = \frac{P_{out}}{P_{in}} = \frac{P_{out}}{P_{out} + P_{Loss}} \cdot 100\,(\%) \tag{3.43}$$

As for a motor, the efficiency indicates how well the motor converts electrical power into mechanical power. The input power of an induction motor is given by

$$P_{in} = 3V_s I_s \cos\theta_s \tag{3.44}$$

As shown in Fig. 3.24, the losses in induction motors are typically grouped into five significant origins: the stator copper loss P_s, the core loss P_f, the rotor copper loss P_r, the windage and friction loss P_v, and the stray-load loss. The stray-load loss indicates residual losses that are difficult to determine by a direct measurement or a calculation. The stator copper loss accounts for a large part of the motor loss. Fig. 3.29 shows the loss distribution rating by rating [3].

The efficiency of induction motors is also dependent on the slip. If all the losses that have little or no relationship to the slip are neglected, then the efficiency can be expressed as a function of the slip as

$$\eta = \frac{P_{out}}{P_{in}} \rightarrow \frac{P_{mech}}{P_{ag}} = \frac{P_{ag} - P_r}{P_{ag}} = 1 - s \tag{3.45}$$

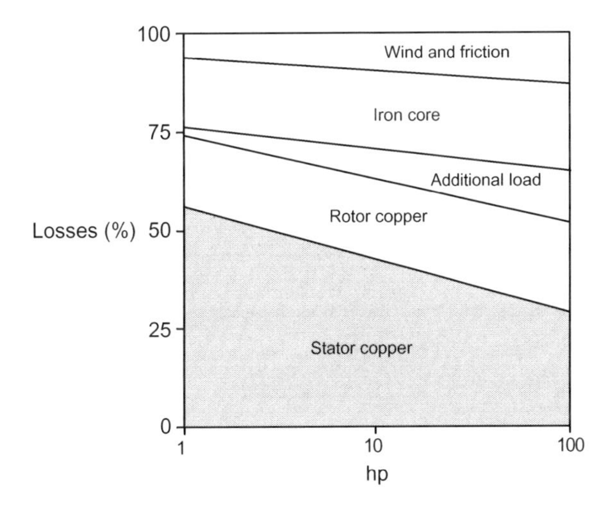

FIGURE 3.29

Loss distribution of induction motor.

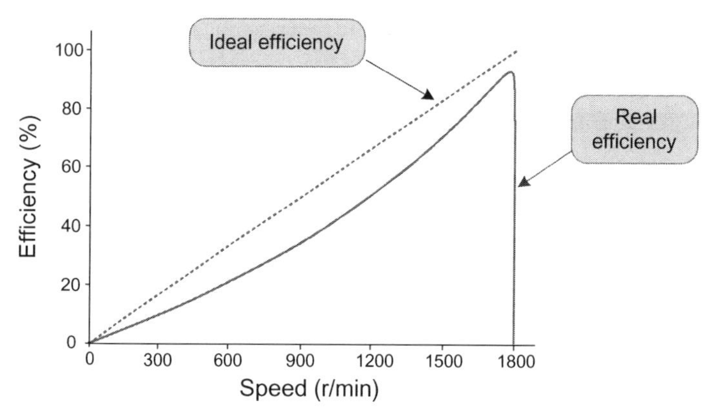

FIGURE 3.30

Actual efficiency and ideal efficiency.

This is called the ideal efficiency or the internal efficiency. The internal efficiency as a function of speed is shown in Fig. 3.30. The internal efficiency gives a measure of how much of the power delivered to the air gap is available for mechanical power. The lower slip operation leads to a higher efficiency because the rotor copper loss is directly proportional to the slip as we can see from Eq. (3.29). In general, larger motors are more efficient than smaller motors. Today's premium efficiency three-phase induction motors have efficiencies ranging from 86.5% at 1 hp to 95.8% at 300 hp.

EXAMPLE 1

A 5 hp, 220 V, 60 Hz, four-pole three-phase induction motor operates at 1740 r/min when fully loaded at its rated voltage and rated frequency. The motor parameters are $R_s = 0.295\ \Omega$, $R_r = 0.379\ \Omega$, $X_m = 22.243\ \Omega$, and $X_{ls} = X_{lr} = 0.676\ \Omega$. Evaluate the following:

1. the slip, power factor, stator current, exciting current, and developed torque when running at rated conditions;
2. the stator current and developed torque when starting at full voltage;
3. the slip at which maximum torque is developed.

Solution

1. The synchronous speed is

$$n_s = \frac{120 f_s}{P} = \frac{120 \cdot 60}{4} = 1800\,\text{r/min}$$

Thus the rated slip is

$$s = \frac{n_s - n}{n_s} = \frac{1800 - 1740}{1800} = 0.0333$$

From the equivalent circuit in Fig. 3.20, the rotor impedance is

$$Z_r = \frac{R_s}{s} + jX_{lr} = 11.37 + j0.679$$

The input impedance is

$$Z_s = R_s + jX_{ls} + \frac{Z \cdot jX_{lr}}{Z_r + jX_{lr}} = 8.887 + j5.599$$

$$= |Z_s| \angle \theta_s = 10.5 \angle 32.212$$

The power factor is

$$\cos \theta_s = 0.846$$

The stator current is

$$I_s = \frac{V_{phase}}{|Z_s|} = \frac{\left(\frac{220}{\sqrt{3}}\right)}{10.5} = 12.092 \text{ A}$$

The rotor current is

$$I_r = I_s \cdot \frac{jX_m}{Z_r + jX_m} = 10.512 \text{ A}$$

The torque is

$$T_{mech} = \frac{1}{\omega_s} I_r^2 \frac{R_r}{s} = \frac{1}{188.5} \cdot 10.512^2 \cdot \frac{0.379}{0.0333} = 20 \text{ Nm}$$

where

$$\omega_s = 2\pi f_s \left(\frac{2}{P}\right) = 188.5 \text{ rad/s}$$

2. Since $s = 1$ at start, the rotor impedance is

$$Z_r = \frac{R_s}{s} + jX_{lr} = 0.379 + j0.679$$

The input impedance is

$$Z_s = R_s + jX_{ls} + \frac{Z \cdot jX_{lr}}{Z_r + jX_{lr}} = 0.652 + j1.343$$

The stator current is

$$I_s = \frac{V_{phase}}{|Z_s|} = \frac{\left(\frac{220}{\sqrt{3}}\right)}{1.493} = 85.08\text{A}$$

and the starting current is seven times larger than the rated current.

The rotor current is

$$I_r = I_s \cdot \frac{jX_m}{Z_r + jX_m} = 82.557\text{A}$$

The starting torque is

$$T_{mech} = \frac{1}{\omega_s} I_r^2 \frac{R_r}{s} = \frac{1}{188.5} \cdot 82.557^2 \cdot \frac{0.379}{1} = 41.1 \text{ Nm}$$

3. The slip at the maximum torque is

$$s_{max} = \frac{R_r}{\sqrt{R_r^2 + (X_{ls} + X_{lr})^2}} = 0.273$$

The maximum torque developed is

$$T_{max} = \frac{3}{2\omega_s} \frac{V_s^2}{R_s + \sqrt{R_s^2 + (X_{ls} + X_{lr})^2}} = 73.28$$

From the above results, the speed–torque curve of this induction motor is shown as

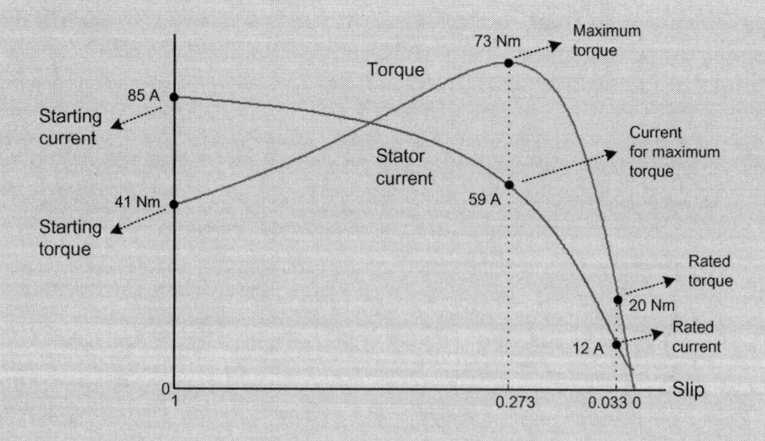

3.1.5 OPERATING MODES OF INDUCTION MOTORS

An induction motor can be operated in three modes according to its operating speed and rotation direction. Fig. 3.31 shows the power, operating speed, and torque in the available three operation modes of an induction motor.

First, a typical operation of an induction motor is the *motoring mode*, in which the rotor rotates in the same direction as the stator rotating magnetic field but at a somewhat slower speed. Thus the slip ranges between 0 and 1 ($0<s\leq 1$), and both the air-gap power and the mechanical power are positive as

$$P_{ag} = I_r^2 \frac{R_r}{s} > 0, \quad P_{mech} = (1-s)P_{ag} > 0 \tag{3.46}$$

This implies that, in the motoring mode, the induction motor converts the applied input power into mechanical power through the air gap.

If an induction motor is driven to a speed higher than the synchronous speed by an external prime mover, then the slip will be negative ($s<0$) and both the air-gap power and the mechanical power are negative as

$$P_{ag} = I_r^2 \frac{R_r}{s} < 0, \quad P_{mech} = (1-s)P_{ag} < 0 \tag{3.47}$$

This implies that the power from the mechanical system flows into the rotor circuit, then across the air gap to the stator circuit and the external electrical system. Thus the induction motor will operate in the *generation mode*.

This mode may occur when the stator frequency applied to an induction motor is lowered to reduce the speed of the rotor. In this process, the

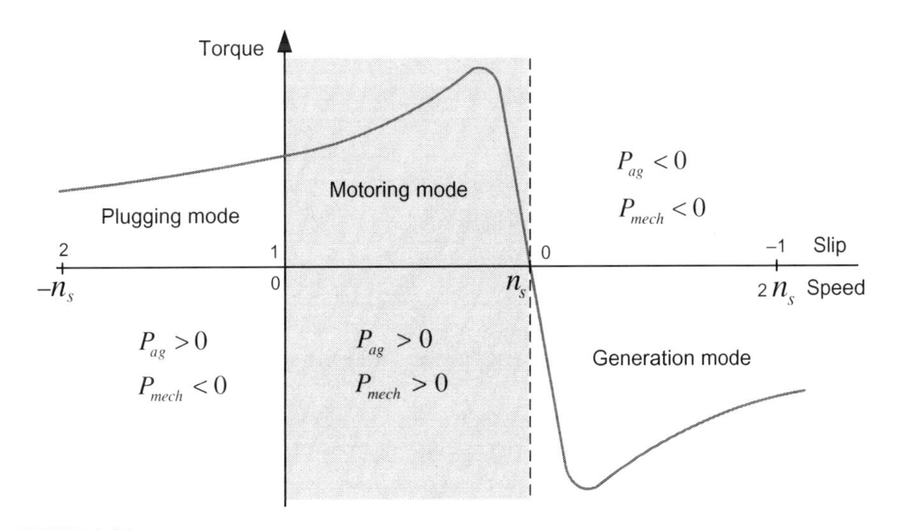

FIGURE 3.31

Three operation modes of an induction motor: plugging mode, motoring mode, and generation mode.

instantaneous speed of the rotor may be higher than the instantaneous synchronous speed because of the inertia of the drive system. This leads to a negative slip, and the direction of the torque developed in the motor will be opposite to the rotation of the rotor. This torque will act as a braking torque that slows down the rotor speed. In this case, the motor operates as a generator, and the generated power is fed back into the power source. This process is called *regenerative braking*.

The last operating mode is the *plugging mode*, in which the rotor rotates in the direction opposite to the stator magnetic field and thus, the slip $s > 1$. In this mode, the air-gap power and the mechanical power are given as

$$P_{ag} = I_r^2 \frac{R_r}{s} > 0, \quad P_{mech} = (1 - s)P_{ag} < 0 \tag{3.48}$$

Both powers flow into the rotor and are dissipated in the rotor as heat. Thus the rotor bar may be overheated. This mode can also be used for braking the motor to stop quickly. When an induction motor is running, the direction of the rotation of the stator magnetic field can be reversed by switching any two of the three-phase windings. In this case, the developed torque will be in the direction of the stator magnetic field but opposite to the rotation of the rotor, and the induction motor will operate in the plugging mode.

3.1.6 EFFECT OF ROTOR RESISTANCE

The resistance in the rotor circuit greatly influences the characteristics of an induction motor. This is because the speed at which the maximum torque occurs depends on the rotor resistance as shown in Eq. (3.4).

Fig. 3.32 shows the speed−torque characteristics for two different values of the rotor resistance. A high rotor resistance will provide a high starting torque, leading to a rapid acceleration of the mechanical load system. This is also desirable for a low starting current and a high power factor during starting. However a high rotor resistance results in a high slip during the normal running operation. This causes increased rotor loss and reduced efficiency. Therefore, during a normal operation, a low rotor resistance is desirable. This results in a lower slip and thereby a higher efficiency.

Therefore we can see that there are conflicting requirements of rotor resistance for a satisfactory performance, i.e., a high resistance at starting and low resistance at the normal operating speed. For this reason, different types of rotors have been developed to achieve a variation in rotor resistance with operating speed.

In a wound-rotor induction motor, external resistance can be connected to the rotor windings through slip rings as shown in Fig. 3.6. Thus a satisfactory performance in both the starting and running conditions can be obtained by varying the value of the external resistance with the rotor speed.

For squirrel-cage induction motors, the rotor resistance depends on the operating speed. This is because the effective resistance of the rotor conductor

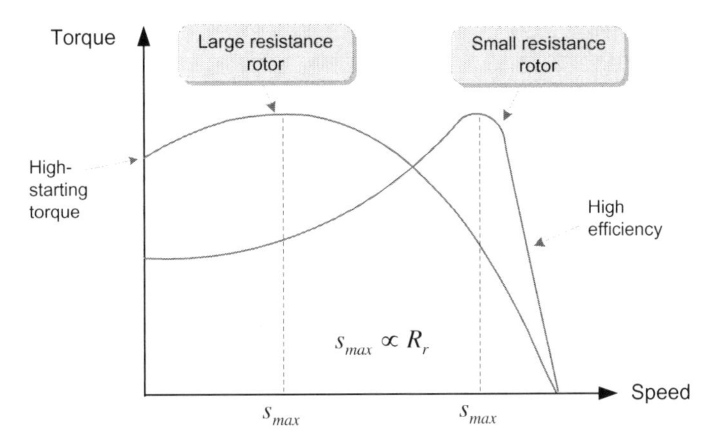

FIGURE 3.32

Torque–speed characteristics for different values of the rotor resistance.

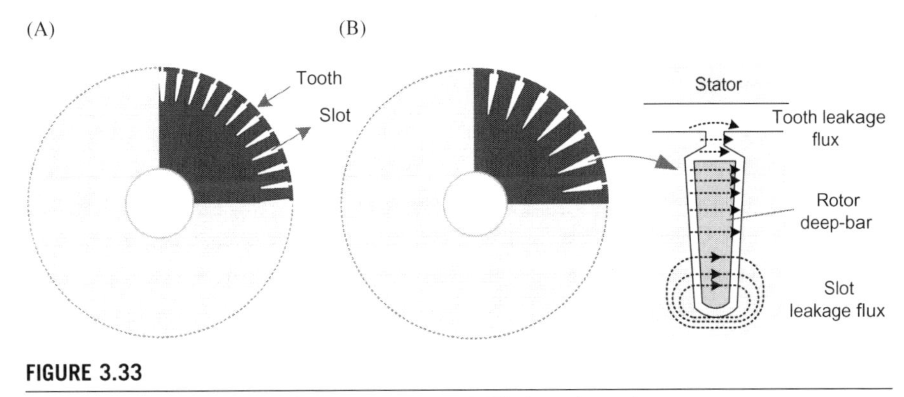

FIGURE 3.33

Comparison of rotor bars. (A) Normal rotor and (B) deep-bar rotor.

may be changed with the rotor frequency due to the skin effect. The skin effect is the tendency of AC current to flow near the surface of the conductor. The skin effect causes the effective area of the cross section of the conductor to reduce at higher frequencies, thus increasing the effective resistance of the conductor. At standstill, the rotor frequency is high since it equals to the stator frequency, whereas at a normal operating speed, it is low at about 1–3 Hz in a 60-Hz motor. As a result, the effective resistance of the rotor circuit decreases as the operating speed increases.

Besides the skin effect, the rotor conductor can be properly designed to have a higher resistance at the starting. In this case, the rotor conductor is made in the form of a *deep-bar* as shown in Fig. 3.33B. In the current-carrying deep-bar, the leakage flux surrounding the bottom part of the deep-bar conductor is greater than that of the upper part. The leakage flux lines are shown by the *dotted lines* on the right of Fig. 3.33B. The leakage inductance of the conductor elements at the bottom part is larger than that of the conductor elements at the upper part. Therefore the AC current in the upper part of the conductor will flow more easily than in the bottom part with a large leakage reactance. This nonuniform current distribution makes the effective resistance of the bar to increase.

Since the leakage reactance is proportional to the frequency, this nonuniform current distribution is more pronounced at the starting of a high rotor frequency. At normal operating speeds, since the rotor frequency is low, the rotor current is almost uniformly distributed over the cross-sectional area of the rotor bar, and thus the effective resistance R_{eff} will be as low as the DC value R_{dc}. For a deep-bar rotor, its effective resistance at starting may be several times greater than at the rated speed as shown in Fig. 3.34.

As explained above, for induction motors with a deep-bar rotor, the rotor resistance varies according to the speed so that a satisfactory performance can be achieved both at the start and normal operations.

A double-cage rotor is another type of rotor whose effective resistance varies as shown in Fig. 3.35 [4]. It consists of two cages to exaggerate its variation more than the deep-bar. The upper cage has smaller cross-sectional areas than the lower cage. In addition, the upper cage is made up of a higher resistance material than the lower cage. Similar to a deep-bar rotor, the rotor bars in the lower cage have a greater leakage flux and thus, higher leakage inductance. At starting, since the rotor frequency is high, the division of the rotor current is mainly decided by the leakage reactance difference between the two cages. Most of the currents will flow in the upper cage that has a lower reactance, but a high resistance.

On the other hand, at a normal operating speed, since the rotor frequency is low, the division of the rotor current is mainly dependent on the resistance of the two cages. Most of the current will flow in the lower cage which has a lower resistance. Therefore, in a double-cage rotor, at the start-up, the current flows mainly in the upper cage which has a higher resistance, while at a normal operating speed, it flows in the lower cage which has a lower resistance.

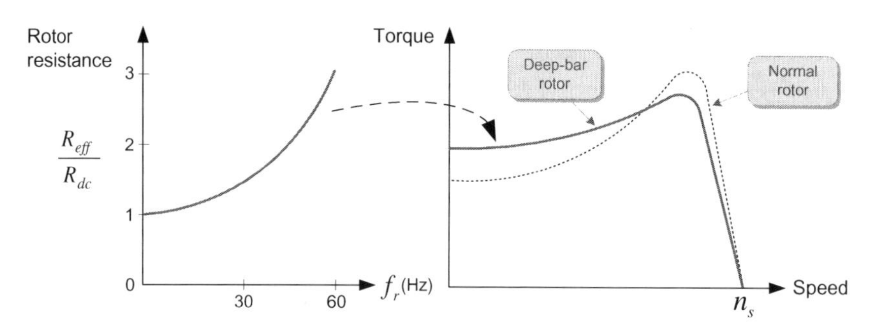

FIGURE 3.34

Effective resistance and output torque in the deep-bar rotor.

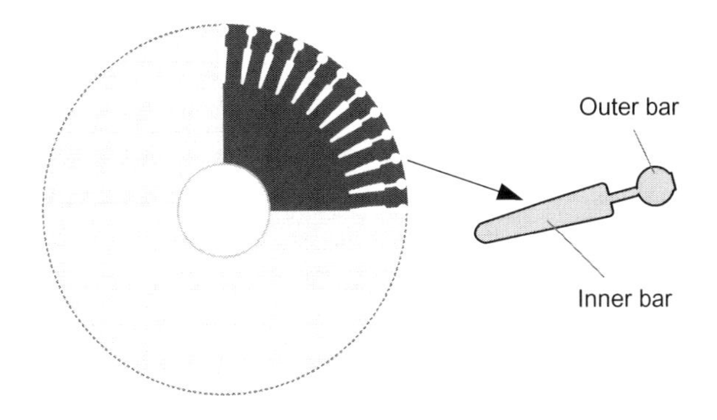

FIGURE 3.35

Double-cage rotor.

3.1.6.1 Design types of induction motors

As we discussed in Section 1.3, there are several types of loads. The required performance characteristics for a driving motor may be different according to the driven load. For example, the performance characteristics of motors for crane or hoist drives may be much different from those of motors for pump and blower drives.

Induction motors can be classified into different designs according to the starting and the normal operating characteristics. For classification of the motor, there are two most widely used standards: National Electrical Manufacture's Association (NEMA) and International Electrotechnical Commission (IEC). The NEMA standards mainly specify four design types: *Design A*, *B*, *C*, and *D*. Of these, the most widely used are Design B (normal torque) and Design C (high torque). Description on these four design types is summarized later.

Design A—It is characterized by normal starting torque (typically 150–170% of the rated), high starting current, low operating slip, and high-breakdown torque (the highest of all the NEMA types); common applications include fans, blowers, and pumps.

Design B—It is characterized by normal starting torque, low starting current, and low operating slip; common applications being the same as Design A. Design B is the most commonly used type for general-purpose applications.

Design C—It is characterized by high starting torque, low starting current, and higher operating slip than Designs A and B; common applications include compressors and conveyors.

Design D—It is characterized by high starting torque (higher than all the NEMA motor types), high starting current, high operating slip (5–13%), and inefficient operation efficiency for continuous loads; it is commonly used for intermittent loads such as a punch press.

Fig. 3.36 shows the speed–torque curve for the four NEMA designs [4].

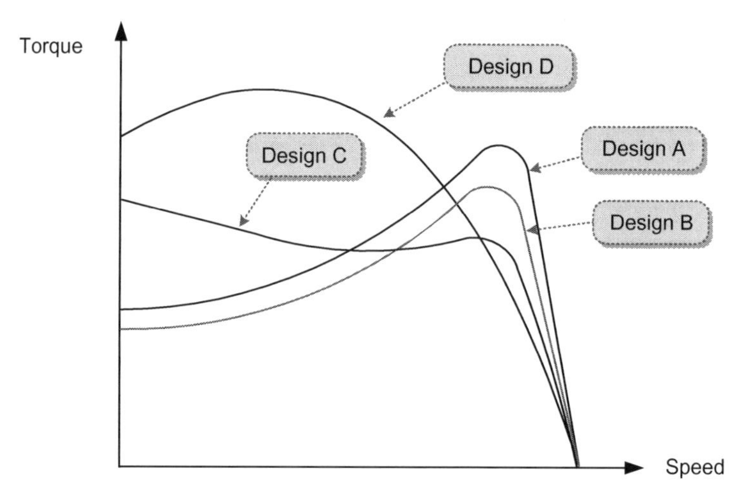

FIGURE 3.36

Torque characteristic according to design types of induction motors.

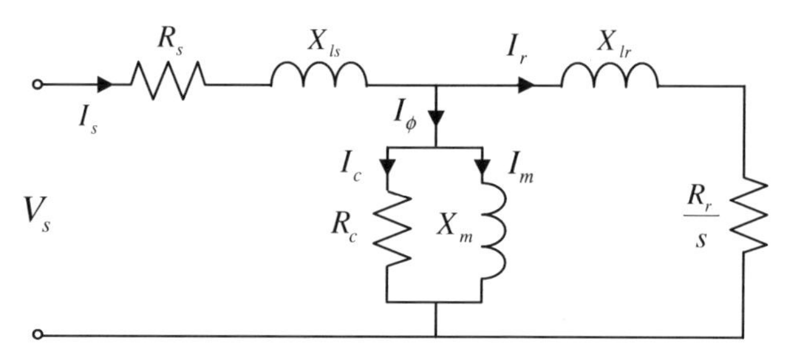

FIGURE 3.37

Equivalent circuit parameters for an induction motor.

3.1.7 DETERMINING EQUIVALENT CIRCUIT PARAMETERS

To analyze the operating and performance characteristics from the equivalent circuit of an induction motor as shown in Fig. 3.37, the parameters for the given induction motor should be determined. In addition, an accurate knowledge of motor parameters is essential for high-performance control of induction motors, which will be discussed in Chapter 5. Now, we will introduce specific tests on the induction motor to determine its equivalent circuit parameters.

3.1.7.1 Measurement of stator resistance R_s

The resistance R between any two terminals of three-phase windings, which may be connected in a delta or in a wye, can be directly measured. We can obtain the

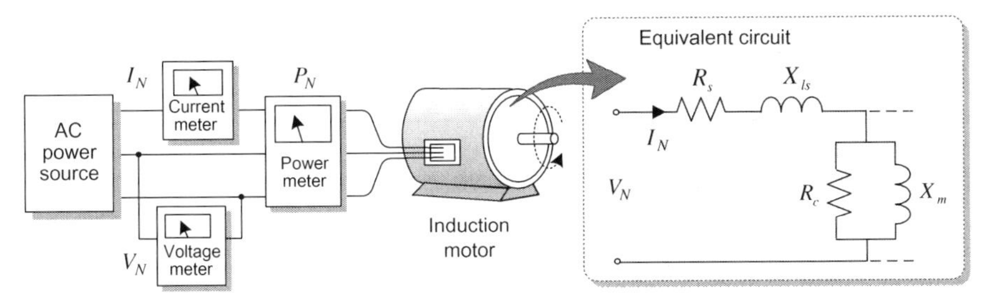

FIGURE 3.38

No-load test.

stator resistance as $R_s = R/2$ for a wye connection or as $R_s = 1.5R$ for a delta connection. This test must be performed with a precise instrument for large-sized motors because their resistance is very small. The stator resistance can be also obtained from the measured current after applying a DC voltage between the two windings.

3.1.7.2 No-load test

This test can determine the stator circuit parameter $L_{ls} + L_m$ and the parameter R_c. Here, R_c represents both the core loss and rotational loss. This test corresponds to the open-circuit test on a transformer. In this test, the rated voltage is applied to the stator terminals without any mechanical load. Then, the voltage V_N, current I_N, and input power P_N are measured at the motor input. The slip of an induction motor at no-load is almost zero, and thus $R_r/s \approx \infty$ and $I_r \approx 0$. Therefore the rotor branch of the equivalent circuit can be neglected as shown in Fig. 3.38.

At no-load test, the input power measured is equal to the losses, which consist of the stator copper loss, the core loss, and the rotational loss of friction and windage losses. The rotor copper loss can be negligible because the rotor current is extremely small. Here, the sum of the core loss and the rotational loss, which is normally almost constant at all operating conditions, is regarded as R_c. With the value of the premeasured stator resistance R_s, the value of R_c may be found by

$$P_N = 3I_N^2(R_s + R_c) \rightarrow R_c = \frac{\left(\frac{P_N}{3}\right)}{I_N^2} - R_s \tag{3.49}$$

From the measurement data P_N, V_N, and I_N, the equivalent impedance and resistance are

$$Z_N = \frac{\left(\frac{V_N}{\sqrt{3}}\right)}{I_N}, \quad R_N = \frac{P_N}{3I_N^2} \tag{3.50}$$

and the equivalent reactance is

$$X_N = \sqrt{Z_N^2 - R_N^2} = X_{ls} + X_m \tag{3.51}$$

The values of X_{ls} and X_m will be separately determined by the following blocked rotor test data.

3.1.7.3 Blocked rotor test

This test determines the rotor resistance R_r and the sum of the leakage inductances $L_{ls} + L_{lr}$, corresponding to the short-circuit test on a transformer. In this test, the rotor is locked so that it cannot rotate, and the current is maintained at the rated value by adjusting the applied voltage. Afterward, the resulting voltage V_B, current I_B, and power P_B are measured at the motor input. Since the rotor is not rotating, the slip $s = 1$, resulting in $X_m \gg R_r + jX_{lr}$. Thus the magnetizing X_m branch of the equivalent circuit can be neglected. The equivalent circuit for the block rotor test is reduced to the circuit as shown in Fig. 3.39.

In this case, the input power measured is equal to the sum of the stator and the rotor copper losses. Thus the rotor resistance is determined by

$$P_B = 3I_B^2(R_r + R_s) \rightarrow R_r = \frac{\left(\frac{P_B}{3}\right)}{I_B^2} - R_s \tag{3.52}$$

From the measurement data V_B and I_B, the equivalent impedance is

$$Z_B = \frac{\left(\frac{V_B}{\sqrt{3}}\right)}{I_B} \tag{3.53}$$

and the reactance X_B, which is the sum $X_{ls} + X_{lr}$ of the stator and rotor leakage reactances, is given by

$$X_B = \sqrt{Z_B^2 - (R_s + R_r)^2} = X_{ls} + X_{lr} \tag{3.54}$$

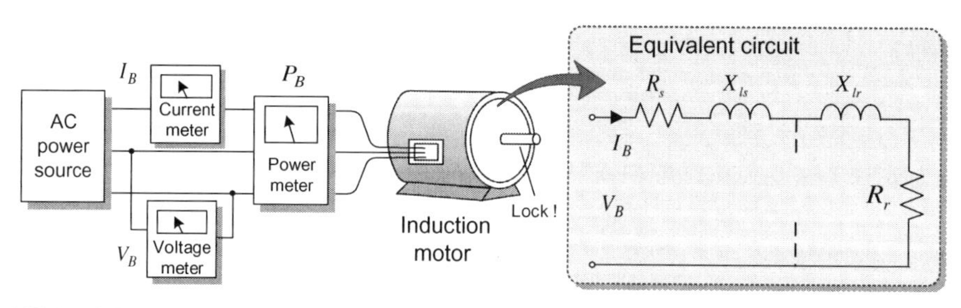

FIGURE 3.39

Blocked rotor test.

As for the proportion of the stator and rotor leakage reactances, empirically, $X_{ls} = X_{lr} = 0.5X_B$ for Design A and D types, $X_{ls} = 0.4X_B$, $X_{lr} = 0.6X_B$ for Design B type, and $X_{ls} = 0.3X_B$, $X_{lr} = 0.7X_B$ for Design C type [4].

Given the value of X_{ls}, the magnetizing reactance can be separated from Eq. (3.51) as

$$X_m = X_N - X_{ls} \tag{3.55}$$

In this test, since the rotor is locked, the rotor frequency is equal to the source frequency of 50 or 60 Hz. However, at normal operating conditions, the rotor frequency is in the range of 1–3 Hz. Since the resistance depends on the operating frequency, an incorrect rotor frequency can lead to misleading results in this test. A typical compromise is to use a frequency 25% or less of the rated frequency [2].

3.1.8 SPEED CONTROL OF INDUCTION MOTORS

Traditionally, induction motors have been used as nearly constant speed motors, which are operated by their direct connection to the power grid. However, because of their low cost and ruggedness, the induction motors are also being used for variable speed drives in many applications.

The operating speed of an induction motor depends on the slip or the synchronous speed of the given load. Therefore the speed control of an induction motor can be classified into two methods as follows:

- Slip control: inefficient, restricted speed control range
- Synchronous speed control: efficient, wide speed control range

3.1.8.1 Slip control

At a constant stator frequency, the speed of an induction motor can be varied according to the slip. The slip can be changed by varying the stator voltage or the rotor resistance.

First, let us take a look at the speed control by changing the stator voltage. As can be seen from Eq. (3.34), the torque developed by an induction motor is proportional to the square of the stator voltage as

$$T_{mech} = \frac{1}{\omega_s} \frac{V_s^2}{\left(R_s + \frac{R_r}{s}\right)^2 + j\left(X_{ls} + X_{lr}\right)^2} \frac{R_r}{s} \propto V_s^2 \tag{3.56}$$

Fig. 3.40 shows the variation of a speed–torque curve with respect to the stator voltage. We can see that the speed control by varying the stator voltage is available only for a limited range. The controllable speed range depends on the value of the slip (i.e., the maximum slip) at which the maximum torque occurs. By comparing Fig. 3.40A with B, we can see that the Design C or D motors have a wider speed control range than the Design A or B motors.

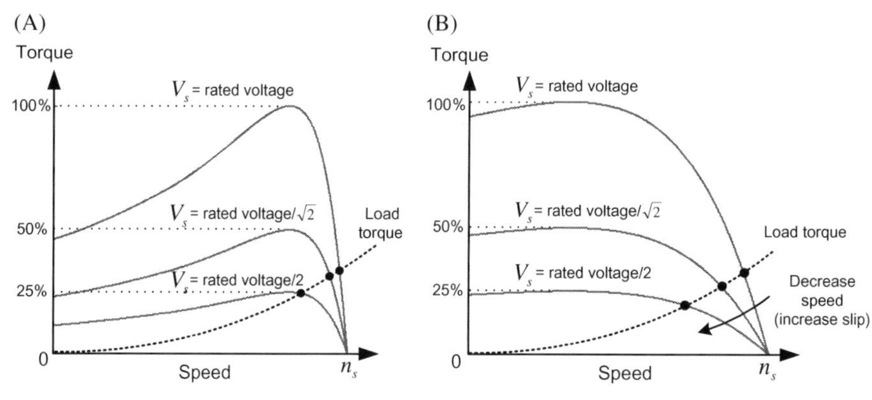

FIGURE 3.40

Torque characteristics according to the stator voltage variation. (A) Design A or B motor and (B) Design C or D motor.

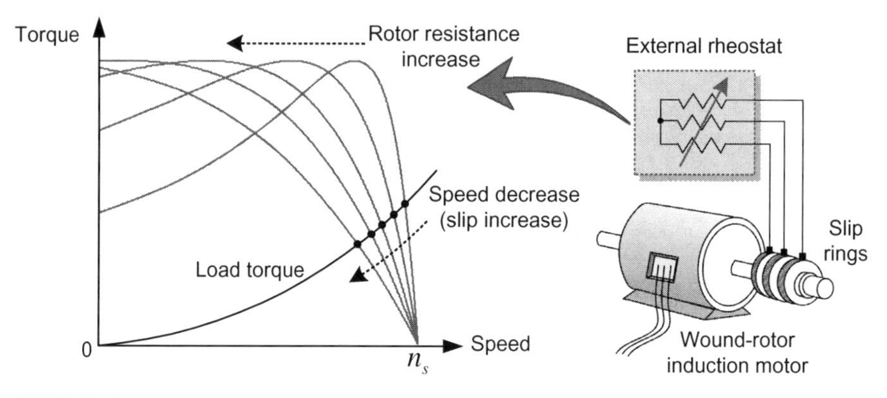

FIGURE 3.41

Speed control by changing the rotor resistance.

Next, let us examine the speed control by changing the rotor resistance. From the Section 3.1.4.3, we can see that the slip at which the maximum torque occurs is proportional to the rotor resistance R_r as

$$s_{max} = \frac{R_r}{\sqrt{R_s^2 + (X_{ls} + X_{lr})^2}} \propto R_r \tag{3.57}$$

Thus, changing the rotor resistance will alter the shape of the speed–torque curve and in turn, adjust the operating speed of the motor. Fig. 3.41 shows the speed–torque curves for different rotor resistances.

To adopt the speed control by changing the rotor resistance, we need to use the wound-rotor type induction motors. The rotor resistance in the wound-rotor type induction motors can be varied by adding an external rheostat or a resistor bank to the rotor windings via the slip rings. Changing the value of the rheostat will change the operating speed of the motor. Similar to the applied voltage varying method, the rotor resistance varying method also gives a limited controllable speed range. Moreover, inserting extra resistances into the rotor circuit seriously reduces the efficiency of the drive.

In the speed control methods of varying the stator voltage or the rotor resistance, the speed is considered to be changed by varying the slip at a constant stator frequency. As we can see from Eq. (3.46), since the efficiency of induction motors depends on the operating slip, the speed reduction by these methods will lead to a reduction in efficiency. Nevertheless, these speed control methods are applicable to small-sized motors that drive loads requiring a torque proportional to the square of the operating speed such as fans or blowers. For such loads, since the power consumption is reduced significantly according to the reduction in the speed, the reduction in efficiency is relatively small.

Instead of wasting energy in the rotor resistance (referred to as the *slip energy*), there is a better approach to improve the efficiency, which is known as a *slip energy recovery system*. In the slip energy recovery system, the slip energy can be returned back to the electric power source. There are two types of the slip energy recovery system: *Kramer* drive system and *Scherbius drive systems*. These systems exploit a power electronic converter connected to the rotor circuit instead of external resistors. Such a system can recover the energy taken from the rotor through the slip rings to decrease the speed and feed it back into the power source.

3.1.8.2 Synchronous speed control

The rotor of an induction motor follows the stator magnetic field which rotates at the synchronous speed proportional to the applied stator frequency. Thus, changing the stator frequency is more fundamental to the speed control. The relationship between the synchronous speed n_s and the stator frequency f_s is

$$n_s = \frac{120 \cdot f_s}{P} \tag{3.58}$$

Though the synchronous speed can be altered by changing the number of poles, this requires a complex motor construction. Moreover, it is not possible to provide a continuous change in the synchronous speed. Thus, it is more effective to change the stator frequency. The speed–torque characteristics for a variation in the stator frequency at the rated stator voltage are shown in Fig. 3.42.

When changing the stator frequency to adjust the speed, if the stator voltage remains constant, then the stator frequency variations lead to the change of the developed torque as well as the operating speed as shown in Fig. 3.42. The explanation is as follows.

At normal operating conditions, since the voltage drop across the stator resistance and leakage reactance are negligible in comparison to the applied voltage, the torque of Eq. (3.34) can be simplified as

$$T_{mech} \cong \frac{V_s^2}{\left(\frac{R_r}{s}\right)^2 + (\omega_s L_{lr})^2} \cdot \frac{R_r}{\omega_s s} = \left(\frac{V_s}{\omega_s}\right)^2 \frac{R_r \omega_{sl}}{R_r^2 + (\omega_{sl} L_{lr})^2} \tag{3.59}$$

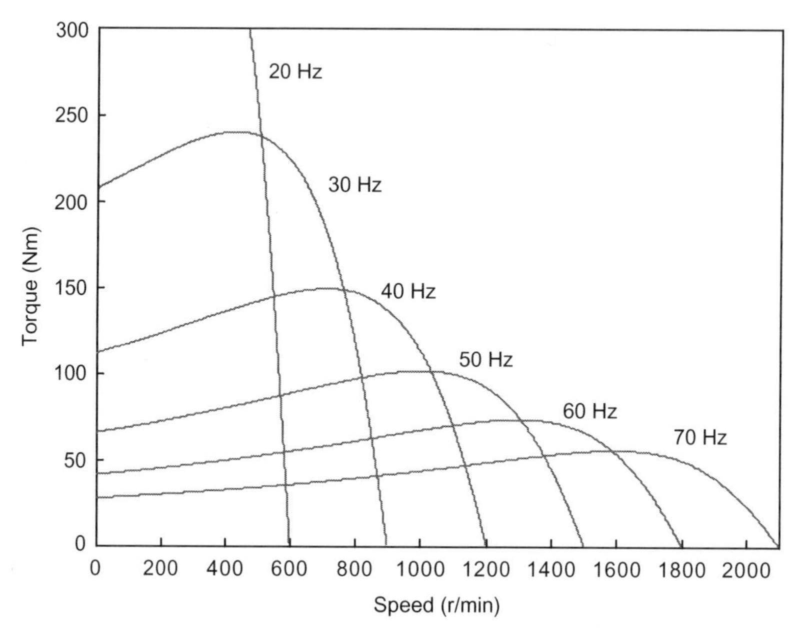

FIGURE 3.42

Speed–torque characteristic curves for several frequencies (5-hp, four-pole motor).

It can be seen that at a fixed stator voltage and slip frequency, the developed torque is inversely proportional to the square of the stator frequency. From Eq. (3.17), we can see that the air-gap flux depends on the ratio of the stator voltage V_s to the stator frequency f_s as

$$\phi \propto \frac{E_s}{f_s} \approx \frac{V_s}{f_s} \quad (\leftarrow E_s = 4.44 f_s N_s \phi K_{ws}) \tag{3.60}$$

Thus, under a fixed stator voltage, the air-gap flux varies inversely with the stator frequency. As can be seen from Eq. (3.59), the developed torque is proportional to the square of the air-gap flux and thus, is inversely proportional to the square of the stator frequency.

The rotor current is also influenced by the variation of the stator frequency. Assuming that at normal operating conditions, the stator resistance and leakage reactance are negligible, the rotor current varies inversely with the stator frequency as

$$I_r = \frac{V_s}{\sqrt{\left(\frac{R_r}{s}\right)^2 + (X_{lr})^2}} = \left(\frac{V_s}{\omega_s}\right) \frac{\omega_{sl}}{\sqrt{R_r^2 + (\omega_{sl} L_{lr})^2}} \propto \frac{1}{f_s} \tag{3.61}$$

When adjusting the operating speed by changing the stator frequency, if the stator voltage is varied linearly with the frequency as shown in Fig. 3.43, then the air-gap flux $\phi \approx V_s/f_s$ will remain constant and thus, the torque and rotor current will remain unchanged. This technique is called the *constant volts*

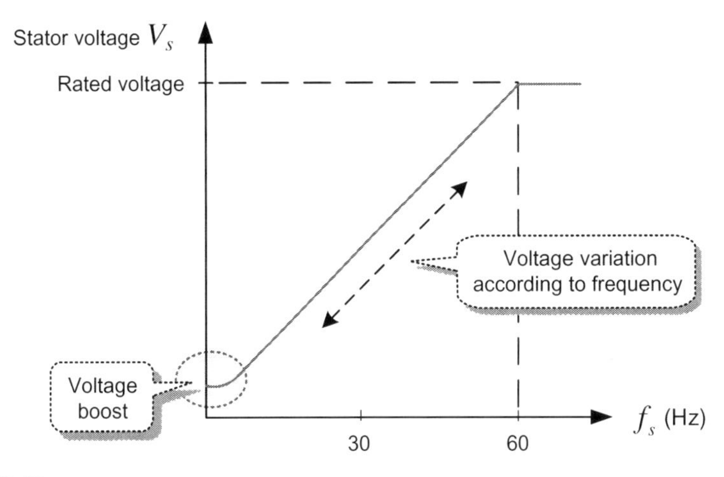

FIGURE 3.43

Constant volts per Hertz (*V/f*) control.

per Hertz (*V/f*) *control*, which has become a popular variable speed drive for induction motors in general-purpose applications today. The *V/f* ratio may be adjusted according to the driven load.

At low frequencies below a few Hertz, due to the influence from the voltage drop across the stator resistance and leakage reactance, the magnitude of the stator voltage determined by this linear *V/f* relationship will be insufficient to keep the magnitude of the air-gap flux constant. This is because the actual air-gap flux ϕ is not proportional to V_s/f_s, but E_s/f_s. This results in reduction of the air-gap flux and in turn, reduction of the output torque. Thus, for the constant *V/f* control at a low operating frequency, we have to boost the stator voltage to compensate for the voltage drop of the stator resistance and leakage reactance as shown in Fig. 3.43.

Fig. 3.44 shows the speed–torque characteristics curves for several frequencies under the constant *V/f* control. To summarize, under the constant *V/f* operation of an induction motor,

- Torque is independent of the stator frequency,
- Rotor current is independent of the stator frequency,
- Synchronous speed is proportional to the stator frequency.

The constant *V/f* control can be utilized for the speed control below the rated speed. If the stator frequency increases more than the rated frequency, the constant *V/f* ratio cannot be maintained due to the limitation of the applied voltage. Thus, for speeds higher than the rated speed, the maximum torque will be reduced due to the reduction in the air-gap flux. The control method for high speeds above the rated speed will be discussed in more detail in Chapter 8.

In the induction motor drives with the constant *V/f* control, it should be noted that the torque and the rotor current are determined only by the slip angular

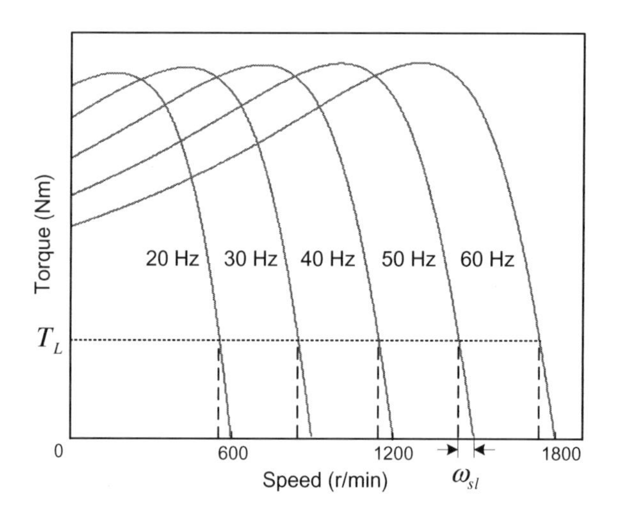

FIGURE 3.44

Speed–torque curves with constant volts per Hertz control (5-hp, four poles).

frequency $\omega_{sl}(=s\omega_s)$. We can see from Fig. 3.44 that the output torque can be produced equally at any speed for the same slip frequency. Thus, even though the speed is reduced by the stator frequency, the operating slip can be kept low so that the efficiency cannot be degraded unlike the slip control. Moreover, at start-up, a high starting torque can be achieved by applying a low stator frequency without requiring a high rotor resistance. After the starting, the stator frequency is increased along with the stator voltage according to the linear *V/f* relationship.

Up to this point, we have discussed the open-loop speed control by the stator frequency. In the open-loop speed control, however, the steady-state speed can be affected by the load variation because the operating slip frequency depends on the load. In particular, in the low-speed region, the variation of the operating slip frequency may bring about a large error in the steady-state speed. Thus this open-loop speed control needs a proper compensation to offset the speed error according to the load variation. More accurate speed control can be achieved by the closed-loop speed control, which adjusts the slip frequency under the constant *V/f* control as the following.

3.1.8.3 Closed-loop speed control by adjusting the slip frequency under constant V/f control

Under the constant *V/f* control, the torque of an induction motor can be controlled by adjusting the slip frequency f_r. The torque of Eq. (3.59) is rewritten in terms of the slip frequency f_r as

$$T_{mech} = \left(\frac{V_s}{\omega_s}\right)^2 \frac{2\pi f_r R_r}{R_r^2 + (2\pi f_r L_{lr})^2} \tag{3.62}$$

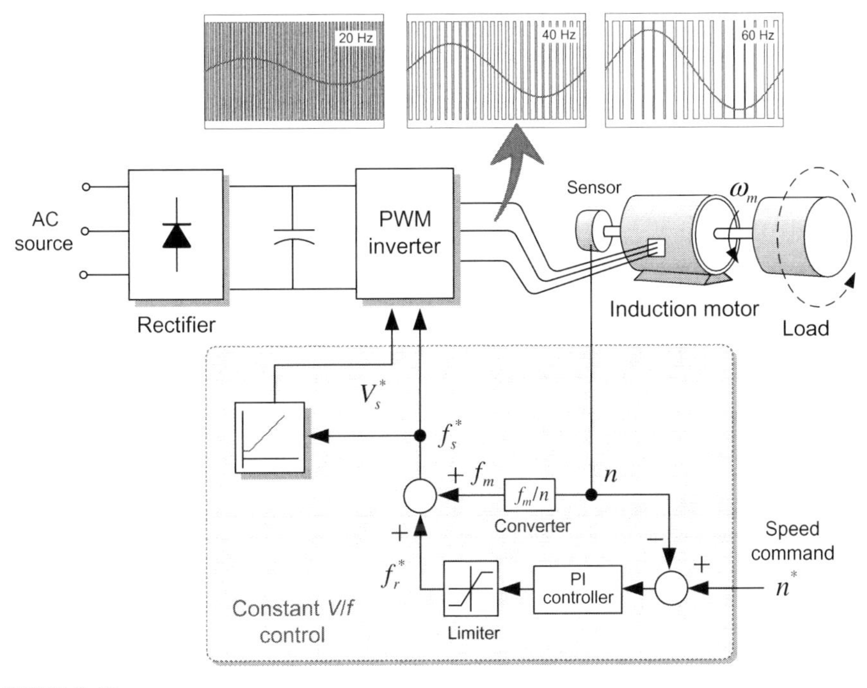

FIGURE 3.45

Speed control by adjusting the slip frequency with constant *V/f* control.

Since the standard induction motors normally have $R_r \gg 2\pi f_r L_{lr}$, if the air-gap flux is constant, then the torque is proportional to the slip frequency f_r as

$$T_{mech} \cong \left(\frac{V_s}{\omega_s}\right)^2 \frac{2\pi}{R_r} f_r \tag{3.63}$$

Fig. 3.45 shows the block diagram of the closed-loop speed control system based on Eq. (3.63).

In this speed control system, with the constant *V/f* operation, the slip frequency f_r is adjusted so that the actual speed n follows the command speed n^*. Since the developed torque of an induction motor can be adjusted by the slip frequency, if the actual speed of the motor is less than its command, then the controller increases the slip frequency to produce a larger torque. On the other hand, if the actual speed is higher than its command, then the controller decreases the slip frequency to reduce the developed torque. In this case, the variation range of the slip frequency should be limited within the maximum slip frequency, at which the breakdown torque occurs.

Once the required slip frequency f_r^* is determined in this manner, the stator frequency command can be determined by the sum of the slip frequency and the rotor speed f_m as

$$f_s^* = f_m \pm f_r^* \tag{3.64}$$

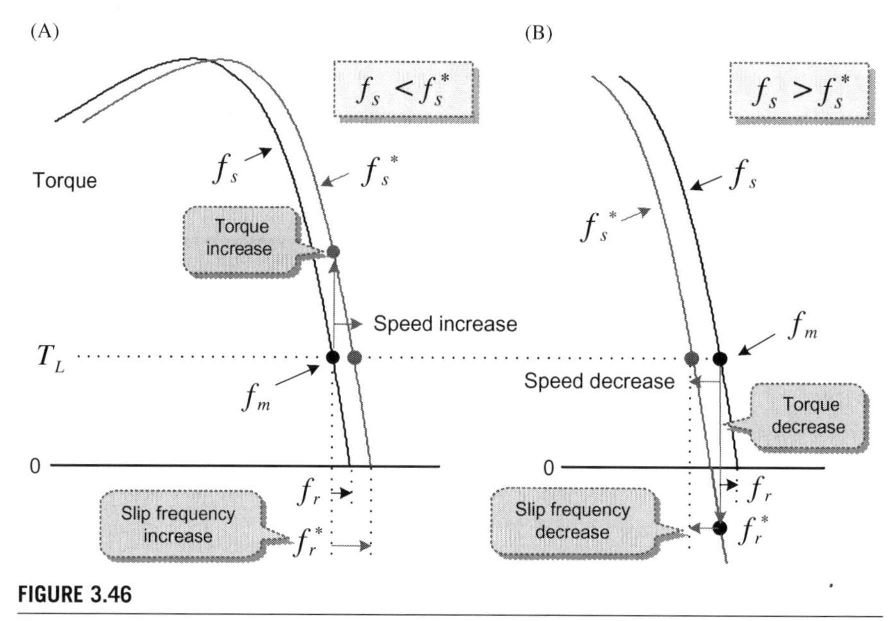

FIGURE 3.46

Operations in the closed-loop speed control system. (A) Speed increase and (B) speed decrease.

From this stator frequency command f_s^*, the voltage command applied to the motor should be determined according to the linear *V/f* relationship shown in Fig. 3.43. This voltage reference V_s^* with the stator frequency will generally be generated by using the PWM inverter, which will be described in Chapter 7. The inverter used for this purpose is called the *Variable Voltage Variable Frequency (VVVF) inverter.*

Fig. 3.46 describes the closed-loop speed control that was shown in Fig. 3.45. Fig. 3.46A shows the operation for the increase in speed. If the stator frequency is increased by the increased slip frequency, then the developed torque becomes greater than the load torque, and thus the speed is increased. On the other hand, Fig. 3.46B shows the operation for the decrease in speed. If the stator frequency is decreased by the decreased slip frequency, then the developed torque becomes smaller than the load torque, and thus the speed is reduced.

This speed control scheme for induction motors is widely used in general-purpose drives such as fans, pumps, and conveyors because it can improve the system efficiency and provide a satisfactory starting torque and a satisfactory steady-state performance. This control can give a speed regulation of 1–2%. However, this control scheme is insufficient to achieve a high dynamic performance. Therefore, we need an advanced control technique such as the *Vector control*, which will be described in Chapter 5.

3.1.9 OPERATION REGIONS OF INDUCTION MOTORS

Induction motors have been traditionally used as constant speed motors supplied by the AC power source. However, recently, it has been increasingly used as variable speed motors driven by an inverter. As stated in the Section 3.1.8.2, the variable speed induction motor drive can be achieved by controlling the frequency and voltage applied to the motor. The operation range of an induction motor can be divided into three regions according to the output torque capability as

- Constant torque region: speed range below the base speed
- Constant power region: speed range above the base speed
- Breakdown torque region: speed range above the base speed maintaining the maximum slip frequency

The speed range above the base speed can be further divided into two subregions: *constant power region* and *breakdown torque region*. Because the field flux is reduced for operation in these regions, they are called *field-weakening regions*. Fig. 3.47 illustrates the output toque characteristic in each region.

3.1.9.1 Constant torque region

In this region, the motor speed is increased by increasing the stator frequency. To maintain the air-gap flux at a constant value, the applied stator voltage will be increased with the stator frequency according to the linear *V*/*f* relationship. This is the constant *V*/*f* control as stated in the Section 3.1.8.2. In this control method, since the output torque is proportional to the slip frequency, the output torque can be produced equally at any speed for the same slip frequency as shown in Fig. 3.48. (The same slip frequency means the same stator current.) Hence, this speed range is called *constant torque region*. The maximum torque developed by the induction motor is limited only by the allowable current rating in this region.

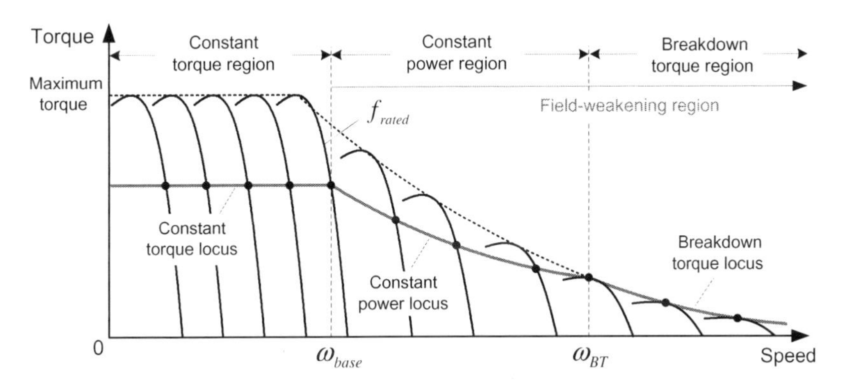

FIGURE 3.47

Operation regions of an induction motor.

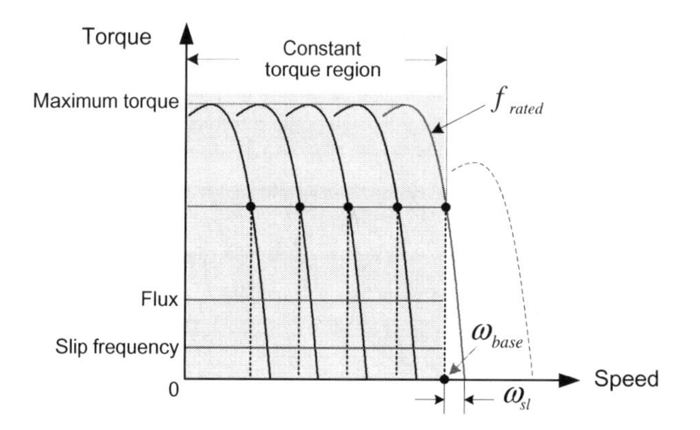

FIGURE 3.48

Operation characteristics in constant torque region.

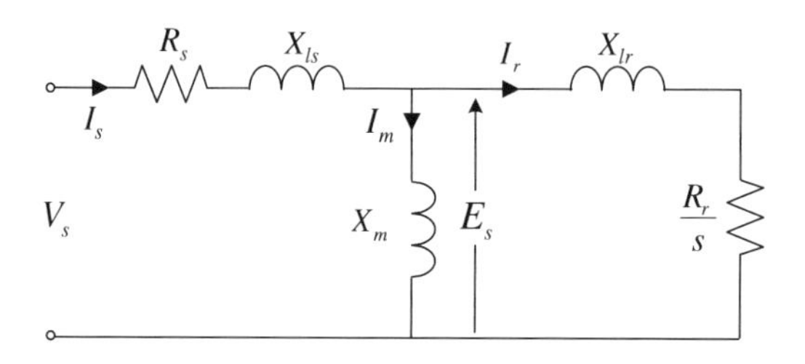

FIGURE 3.49

Steady-state equivalent circuit of an induction motor.

3.1.9.2 Constant power region

Since the back-electromotive force (back-EMF) of the induction motor increases with the speed, the voltage applied to the motor should also be increased with the speed for the stator current to flow properly. However, since the terminal voltage of the motor should not exceed the rated (or nominal) voltage, the stator voltage cannot be increased above the rated voltage despite the speed increase. The onset speed, at which the terminal voltage of the motor becomes the rated voltage, is called the *base speed*. Since the induction motor will be operated with a constant voltage in the speed region above the base speed, the air-gap flux as well as the rotor current will decrease as the operating speed increases.

The steady-state equivalent circuit in Fig. 3.49 implies that under a constant stator voltage V_s, an increase in the operating frequency causes the reactance to increase, resulting in a decrease of both the exciting current I_ϕ and rotor current I_r. Therefore the output torque of an induction motor in this region will be different from that in the constant torque region.

In the Section 3.1.4.3, the output torque and the rotor current of an induction motor were given as

$$T_e = 3\frac{1}{\omega_s} I_r^2 \frac{R_r}{s} \tag{3.65}$$

$$I_r = \frac{E_s}{\left(\frac{R_r}{s}\right) + jX_{lr}} = \left(\frac{E_s}{\omega_s}\right) \frac{\omega_{sl}}{\sqrt{R_r^2 + (\omega_{sl} L_{lr})^2}} \tag{3.66}$$

Substituting Eq. (3.66) into Eq. (3.65), the following torque expression as a function of air-gap flux and rotor current can be obtained.

$$T_e = 3\left(\frac{E_s}{\omega_s}\right) I_r \frac{R_r}{\sqrt{R_r^2 + (\omega_{sl} L_{lr})^2}} \cong K\phi I_r \quad \left(\phi = \frac{E_s}{\omega_s}\right) \tag{3.67}$$

Under a constant stator voltage, an increase in the stator frequency leads to a reduction in both the air-gap flux $\phi(\propto E_s/\omega_s)$ and the rotor current I_r. Therefore the output torque will decrease in proportion to $1/\omega_s^2$. However, if we make the slip frequency ω_{sl} increase as the stator frequency increases, then the rotor current will remain constant due to the reduced impedance R_r/s by the increase in the slip. This can be readily seen in Eq. (3.66) and Fig. 3.49. In this way, if the rotor current remains constant despite the stator frequency increase, then the developed torque decreases in proportion to $1/\omega_s$, not $1/\omega_s^2$, as shown in Fig. 3.50. Thus an enhanced output torque capability can be achieved.

Therefore for operation in this region, the slip frequency ω_{sl} should be increased as the stator frequency increases. In this case, since both the voltage and the current remain constant, this region is called the *constant power region* (strictly speaking, *constant VA*), which corresponds to the field-weakening region of a DC motor.

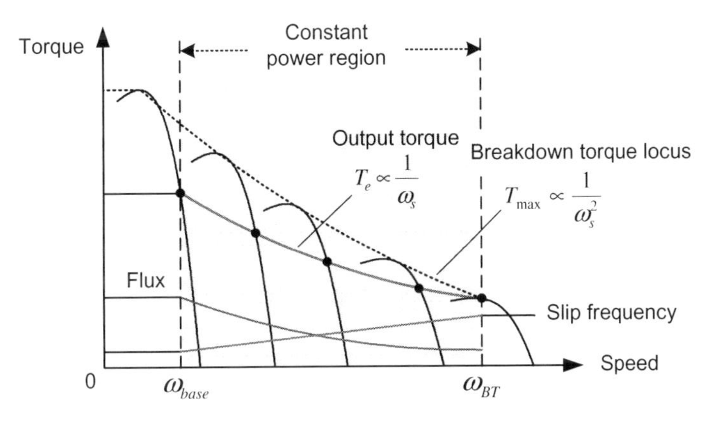

FIGURE 3.50

Operation of constant power regions.

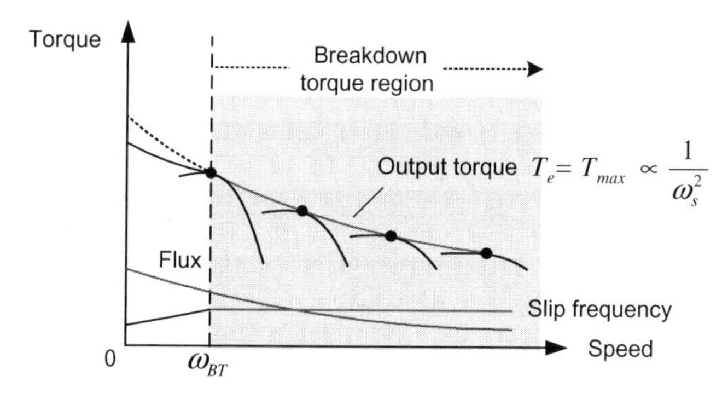

FIGURE 3.51

Operation of breakdown torque regions.

3.1.9.3 Breakdown torque region

In the constant power region, the slip frequency is increased with the speed increase to lower the reduction rate of the output torque. However, there is a limit to increasing the slip frequency because an induction motor has a maximum slip for the stable operation as

$$S_{max} = \frac{R_r}{\sqrt{R_s^2 + (X_{ls} + X_{lr})^2}} \tag{3.68}$$

Once the operating slip frequency reaches its maximum value, the rotor current will decrease and, in turn, the output torque will decrease in proportion to $1/\omega_s^2$ as the stator frequency increases as shown in Fig. 3.51. Since the current will be decreased, the input power will not remain constant. In this case, an induction motor will be operated at the maximum slip for producing the breakdown torque. Thus this region is called the *breakdown torque region*.

Fig. 3.52 depicts characteristics such as voltage, current, developed output torque, flux, and slip frequency over the overall speed range. The curve, which shows the output torque capability versus the speed, is called the *capability curve* of an induction motor.

The base speed, at which the constant power region starts, does not usually coincide with the rated speed and varies with the available voltage, operating current, flux level, and load condition. A detailed explanation on this will be given in Chapter 8. The range of the constant power region depends on the maximum value of the slip frequency. As can be seen from Eq. (3.68), this maximum slip frequency is a function of leakage inductance, and thus, its range depends on the total leakage inductance, $X_{ls} + X_{lr}$. For a standard induction motor, this value is normally 0.2 p.u., producing a constant power range up to two to three times the rated speed. By contrast, induction motors for high-speed drives often have a constant power region range over four times the rated speed.

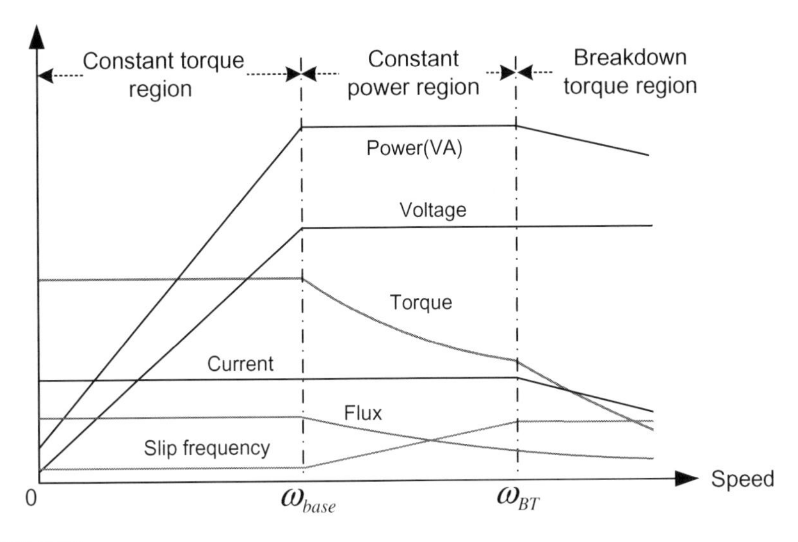

FIGURE 3.52

Capability curve of an induction motor.

3.2 SYNCHRONOUS MOTORS

As stated in Chapter 1, the name *synchronous motors* stems from its characteristic to develop the torque only when the rotor operates at the synchronous speed. Traditionally, large-sized synchronous machines have been mainly used as generators. As motors, they have been used to improve the power factor in the power grid by changing the field current, i.e., excitation. In this case, they are called synchronous condensers. In small sizes, they have been often used in applications requiring a constant speed operation.

However, recently, synchronous motors with a permanent magnet rotor are growing rapidly in use for variable speed drives instead of DC motors. This is because permanent magnet synchronous motors (PMSMs) have many advantages over other motors such as high efficiency, high power density, high dynamic response, and high power factor. However, one disadvantage is its higher cost in comparison to an induction motor. In this section, we will study the basic performance characteristics of the synchronous motor under steady-state conditions, and in Chapter 4 and Chapter 5, we will learn about its dynamic characteristics.

The structure of the stator and rotor of a synchronous motor is shown Fig. 3.53. A synchronous motor has the same stator configuration as an induction motor, but a different rotor structure. Unlike an induction motor, a synchronous motor has independent excitations for the stator and rotor windings,

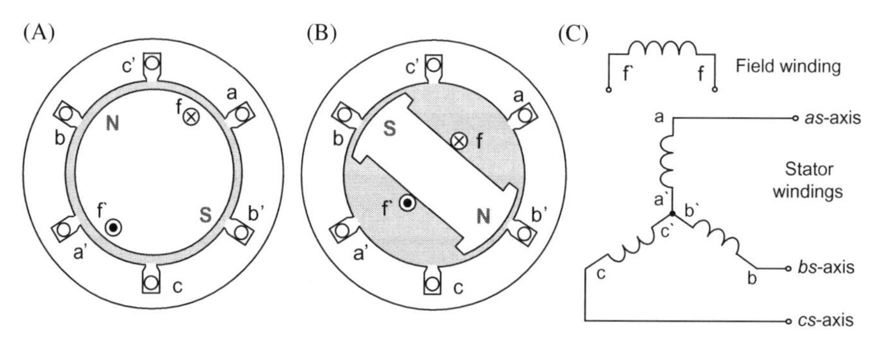

FIGURE 3.53

Stator and rotor structure of a synchronous motor. (A) Cylindrical type, (B) salient pole type, and (C) stator windings.

i.e., AC excitation to the stator winding and DC excitation to the rotor winding. The function of the stator winding, which is commonly called an *armature winding*, is to generate a rotating magnet field through its connection to a three-phase AC power source such as an induction motor. The rotor winding, which is called the *field winding*, generates a field flux with a DC excitation or by a permanent magnet.

There are two types of rotor construction: *cylindrical* (or *nonsalient pole*) *type* and *salient pole type*. A salient pole rotor is normally used for low-speed applications such as hydroelectric generators, while a cylindrical rotor is normally used in high-speed applications such as steam or gas turbine generators. Now, we will explore the steady-state equivalent circuit and the output torque of a synchronous motor.

3.2.1 CYLINDRICAL ROTOR SYNCHRONOUS MOTORS

Since the rotor winding with a DC excitation can be represented as a simple R-L circuit without back-EMF, we need to discuss an equivalent circuit for only the stator winding.

The stator windings of a synchronous motor are the same as those of an induction motor. Thus, from Eq. (3.6), the stator voltage equation of a synchronous motor is given as

$$v_s = R_s i_s + \frac{d\lambda_s}{dt} \tag{3.69}$$

However, since the stator flux linkage of a synchronous motor is different from those of an induction motor, we need to examine the stator flux linkage λ_s.

A synchronous motor has independent excitations for the stator and rotor windings. In the air gap, these excitations generate two magnetic fields which

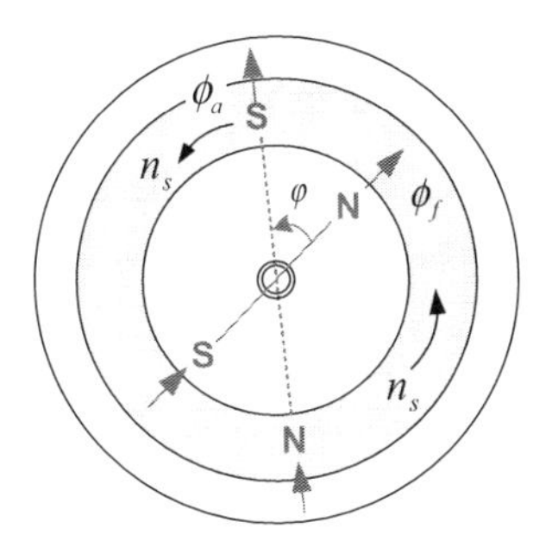

FIGURE 3.54

Rotating field flux of stator and the field flux of rotor.

rotate at the synchronous speed as shown in Fig. 3.54. One is generated by the stator winding currents as in an induction motor, and the other is generated by the field winding current. In synchronous motors, the torque is produced by the interaction of these two magnetic fields.

The rotating field flux ϕ_a generated by the three-phase AC currents I_s flowing in the stator winding consists of the armature reaction flux ϕ_{ar} and the leakage flux ϕ_{al} as

$$\phi_a = \phi_{ar} + \phi_{al} \tag{3.70}$$

The armature reaction flux ϕ_{ar} corresponds to the magnetizing flux in the induction motor. The field flux ϕ_f generated in the field winding is also a rotating magnetic field rotating at the synchronous speed because the rotor itself rotates at the synchronous speed. In the air gap, there is a resultant of these two field fluxes as

$$\phi_s = \phi_f + \phi_{ar} \tag{3.71}$$

Thus the flux linkage λ_s of the stator winding with an effective number of turns N_s will be expressed as

$$\lambda_s = N_s(\phi_f + \phi_{ar}) \tag{3.72}$$

In the stator winding, the voltage is induced by these two fluxes. The voltage E_f induced by the field flux ϕ_f is called the *excitation voltage* as shown in Fig. 3.55. The magnitude of the excitation voltage is proportional to the field current and the rotor speed.

The voltage E_{ar} induced by the armature reaction flux ϕ_{ar} is called the *armature reaction voltage*, whose equation is the same as the Eq. (3.18) of the induced voltage for an induction motor.

The total induced voltage in the stator winding is the sum of these two induced voltages as

$$E_s = E_f + E_{ar} \tag{3.73}$$

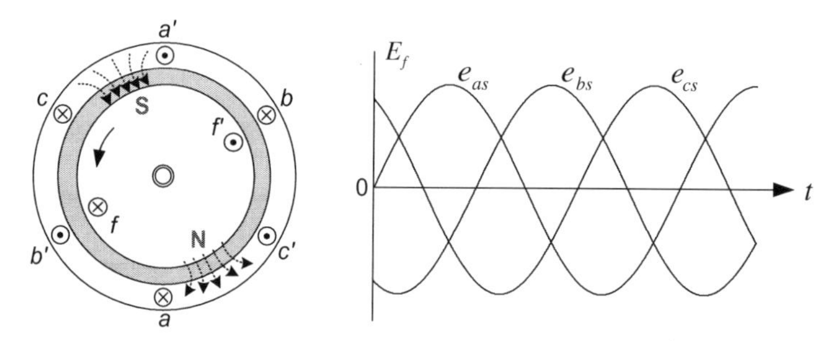

FIGURE 3.55

Excitation voltage induced by the field flux.

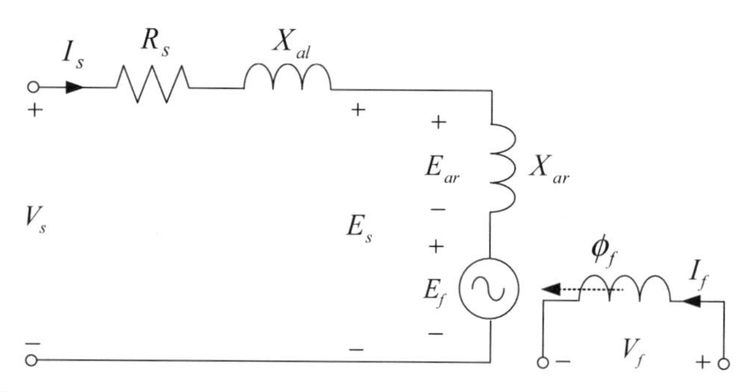

FIGURE 3.56

Per phase equivalent circuit of a cylindrical rotor synchronous motor.

Here, the armature reaction voltage E_{ar} lags the armature reaction flux ϕ_{ar} by 90°, so it can be expressed in terms of the armature reaction reactance X_{ar} as

$$E_{ar} = I_a jX_{ar} \tag{3.74}$$

Considering the total induced voltage, the voltage drop of the stator resistance R_s, and leakage reactance X_{al}, the stator voltage equation of Eq. (3.69) can be expressed as the phasor representation by

$$\begin{aligned} V_s &= I_s R_s + I_s jX_{al} + E_s \\ &= I_s R_s + I_s jX_{al} + I_s jX_{ar} + E_f \\ &= I_s R_s + I_s jX_s + E_f = I_s Z_s + E_f \end{aligned} \tag{3.75}$$

where, $X_s(= X_{ar} + X_{al})$ is the synchronous reactance and $Z_s(= R_s + jX_s)$ is the synchronous impedance.

By using Eq. (3.75), the complete per phase equivalent circuit of a cylindrical rotor synchronous motor under the steady-state condition is shown in Fig. 3.56.

3.2.1.1 *Torque of a cylindrical rotor synchronous motor*

The torque of a cylindrical rotor synchronous motor is obtained by dividing the output power by the synchronous speed ω_s. The input power of a synchronous motor is given by

$$P = 3V_s I_s \cos\theta \tag{3.76}$$

where θ is the phase angle between the stator voltage V_s and the stator current I_s.

Eq. (3.75) of a synchronous motor can be represented by the phasor diagram as shown in Fig. 3.57. The phase angle δ between the stator voltage V_s and the excitation voltage E_f is an important factor of power transfer and stability in synchronous machines and is usually termed the *power angle* (or *load angle*). For motoring operation, this angle δ is also called the *torque angle*, and the stator voltage V_s is ahead of the excitation voltage E_f. The angle δ acts as the slip in an induction motor. The angle δ increases with the increased load. This angle can also be regarded as the difference angle between the stator and the rotor rotating magnetic fields. In motoring operation, the stator magnetic field is ahead of the rotor magnetic field. The power and torque for synchronous machines are usually expressed as a function of the stator voltage, excitation voltage and their angle δ.

Now, we will derive the output torque from the equivalent circuit. From the phasor diagram shown in Fig. 3.57, we have

$$V_s\cos\delta = E_f + I_s X_s \sin(\theta-\delta) + I_s R_s \cos(\theta-\delta) \tag{3.77}$$

$$V_s\sin\delta = I_s X_s \cos(\theta-\delta) - I_s R_s \sin(\theta-\delta) \tag{3.78}$$

If the stator winding resistance R_s is neglected, then we can readily obtain

$$I_s\cos\theta = \frac{E_f\sin\delta}{X_s} \tag{3.79}$$

By substituting this into Eq. (3.76), the power is expressed by

$$P = 3\frac{V_s E_f}{X_s}\sin\delta = P_{max}\sin\delta \tag{3.80}$$

Since the stator losses are neglected by assuming $R_s = 0$, the power of Eq. (3.80) is considered as the output power. Thus the developed torque is obtained by dividing the output power by the synchronous speed ω_s as

$$\text{Torque:}\quad T = \frac{P}{\omega_s} = 3\frac{V_s E_f}{\omega_s X_s}\sin\delta = T_{max}\sin\delta \tag{3.81}$$

Eqs. (3.80) and (3.81) indicate that the output power and the torque vary as a sine of the torque angle δ as shown in Fig. 3.58. The maximum torque occurs at a torque angle of 90 electrical degrees. This is called the *pull-out torque*, which indicates the maximum value of torque that a synchronous motor can develop without pulling out of synchronism. In general, its value varies from 1.25 to 3.5 times the full-load torque. In actual practice, however, the motor will never

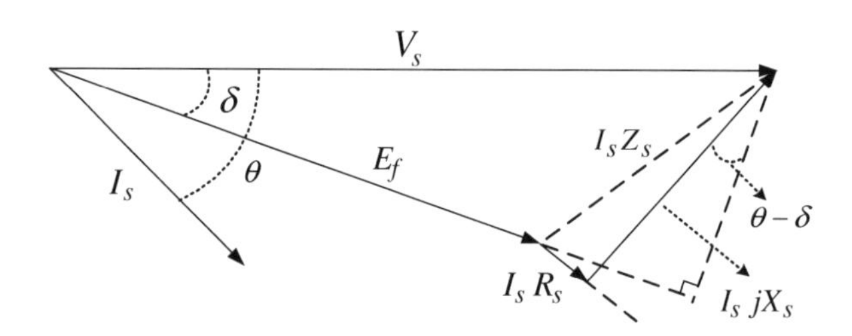

FIGURE 3.57

Phasor diagram for a synchronous motor.

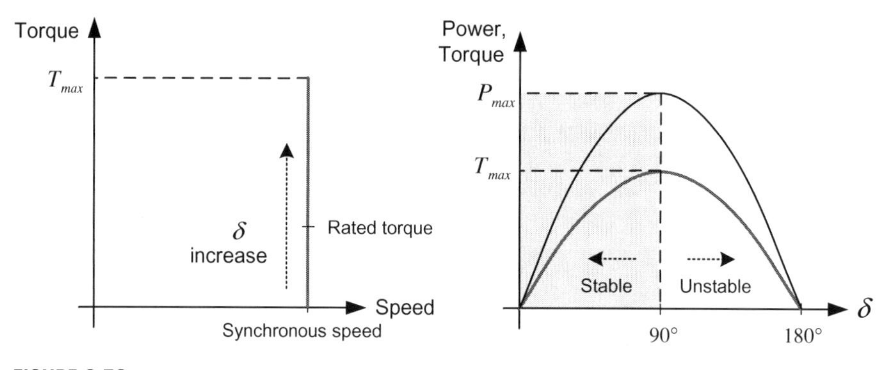

FIGURE 3.58

Output power and torque for a cylindrical rotor synchronous motor.

operate at a torque angle close to 90 electrical degrees because the stator current will be many times its rated value at this operation.

Next, we will discuss the salient pole rotor synchronous motor.

3.2.2 SALIENT POLE ROTOR SYNCHRONOUS MOTORS

As we can see in Fig. 3.59, a salient pole synchronous motor has a nonuniform air gap due to the protruding rotor poles. In this type of motor, the amount of the flux produced by the stator current varies according to the position of the rotor. A higher flux is generated along the poles in comparison to between the poles because the reluctance of the flux path is low along the poles and high between the poles.

To consider the saliency of the rotor in a synchronous motor model, we need to define the d and q axes as the following. The d-axis is defined as the axis along the poles, which is also the direction of the rotor flux produced by the field winding. The reluctance of the d-axis path is low due to the small air gap, and thus the inductance L_d of the d-axis is large. On the other hand, the q-axis is defined as the axis between the poles, which is also the direction of the excitation voltage. The reluctance of the q-axis path is large due to the large air gap, and so the inductance L_q of the q-axis is small. Thus, $L_d > L_q$.

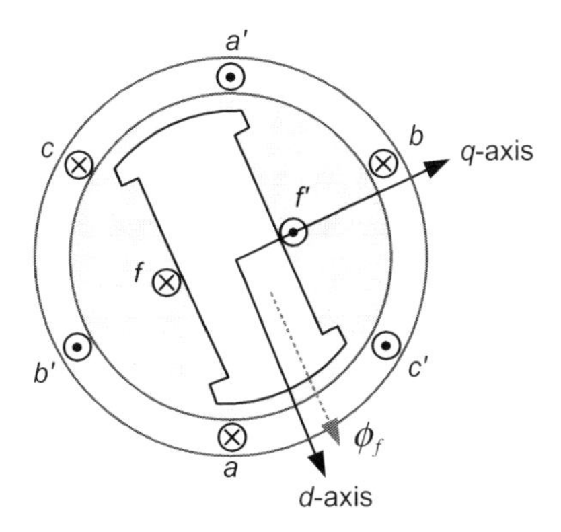

FIGURE 3.59

Salient pole rotor synchronous motor.

In order to consider the flux variation with the position of the rotor, it is more effective to analyze the saliency of the rotor by using the divided stator current into two components of the current I_d aligned with the d-axis and the current I_q aligned with the q-axis. Thus the stator current I_s is expressed as

$$I_s = I_q - jI_d \tag{3.82}$$

where $I_q = I_s\cos\varphi$, $I_d = I_s\sin\varphi$, and φ is the phase angle between I_q and I_a.

According to the d and q axes currents, the armature reaction flux ϕ_{ar} is also divided into two components: the d-axis flux ϕ_d produced by the d-axis current I_d and the q-axis flux ϕ_q by the q-axis current I_q. The relation between ϕ_d and I_d is expressed by the d-axis armature reactance X_{ad}, whereas the relation between ϕ_q and I_q is expressed by the q-axis armature reactance X_{aq}. The d and q axes synchronous reactances are given by

$$X_d = X_{ad} + X_{al} \tag{3.83}$$

$$X_q = X_{aq} + X_{al} \tag{3.84}$$

where X_{al} is the armature leakage reactance. The salient pole synchronous motor always $X_d > X_q$, normally $X_q = (0.5 \sim 0.8)X_d$.

Considering the d and q axes synchronous reactances, and referring to Eq. (3.75), the voltage equation of the salient pole rotor synchronous motor can be expressed by

$$V_s = I_sR_s + I_djX_d + I_qjX_q + E_f \tag{3.85}$$

From this equation, the equivalent circuit of the salient pole rotor synchronous motor is shown in Fig. 3.60.

3.2.2.1 Torque of a salient pole rotor synchronous motor

Now, we will derive the torque of a salient pole rotor synchronous motor. In addition to the torque of a cylindrical rotor, it can be expected that a salient pole rotor synchronous motor has a reluctance torque due to the saliency of the rotor as already discussed in Chapter 1.

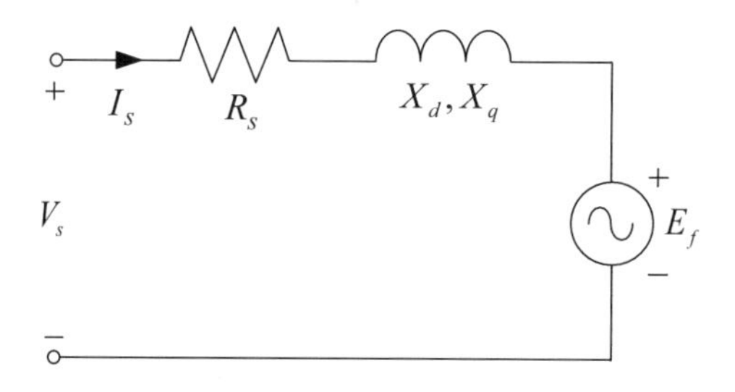

FIGURE 3.60

Equivalent circuit of a salient pole rotor synchronous motor.

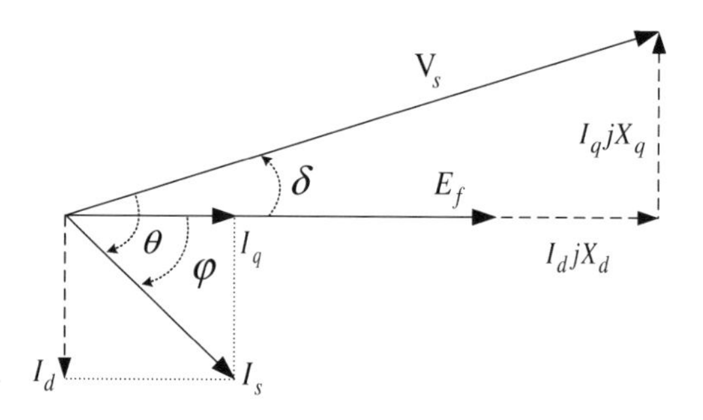

FIGURE 3.61

Phasor diagram of a salient pole synchronous motor.

The input power in terms of *d-q* axes currents, I_d and I_q, can be expressed as

$$\begin{aligned} P &= 3V_sI_s\cos\theta = 3V_sI_s\cos(\varphi+\delta) \\ &= 3V_sI_s(\cos\varphi\cos\delta - \sin\varphi\sin\delta) \\ &= 3V_s(I_q\cos\delta - I_d\sin\delta) \end{aligned} \tag{3.86}$$

From Eq. (3.85), assuming that the stator winding resistance is neglected, the phasor diagram can be expressed as Fig. 3.61.

From this phasor diagram, we have

$$V_s\cos\delta = E_f + I_dX_d \tag{3.87}$$

$$V_s\sin\delta = I_qX_q \tag{3.88}$$

Combining Eqs. (3.86)–(3.88), the output power and the torque are given by

$$P = 3\frac{V_sE_f}{X_s}\sin\delta + 3\frac{V_s^2(X_d - X_q)}{2X_dX_q}\sin 2\delta \tag{3.89}$$

$$\text{Torque: } T = \frac{P}{\omega_s} = \underbrace{3\frac{V_s E_f}{\omega_s X_s}\sin\delta}_{\text{Electromagnetic torque } T_f} + \underbrace{3\frac{V_s^2(X_d - X_q)}{2\omega_s X_d X_q}\sin 2\delta}_{\text{Reluctance torque } T_r} \tag{3.90}$$

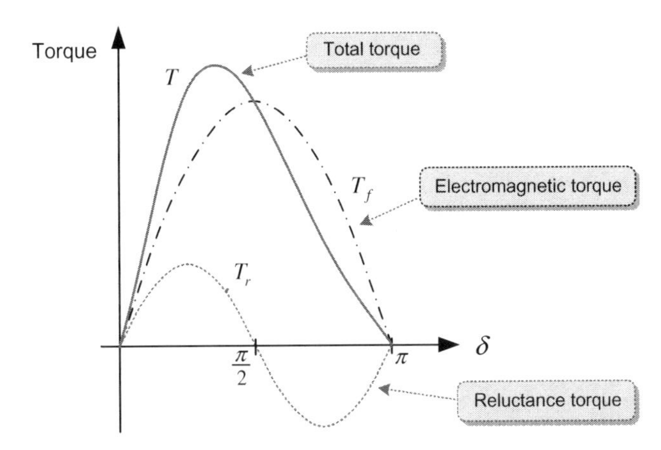

FIGURE 3.62

Torque of a salient pole synchronous motor.

As we can expect, a salient pole synchronous motor has two torque components: *electromagnetic torque* T_f due to the field flux of the rotor (the first term on the right-hand side of Eq. (3.90)) and *reluctance torque* T_r due to the saliency of the rotor (the second term on the right-hand side of Eq. (3.90)). Fig. 3.62 shows the output torque according to the torque angle δ.

Because of this reluctance torque, a synchronous motor with the salient pole rotor can produce more torque than a synchronous motor with the cylindrical rotor. In addition, this motor has the output torque to drive a load even if the field current is zero. Due to the reluctance torque, the torque angle δ at which the pull-out torque occurs is less than 90 electrical degrees.

3.2.3 STARTING OF SYNCHRONOUS MOTORS

Synchronous motors are inherently not able to self-start on an AC power source with the utility frequency of 50 or 60 Hz. As stated in Chapter 1, this is because synchronous motors can develop a torque only when running at the synchronous speed. However, the synchronous speed for the utility frequency is too fast for the rotor to synchronize for starting as shown in Fig. 3.63.

Therefore we need some means for starting synchronous motors. Once the rotor reaches a speed close to the synchronous speed (>95%) through some

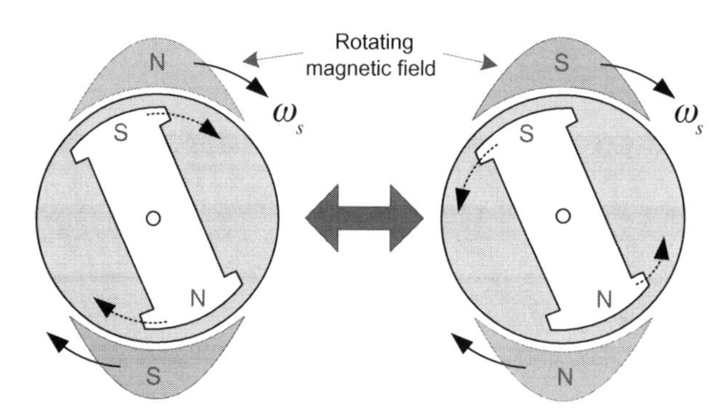

FIGURE 3.63

Synchronization in a synchronous motor.

starting means, the field winding will be excited, so the motor locks into synchronization.

There are some techniques employed to start a synchronous motor. As a simple method, a separate motor is used to drive the rotor to run close to the synchronous speed. A second method is to start the synchronous motor as an induction motor. In this case, the rotor has a special damping winding for the purpose of starting, which is similar to the squirrel-cage arrangements of an induction motor. When the synchronous motor is starting, the torque will be produced by the induced current in the damping winding, so it can start as an induction motor. In addition, the damper winding has another function to help keeping the synchronous motor in synchronism. If the speed of the motor is increased or decreased from the synchronous speed, then a current will be induced in the damper winding and will develop a torque to oppose the change in speed.

As for the starting method of PMSMs used in variable speed drives, we can start the motor slowly at a reduced frequency by using a PWM inverter. In this case, a high starting torque can be developed by using the information of the rotor initial position.

REFERENCES

[1] C.I. Hubert, Electric Machines, Theory, Operation, Applications, Adjustment, and Control, second ed., Prentice Hall, Upper Saddle River, NJ, 2002, p. 135.

[2] P.C. Sen, Principles of Electric Machines and Power Electronics, second ed., John Wiley & Sons, Inc., New York, 1997.

[3] J.F. Funchslocj, W.R. Findley, R.W. Walter, The next generation motor, IEEE Ind. Appl. Mag. 14 (1) (Jan./Feb. 2008) 37–43.

[4] S.J. Chapman, Electric Machinery Fundamentals, third ed., McGrawHill, International editions, Singapore, 1999 (Chapter 7).

CHAPTER

Modeling of alternating current motors and reference frame theory

4

For a proper control of a motor, we need to analyze its dynamic performance as well as steady-state performance. This requires a mathematical model, which describes the behavior of a motor. The motor model can be mainly expressed by the voltage and torque equations. The voltage equation for a motor basically consists of the following components: the voltage drop of the winding resistance R and the induced voltage proportional to the rate of change over time of the winding flux linkage λ as

$$\begin{aligned} v(t) &= Ri(t) + \frac{d\lambda(t)}{dt} \\ &= Ri(t) + \frac{d\left[L(\theta_r)i(t)\right]}{dt} \quad (\theta_r = \omega_r t) \end{aligned} \tag{4.1}$$

As discussed in Chapter 1, the flux linkage can be expressed as the product of inductance and current. For alternating current (AC) motors with windings in the stator and rotor, the mutual-inductance which implies the amount of flux linking between the two windings, is a time-varying parameter. This is because that the mutual-inductance is a function of the rotor position due to the variation in the amount of linking flux with the rotor position. Furthermore, in salient pole synchronous motors, even the self-inductance is a function of the rotor position due to the variation in the air gap according to the rotor position.

As an example, consider the induction motor in Fig. 4.1, in which the stator windings are stationary while the rotor windings are rotating with the rotor. Due to the movement of the rotor, the angular position θ_r between the stator and rotor windings varies over time except for at a standstill. As a result, even if currents (thus, flux) flowing in these two windings remain constant, the amount of the flux linking these windings will vary over time. This results in the mutual-inductance to become time-varying. Due to this time-varying inductance, the voltage equations, which describe the behavior of AC motors, will be expressed by differential equations with time-varying coefficients.

The analysis of such time-varying differential equations will be fairly complex. However, by using the *reference frame transformation*, which will be introduced in this chapter, its complexity can be significantly reduced. This is because the

Electric Motor Control. DOI: http://dx.doi.org/10.1016/B978-0-12-812138-2.00004-0

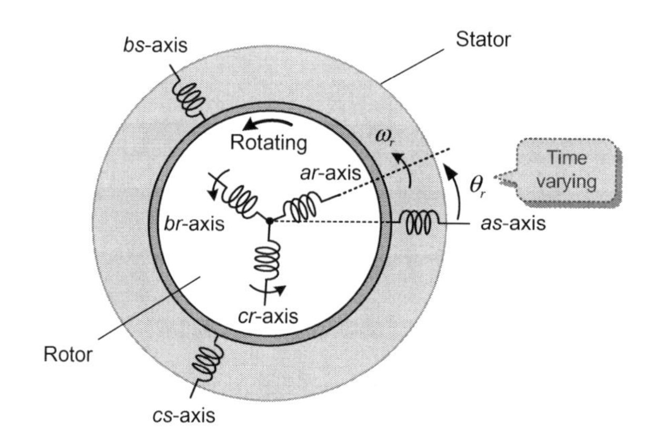

FIGURE 4.1

Angular position between stator and rotor windings.

time-varying inductances can be eliminated in the voltage equations by employing the reference frame transformation. When using this transformation, the three-phase *abc* variables are transformed to *dq* variables in an orthogonal coordinate system. These *dq* orthogonal variables are necessary to employ the *vector control* scheme, which enables a dynamic torque control of AC motors. This scheme will be described in Chapter 5.

In this chapter, we will study the reference frame transformation and introduce the models of AC motors expressed by the *dq* variables in the orthogonal coordinate system. Prior to this, we will start exploring the models of AC motors expressed in the three-phase *abc* variables.

4.1 MODELING OF INDUCTION MOTOR

Unlike the steady-state model of an induction motor shown in Chapter 3, in this section, we will introduce a model to describe the dynamics in the transient-state as well as the steady-state performance.

Consider a three-phase, two-pole induction motor as shown in Fig. 4.2. Assume that the stator windings have a number of effective turns N_s, resistance R_s, leakage inductance L_{ls}, and self-inductance L_s. Similarly, the equivalent rotor windings have a number of effective turns N_r, resistance R_r, leakage inductance L_{lr}, and self-inductance L_r. θ_r is the angular displacement between the stator and rotor axes. Here, we will not consider the nonideal characteristics such as the saturation of iron core, slots effect, and cogging torque. Although these are identical sinusoidal distributed windings, each displaced by 120°, we will consider only the centered coil of each phase for simplicity.

Now, let us derive the voltage equations of an induction motor. Since an induction motor has six windings, we can express it by six voltage equations. The

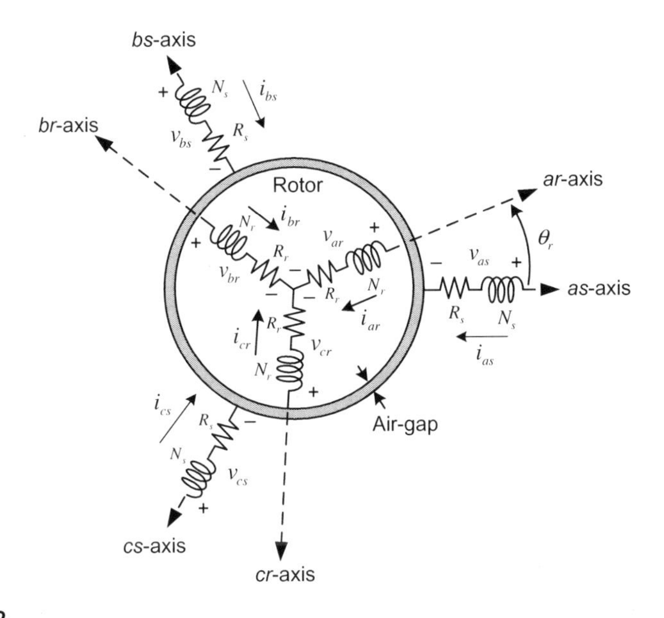

FIGURE 4.2

Stator and rotor windings of an induction motor.

winding voltage is composed of the voltage drop of the winding resistance and the voltage induced by the flux linkage variation of the winding. Thus the six voltage equations for the stator and rotor windings can be written by

$$v_{as} = R_s i_{as} + \frac{d\lambda_{as}}{dt} \tag{4.2}$$

$$v_{bs} = R_s i_{bs} + \frac{d\lambda_{bs}}{dt} \tag{4.3}$$

$$v_{cs} = R_s i_{cs} + \frac{d\lambda_{cs}}{dt} \tag{4.4}$$

$$v'_{ar} = R' i'_{ar} + \frac{d\lambda'_{ar}}{dt} \tag{4.5}$$

$$v'_{br} = R' i'_{br} + \frac{d\lambda'_{br}}{dt} \tag{4.6}$$

$$v'_{cr} = R' i'_{cr} + \frac{d\lambda'_{cr}}{dt} \tag{4.7}$$

where v_{as}, v_{bs}, v_{cs} are the stator voltages, i_{as}, i_{bs}, i_{cs} are the stator currents, v'_{ar}, v'_{br}, v'_{cr} are the rotor voltages, i'_{ar}, i'_{br}, i'_{cr} are the rotor currents, λ_{as}, λ_{bs}, λ_{cs} are the stator flux linkages, and λ_{ar}, λ_{br}, λ_{cr} are the rotor flux linkages. Here, the primed variables denote the rotor variables that are referred to the stator by considering the turns ratio of the two windings. From this point on, we will skip the prime symbol (′).

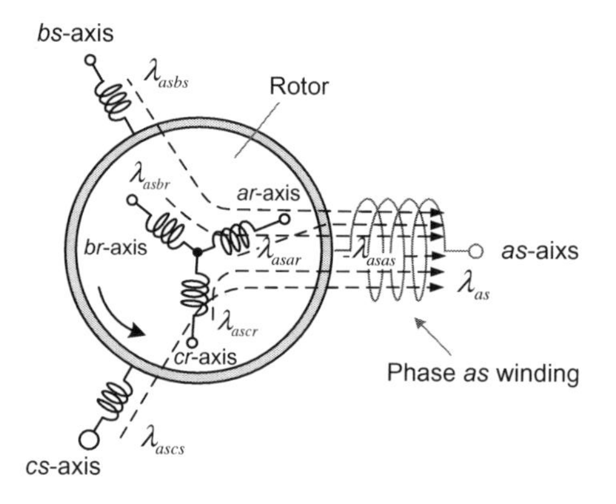

FIGURE 4.3

Flux linkage λ_{as} of the phase *as* winding.

To solve these voltage equations, we need to know the flux linkage of each winding. Since there are six windings, one winding will be linked by not only the flux produced by its own winding current but also the flux produced by the current flowing in the other five windings. For example, the total flux linkage λ_{as} of the phase *as* winding shown in Fig. 4.3 has the following six components.

$$\begin{aligned}\lambda_{as} &= \lambda_{asas} + \lambda_{asbs} + \lambda_{ascs} + \lambda_{asar} + \lambda_{asbr} + \lambda_{ascr} \\ &= L_{asas}i_{as} + L_{asbs}i_{bs} + L_{ascs}i_{cs} + L_{asar}i_{ar} + L_{asbr}i_{br} + L_{ascr}i_{cr}\end{aligned} \tag{4.8}$$

where the flux linkage λ_{xsys} represents the flux, which is produced by the current i_{ys} flowing in the *ys* winding and links the *xs* winding. The flux linkage λ_{xsys} can be expressed as the product of the related current and inductance. In this case inductance L_{xsys} is defined as the ratio of the flux linking the *xs* winding to the current i_{ys} generating the flux, i.e., $L_{xsys} = \lambda_{xsys}/i_{ys}$.

Six flux linkages in the stator and rotor windings can be written by

$$\begin{aligned}\begin{bmatrix}\lambda_{as}\\ \lambda_{bs}\\ \lambda_{cs}\\ \hdashline \lambda_{ar}\\ \lambda_{br}\\ \lambda_{cr}\end{bmatrix} &= \left[\begin{array}{ccc:ccc}\lambda_{asas} & \lambda_{asbs} & \lambda_{ascs} & \lambda_{asar} & \lambda_{asr} & \lambda_{ascr}\\ \lambda_{bsas} & \lambda_{bsbs} & \lambda_{bscs} & \lambda_{bsar} & \lambda_{bsbr} & \lambda_{bscr}\\ \lambda_{csas} & \lambda_{csbs} & \lambda_{csbs} & \lambda_{csar} & \lambda_{csbr} & \lambda_{cscr}\\ \hdashline \lambda_{aras} & \lambda_{arbs} & \lambda_{arcs} & \lambda_{arbr} & \lambda_{arbr} & \lambda_{arcr}\\ \lambda_{bras} & \lambda_{brbs} & \lambda_{brcs} & \lambda_{brar} & \lambda_{brbr} & \lambda_{brcr}\\ \lambda_{cras} & \lambda_{crbs} & \lambda_{crcs} & \lambda_{crar} & \lambda_{crbr} & \lambda_{crbcr}\end{array}\right] \\ &= \left[\begin{array}{ccc:ccc}L_{asas} & L_{asbs} & L_{ascs} & L_{asar} & L_{asr} & L_{ascr}\\ L_{bsas} & L_{bsbs} & L_{bscs} & L_{bsar} & L_{bsbr} & L_{bscr}\\ L_{csas} & L_{csbs} & L_{csbs} & L_{csar} & L_{csbr} & L_{cscr}\\ \hdashline L_{aras} & L_{arbs} & L_{arcs} & L_{arbr} & L_{arbr} & L_{arcr}\\ L_{bras} & L_{brbs} & L_{brcs} & L_{brar} & L_{brbr} & L_{brcr}\\ L_{cras} & L_{crbs} & L_{crcs} & L_{crar} & L_{crbr} & L_{crbcr}\end{array}\right]\begin{bmatrix}i_{as}\\ i_{bs}\\ i_{cs}\\ i_{ar}\\ i_{br}\\ i_{cr}\end{bmatrix}\end{aligned} \tag{4.9}$$

These inductances are largely divided into four groups as

$$\begin{bmatrix} \boldsymbol{\lambda}_{abcs} \\ \boldsymbol{\lambda}_{abcr} \end{bmatrix} = \begin{bmatrix} \boldsymbol{L}_s & \boldsymbol{L}_{sr} \\ (\boldsymbol{L}_{sr})^T & \boldsymbol{L}_r \end{bmatrix} \begin{bmatrix} \boldsymbol{i}_{abcs} \\ \boldsymbol{i}_{abcr} \end{bmatrix} \tag{4.10}$$

where $\boldsymbol{L}_s$ and $\boldsymbol{L}_r$ represent the inductance matrix of the stator and rotor windings, respectively, and $\boldsymbol{L}_{sr}$ represents the mutual-inductance matrix between the stator and rotor windings.

Next, we will determine these inductances.

4.1.1 STATOR WINDINGS

The stator inductance matrix $\boldsymbol{L}_s$ as shown in Fig. 4.4 consists of the self-inductances of each stator winding and the mutual-inductance between the stator windings as

$$\boldsymbol{L}_s = \begin{bmatrix} L_{asas} & L_{asbs} & L_{ascs} \\ L_{bsas} & L_{bsbs} & L_{bscs} \\ L_{csas} & L_{csbs} & L_{cscs} \end{bmatrix} \tag{4.11}$$

Stator self-inductance

Stator mutual-inductance

We can obtain expressions for the self-inductances and the mutual-inductances as follows.

The flux produced by each winding can be separated into two components: the *leakage flux* which links only its own winding and the *magnetizing flux* which links both its own winding and the other windings. These can be expressed

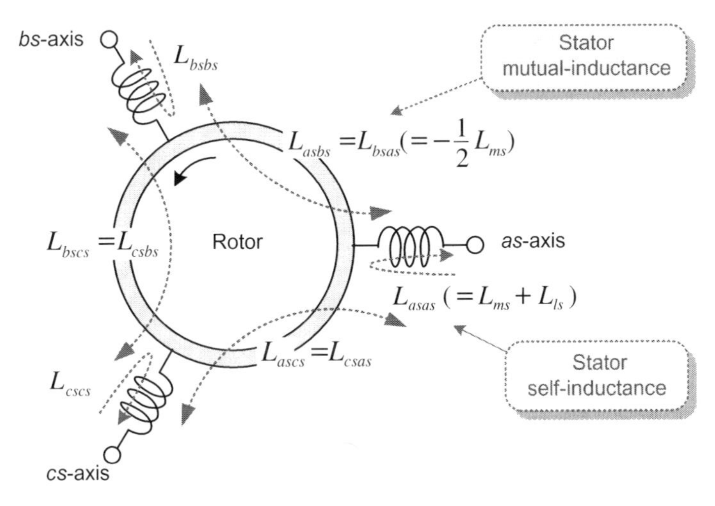

FIGURE 4.4

Stator inductances.

in terms of the leakage inductance and the magnetizing inductance, respectively. Thus the stator self-inductances L_{asas}, L_{bsbs}, L_{cscs} consist of the leakage inductance L_{ls} and the magnetizing inductance L_{ms} as

$$L_{asas} = L_{bsbs} = L_{cscs} = L_{ls} + L_{ms} \tag{4.12}$$

where $L_{ms} = \mu_o N_s^2 (rl/g)(\pi/4)$, μ_o is the permeability of air, l is the axial length of the air gap, and r is the radius to the mean of the air gap [1].

The mutual-inductances L_{asbs}, L_{ascs}, L_{bsas}, L_{bscs}, L_{csas}, L_{csbs} between the two stator windings, which are displaced from each other by 120°, are all the same and are related to the magnetizing inductance as

$$\begin{aligned} L_{asbs} &= L_{ascs} = L_{bsas} = L_{bscs} = L_{csas} = L_{csbs} \\ &= L_{ms} \cos\left(\frac{2\pi}{3}\right) = -\frac{1}{2} L_{ms} \end{aligned} \tag{4.13}$$

Here, the negative sign in the mutual-inductance indicates that each winding produces the flux in a direction opposite to one another. From the inductances of Eqs. (4.12) and (4.13), the inductances of the stator windings are given by

$$\mathbf{L}_s = \begin{bmatrix} L_{asas} & L_{asbs} & L_{ascs} \\ L_{bsas} & L_{bsbs} & L_{bscs} \\ L_{csas} & L_{csbs} & L_{cscs} \end{bmatrix} = \begin{bmatrix} L_{ls} + L_{ms} & -\dfrac{L_{ms}}{2} & -\dfrac{L_{ms}}{2} \\ -\dfrac{L_{ms}}{2} & L_{ls} + L_{ms} & -\dfrac{L_{ms}}{2} \\ -\dfrac{L_{ms}}{2} & -\dfrac{L_{ms}}{2} & L_{ls} + L_{ms} \end{bmatrix} \tag{4.14}$$

Because the length of the air gap is constant, these inductances become a constant value, irrespective of the rotor position.

4.1.2 ROTOR WINDINGS

The rotor inductance matrix $\mathbf{L}_r$ as shown in Fig. 4.5 consists of the self-inductances of each rotor windings and the mutual-inductance between these windings as

$$\mathbf{L}_r = \begin{bmatrix} L_{arar} & L_{arbr} & L_{arcr} \\ L_{brar} & L_{brbr} & L_{brcr} \\ L_{crar} & L_{crbr} & L_{crcr} \end{bmatrix} \tag{4.15}$$

Similar to the stator self-inductances, the rotor self-inductances L_{arar}, L_{brbr}, L_{crcr} consist of the leakage inductance L_{lr} and the magnetizing inductance L_{mr} as

$$L_{arar} = L_{brbr} = L_{crcr} = L_{lr} + L_{mr} \tag{4.16}$$

where $L_{mr} = \mu_o N_r^2 (rl/g)(\pi/4) = (N_r/N_s)^2 L_{ms}$.

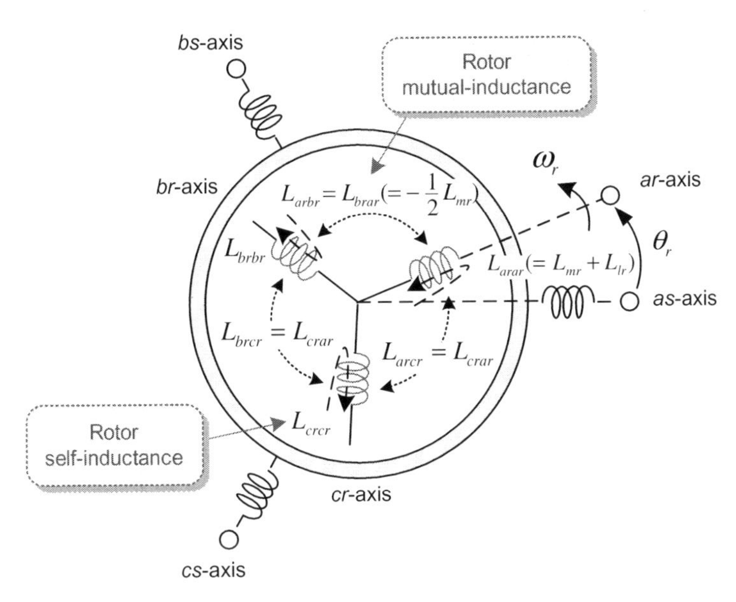

FIGURE 4.5

Inductances of the rotor windings.

The mutual-inductances L_{arbr}, L_{arcr}, L_{brar}, L_{brcr}, L_{crar}, L_{crbr} between the two rotor windings, which are displaced from each other by 120°, are all the same and related to the magnetizing inductance as

$$\begin{aligned} L_{arbr} = L_{arcr} = L_{brar} = L_{brcr} = L_{crar} = L_{crbr} = L_{mr}\cos\left(\frac{2\pi}{3}\right) &= -\frac{1}{2}L_{mr} \\ &= -\frac{1}{2}\left(\frac{N_r}{N_s}\right)^2 L_{ms} \end{aligned} \tag{4.17}$$

where N_r/N_s is the turns ratio of the stator and rotor windings. With these inductances, the inductances of the rotor windings are given by

$$\mathbf{L}_r = \begin{bmatrix} L_{arar} & L_{arbr} & L_{arcr} \\ L_{brar} & L_{brbr} & L_{brcr} \\ L_{crar} & L_{crbr} & L_{crcr} \end{bmatrix} = \begin{bmatrix} L_{ls} + n^2 L_{ms} & -n^2\dfrac{L_{ms}}{2} & -n^2\dfrac{L_{ms}}{2} \\ -\dfrac{L_{ms}}{2} & L_{ls} + n^2 L_{ms} & -n^2\dfrac{L_{ms}}{2} \\ -n^2\dfrac{L_{ms}}{2} & -n^2\dfrac{L_{ms}}{2} & L_{ls} + n^2 L_{ms} \end{bmatrix}, \quad n = \left(\frac{N_r}{N_s}\right) \tag{4.18}$$

These inductances are a constant value, irrespective of the rotor position.

4.1.3 INDUCTANCE BETWEEN THE STATOR AND ROTOR WINDINGS

Next, let us examine the mutual-inductance of the stator and rotor windings. First, we will explore the mutual-inductance matrix $\boldsymbol{L}_{sr}$. This is related to the amount of flux, which is produced by the rotor windings and links the stator windings. The mutual-inductance consists of nine components as

$$L_{sr} = \begin{bmatrix} L_{asar} & L_{asbr} & L_{ascr} \\ L_{bsar} & L_{bsbr} & L_{bscr} \\ L_{csar} & L_{csbr} & L_{cscr} \end{bmatrix} \tag{4.19}$$

As an example, examine the mutual-inductance L_{asar}, which represents the ratio of the flux linking in the stator *as* winding to the rotor *ar* winding current generating the flux. If the rotor winding is rotating at a speed ω_r, the relative position θ_r between the two windings will vary over time. Thus the mutual-inductance vary sinusoidally with respect to the displacement angle θ_r of the rotor as shown in Fig. 4.6.

Considering the turns ratio between the two windings, L_{asar} is given by

$$L_{asar} = L_{mr}\left(\frac{N_s}{N_r}\right)\cos\theta_r = L_{ms}\left(\frac{N_r}{N_s}\right)\cos\theta_r \quad \left(\theta_r = \int \omega_r dt\right) \tag{4.20}$$

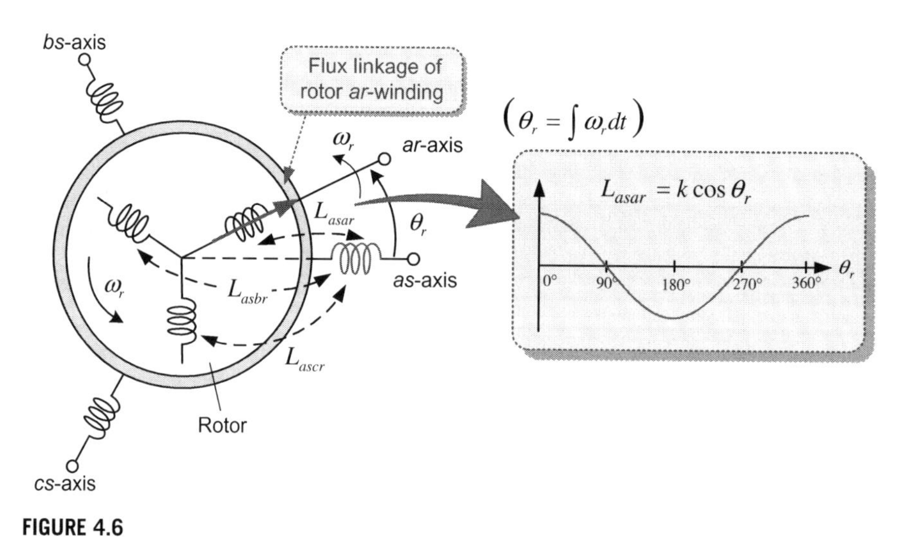

FIGURE 4.6

Mutual-inductance between the stator *as* winding and the rotor *ar* winding.

Likewise, the other mutual-inductances are given by

$$L_{asar} = L_{bsbr} = L_{cscr} = \left(\frac{N_r}{N_s}\right) L_{ms} \cos\theta_r \tag{4.21}$$

$$L_{asbr} = L_{bscr} = L_{csar} = \left(\frac{N_r}{N_s}\right) L_{ms} \cos\left(\theta_r + \frac{2\pi}{3}\right) \tag{4.22}$$

$$L_{ascr} = L_{bsar} = L_{csbr} = \left(\frac{N_r}{N_s}\right) L_{ms} \cos\left(\theta_r - \frac{2\pi}{3}\right) \tag{4.23}$$

With these inductances, the mutual-inductance matrix $\boldsymbol{L}_{sr}$ is given by

$$\boldsymbol{L}_{sr} = \begin{bmatrix} L_{asar} & L_{asbr} & L_{ascr} \\ L_{bsar} & L_{bsbr} & L_{bscr} \\ L_{csar} & L_{csbr} & L_{cscr} \end{bmatrix} = nL_{ms} \begin{bmatrix} \cos\theta_r & \cos\left(\theta_r + \frac{2\pi}{3}\right) & \cos\left(\theta_r - \frac{2\pi}{3}\right) \\ \cos\left(\theta_r - \frac{2\pi}{3}\right) & \cos\theta_r & \cos\left(\theta_r + \frac{2\pi}{3}\right) \\ \cos\left(\theta_r + \frac{2\pi}{3}\right) & \cos\left(\theta_r - \frac{2\pi}{3}\right) & \cos\theta_r \end{bmatrix} \tag{4.24}$$

From Eq. (4.24), we can readily see that the mutual-inductances are a function of the rotor position and thus vary over time except for at a standstill. Consequently, the time-varying coefficients will appear in the stator voltage equations of Eqs. (4.2)–(4.4).

The flux linkage of the stator windings, from Eqs. (4.14) and (4.24), is given by

$$\begin{aligned} \boldsymbol{\lambda}_{abcs} &= \boldsymbol{L}_s \boldsymbol{i}_{abcs} + \boldsymbol{L}_{sr} \boldsymbol{i}_{abcr} \\ &= \begin{bmatrix} L_{ls} + L_{ms} & -\frac{L_{ms}}{2} & -\frac{L_{ms}}{2} \\ -\frac{L_{ms}}{2} & L_{ls} + L_{ms} & -\frac{L_{ms}}{2} \\ -\frac{L_{ms}}{2} & -\frac{L_{ms}}{2} & L_{ls} + L_{ms} \end{bmatrix} \begin{bmatrix} i_{as} \\ i_{bs} \\ i_{cs} \end{bmatrix} \\ &\quad + nL_{ms} \begin{bmatrix} \cos\theta_r & \cos\left(\theta_r + \frac{2\pi}{3}\right) & \cos\left(\theta_r - \frac{2\pi}{3}\right) \\ \cos\left(\theta_r - \frac{2\pi}{3}\right) & \cos\theta_r & \cos\left(\theta_r + \frac{2\pi}{3}\right) \\ \cos\left(\theta_r + \frac{2\pi}{3}\right) & \cos\left(\theta_r - \frac{2\pi}{3}\right) & \cos\theta_r \end{bmatrix} \begin{bmatrix} i_{ar} \\ i_{br} \\ i_{cr} \end{bmatrix} \end{aligned} \tag{4.25}$$

Next, examine the mutual-inductance $\boldsymbol{L}_{rs}$. This is related to the amount of flux, which is produced by the stator windings and links the rotor windings. This mutual-inductance $\boldsymbol{L}_{rs}$ is equal to the transpose of $\boldsymbol{L}_{rs}$ as

$$\boldsymbol{L}_{rs} = \boldsymbol{L}_{sr}^T = \begin{bmatrix} L_{aras} & L_{arbs} & L_{arcs} \\ L_{bras} & L_{brbs} & L_{brcs} \\ L_{cras} & L_{crbs} & L_{crcs} \end{bmatrix}$$

$$= nL_{ms} \begin{bmatrix} \cos\theta_r & \cos\left(\theta_r - \frac{2\pi}{3}\right) & \cos\left(\theta_r + \frac{2\pi}{3}\right) \\ \cos\left(\theta_r + \frac{2\pi}{3}\right) & \cos\theta_r & \cos\left(\theta_r - \frac{2\pi}{3}\right) \\ \cos\left(\theta_r - \frac{2\pi}{3}\right) & \cos\left(\theta_r + \frac{2\pi}{3}\right) & \cos\theta_r \end{bmatrix} \tag{4.26}$$

Like $\boldsymbol{L}_{sr}$, the mutual-inductance $\boldsymbol{L}_{rs}$ also becomes time-varying except at a standstill. Therefore the time-varying coefficients will appear in the voltage equations of the rotor windings.

The flux linkage of the rotor windings, from Eqs. (4.18) and (4.26), is given by

$$\boldsymbol{\lambda}_{abcr} = \boldsymbol{L}_r \boldsymbol{i}_{abcr} + \boldsymbol{L}_{rs} \boldsymbol{i}_{abcs}$$

$$= \begin{bmatrix} L_{ls} + n^2 L_{ms} & -n^2 \frac{L_{ms}}{2} & -n^2 \frac{L_{ms}}{2} \\ -\frac{L_{ms}}{2} & L_{ls} + n^2 L_{ms} & -n^2 \frac{L_{ms}}{2} \\ -n^2 \frac{L_{ms}}{2} & -n^2 \frac{L_{ms}}{2} & L_{ls} + n^2 L_{ms} \end{bmatrix} \begin{bmatrix} i_{ar} \\ i_{br} \\ i_{cr} \end{bmatrix}$$

$$+ nL_{ms} \begin{bmatrix} \cos\theta_r & \cos\left(\theta_r - \frac{2\pi}{3}\right) & \cos\left(\theta_r + \frac{2\pi}{3}\right) \\ \cos\left(\theta_r + \frac{2\pi}{3}\right) & \cos\theta_r & \cos\left(\theta_r - \frac{2\pi}{3}\right) \\ \cos\left(\theta_r - \frac{2\pi}{3}\right) & \cos\left(\theta_r + \frac{2\pi}{3}\right) & \cos\theta_r \end{bmatrix} \begin{bmatrix} i_{as} \\ i_{bs} \\ i_{cs} \end{bmatrix} \tag{4.27}$$

Finally, from Eqs. (4.25) and (4.27), the total flux linkages of the induction motor are given by

$$\begin{bmatrix}\lambda_{as}\\ \lambda_{bs}\\ \lambda_{cs}\\ \hline \lambda_{ar}\\ \lambda_{br}\\ \lambda_{cr}\end{bmatrix}=\left[\begin{array}{ccc|ccc} L_{ls}+L_{ms} & -\frac{L_{ms}}{2} & -\frac{L_{ms}}{2} & nL_{ms}\cos\theta_r & nL_{ms}\cos(\theta_r+\frac{2\pi}{3}) & nL_{ms}\cos(\theta_r-\frac{2\pi}{3})\\ -\frac{L_{ms}}{2} & L_{ls}+L_{ms} & -\frac{L_{ms}}{2} & nL_{ms}\cos(\theta_r-\frac{2\pi}{3}) & nL_{ms}\cos\theta_r & nL_{ms}\cos(\theta_r+\frac{2\pi}{3})\\ -\frac{L_{ms}}{2} & -\frac{L_{ms}}{2} & L_{ls}+L_{ms} & nL_{ms}\cos(\theta_r+\frac{2\pi}{3}) & nL_{ms}\cos(\theta_r-\frac{2\pi}{3}) & nL_{ms}\cos\theta_r\\ \hline nL_{ms}\cos\theta_r & nL_{ms}\cos(\theta_r-\frac{2\pi}{3}) & nL_{ms}\cos(\theta_r+\frac{2\pi}{3}) & L_{lr}+n^2L_{ms} & -\frac{1}{2}n^2L_{ms} & -\frac{1}{2}n^2L_{ms}\\ nL_{ms}\cos(\theta_r+\frac{2\pi}{3}) & nL_{ms}\cos\theta_r & nL_{ms}\cos(\theta_r-\frac{2\pi}{3}) & -\frac{1}{2}n^2L_{ms} & L_{lr}+n^2L_{ms} & -\frac{1}{2}n^2L_{ms}\\ nL_{ms}\cos(\theta_r-\frac{2\pi}{3}) & nL_{ms}\cos(\theta_r+\frac{2\pi}{3}) & nL_{ms}\cos\theta_r & -\frac{1}{2}n^2L_{ms} & -\frac{1}{2}n^2L_{ms} & L_{lr}+n^2L_{ms}\end{array}\right]\begin{bmatrix}i_{as}\\ i_{bs}\\ i_{cs}\\ \hline i_{ar}\\ i_{br}\\ i_{cr}\end{bmatrix}$$

(4.28)

By substituting these flux linkages into Eqs. (4.2)–(4.7), the voltage equations of the induction motor are completed.

4.2 MODELING OF PERMANENT MAGNET SYNCHRONOUS MOTOR

The stator configuration of a synchronous motor is the same as that of an induction motor. The stator windings are connected to a three-phase AC source to generate a rotating magnetic field. However, unlike an induction motor having three-phase windings, the rotor of a synchronous motor has a field winding or a permanent magnet to generate a magnetic field flux on it. In recent years the magnetic field for synchronous motors has been generated mostly by using a permanent magnet rather than a field winding.

In addition to the electromagnetic torque produced by the magnetic fields of the stator and rotor, synchronous motors with a salient rotor can exploit reluctance torque. Synchronous motors can be divided into several categories according to the ratio of electromagnet torque to reluctance torque as shown in Fig. 4.7 [2].

In this book we will limit the scope of our study only to permanent magnet synchronous motors (PMSMs) except for synchronous reluctance motors, which only exploit reluctance torque.

PMSMs offer many advantages over other motors. They have a higher efficiency than induction motors because there are no copper losses by the rotor. In addition, they exhibit a high power density, high torque-to-volume ratios, and a fast dynamic response because of the availability of cost-effective powerful permanent magnet materials such as NdFeB.

Besides these advantages, PMSMs with magnetic rotor saliency use reluctance torque as well as magnet torque. Owing to these attractive characteristics, PMSMs are gaining more attention for a wide variety of residential and industrial drive applications, ranging from general-purpose drives to high-performance drives.

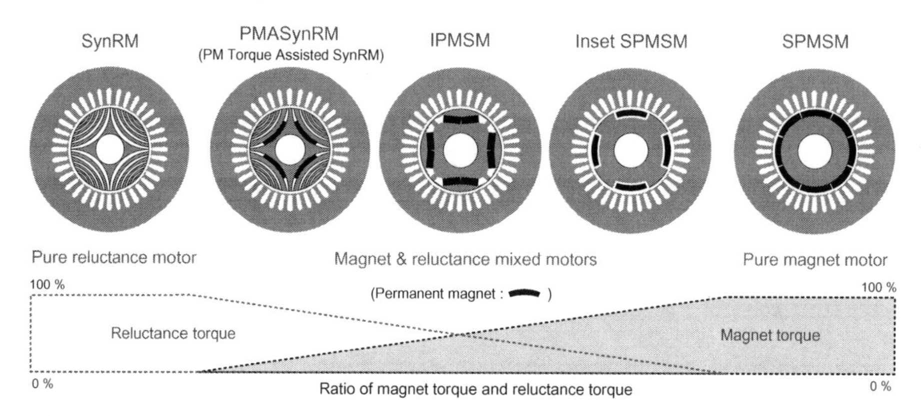

FIGURE 4.7

Several categories of synchronous motors.

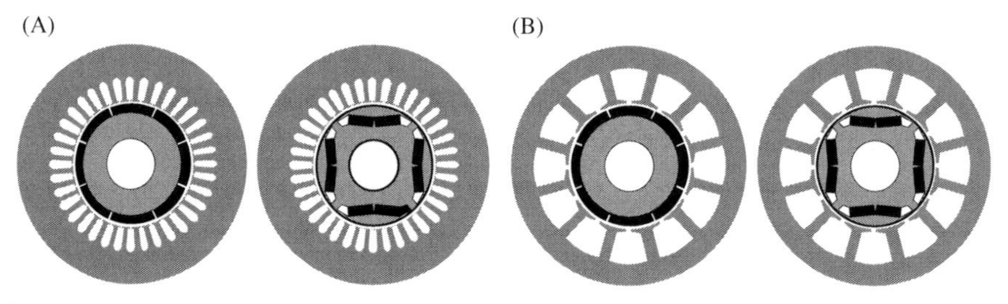

FIGURE 4.8

Different structures for the stator and rotor of PMSMs. (A) Distributed winding and (B) concentrated winding.

4.2.1 STRUCTURE OF PERMANENT MAGNET SYNCHRONOUS MOTORS

4.2.1.1 Stator structure

A distributed winding has been typically used as a three-phase stator winding configuration to produce a sinusoidal flux in the air gap as shown in Fig. 4.8A. However, recently, a concentrated winding has been increasingly used in PMSMs as shown in Fig. 4.8B.

Although a concentrated winding gives less sinusoidal magnetomotive force distribution compared to a distributed winding, this type of winding can provide several advantages such as high power density, high efficiency, short end turns, high slot-fill factor, low manufacturing costs by segmented stator structures, low cogging torque, flux-weakening capability, and fault tolerance [3].

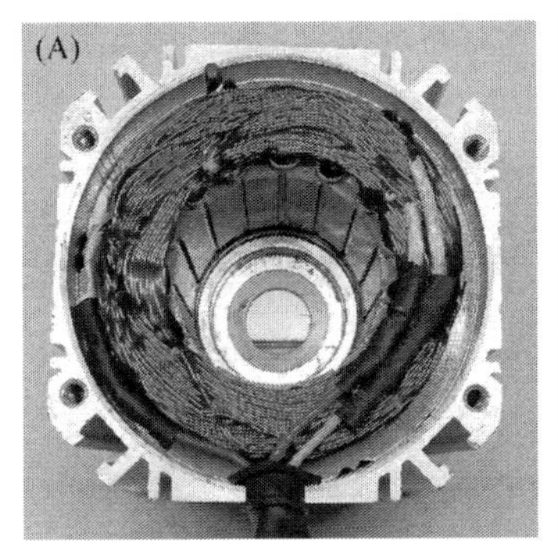

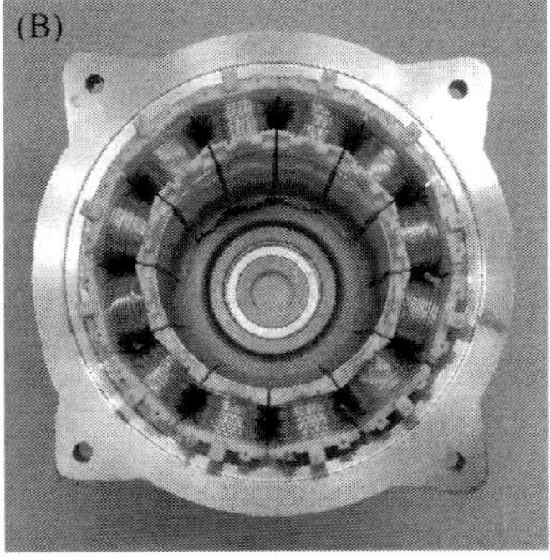

FIGURE 4.9

Comparison of distributed and concentrated windings. (A) Distributed winding and (B) concentrated winding.

Fig. 4.9 compares the distributed winding and concentrated winding in the stator of a PMSM. There are two main types of a concentrated winding: *single-layer winding* configuration, where each slot is occupied by the coil sides of one phase, and *double-layer winding* configuration, where each slot is occupied equally by the coil sides of two phases [4].

4.2.1.2 Rotor structure

The rotor of PMSMs has a permanent magnet to generate a magnetic field flux on it. The characteristics of PMSMs such as the power density, torque density, weight, and size may vary according to the chosen permanent magnet material. These days, a neodymium magnet (also known as NdFeB) is the most commonly employed for PMSMs due to its excellent magnetic characteristics and low cost. Even though we use the same permanent magnet material, many characteristics of PMSMs are heavily dependent on rotor configurations involving the shape and arrangement of the permanent magnets.

PMSMs can be divided into two types according to the direction of the magnetic flux produced from the permanent magnets as shown in Fig. 4.10: *radial-flux type* and *axial-flux type*. In the radial-flux type, the rotor which is cylindrical in shape rotates inside the stator and the magnetic flux crosses the air gap in a radial direction. On the other hand, in the axial-flux type, the rotor which resembles a pancake rotates beside the stator and the magnetic flux crosses the air gap in an axial direction [5].

Out of these two types, the radial-flux type is more commonly used in many applications. The radial-flux type PMSMs can also be further classified into two

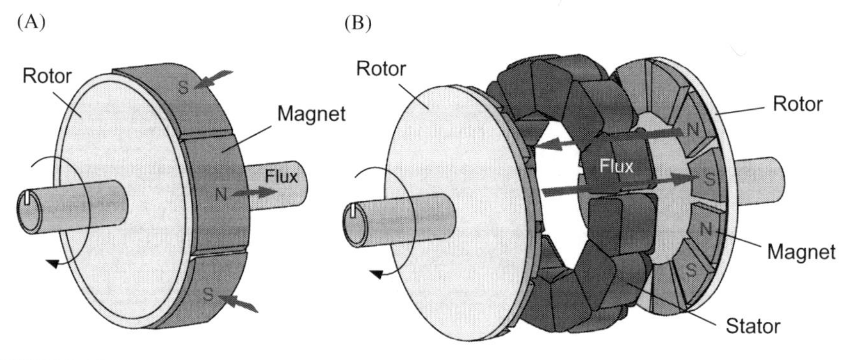

FIGURE 4.10

PMSM configurations. (A) Radial-flux type and (B) axial-flux type.

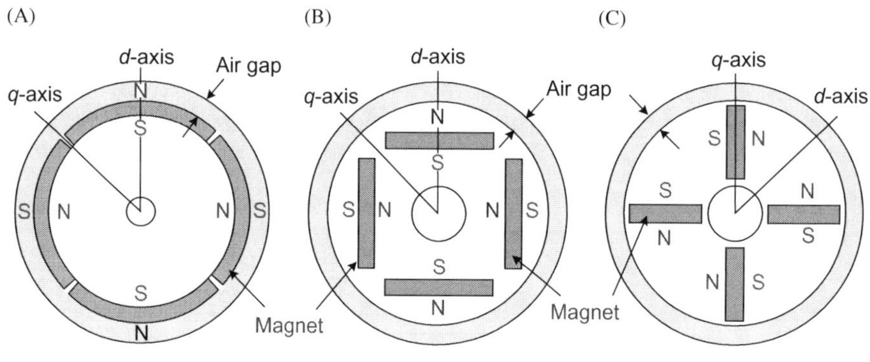

FIGURE 4.11

Rotor topologies of PMSMs (four-pole). (A) Surface mounted, (B) interior (parallel), and (C) interior (perpendicular).

categories based on the position of the permanent magnets as shown in Fig. 4.11: *surface-mounted permanent magnet synchronous motor* (SPMSM) with magnets mounted on the surface of the rotor and *interior permanent magnet synchronous motor* (IPMSM) with magnets buried inside the rotor iron core.

As for IPMSM, there are two topologies according to the arrangement of the permanent magnets: *parallel topology*, in which the magnetic poles are located parallel to the circumference of the rotor as shown in Fig. 4.11B and *perpendicular topology* (also called *spoke type*), in which the magnetic poles are located perpendicularly to the circumference as shown in Fig. 4.11C.

IPMSMs with the parallel topology are more commonly used. The perpendicular topology can concentrate the magnetic flux of the magnets and achieve a high pole number, so that the flux density in the air gap can be higher than the flux density of the magnet itself. Thus although low cost and low flux density magnet

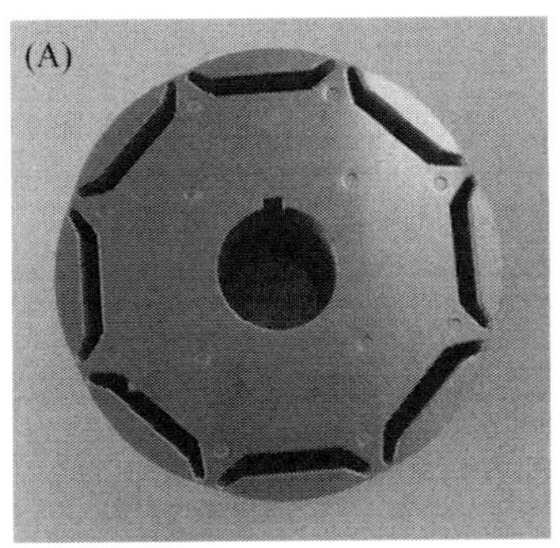

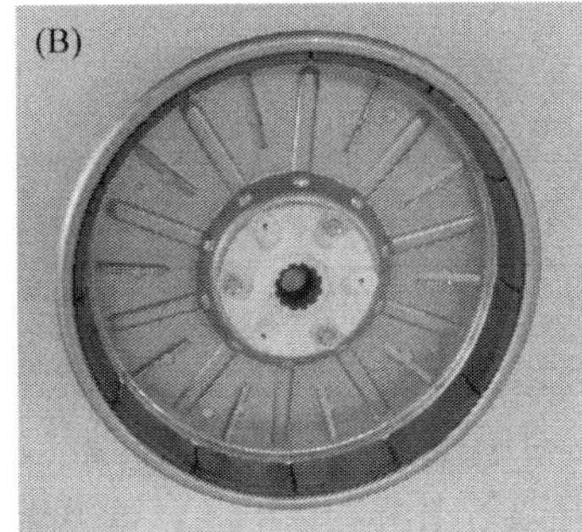

FIGURE 4.12

Rotor configurations of IPMSMs. (A) Inner rotor type and (B) outer rotor type.

material such as ferrite is employed, the motor can exhibit a high torque density. However, due to the flux concentration at the pole edges, there may be a high harmonic distortion in the air-gap flux density near the pole tips and thereby in the back-EMF profile. This also leads to a high ripple in the developed torque.

The rotor can also be constructed as an *inner rotor type* rotating inside the stator windings or an *outer rotor type* rotating outside the stator windings as shown in Fig. 4.12. The inner rotor type has lower inertia, and thus, results in a quick speed response. On the other hand, the outer rotor type has relatively high inertia and, thus, is favorable for constant speed operations.

PERMANENT MAGNETS USED IN ELECTRIC MOTORS

The permanent magnet used in motors can be mainly divided into the following three categories according to the type of magnetic materials: *Alnico, Ferrite, and Rare-earth* (SmCo, NdFeB).

1. Alnico magnets

   Alnico gives a high residual flux density (known as *remanence*) but a low coercivity. It also has a low temperature coefficient of remanence. However, due to a low coercivity, its magnetization is easily reduced by applied external magnetic field (this is called *demagnetization*).

2. Ferrite (or ceramic) magnets

   Ferrite offers a higher coercivity but a lower residual flux density than Alnico. Because of their cheap and moderate magnetic property, Ferrite magnets are widely used in low-performance applications.

3. Rare-earth magnets

   a. Samarium–cobalt (SmCo) magnets

      SmCo gives a higher energy product (BH_{max}) caused by its high residual flux density and coercivity. The Sm-Co magnets have higher Curie temperature (at which the material loses its magnetism) and a low temperature coefficient of coercivity and remanence but are very expensive and brittle.

(*Continued*)

PERMANENT MAGNETS USED IN ELECTRIC MOTORS (CONTINUED)

b. Neodymium–iron–boron (NdFeB) magnets

NdFeB is considered as one of the best magnetic materials at present since it gives a much higher residual flux density and coercivity. However, Neodymium magnets are sensitive to temperature and can even lose magnet properties at a high temperature. This drawback is steadily improving.

Among these, NdFeB is the most commonly employed permanent magnet material for PMSMs due to its excellent magnetic characteristics and its recent price reductions. The following figure shows B–H characteristic curves of these magnetic materials.

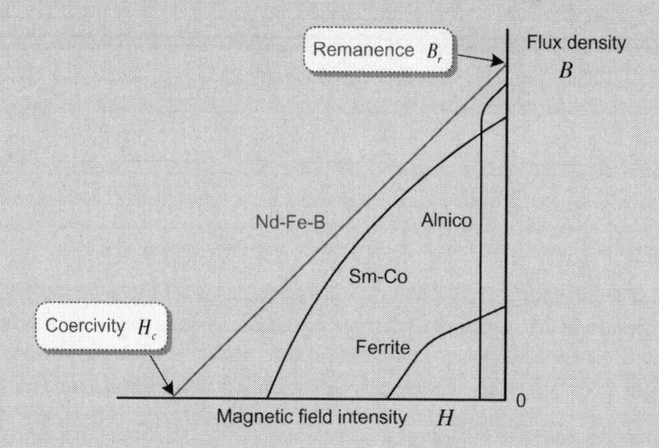

- Remanence B_r and coercivity H_c

  Remanence or residual flux density indicates the magnetic flux density remaining in a magnetic material after an external magnetic field is removed. Magnetic materials with a large residual flux density are desirable for strong permanent magnets. Coercivity indicates the intensity of the magnetic field required to reduce the magnetization of the material to zero. Magnetic materials with high coercivity are commonly used to make permanent magnets for electric motors.

- Temperature effects on permanent magnets

  The permanent magnets are sensitive to temperature. The residual flux density decreases with the increase in temperature. For example, NdFeB has the temperature coefficient of about 0.1% per degree celsius. The initial flux of the magnet can be restored when it returns to the original temperature. However, if the magnet is exposed to temperatures above the maximum working temperature and experiences a big loss of magnetism, irreversible changes may occur and its original flux level cannot be restored when it returns to the original temperature. Therefore it is advisable to operate magnets below the maximum working temperature. This limits the maximum torque capability of motors because the maximum current should be limited to avoid the demagnetization of magnets.

All the rotors shown in Fig. 4.11 are cylindrical in shape, and thus the physical air gap is constant. However the effective air gap may vary with the arrangement of the magnets. Since the permeability of a magnet approximates to that of air, the magnets placed in the flux path may be regarded as air. Thus, from the viewpoint of the magnetic field, the effective air gap will be the sum of the actual air-gap length and the radial thickness of the magnets.

For SPMSMs, which have magnets of uniform thickness on the surface of the rotor, the effective air gap is increased by the radial thickness of the magnets. Thus SPMSMs have a small stator winding inductance, and the d-axis inductance equals the q-axis inductance.

For IPMSMs, the $d-q$ axes inductances may be different depending on the arrangement of the magnets. On the one hand, in parallel topology IPMSMs the magnets are placed along with the d-axis so that the q-axis inductance becomes higher than the d-axis inductance. On the other hand, in perpendicular topology IPMSMs, the q-axis inductance is smaller than the d-axis inductance. Because of the magnetic saliency resulting from the inductance difference between the d- and q-axes, IPMSMs can exploit the reluctance torque.

The SPMSMs are unsuitable for high-speed operations producing a lager centrifugal force because of the magnets stuck on the surface of the rotor. Furthermore, because of the small inductance resulting from the increased effective air gap, SPMSMs have a relatively limited flux-weakening capability for high-speed operations. On the other hand, IPMSMs have more mechanical strength for high-speed operations because the magnets are inserted inside the rotor. Furthermore, the d-axis inductance of IPMSMs is high compared to that of an equivalent SPMSMs, and this is of a great advantage to IPMSMs for flux-weakening operation, which will be discussed in Chapter 8. Therefore IPMSMs are appropriate for high-speed operations.

As explained above, SPMSMs and IPMSMs are different in their characteristics, requiring different control methods. Now, we will find the model of the PMSMs. Initially, we will introduce the model of an IPMSM with a rotor that has a magnetic saliency. The model of an SPMSM can be derived easily from the IPMSM model. As stated earlier, an IPMSM has parallel and perpendicular topologies. These two topologies can be expressed as the same models but are different in the values of their $d-q$ axes inductances. In this section the most commonly used parallel topology IPMSM will be discussed.

4.2.2 MODEL OF PERMANENT MAGNET SYNCHRONOUS MOTORS

For a parallel topology IPMSM, the q-axis inductance is higher than the d-axis inductance because of the magnets placed along with the d-axis. Although the rotor of an IPMSM is cylindrical in shape, we will express it as a magnetic equivalent salient rotor as shown in Fig. 4.13 to simplify the analysis of the IPMSM.

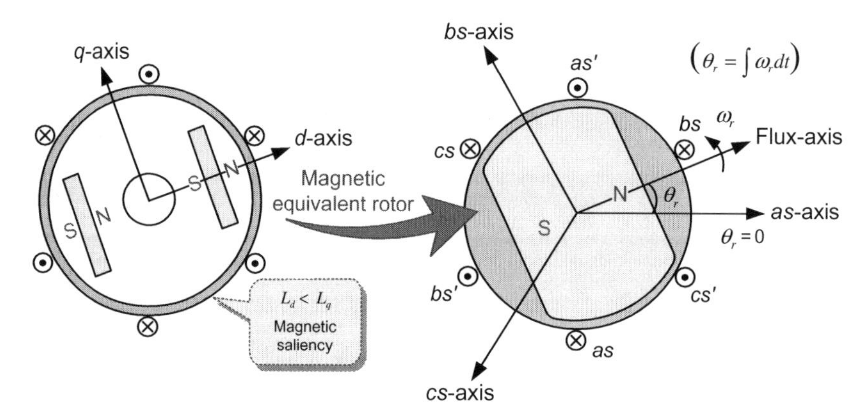

FIGURE 4.13

Rotor and its equivalent of an IPMSM.

As a PMSM has no rotor circuits, we can consider only the voltage equations for the stator windings. The stator windings of a synchronous motor are the same as those of an induction motor. Thus the stator voltage equations will be

$$\boldsymbol{v}_{abcs} = \boldsymbol{R}_s \boldsymbol{i}_{abcs} + \frac{d\boldsymbol{\lambda}_{abcs}}{dt} \tag{4.29}$$

where $\boldsymbol{v}_{abcs} = [v_{as} v_{bs} v_{cs}]^T$, $\boldsymbol{i}_{abcs} = [i_{as} i_{bs} i_{cs}]^T$, $\boldsymbol{\lambda}_{abcs} = [\lambda_{as} \lambda_{bs} \lambda_{cs}]^T$.

Since a synchronous motor has a different rotor configuration from that of an induction motor, the components of the flux linkage of the stator windings are different.

The flux linkage of an IPMSM has the following four components: the flux linkages due to the stator currents and the flux linkage due to the permanent magnet as

$$\lambda_{as} = \lambda_{asas} + \lambda_{asbs} + \lambda_{ascs} + \phi_{asf} \tag{4.30}$$

$$\lambda_{bs} = \lambda_{bsas} + \lambda_{bsbs} + \lambda_{bscs} + \phi_{bsf} \tag{4.31}$$

$$\lambda_{cs} = \lambda_{csas} + \lambda_{csbs} + \lambda_{cscs} + \phi_{csf} \tag{4.32}$$

These flux linkages can be expressed as the product of the related current and inductance as

$$\lambda_{as} = L_{asas} i_{as} + L_{asbs} i_{bs} + L_{ascs} i_{cs} + L_{asf} I_f \tag{4.33}$$

$$\lambda_{bs} = L_{bsas} i_{as} + L_{bsbs} i_{bs} + L_{bscs} i_{cs} + L_{bsf} I_f \tag{4.34}$$

$$\lambda_{cs} = L_{csas} i_{as} + L_{csbs} i_{bs} + L_{cscs} i_{cs} + L_{csf} I_f \tag{4.35}$$

Here, the flux linkages ϕ_{asf}, ϕ_{bsf}, ϕ_{csf} due to the permanent magnet are expressed as the product of the equivalent field current I_f and the related inductance L_{asf}, L_{bsf}, L_{csf}.

Now, we will determine the stator inductances of

$$\boldsymbol{L}_s = \begin{bmatrix} L_{asas} & L_{asbs} & L_{ascs} \\ L_{bsas} & L_{bsbs} & L_{bscs} \\ L_{csas} & L_{csbs} & L_{cscs} \end{bmatrix} \tag{4.36}$$

As can be seen in Section 4.1.1, the stator inductances consist of the self-inductances of the stator windings and the mutual-inductances between the stator windings. Furthermore, the stator self-inductances L_{asas}, L_{bsbs}, L_{cscs} consist of the leakage inductance and the magnetizing inductance.

The self-inductance of an IPMSM, unlike an induction motor, may be different depending on the angular position of the rotor because the effective air gap may vary with the position of the rotor. Consider an example of the phase *as* winding as shown in Fig. 4.14. In this case, the self-inductance becomes the maximum value at the rotor positions of both 90° and 270° (the smallest reluctance position), where the flux produced by the phase *as* current becomes the maximum value. This, in turn, makes the flux linking its own winding to become the maximum value. On the other hand, the self-inductance becomes the minimum value at the rotor positions of both 0 and 180° (the largest reluctance position).

Thus, the self-inductances vary sinusoidally with respect to the rotor angle θ_r and can be expressed as

$$L_{asas} = L_{ls} + L_A - L_B\cos 2\theta_r \tag{4.37}$$

$$L_{bsbs} = L_{ls} + L_A - L_B\cos 2\left(\theta_r - \frac{2}{3}\pi\right) \tag{4.38}$$

$$L_{cscs} = L_{ls} + L_A - L_B\cos 2\left(\theta_r + \frac{2}{3}\pi\right) \tag{4.39}$$

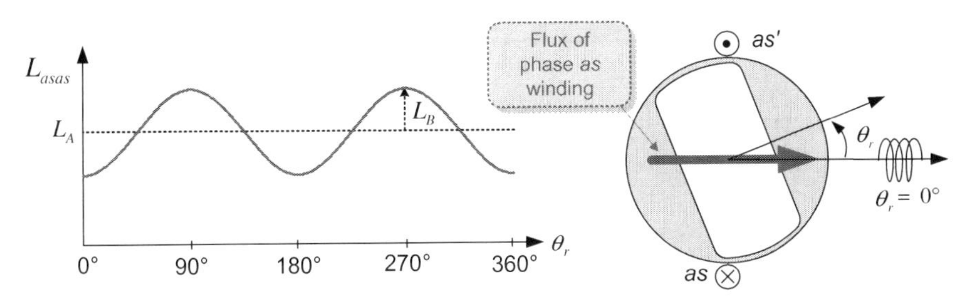

FIGURE 4.14

Self-inductance of phase *as* winding with respect to rotor positions.

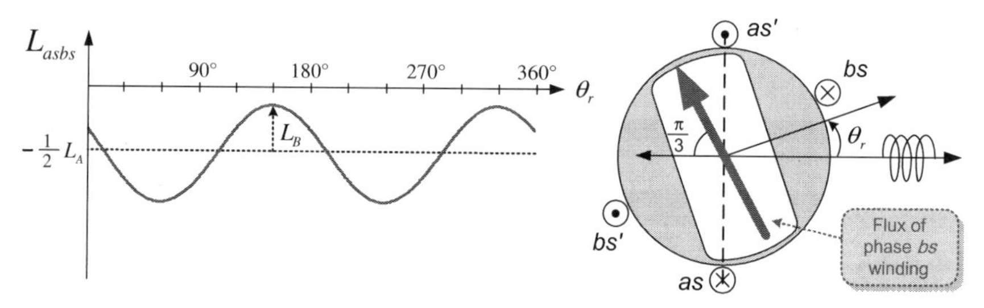

FIGURE 4.15

mutual-inductance between the phase *as* and *bs* stator windings.

where L_{ls} represents the leakage inductance, L_A represents the average value of the magnetizing inductance, and L_B represents the variation in value of the magnetizing inductance.

Similar to those of an induction motor, the mutual-inductances between the stator windings also vary sinusoidally with respect to the angle θ_r as shown in Fig. 4.15.

Thus they are given by

$$L_{asbs} = L_{bsas} = -\frac{1}{2}L_A - L_B\cos 2\left(\theta_r - \frac{\pi}{3}\right) \tag{4.40}$$

$$L_{ascs} = L_{csas} = -\frac{1}{2}L_A - L_B\cos 2\left(\theta_r + \frac{\pi}{3}\right) \tag{4.41}$$

$$L_{bscs} = L_{csbs} = -\frac{1}{2}L_A - L_B\cos 2\theta_r \tag{4.42}$$

With these inductances, the stator inductance is given by

$$\mathbf{L}_s = \begin{bmatrix} L_{ls} + L_A - L_B\cos 2\theta_r & -\frac{1}{2}L_A - L_B\cos 2\left(\theta_r - \frac{\pi}{3}\right) & -\frac{1}{2}L_A - L_B\cos 2\left(\theta_r + \frac{\pi}{3}\right) \\ -\frac{1}{2}L_A - L_B\cos 2\left(\theta_r - \frac{\pi}{3}\right) & L_{ls} + L_A - L_B\cos 2\left(\theta_r - \frac{2}{3}\pi\right) & -\frac{1}{2}L_A - L_B\cos 2\theta_r \\ -\frac{1}{2}L_A - L_B\cos 2\left(\theta_r + \frac{\pi}{3}\right) & -\frac{1}{2}L_A - L_B\cos 2\theta_r & L_{ls} + L_A - L_B\cos 2\left(\theta_r - \frac{4}{3}\pi\right) \end{bmatrix} \tag{4.43}$$

Next, let us determine the mutual-inductances L_{asf} L_{bsf} L_{csf} relevant to the amount of the permanent magnet flux linking the stator windings. These mutual-inductance are different depending on the position of the rotor because the magnet flux linking to the stator windings varies with the position of the rotor. As an example of the phase *as* winding, from Fig. 4.16, the mutual-inductance becomes a maximum value at the rotor position of 0°, while it becomes a minimum value at the rotor position of 180°.

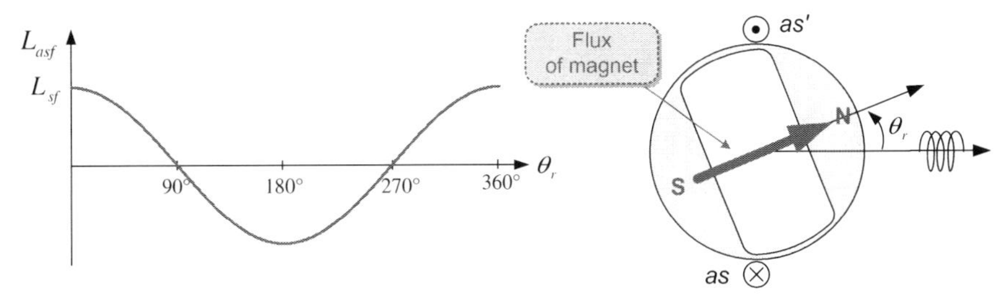

FIGURE 4.16

Mutual-inductance between stator *as* winding and magnet.

Unlike self-inductances, the mutual-inductances vary by $\cos\theta_r$ per one mechanical revolution of the rotor. So we have

$$L_{asf} = L_{sf}\cos\theta_r \tag{4.44}$$

$$L_{bsf} = L_{sf}\cos\left(\theta_r - \frac{2}{3}\pi\right) \tag{4.45}$$

$$L_{csf} = L_{sf}\cos\left(\theta_r + \frac{2}{3}\pi\right) \tag{4.46}$$

where L_{sf} depends on the amplitude of the flux linkage established by the permanent magnet.

From these inductances, the total flux linkages of an IPMSM are given by

$$\begin{aligned}\boldsymbol{\lambda}_{abcs} &= \boldsymbol{L}_s\boldsymbol{i}_{abcs} + \boldsymbol{L}_f I_f \\ &= \begin{bmatrix} L_{ls}+L_A-L_B\cos 2\theta_r & -\frac{1}{2}L_A - L_B\cos 2\left(\theta_r-\frac{\pi}{3}\right) & -\frac{1}{2}L_A - L_B\cos 2\left(\theta_r+\frac{\pi}{3}\right) \\ -\frac{1}{2}L_A - L_B\cos 2\left(\theta_r-\frac{\pi}{3}\right) & L_{ls}+L_A-L_B\cos 2\left(\theta_r-\frac{2}{3}\pi\right) & -\frac{1}{2}L_A - L_B\cos 2\theta_r \\ -\frac{1}{2}L_A - L_B\cos 2\left(\theta_r+\frac{\pi}{3}\right) & -\frac{1}{2}L_A - L_B\cos 2\theta_r & L_{ls}+L_A-L_B\cos 2\left(\theta_r-\frac{4}{3}\pi\right) \end{bmatrix}\begin{bmatrix} i_{as} \\ i_{bs} \\ i_{cs}\end{bmatrix} \\ &\quad + L_{sf}\begin{bmatrix}\cos\theta_r \\ \cos\left(\theta_r-\frac{2}{3}\pi\right) \\ \cos\left(\theta_r-\frac{4}{3}\pi\right)\end{bmatrix} I_f \end{aligned} \tag{4.47}$$

All the inductances in the flux linkages of an IPMSM are time varying except for at a standstill of the motor. Therefore the time-varying coefficients will appear in the voltage equations of an IPMSM. In the case of an SPMSM, we can easily obtain the flux linkages by letting $L_B = 0$ in Eq. (4.47).

From the discussion so far, we can see that the voltage equations of an induction motor and a synchronous motor are expressed as the differential equations with time-varying coefficients. Fortunately, by employing the reference frame transformation, these time-varying inductances in the voltage equations can be eliminated. Now, we will study the reference frame transformation and the AC motor models expressed by the *dq* variables.

4.3 REFERENCE FRAME TRANSFORMATION

Reference frame transformation refers to a transformation of the three-phase *abc* variables commonly used in AC systems to *dqn* variables orthogonal to each other. This transformation is often known as *d–q transformation*. Fig. 4.17 compares these variables. Here, the direction of the *abc* variables is considered to be the direction of the magnetic axes of their windings.

The direction of *d*, *q*, and *n* axes in orthogonal coordinates will be defined as the following.

- *d*-axis (direct axis)
  The direction of the *d* variable, which is called the *d*-axis, is normally chosen as the direction of the magnetic flux in the AC motor. In the vector control for AC motors, the *d*-axis is regarded as the reference axis, and the flux-producing component of motor current is aligned along the *d*-axis.
- *q*-axis (quadrature axis)
  The direction of the *q* variable, which is called the *q*-axis, is defined as the direction 90° ahead of the *d*-axis. In the vector control for AC motors,

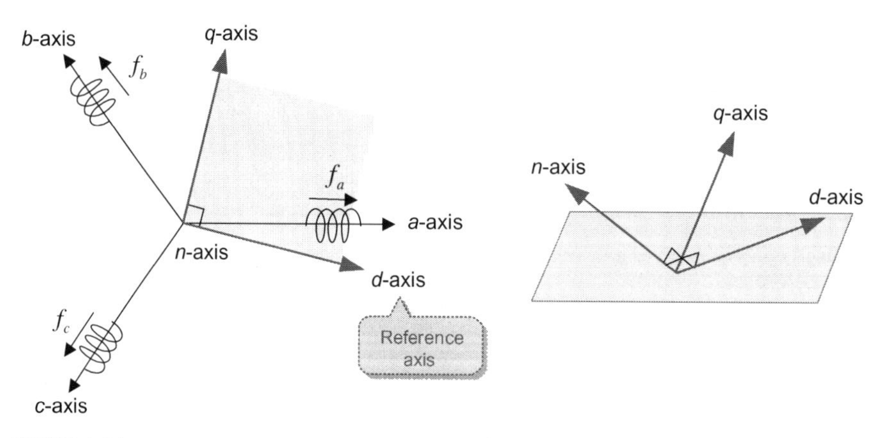

FIGURE 4.17

abc coordinate and *d–q* axes coordinate.

the torque-producing component of motor current or the back-EMF is aligned along the q-axis.

- n-axis (neutral axis)

  The direction of the n variable, which is called the n-axis, is defined as the direction that is orthogonal to both the d- and q-axes. The n-axis has nothing to do with the mechanical output power of the AC motor but is related to the losses.

4.3.1 TYPES OF THE *d–q* REFERENCE FRAME

The frame of reference, consisting of d, q, and n axes, may rotate at any speed or remain stationary. Therefore, it can be mainly divided into *stationary reference frame* and *rotating reference frame* according to whether or not it will rotate as shown in Fig. 4.18.

- Stationary reference frame

  This frame of reference remains stationary. In other words, in the stationary reference frame, the $d-q$ coordinate system does not rotate. In this book, this will be denoted by $d^s - q^s$ axes. Normally, in AC motor drives, d^s-axis is chosen as the axis of phase *as*. This reference frame is also called *stator reference frame*.

- Rotating reference frame

  This frame of reference rotates at an angular speed ω. In other words, in the rotating reference frame, the $d-q$ coordinate system rotates at a speed ω. The speed of rotation can be chosen arbitrarily. In this book this reference frame will be denoted by $d^\omega - q^\omega$ axes.

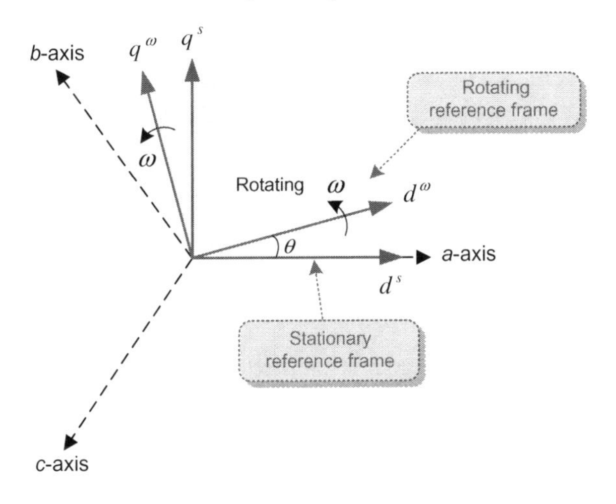

FIGURE 4.18

d–q axes stationary reference frame and rotating reference frame.

Although the rotation of the $d-q$ coordinate system can be arbitrarily set to any speed, there are two widely used speeds or reference frames. The first one is a *synchronously rotating reference frame* or *synchronous reference frame*, which rotates at the speed of the rotating magnetic field and will be denoted by d^e-q^e axes. The second one is a *rotor reference frame*, which rotates at the speed of the rotor and will be denoted by d^r-q^r axes.

The angle between the rotating reference frame and the stationary reference frame may vary over time. This angle θ is given by an integral of the angular velocity ω of the rotating reference frame as

$$\theta = \int \omega(\tau)d\tau + \theta(0) \tag{4.48}$$

where $\theta(0)$ means the initial angle at time $t=0$, and commonly, $\theta(0)=0$.

In this book we will use the notation "f_{AB}^{ω}" to represent the variables employed in various reference frames. Here, "f" represents variables such as voltage, current, and flux linkage. A superscript "ω" denotes the angular speed of the reference frame. First, the subscript "A" denotes the type of the axis. $A=d,q,n$ for dqn variables, and $A=a,\ b,\ c$ for abc variables. Second, the subscript "B" denotes where the variable is. $B=s$ for the stator, and $B=r$ for the rotor. For example, "i_{ds}^{e}" represents the d-axis stator current in the synchronously rotating reference frame.

The reference frame transformation was originally proposed by R.H. Park in the late 1920s. This transformation is used only for synchronous machines in the rotor reference frame, which is commonly known as *Park's transformation*. Since then, several reference frame transformations have been developed for induction and synchronous machines. Later, it was found that all known types of reference frames can be obtained from arbitrary reference frames by simply changing the rotation speed. This arbitrary reference frame transformation is referred to as the *generalized rotating transformation* [1].

As mentioned earlier, the reference frame transformation to eliminate the time-varying coefficients refers to the transformation of the three-phase *abc* variables to *dqn* variables. The reference frame transformation can be easily carried out by using the *matrix forms* or *complex vectors*. First, let us take a look at the reference frame transformation by the matrix equations.

4.3.2 REFERENCE FRAME TRANSFORMATION BY MATRIX EQUATIONS

Fig. 4.19 demonstrates the transformation of the three-phase *abc* variables into *dqn* variables in the stationary reference frame.

The reference frame transformation can be simply considered as the orthogonal projection of the three-phase *abc* variables f_a, f_b, f_c onto the $d-q$ axis

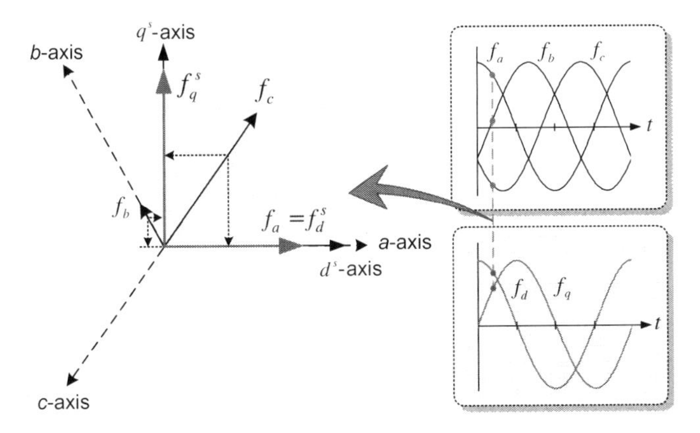

FIGURE 4.19

Transformation into the stationary reference frame.

in the stationary reference frame ($d^s - q^s$ axes). For the example in Fig. 4.19, using trigonometric relationships, the d- and q-axes variables becomes

$$f_d^s = k\left[f_a \cos(0) + f_b \cos\left(-\frac{2}{3}\pi\right) + f_c \cos\left(\frac{2}{3}\pi\right)\right] \tag{4.49}$$

$$f_q^s = k\left[f_a \sin(0) + f_b \sin\left(-\frac{2}{3}\pi\right) + f_c \sin\left(\frac{2}{3}\pi\right)\right] \tag{4.50}$$

Here, the coefficient k can be chosen arbitrarily.

The frame of reference can rotate at an angular velocity ω. So, the transformation of the three-phase stationary abc variables into dqn variables in the arbitrary reference frame rotating at a speed ω can be generally formulated as

$$\boldsymbol{f}_{dqn}^{\omega} = \boldsymbol{T}(\theta)\boldsymbol{f}_{abc} \tag{4.51}$$

where $\boldsymbol{f}_{dqn} = \left[f_d f_q f_n\right]^T$, $\boldsymbol{f}_{abc} = [f_a f_b f_c]^T$, and $[]^T$ denotes the transpose of a matrix. f_x can represent the variables of an AC motor such as voltage, current, and flux linkage.

The transformation matrix $\boldsymbol{T}(\theta)$ is defined as

$$\boldsymbol{T}(\theta) = \frac{2}{3}\begin{bmatrix} \cos\theta & \cos\left(\theta - \frac{2}{3}\pi\right) & \cos\left(\theta + \frac{2}{3}\pi\right) \\ -\sin\theta & -\sin\left(\theta - \frac{2}{3}\pi\right) & -\sin\left(\theta + \frac{2}{3}\pi\right) \\ \frac{1}{2} & \frac{1}{2} & \frac{1}{2} \end{bmatrix} \tag{4.52}$$

where the angular displacement $\theta = \int \omega(\tau)d\tau + \theta(0)$.

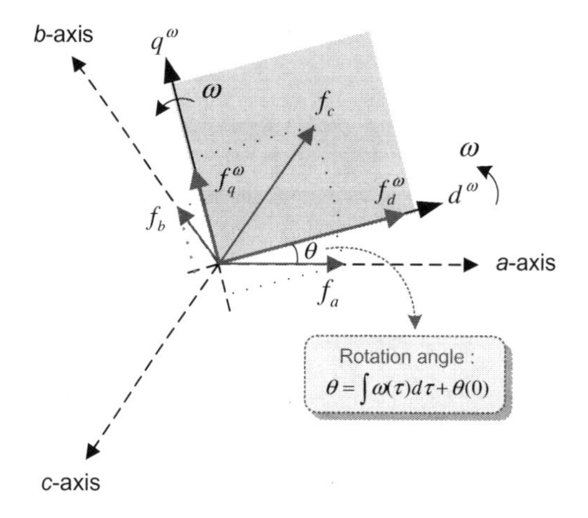

FIGURE 4.20

Transformation into the arbitrary rotating reference frame.

From Eq. (4.52), it can be seen that the f_n variable is independent of the reference frame but is related arithmetically to the *abc* variables. Fig. 4.20 demonstrates the resolution of the three-phase *abc* variables f_a, f_b, f_c into the d- and q-axes in the stationary reference frame ($d^s - q^s$ axes).

As mentioned previously, when transforming, the coefficient k can be chosen arbitrarily. In the transformation matrix $\boldsymbol{T}(\theta)$ of Eq. (4.52), the coefficient k is chosen as 2/3. In this case the magnitude of the *dq* variables is exactly identical to that of the *abc* variables, so, this transformation is called *magnitude invariance transformation*. However, the power and the torque evaluated in the *dqn* variables become 2/3 less than those evaluated in the *abc* variables, which will be shown later. On the other hand, when using the coefficient of $\sqrt{2/3}$, the power remains the same value in the two reference frames, i.e., $P_{dqn} = P_{abc}$. So, this transformation is called *power invariance transformation*. However, the magnitude of the *dq* variables is not equal to the magnitude of the *abc* variables. The coefficient of 2/3 is normally used when applying the transformation to motor variables. So, in this book, we always use the coefficient of 2/3 for magnitude invariance transformation.

4.3.2.1 *Transformation of* abc *variables into* dqn *variables in the stationary reference frame*

From setting $\theta = 0$ in Eq. (4.51), the transformation of the three-phase *abc* variables into *dqn* variables in the stationary reference frame is given by

$$\boldsymbol{f}^s_{dqn} = \boldsymbol{T}(0)\boldsymbol{f}_{abc} = \frac{2}{3}\begin{bmatrix} 1 & -\frac{1}{2} & -\frac{1}{2} \\ 0 & \frac{\sqrt{3}}{2} & -\frac{\sqrt{3}}{2} \\ \frac{1}{\sqrt{2}} & \frac{1}{\sqrt{2}} & \frac{1}{\sqrt{2}} \end{bmatrix}\begin{bmatrix} f_a \\ f_b \\ f_c \end{bmatrix} \tag{4.53}$$

This is known as *Clark's transformation*.

Eq. (4.53) can be disassembled as the following.

- Transformation of *abc* variables into *dqn* variables in the stationary reference frame

$$f_d^s = \frac{2f_a - f_b - f_c}{3} \tag{4.54}$$

$$f_q^s = \frac{1}{\sqrt{3}}(f_b - f_c) \tag{4.55}$$

$$f_n^s = \frac{2(f_a + f_b + f_c)}{3} \tag{4.56}$$

In the stationary reference frame, the *dqn* variables are arithmetically related to the *abc* variables. In particular, if the sum of the *abc* variables in a balanced three-phase system with no neutral connection is zero, i.e., $f_a + f_b + f_c = 0$, then the *n*-axis variable is zero, i.e., $f_n^s = 0$. Thus, $f_d^s = f_a$, i.e., the d^s-axis variable is always equal to the phase *a*-axis variable. In this case the *dq* variables are reduced as following.

$$f_d^s = f_a \quad (f_n^s = 0) \tag{4.57}$$

$$f_q^s = \frac{1}{\sqrt{3}}(f_b - f_c) \tag{4.58}$$

The inverse transformation of the *dqn* variables in the stationary reference frame into the *abc* variables is as following.

- Inverse transformation ($f_n^s = 0$)

$$f_a = f_d^s \tag{4.59}$$

$$f_b = -\frac{1}{2}f_d^s + \frac{\sqrt{3}}{2}f_q^s \tag{4.60}$$

$$f_c = -\frac{1}{2}f_d^s - \frac{\sqrt{3}}{2}f_q^s \tag{4.61}$$

4.3.2.2 Transformation between reference frames

Consider a transformation between reference frames. For the analysis of AC motors, it is often required to transform the variables in one reference frame to variables in another reference frame as shown in Fig. 4.21.

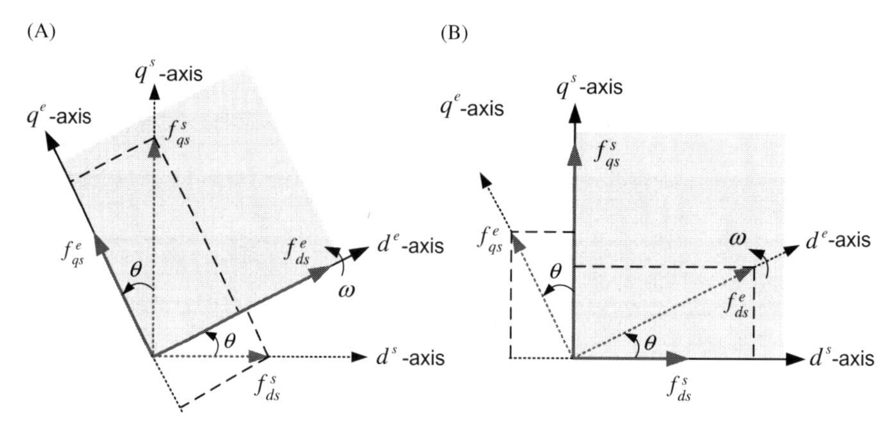

FIGURE 4.21

Transformation between reference frames. (A) Stationary into rotating frame and (B) rotating into stationary frame.

As a widely used transformation, the transformation of the stationary reference frame into the rotating reference frame can be formulated as

$$\boldsymbol{f}^{e}_{dqn} = \boldsymbol{R}(\theta)\boldsymbol{f}^{s}_{dqn} = \begin{bmatrix} \cos\theta & \sin\theta & 0 \\ -\sin\theta & \cos\theta & 0 \\ 0 & 0 & 1 \end{bmatrix} \begin{bmatrix} f^s_d \\ f^s_q \\ f^s_n \end{bmatrix} \tag{4.62}$$

The angular displacement $\theta = \int \omega(\tau)d\tau + \theta(0)$.

This is known as *Park's transformation.*

In the case where the n variable is zero, i.e., $f^s_n = 0$, Eq. (4.62) can be separated into Eqs. (4.63) and (4.64).

- Transformation of stationary reference frame into rotating reference frame

$$f^e_d = f^s_d \cos\theta + f^s_q \sin\theta \tag{4.63}$$

$$f^e_q = -f^s_d \sin\theta + f^s_q \cos\theta \tag{4.64}$$

Likewise, for the inverse transformation, it will be Eqs. (4.65) and (4.66).

- Inverse transformation of stationary reference frame into rotating reference frame

$$f^s_d = f^e_d \cos\theta - f^e_q \sin\theta \tag{4.65}$$

$$f^s_q = f^e_d \sin\theta + f^e_q \cos\theta \tag{4.66}$$

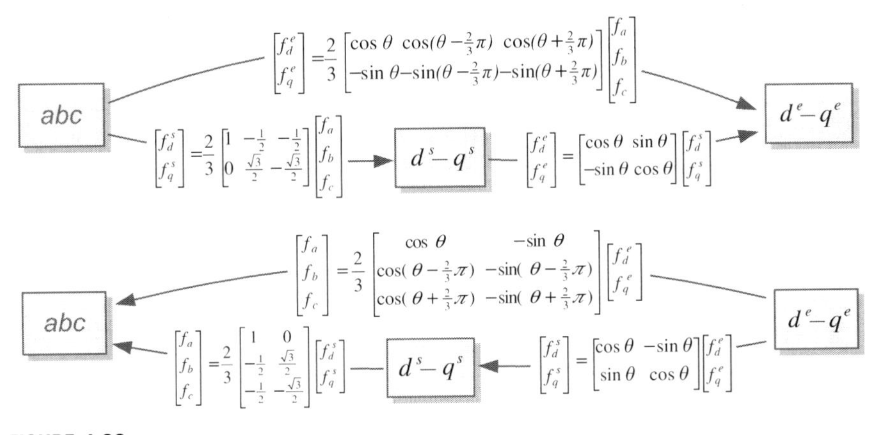

FIGURE 4.22

Reference frame transformations.

The procedure for all the transformations just mentioned is summarized in Fig. 4.22.

Next, we will find the instantaneous power expressed in *dqn* variables. The instantaneous power may be expressed in *abc* variables as

$$P = v_a i_a + v_b i_b + v_c i_c = \boldsymbol{V}_{abc}^T \boldsymbol{I}_{abc} \tag{4.67}$$

where $\boldsymbol{V}_{abc}^T = [v_a v_b v_c]^T$, $\boldsymbol{I}_{abc} = [i_a i_b i_c]$.

By transforming the above equation into the arbitrary reference frame rotating at any angular velocity ω, the instantaneous power is given by

$$\begin{aligned} P = \boldsymbol{V}_{abc}^T \boldsymbol{I}_{abc} &= \left[\boldsymbol{T}^{-1}(\theta)\boldsymbol{V}_{dqn}^{\omega}\right]^T \left[\boldsymbol{T}^{-1}(\theta)\boldsymbol{I}_{dqn}^{\omega}\right] \\ &= \frac{3}{2}(\boldsymbol{V}_{dqn}^{\omega})^T \boldsymbol{I}_{dqn}^{\omega} = \frac{3}{2}(v_d^{\omega} i_d^{\omega} + v_q^{\omega} i_q^{\omega} + v_n^{\omega} i_n^{\omega}) \end{aligned} \tag{4.68}$$

where $\boldsymbol{V}_{dqn}^{\omega} = \left[v_d^{\omega} v_q^{\omega} v_n^{\omega}\right]^T$, $\boldsymbol{I}_{dqn}^{\omega} = \left[i_d^{\omega} i_q^{\omega} i_n^{\omega}\right]^T$, $\boldsymbol{T}^{-1}(\theta) = \frac{3}{2}\boldsymbol{T}^T(\theta)$.

Although the power of Eq. (4.68) has a 3/2 factor due to the choice of the constant used in the transformation matrix $\boldsymbol{T}(\theta)$, it is expressed as the sum of the product of the voltage and the current in each axis like the power expression in the three-phase system. The waveforms of the voltage, current, and flux linkage vary according to the angular velocity of the reference frame, whereas the waveform of the power remains unchanged.

EXAMPLE 1

Transform the following balanced three-phase *abc* voltages into *dqn* voltages in the stationary and rotating reference frames.

$$v_{as} = V_m\cos\omega_e t,\quad v_{bs} = V_m\cos(\omega_e t - 120°),\quad v_{cs} = V_m\cos(\omega_e t - 240°),$$

Solution

From Eq. (4.53), *dqn* voltages in the stationary reference frame are given as

$$v_{ds}^s = \frac{2v_{as} - v_{bs} - v_{cs}}{3} = v_{as} = V_m \cos\omega_e t$$

$$v_{qs}^s = \frac{1}{\sqrt{3}}(v_{bs} - v_{cs}) = V_m \sin\omega_e t$$

$$v_{ns}^s = \frac{\sqrt{2}(v_{as} + v_{bs} + v_{cs})}{3} = 0$$

Therefore, the three-phase *abc* variables, each displaced by 120°, are transformed into two-phase *dq* variables, each displaced by 90°, in the stationary reference frame. These amplitudes are equal to each other but the variable of d-axis is ahead of the variable of q-axis by 90°.

From Eq. (4.62), *dq* voltages in the synchronously rotating reference frame ($d^e - q^e$ axes) are given as

$$\begin{aligned} v_{ds}^e &= v_{ds}^s\cos\theta_e + v_{qs}^s\sin\theta_e \\ &= V_m\cos\omega_e t\cos\theta_e + V_m\sin\omega_e t\sin\theta_e = V_m \end{aligned}$$

$$\begin{aligned} v_{qs}^e &= -v_{ds}^s\sin\theta_e + v_{qs}^s\cos\theta_e \\ &= -V_m\cos\omega_e t\sin\theta_e + V_m\sin\omega_e t\cos\theta_e = 0 \end{aligned}$$

Here, the initial position of the reference frame is set to zero, thus $\theta_e = \omega_e t$. It should be noted that the three-phase *abc* variables are expressed as steady-state direct current (DC) values in the *dq* reference frame rotating at the angular velocity ω_e same as that of *abc* variables. The amplitude of the DC value is equal to the peak value of the *abc* variables.

MATLAB/SIMULINK SIMULATION: REFERENCE FRAME TRANSFORMATIONS

- Simulation conditions: $V_m = 100$ V, frequency $= 100$ Hz,
  Phase *as* voltage $v_{as} = V_m \cos \omega_e t = 100 \cos(2\pi 100t)$
- Whole simulation block diagram

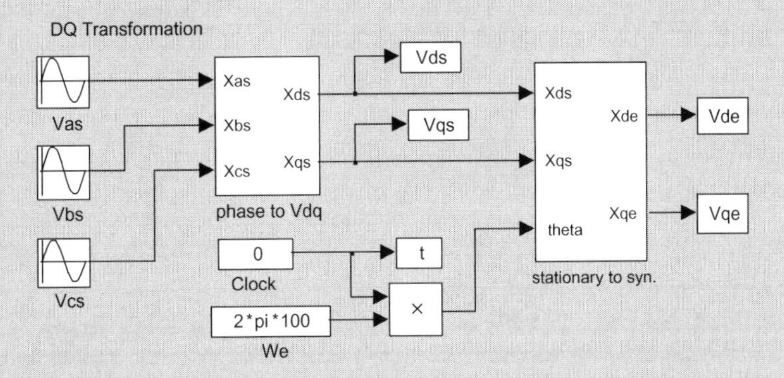

- Transformation of the three-phase *abc* variables into *dq* variables in the stationary reference frame (phase to V_{dq} block)

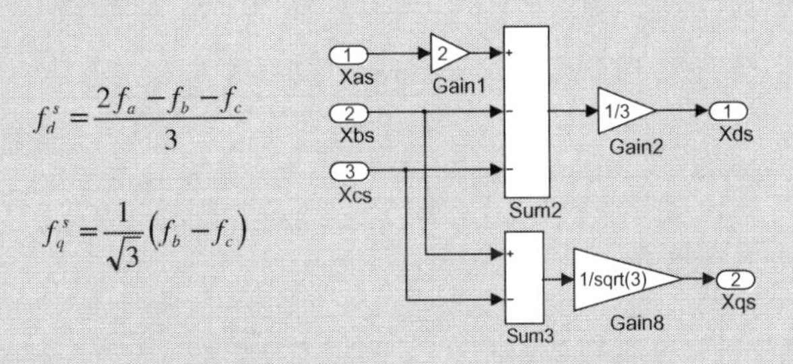

- Transformation of the stationary reference frame into the rotating reference frame (stationary to syn. block)

$$f_d^e = f_d^s \cos\theta + f_q^s \sin\theta, \quad f_q^e = -f_d^s \sin\theta + f_q^s \cos\theta$$

  - Angular displacement: $\theta = \int \omega_e(t)dt = \int (2\pi f)dt = 2\pi 100t = 2\pi 100t$

(Continued)

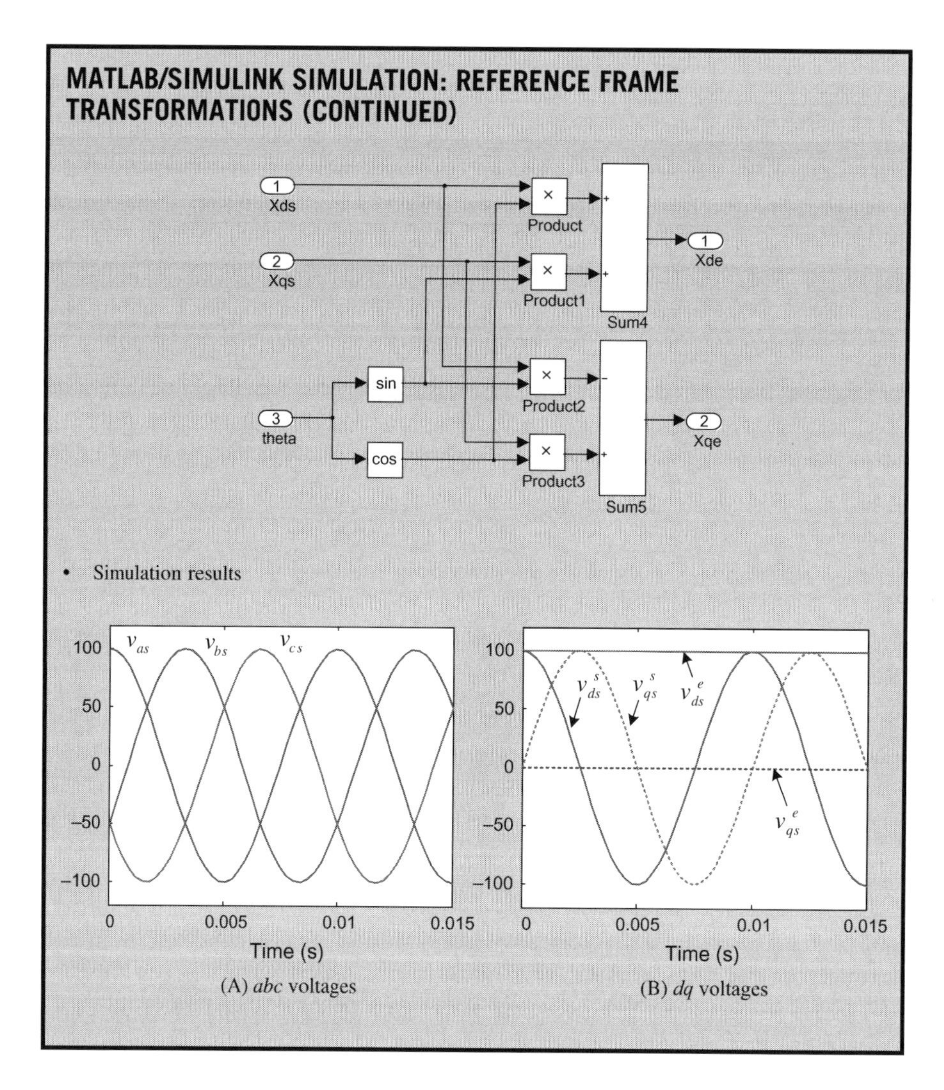

(A) *abc* voltages

(B) *dq* voltages

We have just studied the reference frame transformation by the matrix equations. We can carry out this transformation more efficiently by using a complex vector, which is a representation of the three quantities in the three-phase system as one quantity. Moreover, by treating three-phase AC motor quantities as complex vectors, the work required in the calculations for a three-phase AC motor can be simplified. Now, we will take a look at the reference frame transformation by the complex vector.

4.3.3 REFERENCE FRAME TRANSFORMATION BY COMPLEX VECTOR

Any three-phase quantity can be represented by a vector $\boldsymbol{f}_{abc}$ in the complex plane as shown in Fig. 4.23.

This vector is called the *complex space vector*, or in short, *space vector*. The complex vector representation allows us to analyze the three-phase system as a whole instead of looking at each phase individually. A space vector is different from a *phasor*, which is a representation of the amplitude and phase for a sine wave in the steady-state condition.

A complex space vector is defined as

$$\boldsymbol{f}_{abc} \equiv \frac{2}{3}(f_a + af_b + a^2 f_c) \quad \left(a = e^{j(2\pi/3)},\ a^2 = e^{j(4\pi/3)}\right) \tag{4.69}$$

This space vector rotates at the same angular velocity ω_e as that of the three-phase quantities as shown in Fig. 4.24.

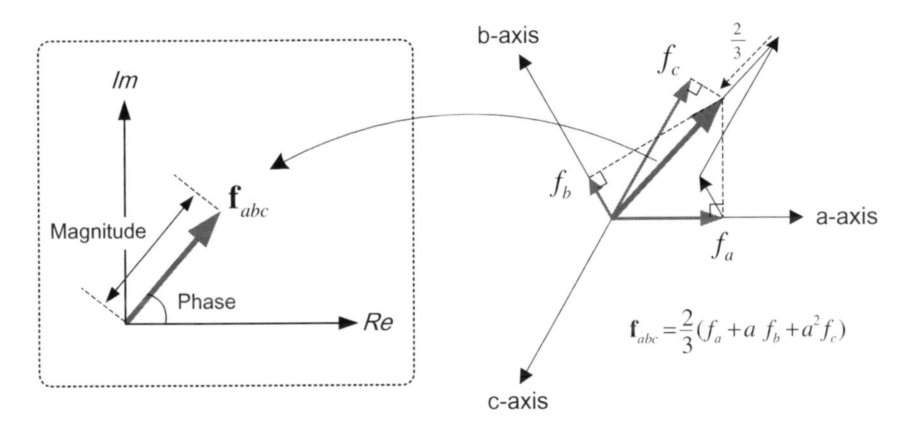

FIGURE 4.23

Complex space vector.

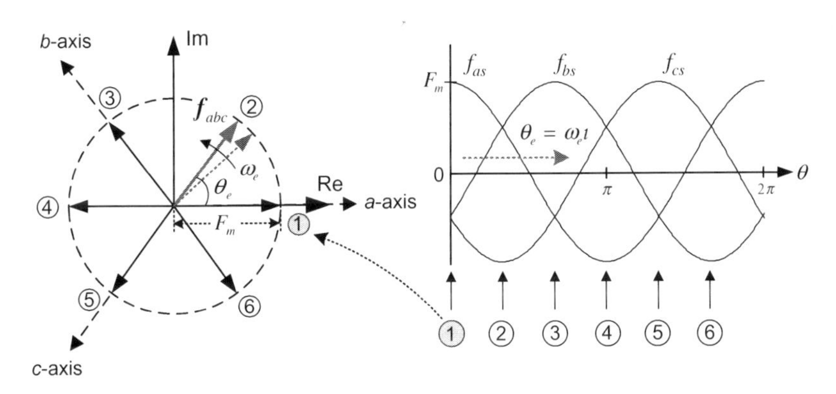

FIGURE 4.24

Three-phase quantities and complex space vector.

The amplitude of a space vector is equal to the peak value of the phase quantity. One example of a space vector is the rotating magnetic field as mentioned in Chapter 3. The rotating magnetic field is identical to the space vector of magnetic fields produced by currents flowing in three-phase windings.

Example 2

Find the space vector representation of three-phase voltages as

$$v_{as} = V_m \cos \omega_e t, \quad v_{bs} = V_m \cos(\omega_e t - 120°), \quad v_{cs} = V_m \cos(\omega_e t - 240°)$$

Solution

According to the definition, voltage space vector $\boldsymbol{V}_{abc}$ is given as

$$\boldsymbol{V}_{abc} = \frac{2}{3}(v_{as} + a v_{bs} + a^2 v_{cs})$$

From Euler formula $e^{j\theta} = \cos\theta + j\sin\theta$, we have $\cos\omega_e t = (e^{j\omega_e t} + e^{-j\omega_e t})/2$, $\sin\omega_e t = (e^{j\omega_e t} - e^{-j\omega_e t})/2j$. Thus,

$$\begin{aligned}\boldsymbol{V}_{abc} &= \frac{2V_m}{3}\left(\frac{e^{j\omega_e t} + e^{-j\omega_e t}}{2} + a\frac{e^{j(\omega_e t - 2\pi/3)} + e^{-j(\omega_e t - 2\pi/3)}}{2} + a^2\frac{e^{j(\omega_e t - 4\pi/3)} + e^{-j(\omega_e t - 4\pi/3)}}{2}\right) \\ &= \frac{V_m}{3}(3e^{j\omega_e t} + e^{-j\omega_e t}(1 + a + a^2)) = V_m e^{j\omega_e t}\end{aligned}$$

where $a = e^{j(2\pi/3)} = \cos\frac{2}{3}\pi + j\sin\frac{2}{3}\pi = -\frac{1}{2} + j\frac{\sqrt{3}}{2}$,

$$a^2 = e^{j(4\pi/3)} = \cos\frac{4}{3}\pi + j\sin\frac{4}{3}\pi = -\frac{1}{2} - j\frac{\sqrt{3}}{2}$$

$$\therefore \quad \boldsymbol{V}_{abc} = V_m e^{j\omega_e t} = V_m e^{j\theta_e} \quad \left(\theta_e = \int \omega_e(t)dt + \theta(0)\right)$$

Inversely, the quantities of a three-phase system from the space vector $\boldsymbol{f}_{abc}$ are given by

$$\begin{aligned}f_a &= \text{Re}\left[\boldsymbol{f}_{abc}\right] + f_n^{\omega} \\ &= \text{Re}\left[\frac{2}{3}(f_a + af_b + a^2 f_c)\right] + \frac{1}{3}(f_a + f_b + f_c) = f_a\end{aligned} \tag{4.70}$$

$$f_b = \text{Re}[a^2 \boldsymbol{f}_{abc}] + f_n^{\omega} \tag{4.71}$$

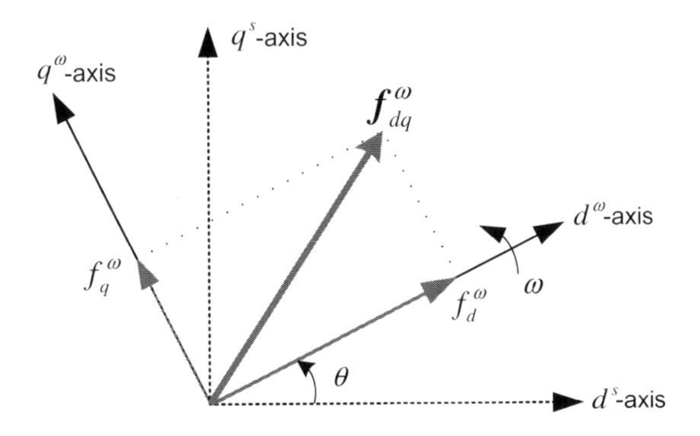

FIGURE 4.25

Vector $\boldsymbol{f}_{dq}^{\omega}$ in the arbitrary reference frame rotating at an angular velocity.

$$f_c = \mathrm{Re}[a\boldsymbol{f}_{abc}] + f_n^{\omega} \tag{4.72}$$

where "Re" represents the real part of the complex vector.

When comparing the complex plane with the d–q coordinate system mentioned in the previous section, we can see that these two systems have similar orthogonal coordinates, that is, the real-axis and the imaginary-axis correspond to the d-axis and the q-axis in the stationary reference frame, respectively. Thus, we can consider that the space vector $\boldsymbol{f}_{abc}$ equals to the vector $\boldsymbol{f}_{dq}^{s}$ in the stationary reference frame, which has the d- and q-axes components of Eqs. (4.49) and (4.50) as

$$\boldsymbol{f}_{dq}^{s} = f_d^s + jf_q^s \tag{4.73}$$

However, since the d–q axis may rotate at any angular velocity, we need to take a look at the relationship between a space vector and a vector in the d–q axes rotating at an angular velocity. A vector $\boldsymbol{f}_{dq}^{\omega}$ in the arbitrary reference frame rotating at a speed ω in Fig. 4.25 can be expressed as

$$\boldsymbol{f}_{dq}^{\omega} = f_d^{\omega} + jf_q^{\omega} \tag{4.74}$$

Here, from Eqs. (4.49) and (4.50), the d- and q-axes components of this vector are given by

$$f_d^{\omega} = \frac{2}{3}\left[f_a\cos\theta + f_b\cos\left(\theta - \frac{2}{3}\pi\right) + f_c\cos\left(\theta - \frac{4}{3}\pi\right)\right]$$

$$f_q^{\omega} = -\frac{2}{3}\left[f_a\sin\theta + f_b\sin\left(\theta - \frac{2}{3}\pi\right) + f_c\sin\left(\theta - \frac{4}{3}\pi\right)\right]$$

Using Euler formula, Eq. (4.74) becomes

$$
\begin{aligned}
\boldsymbol{f}_{dq}^{\omega} &= \frac{2}{3}\left[f_a(\cos\theta - \mathrm{j}\sin\theta) + f_b\left(\cos\left(\theta - \frac{2}{3}\pi\right) - \mathrm{j}\sin\left(\theta - \frac{2}{3}\pi\right)\right)\right. \\
&\quad \left. + f_c\left(\cos\left(\theta - \frac{4}{3}\pi\right) - \mathrm{j}\sin\left(\theta - \frac{4}{3}\pi\right)\right)\right] \\
&= \frac{2}{3}\left[f_a e^{-\mathrm{j}\theta} + f_a e^{-\mathrm{j}\left(\theta - \frac{2}{3}\pi\right)} + f_a e^{-\mathrm{j}\left(\theta - \frac{4}{3}\pi\right)} \right] = \frac{2}{3}\left[f_a + af_a + a^2 f_a\right]e^{-\mathrm{j}\theta} \\
&= \boldsymbol{f}_{abc} e^{-\mathrm{j}\theta}
\end{aligned}
\tag{4.75}
$$

From Eq. (4.75), we can see that there is only a phase difference of θ between $\boldsymbol{f}_{abc}$ and $\boldsymbol{f}_{dq}^{\omega}$. This phase difference comes from the angular displacement between the complex plane and the d–q axes rotating at an angular velocity.

By letting $\theta = 0$ in Eq. (4.75), we can identify that the space vector $\boldsymbol{f}_{abc}$ equals to the vector $\boldsymbol{f}_{dq}^{s}$ in the d–q axis of the stationary reference frame as

$$
\boldsymbol{f}_{dq}^{s} = \boldsymbol{f}_{abc} e^{-\mathrm{j}0} = \boldsymbol{f}_{abc} \tag{4.76}
$$

A vector $\boldsymbol{f}_{dq}^{\omega}$ in the rotating reference frame with the angular velocity $\omega = \omega_e$ can be expressed as

$$
\boldsymbol{f}_{dq}^{e} = \boldsymbol{f}_{abc} e^{-\mathrm{j}\theta_e} = \boldsymbol{f}_{dq}^{s} e^{-\mathrm{j}\theta_e} \quad \left(\theta_e = \int \omega_e(t)dt + \theta(0)\right) \tag{4.77}
$$

The instantaneous power is expressed in terms of the complex vector as

$$
P = \frac{3}{2}\mathrm{Re}\left[\boldsymbol{V}_{abc}\boldsymbol{I}_{abc}^{*}\right] = \frac{3}{2}\mathrm{Re}\left[\boldsymbol{V}_{dq}^{\omega}\boldsymbol{I}_{dq}^{\omega *}\right] \tag{4.78}
$$

where $\boldsymbol{I}_{abc}^{*}$, $\boldsymbol{I}_{dq}^{*}$ are the complex conjugates of $\boldsymbol{I}_{abc}$, $\boldsymbol{I}_{dq}$, respectively.

We have just studied the transformation changing the reference frame of variables. Now, we are going to find out the model of AC motors expressed in the d–q reference frame. Then, it will be found that the time-varying inductances in the voltage equations of AC motors can be eliminated by employing the reference frame transformation. First, let us find out the d–q axes model for an induction motor.

4.4 *d–q* AXES MODEL OF AN INDUCTION MOTOR

From Eqs. (4.2)–(4.7), the voltage equations for the stator and rotor windings of an induction motor and the flux linkages are rewritten as

$$
\boldsymbol{v}_{abcs} = \boldsymbol{R}_s \boldsymbol{i}_{abcs} + \frac{d\boldsymbol{\lambda}_{abcs}}{dt} \tag{4.79}
$$

$$\boldsymbol{v}_{abcr} = \boldsymbol{R}_r \boldsymbol{i}_{abcr} + \frac{d\boldsymbol{\lambda}_{abcr}}{dt} \tag{4.80}$$

$$\begin{bmatrix} \boldsymbol{\lambda}_{abcs} \\ \boldsymbol{\lambda}_{abcr} \end{bmatrix} = \begin{bmatrix} \boldsymbol{L}_s & \boldsymbol{L}_{sr} \\ (\boldsymbol{L}_{sr})^T & \boldsymbol{L}_r \end{bmatrix} \begin{bmatrix} \boldsymbol{i}_{abcs} \\ \boldsymbol{i}_{abcr} \end{bmatrix} \tag{4.81}$$

Combining Eqs. (4.79) and (4.80) with (4.81), we have the voltage equations expressed in currents and inductances as

$$\begin{bmatrix} \boldsymbol{v}_{abcs} \\ \boldsymbol{v}_{abcr} \end{bmatrix} = \begin{bmatrix} R_s + p\boldsymbol{L}_s & \boldsymbol{L}_{sr} \\ p(\boldsymbol{L}_{sr})^T & R_r + p\boldsymbol{L}_r \end{bmatrix} \begin{bmatrix} \boldsymbol{i}_{abcs} \\ \boldsymbol{i}_{abcr} \end{bmatrix} \tag{4.82}$$

where p is the time derivative operator d/dt

Now, let us transform these equations expressed as *abc* variables into the d–q coordinate system.

4.4.1 VOLTAGE EQUATIONS IN THE d–q AXES

The transformation of the stator voltage equations of Eq. (4.79) into the d–q axes rotating at an arbitrary speed ω is as follows:

$$\boldsymbol{v}_{abcs} = \boldsymbol{R}_s \boldsymbol{i}_{abcs} + \frac{d\boldsymbol{\lambda}_{abcs}}{dt}$$

$$\rightarrow \boldsymbol{T}(\theta)\boldsymbol{v}_{abcs} = \boldsymbol{T}(\theta)\boldsymbol{R}_s \boldsymbol{i}_{abcs} + \boldsymbol{T}(\theta)\frac{d\boldsymbol{\lambda}_{abcs}}{dt} \quad (\text{where } \boldsymbol{T}(\theta)\boldsymbol{R}_s\boldsymbol{T}(\theta)^{-1} = \boldsymbol{R}_s)$$

$$\rightarrow \boldsymbol{v}^{\omega}_{dqns} = \boldsymbol{R}_s \boldsymbol{i}^{\omega}_{dqns} + \boldsymbol{T}(\theta)\frac{d\boldsymbol{T}(\theta)^{-1}}{dt}\boldsymbol{\lambda}^{\omega}_{dqns} + \boldsymbol{T}(\theta)\boldsymbol{T}(\theta)^{-1} \quad \left(\text{where } \boldsymbol{T}(\theta)\frac{d\boldsymbol{T}(\theta)^{-1}}{dt} = \begin{bmatrix} 0 & -\omega & 0 \\ \omega & 0 & 0 \\ 0 & 0 & 0 \end{bmatrix}\right)$$

$$\rightarrow \boldsymbol{v}^{\omega}_{dqns} = \boldsymbol{R}_s \boldsymbol{i}^{\omega}_{dqns} + \begin{bmatrix} 0 & -\omega & 0 \\ \omega & 0 & 0 \\ 0 & 0 & 0 \end{bmatrix}\boldsymbol{\lambda}^{\omega}_{dqns} + \frac{d\boldsymbol{\lambda}^{\omega}_{dqns}}{dt}$$

This can be resolved into as

$$v^{\omega}_{ds} = R_s i^{\omega}_{ds} + \frac{d\lambda^{\omega}_{ds}}{dt} - \omega\lambda^{\omega}_{qs} \tag{4.83}$$

$$v^{\omega}_{qs} = R_s i^{\omega}_{qs} + \frac{d\lambda^{\omega}_{qs}}{dt} + \omega\lambda^{\omega}_{ds} \tag{4.84}$$

$$v^{\omega}_{ns} = R_s i^{\omega}_{ns} + \frac{d\lambda^{\omega}_{ns}}{dt} \tag{4.85}$$

Comparing the voltage equations in the *abc* coordinate system, we can see that there are *speed voltage* terms, $\omega\lambda^{\omega}_{qs}$ and $\omega\lambda^{\omega}_{ds}$, in the d–q coordinate system due to the rotation of the axes. In addition, it should be noted that the flux linkages in these speed voltages are cross-coupled.

Next, we will transform the rotor voltage equations of Eq. (4.80) into d–q axes rotating at an arbitrary speed ω. In this case we should be careful when selecting the angle used in the transformation. From the standpoint of the rotor, which rotates at the speed ω_r, the d–q axes will rotate at the difference speed

$\omega - \omega_r$. Thus the angle used in the transformation should be $\theta - \theta_r(=\beta)$. Here, $\theta\left(=\int \omega_e(t)dt\right)$ is the displacement angle of the d–q axes and $\theta_r\left(=\int \omega_r(t)dt\right)$is the displacement angle of the rotor.

With the angle β, we can transform the rotor voltage equations into the d–q axes rotating at an arbitrary speed ω as

$$\boldsymbol{v}_{abcr} = \boldsymbol{R}_r\boldsymbol{i}_{abcr} + \frac{d\boldsymbol{\lambda}_{abcr}}{dt}$$

$$\rightarrow \boldsymbol{T}(\beta)\boldsymbol{v}_{abcr} = \boldsymbol{T}(\beta)\boldsymbol{R}_r\boldsymbol{i}_{abc} + \boldsymbol{T}(\beta)\frac{d\boldsymbol{\lambda}_{abcr}}{dt}$$

$$\rightarrow \boldsymbol{v}^{\omega}_{dqnr} = \boldsymbol{T}(\beta)\boldsymbol{R}_r(\boldsymbol{T}(\beta)^{-1}\boldsymbol{i}^{\omega}_{dqnr}) + \boldsymbol{T}(\beta)\frac{d[\boldsymbol{T}(\beta)^{-1}\boldsymbol{\lambda}^{\omega}_{dqnr}]}{dt} \quad \text{where } (\boldsymbol{T}(\beta)\boldsymbol{R}_s\boldsymbol{T}(\beta)^{-1} = \boldsymbol{R}_r)$$

$$\rightarrow \boldsymbol{v}^{\omega}_{dqnr} = \boldsymbol{R}_r\boldsymbol{i}^{\omega}_{dqnr} + \boldsymbol{T}(\beta)\frac{d\boldsymbol{T}(\beta)^{-1}}{dt}\boldsymbol{\lambda}^{\omega}_{dqnr} + \boldsymbol{T}(\beta)\boldsymbol{T}(\beta)^{-1}\frac{d\boldsymbol{\lambda}^{\omega}_{dqnr}}{dt}$$

$$\rightarrow \boldsymbol{v}^{\omega}_{dqnr} = \boldsymbol{R}_r\boldsymbol{i}^{\omega}_{dqnr} + \begin{bmatrix} 0 & -(\omega-\omega_r) & 0 \\ \omega-\omega_r & 0 & 0 \\ 0 & 0 & 0 \end{bmatrix}\boldsymbol{\lambda}^{\omega}_{dqrs} + \frac{d\boldsymbol{\lambda}^{\omega}_{dqnr}}{dt}$$

This can be resolved into

$$v^{\omega}_{dr} = R_r i^{\omega}_{dr} + \frac{d\lambda^{\omega}_{dr}}{dt} - (\omega - \omega_r)\lambda^{\omega}_{qr} \tag{4.86}$$

$$v^{\omega}_{qr} = R_r i^{\omega}_{qr} + \frac{d\lambda^{\omega}_{qr}}{dt} + (\omega - \omega_r)\lambda^{\omega}_{dr} \tag{4.87}$$

$$v^{\omega}_{nr} = R_r i^{\omega}_{nr} + \frac{d\lambda^{\omega}_{nr}}{dt} \tag{4.88}$$

For squirrel-cage rotor induction motors, since the rotor bars are short-circuited by the end rings, the rotor voltage is zero, and thus, $v^{\omega}_{dr} = 0$, $v^{\omega}_{qr} = 0$, $v^{\omega}_{nr} = 0$.

To complete these stator and rotor voltage equations, we need the flux linkages expressed as dq variables.

4.4.2 FLUX LINKAGE EQUATIONS IN THE d–q AXES

The transformation of the stator flux linkage of Eq. (4.81) into the d–q axes rotating at an arbitrary speed ω is as follows:

$$\boldsymbol{\lambda}_{abcs} = \boldsymbol{L}_s\boldsymbol{i}_{abcs} + \boldsymbol{L}_{sr}\boldsymbol{i}_{abcr}$$

$$\rightarrow \boldsymbol{T}(\theta)\boldsymbol{\lambda}_{abcs} = \boldsymbol{T}(\theta)\boldsymbol{L}_s\boldsymbol{i}_{abcs} + \boldsymbol{T}(\theta)\boldsymbol{L}_{sr}\boldsymbol{i}_{abcr}$$

$$\rightarrow \boldsymbol{\lambda}^{\omega}_{dqns} = \boldsymbol{T}(\theta)\boldsymbol{L}_s(\boldsymbol{T}^{-1}(\theta)\boldsymbol{i}^{\omega}_{dqns}) + \boldsymbol{T}(\theta)\boldsymbol{L}_{sr}(\boldsymbol{T}^{-1}(\beta)\boldsymbol{i}^{\omega}_{dqnr})$$

$$\rightarrow \boldsymbol{\lambda}^{\omega}_{dqns} = \begin{bmatrix} L_{ls} + \frac{3}{2}L_{ms} & 0 & 0 \\ 0 & L_{ls} + \frac{3}{2}L_{ms} & 0 \\ 0 & 0 & L_{ls} \end{bmatrix}\boldsymbol{i}^{\omega}_{dqns} + \begin{bmatrix} \frac{3}{2}L_{ms} & 0 & 0 \\ 0 & \frac{3}{2}L_{ms} & 0 \\ 0 & 0 & 0 \end{bmatrix}\boldsymbol{i}^{\omega}_{dqnr}$$

where

$$T(\theta)L_sT^{-1}(\theta)=\begin{bmatrix} L_{ls}+\frac{3}{2}L_{ms} & 0 & 0 \\ 0 & L_{ls}+\frac{3}{2}L_{ms} & 0 \\ 0 & 0 & L_{ls} \end{bmatrix},\quad T(\theta)L_{sr}T^{-1}(\beta)=\begin{bmatrix} \frac{3}{2}L_{ms} & 0 & 0 \\ 0 & \frac{3}{2}L_{ms} & 0 \\ 0 & 0 & 0 \end{bmatrix}$$

By letting $L_m=\frac{3}{2}L_{ms}$, $L_s=L_{ls}+L_m$ in the above equations, we will have the stator flux linkages in the $d-q$ axes as

$$\lambda_{ds}^{\omega}=L_{ls}i_{ds}^{\omega}+L_m(i_{ds}^{\omega}+i_{dr}^{\omega})=L_si_{ds}^{\omega}+L_mi_{dr}^{\omega} \tag{4.89}$$

$$\lambda_{qs}^{\omega}=L_{ls}i_{qs}^{\omega}+L_m(i_{qs}^{\omega}+i_{qr}^{\omega})=L_si_{qs}^{\omega}+L_mi_{qr}^{\omega} \tag{4.90}$$

$$\lambda_{ns}^{\omega}=L_{ls}i_{ns}^{\omega} \tag{4.91}$$

Next, we transform the rotor flux linkage of Eq. (4.81) into the $d-q$ axes rotating at an arbitrary speed ω as follows.

$$\boldsymbol{\lambda}_{abcr}=\boldsymbol{L}_r\boldsymbol{i}_{abcr}+\boldsymbol{L}_{rs}\boldsymbol{i}_{abcs}$$

$$\rightarrow \boldsymbol{T}(\beta)\boldsymbol{\lambda}_{abcr}=\boldsymbol{T}(\beta)\boldsymbol{L}_r\boldsymbol{i}_{abcr}+\boldsymbol{T}(\beta)\boldsymbol{L}_{sr}^T\boldsymbol{i}_{abcs}$$

$$\rightarrow \boldsymbol{\lambda}_{dqnr}^{\omega}=\boldsymbol{T}(\beta)\boldsymbol{L}_r(\boldsymbol{T}^{-1}(\beta)\boldsymbol{i}_{dqnr}^{\omega})+\boldsymbol{T}(\beta)\boldsymbol{L}_{sr}^T(\boldsymbol{T}^{-1}(\theta)\boldsymbol{i}_{dqns}^{\omega})$$

$$\rightarrow \boldsymbol{\lambda}_{dqnr}^{\omega}=\begin{bmatrix} L_{ls}+\frac{3}{2}L_{mr} & 0 & 0 \\ 0 & L_{ls}+\frac{3}{2}L_{mr} & 0 \\ 0 & 0 & L_{ls} \end{bmatrix}\boldsymbol{i}_{dqnr}^{\omega}+\begin{bmatrix} \frac{3}{2}L_{ms} & 0 & 0 \\ 0 & \frac{3}{2}L_{ms} & 0 \\ 0 & 0 & 0 \end{bmatrix}\boldsymbol{i}_{dqns}^{\omega}$$

Assuming the turns ratio $N_s/N_r=1$ and letting $L_m=\frac{3}{2}L_{mr}=\frac{3}{2}L_{ms}$, $L_r=L_{lr}+L_m$ in the above equations, we will have rotor flux linkages in the $d-q$ axes as

$$\lambda_{dr}^{\omega}=L_{lr}i_{dr}^{\omega}+L_m(i_{dr}^{\omega}+i_{ds}^{\omega})=L_ri_{dr}^{\omega}+L_mi_{ds}^{\omega} \tag{4.92}$$

$$\lambda_{qr}^{\omega}=L_{lr}i_{qr}^{\omega}+L_m(i_{qr}^{\omega}+i_{qs}^{\omega})=L_ri_{qr}^{\omega}+L_mi_{qs}^{\omega} \tag{4.93}$$

$$\lambda_{nr}^{\omega}=L_{lr}i_{nr}^{\omega} \tag{4.94}$$

From the flux linkage expressions of the stator and rotor in the $d-q$ axes, it is found that the main purpose of the transformation has been achieved. In other words, by employing the reference frame transformation, the time-varying mutual-inductances due to the relative motion between the stator and rotor windings become constant in the $d-q$ reference frame. This result can be valid regardless of the rotating speed of the reference frame. Furthermore, the flux linkages become magnetically decoupled between the d- and q-axes. Fig. 4.26 shows the $d-q$ axes equivalent circuits.

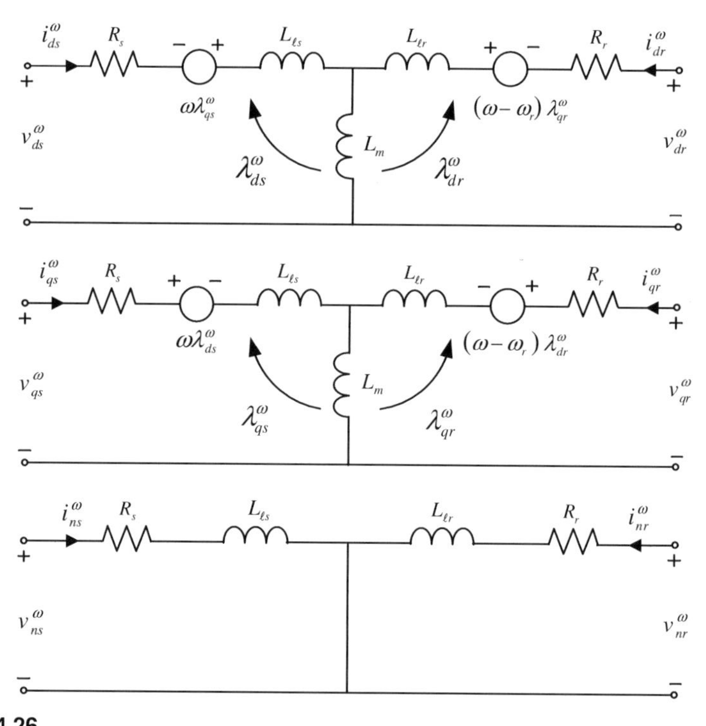

FIGURE 4.26

d–q axes equivalent circuit of an induction motor.

We can obtain voltage and flux linkage equations in the different reference frames by setting $\omega = 0$ for the stationary reference frame, $\omega = \omega_e$ for synchronously rotating reference frame, and $\omega = \omega_r$ for rotor reference frame.

- Stationary reference frame ($\omega = 0$)

$$v_{ds}^s = R_s i_{ds}^s + p\lambda_{ds}^s \qquad \lambda_{ds}^s = L_s i_{ds}^s + L_m i_{dr}^s$$

$$v_{qs}^s = R_s i_{qs}^s + p\lambda_{qs}^s \qquad \lambda_{qs}^s = L_s i_{qs}^s + L_m i_{qr}^s$$

$$0 = R_r i_{dr}^s + p\lambda_{dr}^s + \omega_r \lambda_{qr}^s \qquad \lambda_{dr}^s = L_r i_{dr}^s + L_m i_{ds}^s$$

$$0 = R_r i_{qr}^s + p\lambda_{qr}^s - \omega_r \lambda_{dr}^s \qquad \lambda_{qr}^s = L_r i_{qr}^s + L_m i_{qs}^s$$

- Synchronously rotating reference frame ($\omega = \omega_e$)

$$v_{ds}^e = R_s i_{ds}^e + p\lambda_{ds}^e - \omega_e \lambda_{qs}^e \qquad \lambda_{ds}^e = L_s i_{ds}^e + L_m i_{dr}^e$$

$$v_{qs}^e = R_s i_{qs}^e + p\lambda_{qs}^e + \omega_e \lambda_{ds}^e \qquad \lambda_{qs}^e = L_s i_{qs}^e + L_m i_{qr}^e$$

$$0 = R_r i_{dr}^e + p\lambda_{dr}^e - (\omega_e - \omega_r)\lambda_{qr}^e \qquad \lambda_{dr}^e = L_r i_{dr}^e + L_m i_{ds}^e$$

$$0 = R_r i_{qr}^e + p\lambda_{qr}^e + (\omega_e - \omega_r)\lambda_{dr}^e \qquad \lambda_{qr}^e = L_r i_{qr}^e + L_m i_{qs}^e$$

- Voltage equations of an induction motor in the d–q axes

$$v_{ds}^\omega = R_s i_{ds}^\omega + \frac{d\lambda_{ds}^\omega}{dt} - \omega\lambda_{qs}^\omega$$

$$v_{qs}^\omega = R_s i_{qs}^\omega + \frac{d\lambda_{qs}^\omega}{dt} + \omega\lambda_{ds}^\omega$$

$$v_{ns}^\omega = R_s i_{ns}^\omega + \frac{d\lambda_{ns}^\omega}{dt}$$

$$v_{dr}^\omega = R_r i_{dr}^\omega + \frac{d\lambda_{dr}^\omega}{dt} - (\omega - \omega_r)\lambda_{qr}^\omega$$

$$v_{qr}^\omega = R_r i_{qr}^\omega + \frac{d\lambda_{qr}^\omega}{dt} + (\omega - \omega_r)\lambda_{dr}^\omega$$

$$v_{nr}^\omega = R_r i_{nr}^\omega + \frac{d\lambda_{nr}^\omega}{dt}$$

- Flux linkages of an induction motor in the d–q axes

$$\lambda_{ds}^\omega = L_{ls} i_{ds}^\omega + L_m(i_{ds}^\omega + i_{dr}^\omega) = L_s i_{ds}^\omega + L_m i_{dr}^\omega$$

$$\lambda_{qs}^\omega = L_{ls} i_{qs}^\omega + L_m(i_{qs}^\omega + i_{qr}^\omega) = L_s i_{qs}^\omega + L_m i_{qr}^\omega$$

$$\lambda_{ns}^\omega = L_{ls} i_{ns}^\omega$$

$$\lambda_{dr}^\omega = L_{lr} i_{dr}^\omega + L_m(i_{dr}^\omega + i_{ds}^\omega) = L_r i_{dr}^\omega + L_m i_{ds}^\omega$$

$$\lambda_{qr}^\omega = L_{lr} i_{qr}^\omega + L_m(i_{qr}^\omega + i_{qs}^\omega) = L_r i_{qr}^\omega + L_m i_{qs}^\omega$$

$$\lambda_{nr}^\omega = L_{lr} i_{nr}^\omega$$

4.4.3 TORQUE EQUATION IN THE d–q AXES

As it can be seen in Chapter 1, the torque of double-fed machines such as induction motors is given by

$$T_e = \left(\frac{P}{2}\right)(\boldsymbol{i}_{abcs})^T \frac{\partial \boldsymbol{L}_{sr}}{\partial \theta_r} \boldsymbol{i}_{abcr} \tag{4.95}$$

The transformation of this torque into the $d-q$ axes rotating at an arbitrary speed ω is as follows:

$$
\begin{aligned}
T_e &= \left(\frac{P}{2}\right)(\boldsymbol{i}_{abcs})^T \frac{\partial \boldsymbol{L}_{sr}}{\partial \theta_r}\boldsymbol{i}_{abcr} \\
&= \left(\frac{P}{2}\right)(T(\theta)^{-1}\boldsymbol{i}^{\omega}_{dqns})^T \frac{\partial \boldsymbol{L}_{sr}}{\partial \theta_r}(T(\beta)^{-1}\boldsymbol{i}^{\omega}_{dqnr}) \\
&= \left(\frac{P}{2}\right)(\boldsymbol{i}^{\omega}_{dqns})^T \left(\frac{3}{2}T(\theta)\frac{\partial \boldsymbol{L}_{sr}}{\partial \theta_r}T(\beta)^{-1}\right)\boldsymbol{i}^{\omega}_{dqnr} \qquad (4.96) \\
&= \left(\frac{P}{2}\right)(\boldsymbol{i}^{\omega}_{dqns})^T \frac{3}{2}\begin{bmatrix} 0 & \frac{3}{2}L_{ms} & 0 \\ -\frac{3}{2}L_{ms} & 0 & 0 \\ 0 & 0 & 0 \end{bmatrix}\boldsymbol{i}^{\omega}_{dqnr}
\end{aligned}
$$

where $\boldsymbol{T}^{-1}(\theta) = \frac{3}{2}\boldsymbol{T}(\theta)^T$.

This torque can be resolved into

$$\therefore \quad T_e = \frac{3}{2}\frac{P}{2}L_m(i^{\omega}_{qs}i^{\omega}_{dr} - i^{\omega}_{ds}i^{\omega}_{qr}) \qquad (4.97)$$

where $L_m = \frac{3}{2}L_{ms}$.

The torque can be also expressed in terms of different motor variables but is independent of the speed of the reference frame as following.

$$
\begin{aligned}
T_e &= \frac{3}{2}\frac{P}{2}L_m \mathrm{Im}\left[i^*_{dqr}i_{dqs}\right] = \frac{3}{2}\frac{P}{2}L_m(i_{qs}i_{dr} - i_{ds}i_{qr}) && (4.98) \\
&= \frac{3}{2}\frac{P}{2}\frac{L_m}{L_r}\mathrm{Im}\left[\lambda^*_{dqr}i_{dqs}\right] = \frac{3}{2}\frac{P}{2}\frac{L_m}{L_r}(\lambda_{dr}i_{qs} - \lambda_{qr}i_{ds}) && (4.99) \\
&= \frac{3}{2}\frac{P}{2}\mathrm{Im}\left[\lambda^*_{dqm}i_{dqs}\right] = \frac{3}{2}\frac{P}{2}(\lambda_{dm}i_{qs} - \lambda_{qm}i_{ds}) && (4.100) \\
&= \frac{3}{2}\frac{P}{2}\mathrm{Im}\left[\lambda^*_{dqs}i_{dqs}\right] = \frac{3}{2}\frac{P}{2}(\lambda_{ds}i_{qs} - \lambda_{qs}i_{ds}) && (4.101) \\
&= \frac{3}{2}\frac{P}{2}\mathrm{Im}\left[i^*_{dqr}\lambda^*_{dqm}\right] = \frac{3}{2}\frac{P}{2}(i_{dr}\lambda_{qm} - i_{qr}\lambda_{dm}) && (4.102) \\
&= \frac{3}{2}\frac{P}{2}\mathrm{Im}\left[i^*_{dqr}\lambda^*_{dqs}\right] = \frac{3}{2}\frac{P}{2}(i_{dr}\lambda_{qr} - i_{qr}\lambda_{dr}) && (4.103) \\
&= \frac{3}{2}\frac{P}{2}\frac{L_m}{L_r}\mathrm{Im}\left[i^*_{dqr}\lambda_{dqs}\right] = \frac{3}{2}\frac{P}{2}\frac{L_m}{L_r}(i_{dr}\lambda_{qs} - i_{qr}\lambda_{ds}) && (4.104) \\
&= \frac{3}{2}\frac{P}{2}\frac{L_m}{L_\sigma L_r}\mathrm{Im}\left[\lambda^*_{dqr}\lambda_{dqs}\right] = \frac{3}{2}\frac{P}{2}\frac{L_m}{L_\sigma L_r}(\lambda_{dr}\lambda_{qs} - \lambda_{qr}\lambda_{ds}) && (4.105)
\end{aligned}
$$

Among these, the torque of Eq. (4.99) is widely used for the vector control, especially the rotor flux oriented control.

MATLAB/SIMULINK SIMULATION: AN INDUCTION MOTOR

With the model of an induction motor in the d–q axes, a 3-hp, 4-pole, 220 V, 60 Hz, three-phase induction motor will be simulated for acceleration characteristics when a full voltage is applied at a standstill.

- Overall diagram for simulation

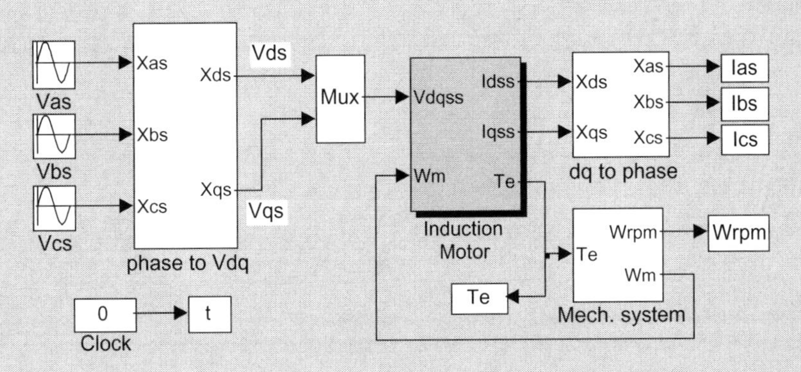

- Applied voltages

$$v_{as} = V_m \cos \omega_e t$$

$$v_{bs} = V_m \cos(\omega_e t - 120°)$$

$$v_{cs} = V_m \cos(\omega_e t - 240°)$$

$$\left(V_m = \sqrt{2}\frac{220}{\sqrt{3}}, \quad \omega_e = 2\pi 60 \right)$$

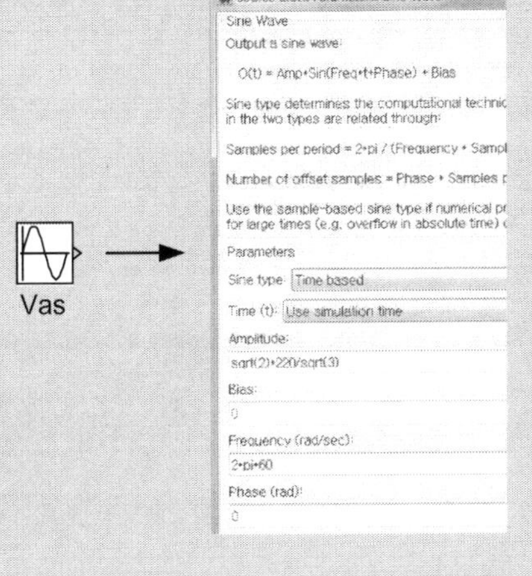

(Continued)

MATLAB/SIMULINK SIMULATION: AN INDUCTION MOTOR (CONTINUED)

- Induction motor $d-q$ equations in the stationary reference frame

$$v_{ds}^s = R_s i_{ds}^s + p\lambda_{ds}^s \rightarrow \lambda_{ds}^s = \int (v_{ds}^s - R_s i_{ds}^s)dt$$

$$v_{qs}^s = R_s i_{qs}^s + p\lambda_{qs}^s \rightarrow \lambda_{qs}^s = \int (v_{qs}^s - R_s i_{qs}^s)dt$$

$$0 = R_r i_{dr}^s + p\lambda_{dr}^s + \omega_r \lambda_{qr}^s \rightarrow \lambda_{dr}^s = \int (-R_r i_{dr}^s - \omega_r \lambda_{qr}^s)dt$$

$$0 = R_r i_{qr}^s + p\lambda_{qr}^s - \omega_r \lambda_{dr}^s \rightarrow \lambda_{qr}^s = \int (-R_r i_{qr}^s + \omega_r \lambda_{dr}^s)dt$$

$$i_{ds}^s = \frac{L_r \lambda_{ds}^s - L_m \lambda_{dr}^s}{L_s L_r - L_m^2}, \quad i_{qs}^s = \frac{L_r \lambda_{qs}^s - L_m \lambda_{qr}^s}{L_s L_r - L_m^2}$$

$$i_{dr}^s = \frac{L_s \lambda_{dr}^s - L_m \lambda_{ds}^s}{L_s L_r - L_m^2}, \quad i_{qr}^s = \frac{L_s \lambda_{qr}^s - L_m \lambda_{qs}^s}{L_s L_r - L_m^2}$$

$$T_e = \frac{3}{2}\frac{P}{2} L_m (i_{qs}^s i_{dr}^s - i_{ds}^s i_{qr}^s)$$

- Induction motor block diagram (Induction Motor Block)

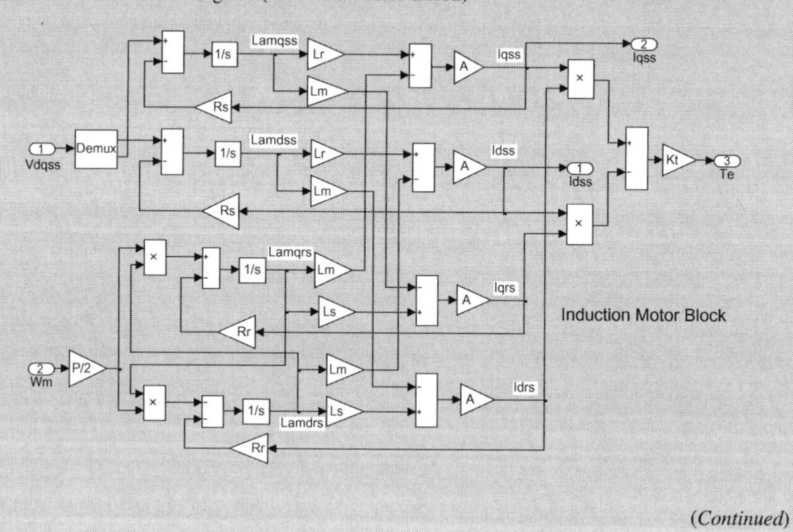

(Continued)

MATLAB/SIMULINK SIMULATION: AN INDUCTION MOTOR (CONTINUED)

- Simulation results

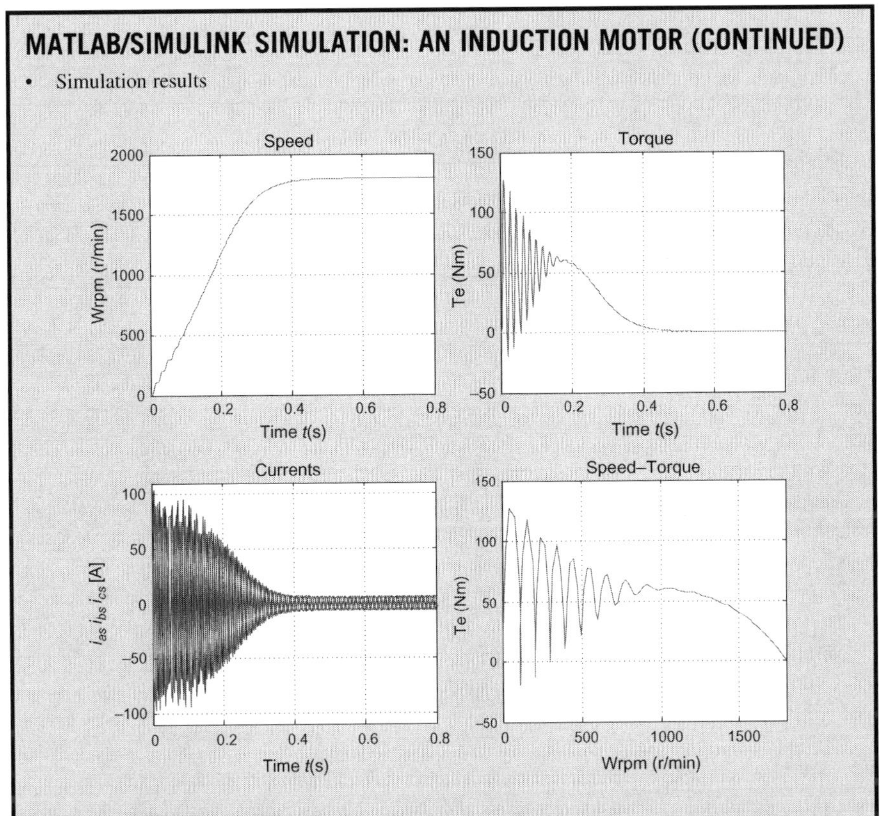

At a standstill the input impedance of an induction motor is small and the back-EMF is zero. Thus, when starting with a full voltage, the starting current becomes large, even more than several times the rated value as can be seen the simulation result.

We can see that, from the speed result, the induction motor reaches the final speed without oscillations. However, in the case of higher power induction motors, there may be large overshoots and oscillations in speed and torque. This depends on the parameters of the motor.

4.5 $d\text{–}q$ AXES MODEL OF A PERMANENT MAGNET SYNCHRONOUS MOTOR

Now, let us transform the model of an IPMSM expressed as *abc* variables into the $d\text{–}q$ coordinate system.

4.5.1 VOLTAGE EQUATIONS IN THE $d-q$ AXES

The stator windings of a PMSM are the same as those of an induction motor. Thus, they can be expressed by the same the $d-q$ axes stator voltage equations, Eqs. (4.83)–(4.85), of an induction motor, and are rewritten as

$$v_{ds}^{\omega} = R_s i_{ds}^{\omega} + \frac{d\lambda_{ds}^{\omega}}{dt} - \omega\lambda_{qs}^{\omega} \tag{4.106}$$

$$v_{qs}^{\omega} = R_s i_{qs}^{\omega} + \frac{d\lambda_{qs}^{\omega}}{dt} + \omega\lambda_{ds}^{\omega} \tag{4.107}$$

$$v_{ns}^{\omega} = R_s i_{ns}^{\omega} + \frac{d\lambda_{ns}^{\omega}}{dt} \tag{4.108}$$

However the flux linkages of a PMSM are different from those of an induction motor. Now, let us find out the flux linkages in the $d-q$ axes required to complete the above voltage equations.

4.5.2 FLUX LINKAGE EQUATIONS IN THE $d-q$ AXES

The transformation of the stator flux linkage of an IMPSM (Eq. 4.47) into the $d-q$ axes rotating at an arbitrary speed ω is as follows:

$$\begin{aligned}
&\boldsymbol{\lambda}_{abcs} = \boldsymbol{L}_s \boldsymbol{i}_{abcs} + \boldsymbol{L}_f \boldsymbol{i}_f \\
\rightarrow\ &\boldsymbol{T}(\theta)\boldsymbol{\lambda}_{abcs} = \boldsymbol{T}(\theta)\boldsymbol{L}_s \boldsymbol{i}_{abcs} + \boldsymbol{T}(\theta)\boldsymbol{L}_f \boldsymbol{i}_f \\
\rightarrow\ &\boldsymbol{\lambda}_{dqns}^{\omega} = \boldsymbol{T}(\theta)\boldsymbol{L}_s(\boldsymbol{T}^{-1}(\theta)\boldsymbol{i}_{dqns}^{\omega}) + \boldsymbol{T}(\theta)\boldsymbol{L}_f \boldsymbol{i}_f
\end{aligned}$$

Transformations of inductances needed by the above equation are as follows. From the following stator inductance

$$\boldsymbol{L}_s = \begin{bmatrix} L_{ls} + L_A & -\frac{1}{2}L_A & -\frac{1}{2}L_A \\ -\frac{1}{2}L_A & L_{ls} + L_A & -\frac{1}{2}L_A \\ -\frac{1}{2}L_A & -\frac{1}{2}L_A & L_{ls} + L_A \end{bmatrix} - L_B \begin{bmatrix} \cos 2\theta_r & \cos 2(\theta_r - \frac{\pi}{3}) & \cos 2(\theta_r + \frac{\pi}{3}) \\ \cos 2(\theta_r - \frac{\pi}{3}) & \cos 2(\theta_r - \frac{2}{3}\pi) & \cos 2\theta_r \\ \cos 2(\theta_r + \frac{\pi}{3}) & \cos 2\theta_r & \cos 2(\theta_r - \frac{4}{3}\pi) \end{bmatrix}$$

we have

$$\boldsymbol{T}(\theta)\boldsymbol{L}_s\boldsymbol{T}^{-1}(\theta) = \begin{bmatrix} L_{ls} + \frac{3}{2}L_A & 0 & 0 \\ 0 & L_{ls} + \frac{3}{2}L_A & 0 \\ 0 & 0 & L_{ls} \end{bmatrix} + \frac{3}{2}L_B \begin{bmatrix} -\cos 2(\theta - \theta_r) & \sin 2(\theta - \theta_r) & 0 \\ \sin 2(\theta - \theta_r) & \cos 2(\theta - \theta_r) & 0 \\ 0 & 0 & 0 \end{bmatrix}$$

From the inductance

$$\boldsymbol{L}_f = L_{sf}\begin{bmatrix} \cos\theta_r \\ \cos(\theta_r - \frac{2}{3}\pi) \\ \cos(\theta_r - \frac{4}{3}\pi) \end{bmatrix}$$

we have

$$\boldsymbol{T}(\theta)\boldsymbol{L}_f = L_{sf}\begin{bmatrix} \cos(\theta-\theta_r) \\ -\sin(\theta-\theta_r) \\ 0 \end{bmatrix}$$

Thus the stator flux linkage in the d–q axes rotating at an arbitrary speed ω is given as

$$\boldsymbol{\lambda}_{dqs}^{\omega} = \begin{bmatrix} L_{ls}+\frac{3}{2}(L_A - L_B\cos 2(\theta-\theta_r)) & \frac{3}{2}\sin 2(\theta-\theta_r) & 0 \\ \frac{3}{2}\sin 2(\theta-\theta_r) & L_{ls}+\frac{3}{2}(L_A + L_B\cos 2(\theta-\theta_r)) & 0 \\ 0 & 0 & L_{ls} \end{bmatrix}\boldsymbol{i}_{dqs}^{\omega} + \begin{bmatrix} \cos(\theta-\theta_r) \\ -\sin(\theta-\theta_r) \\ 0 \end{bmatrix}\phi_f \tag{4.109}$$

where $\phi_f = L_{sf} i_f$.

By letting $L_{ds} = L_{ls} + \frac{3}{2}(L_A - L_B)$, $L_{qs} = L_{ls} + \frac{3}{2}(L_A + L_B)$, Eq. (4.109) becomes

$$\boldsymbol{\lambda}_{dqs}^{\omega} = \begin{bmatrix} \frac{L_{ds}+L_{qs}}{2} + \frac{L_{ds}-L_{qs}}{2}\cos 2(\theta-\theta_r) & \frac{L_{ds}-L_{qs}}{2}\sin 2(\theta-\theta_r) & 0 \\ \frac{L_{ds}-L_{qs}}{2}\sin 2(\theta-\theta_r) & \frac{L_{ds}+L_{qs}}{2} - \frac{L_{ds}-L_{qs}}{2}\cos 2(\theta-\theta_r) & 0 \\ 0 & 0 & L_{ls} \end{bmatrix}\boldsymbol{i}_{dqs}^{\omega} + \begin{bmatrix} \cos(\theta-\theta_r) \\ -\sin(\theta-\theta_r) \\ 0 \end{bmatrix}\phi_f \tag{4.110}$$

This stator flux linkage can be expressed in the stationary reference frame ($\theta = 0$) as

$$\boldsymbol{\lambda}_{dqs}^{s} = \begin{bmatrix} \frac{L_{ds}+L_{qs}}{2} + \frac{L_{ds}-L_{qs}}{2}\cos 2\theta_r & \frac{L_{ds}-L_{qs}}{2}\sin 2\theta_r & 0 \\ \frac{L_{ds}-L_{qs}}{2}\sin 2\theta_r & \frac{L_{ds}+L_{qs}}{2} - \frac{L_{ds}-L_{qs}}{2}\cos 2\theta_r & 0 \\ 0 & 0 & L_{ls} \end{bmatrix}\boldsymbol{i}_{dqs}^{s} + \begin{bmatrix} \cos\theta_r \\ -\sin\theta_r \\ 0 \end{bmatrix}\phi_f \tag{4.111}$$

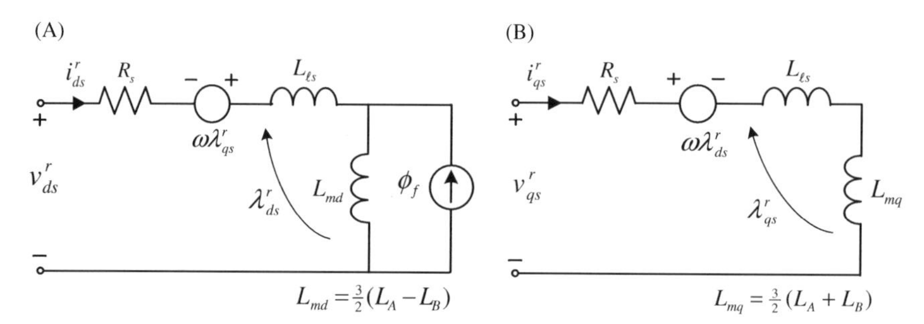

FIGURE 4.27

$d^r - q^r$ axes equivalent circuit of an IPMSM. (A) d^r-axis and (B) q^r-axis.

In this case it is found that there are still time-varying inductances in the flux linkages. Unlike the case of an induction motor, it should be noted that the time-varying inductances of an IPMSM can be eliminated by employing the transformation into only the rotor reference frame ($d^r - q^r$ axes) rotating at the rotation speed ω_r of the rotor. Thus, letting $\theta = \theta_r$ in Eq. (4.110), we have the stator flux linkages with a constant inductance in the rotor reference frame as

$$\boldsymbol{\lambda}^r_{dqn} = \begin{bmatrix} L_{ds} & 0 & 0 \\ 0 & L_{qs} & 0 \\ 0 & 0 & L_{ls} \end{bmatrix} \begin{bmatrix} i^r_{ds} \\ i^r_{qs} \\ i^r_{ns} \end{bmatrix} + \begin{bmatrix} \phi_f \\ 0 \\ 0 \end{bmatrix} \tag{4.112}$$

This can be resolved into

$$\lambda^r_{ds} = L_{ds} i^r_{ds} + \phi_f \tag{4.113}$$

$$\lambda^r_{qs} = L_{qs} i^r_{ds} \tag{4.114}$$

$$\lambda^r_{ns} = L_{ls} i^r_{ns} \tag{4.115}$$

By letting $L_{ds} = L_{qs}$ in the above equations, we can obtain the flux linkages of an SPMSM with a cylindrical rotor. Fig. 4.27 shows the $d^r - q^r$ axes equivalent circuit of an IPMSM.

4.5.3 TORQUE EQUATION IN THE d–q AXES

We can obtain the output torque of an IPMSM from its output power. As we can see from Eq. (4.68), the input power may be expressed in the d–q axes as

$$P_{in} = \frac{3}{2}(v^r_{ds} i^r_{ds} + v^r_{qs} i^r_{qs}) \tag{4.116}$$

By substituting the voltages and flux linkages into Eq. (4.116), we have

$$P_{in} = \frac{3}{2}\left(\left(R_s i^r_{ds} + \frac{d\lambda^r_{ds}}{dt} - \omega_r \lambda^r_{qs}\right) i^r_{ds} - \left(R_s i^r_{qs} + \frac{d\lambda^r_{qs}}{dt} + \omega_r \lambda^r_{ds}\right) i^r_{qs}\right)$$
$$= \frac{3}{2}\left(R_s(i^{r2}_{ds} + i^{r2}_{qs}) + i^r_{ds}\frac{d\lambda^r_{ds}}{dt} + i^r_{qs}\frac{d\lambda^r_{qs}}{dt} + \omega_r \phi_f i^r_{qs} + \omega_r(L_{ds} - L_{qs}) i^r_{ds} i^r_{qs}\right) \tag{4.117}$$

Stator copper losses — Variation of magnetic energy — Mechanical output power P_{out}

The first term on the right-hand side of Eq. (4.117) represents the stator copper losses. The second two terms represent the variation in the magnetic energy. Thus the last two terms represent the mechanical output power that develops the output toque. The output torque is obtained by dividing the output power by the rotor speed ω_r as

$$T_e = \frac{P}{2}\frac{3}{2}\left[\phi_f i^r_{qs} + (L_{ds} - L_{qs}) i^r_{ds} i^r_{qs}\right] \tag{4.118}$$

Magnet torque — Reluctance torque

This equation implies that the developed torque of an IPMSM consists of two components. The first term represents the magnet torque produced by the permanent magnet, and the second term represents the reluctance torque due to a saliency.

For the torque of an SPMSM with a cylindrical rotor, by letting $L_{ds} = L_{qs} = L_s$ in Eq. (4.118), we can obtain

$$T_e = \frac{P}{2}\frac{3}{2}\phi_f i^r_{qs} \tag{4.119}$$

- Voltage and flux linkage equations of an IPMSM in the $d^r - q^r$ axes

$$v^r_{ds} = R_s i^r_{ds} + \frac{d\lambda^r_{ds}}{dt} - \omega_r \lambda^r_{qs}$$

$$v^r_{qs} = R_s i^r_{qs} + \frac{d\lambda^r_{qs}}{dt} + \omega_r \lambda^r_{ds}$$

$$\lambda^r_{ds} = L_{ds} i^r_{ds} + \phi_f$$

$$\lambda^r_{qs} = L_{qs} i^r_{ds}$$

$$T_e = \frac{P}{2}\frac{3}{2}\left[\phi_f i^r_{qs} + (L_{ds} - L_{qs}) i^r_{ds} i^r_{qs}\right]$$

REFERENCES

[1] P.C. Krause, et al., Analysis of Electric Machinery and Drive System, second ed., Wiley-Interscience, New York, NY, 2002.

[2] S. Morimoto, Trend of permanent magnet synchronous machines, IEEJ Trans. Electr. Electron. Eng. 2 (2007) 101–108.

[3] A.M. EL-Refaie, Fractional-slot concentrated-windings synchronous permanent magnet machines: opportunities and challenges, IEEE Trans. Ind. Electron. 57 (1) (2010) 107–121.

[4] F. Magnussen, C. Sadrangani, Winding factors and joule losses of permanent magnet machines with concentrated windings, in: Conf. Rec. IEEE IEMDC'03, 2003, pp. 333–339.

[5] T.J. Woolmer, M.D. McCulloch, Analysis of the yokeless and segmented armature machine, in: Conf. Rec. IEEE IEMDC'07, 2007, pp. 704–708.

CHAPTER

5 Vector control of alternating current motors

The output torque of a motor is the primary control object in motor drive systems. This is because the position or speed of loads can be controlled by controlling the torque of the driving motor. There are two methods of the torque control of alternating current (AC) motors: *average torque control* and *instantaneous torque control*.

The average torque control is a cost-effective variable speed control solution for general-purpose applications, such as fans, blowers, and pump drives, which do not require an accurate speed or torque control. In these applications the main objective for control is the average speed of the motor/load, which is usually regulated by controlling the average torque of the motor. A typical example of the average torque control technique (also called the *scalar control method*) is the constant V/f control in the induction motor drives, which was stated in Chapter 3. However, this method can control the motor torque only in the steady-state condition and thus cannot be used to control the dynamic behavior of the motor.

On the other hand, the instantaneous torque control is needed for high-performance applications such as robots, elevators, CNC machine tools, and automation line drives. In these applications a precise speed/torque control and a fast dynamic response are required. To achieve these, it is necessary to control the instantaneous torque of the motor. For direct current (DC) motors, the instantaneous torque control can be readily achieved through the control of the armature current as discussed in Chapter 2. However, for AC motors, a complicated technique called the *vector control method* is required. The *field-oriented control technique* is a typical vector control method. The direct torque control is also another method for the instantaneous torque. This technique is less complicated than the vector control method, but is not as popular.

A scalar control method based on the control of the average current can control the average torque of a motor in the steady-state condition. On the other hand, a vector control method based on the control of both the magnitude and the direction (phase) of the motor current can control its dynamics at the transient condition as well as the torque of a motor in the steady-state condition.

Electric Motor Control. DOI: http://dx.doi.org/10.1016/B978-0-12-812138-2.00005-2

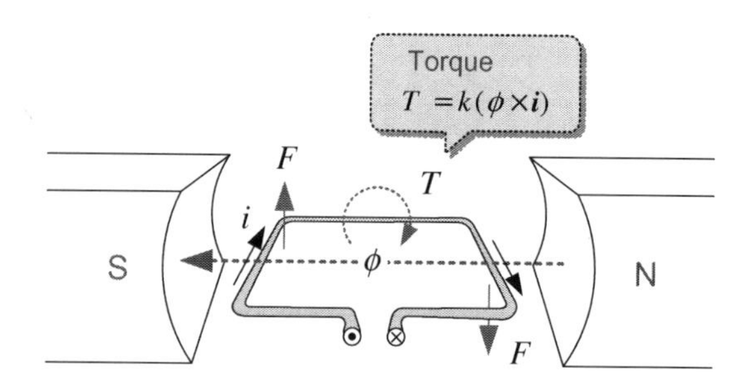

FIGURE 5.1

Torque production in the motor.

In this chapter we will explain the vector control method of AC motors in detail. Prior to this, we need to understand the instantaneous torque. As shown in Fig. 5.1, the instantaneous torque T of a motor is expressed as the cross-product of the magnetic flux vector $\boldsymbol{\Phi}$ and current vector $\boldsymbol{i}$, and is denoted by

$$T = k(\boldsymbol{\Phi} \times \boldsymbol{i}) \tag{5.1}$$

$$T = k|\phi||i|\sin\theta \tag{5.2}$$

From this, it can be seen that the instantaneous torque of a motor depends on not only the magnitude of both the magnetic flux and current vectors but also the space angle θ between two vectors. Hence, to control the instantaneous torque, we need to control the directions of the flux and the current as well as their magnitudes. In general, the direction of the current vector is controlled based on the flux vector $\boldsymbol{\Phi}$. Such method for the instantaneous torque control of AC motors is called *vector control* or *field-oriented control*.

To understand the vector control of AC motors, first, we need to take a close look at the principle of torque production in a separately excited DC motor. This is because the vector control of AC motors imitates the torque control of a separately excited DC motor. A separately excited DC motor has a desirable mechanical structure, which makes it possible to control its instantaneous torque by controlling only its armature current without using any special control technique. Thus a separately excited DC motor can be considered as a role model for the instantaneous torque control of AC motors.

5.1 CONDITIONS FOR INSTANTANEOUS TORQUE CONTROL OF MOTORS

As we discussed in Chapter 2, in DC motors, the field flux ϕ_f is produced by a field winding or a permanent magnet in the stator. Under this field flux, a torque on the conductors of the armature winding carrying the current i_a is produced. A separately excited DC motor has a special mechanical structure, as shown in

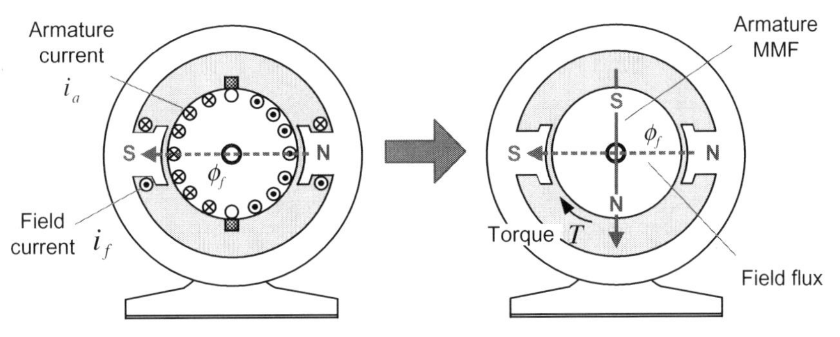

FIGURE 5.2

Separately excited DC motor.

Fig. 5.2, where the direction of the armature current i_a (i.e., magnetomotive force (mmf)) is always 90 electrical degrees with regard to the field flux ϕ_f regardless of rotor rotation.

Like this, since the space angle between the armature current and the field flux always remains at 90 electrical degrees without using any particular control technique, the developed torque can be maximized under a given flux and current. Moreover, as a result of this orthogonality, there is no mutual coupling between the field flux and the armature current, though there is an effect of armature reaction.

In addition, this orthogonality makes the instantaneous torque of Eq. (5.2) for a separately excited DC motor to be expressed simply as

$$T = k|\phi_f||i_a| \tag{5.3}$$

In this condition, if we can control the magnitudes of the field flux and the current independently, the instantaneous torque control will become simpler. When using a permanent magnet or making the field flux constant with a separate DC voltage source, the torque of Eq. (5.3) becomes

$$T = k'|i_a|, \quad k' = k|\phi_f| \tag{5.4}$$

This implies that it is possible to control the instantaneous torque of a DC motor by controlling only the magnitude of the armature current $|i_a|$. If the armature current can be controlled rapidly, the instantaneous torque can also be controlled rapidly to that extent. By using a power electronic converter such as H-bridge circuit with a current controller as stated in Chapter 2, the armature current can be controlled fast enough with a control bandwidth up to several hundred Hertz. Alternatively, we can consider regulating the field flux to control the instantaneous torque under a constant armature current. This, however, may result in a sluggish response and cause a saturation problem. Thus it is desirable to not use the field flux control except in high-speed operations.

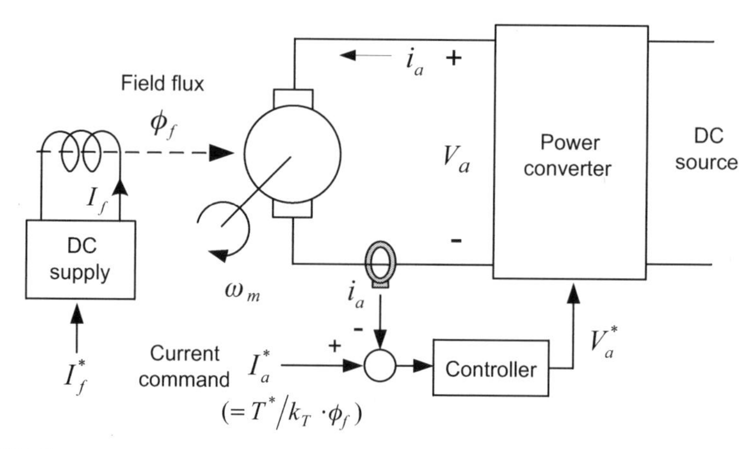

FIGURE 5.3

Torque control system of a DC motor.

On the basis of the analysis stated earlier, we can find out the following requirements for the instantaneous torque control of a DC motor.

The requirements for the instantaneous torque of a DC motor:

1. The space angle between the field flux and the armature current (i.e., mmf) always has to be 90 electrical degrees.
2. Both the field flux and the armature current should be controlled independently.
3. The armature current can be controlled instantaneously.

When the above three requirements are satisfied instantaneously, the instantaneous torque control for a DC motor can be achieved. Separately excited DC motors inherently satisfy the requirements 1 and 2 due to their special mechanical structure. Thus only the requirement 3 remains to be met.

A typical system for the torque control of a DC motor is shown in Fig. 5.3. In such a system, the field flux ϕ_f is usually controlled constantly and the instantaneous torque is controlled by the armature current i_a. The armature current command I_a^* is given from the torque command T_e for driving the load. If the armature current i_a is instantaneously controlled to follow the current command by using a power electronic converter such as H-bridge circuit, the instantaneous torque of the DC motor can follow the command torque.

5.2 VECTOR CONTROL OF AN INDUCTION MOTOR

Even though the output torque of induction motors is not easy to control, it can be controlled instantaneously like DC motors. In induction motors, unlike DC

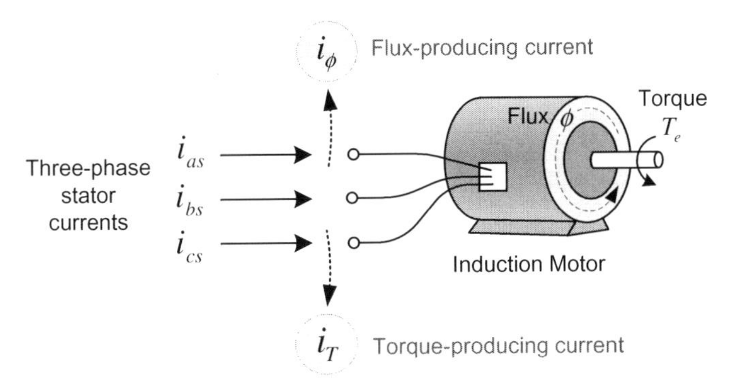

FIGURE 5.4

Currents of an induction motor.

motors, the field flux–producing current and the torque-producing current (equivalent to armature current) are provided all together through the three-phase stator windings as shown in Fig. 5.4.

Therefore it appears that the instantaneous torque control of an induction motor is not as simple as that of a DC motor. The three requirements for controlling the instantaneous torque of a DC motor as mentioned earlier can also be applied to an induction motor as follows:

1. The space angle between the flux-producing current (or the field flux) and the torque-producing current has to be always be 90 degrees.
2. Both the flux-producing current (or the field flux) and the torque-producing current should be controlled independently.
3. The torque-producing current can be controlled instantaneously.

Now we will take a close look at how we can meet these three requirements for the instantaneous torque control of an induction motor.

5.2.1 INSTANTANEOUS TORQUE CONTROL OF AN INDUCTION MOTOR

In an induction motor the currents for producing both the field flux and the torque are provided through the three-phase stator windings. Thus before applying the above three requirements, we first need to know how much the field flux- and torque-producing components are included in the three-phase stator currents i_{as}, i_{bs}, i_{cs}.

This knowledge can be obtained by resolving the three-phase currents into two orthogonal currents, which are utilized as flux-producing current and torque-producing current. This resolution can be easily achieved by using the reference frame transformation as mentioned in Chapter 4. When we transform the

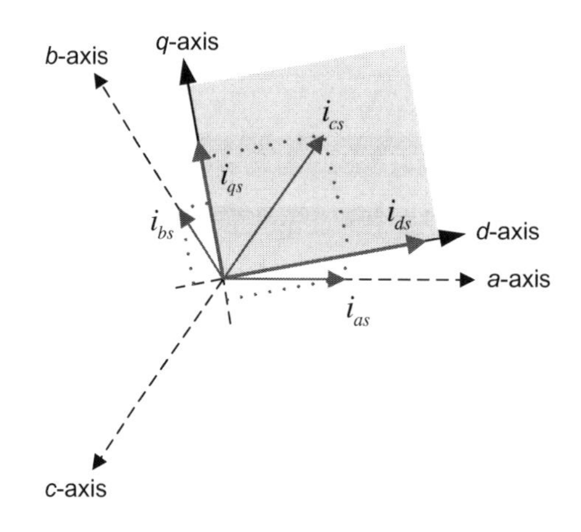

FIGURE 5.5

Reference frame transformation for currents.

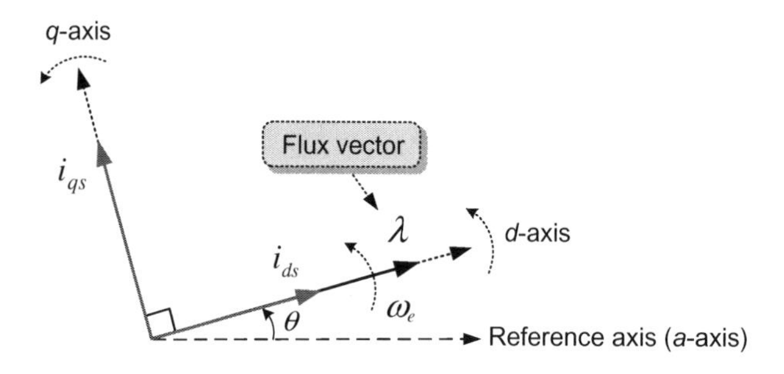

FIGURE 5.6

Assignment of d–q axes based on the flux vector λ.

three-phase *abc* stator currents of an induction motor into the d–q axes currents by using the transformation as shown in Fig. 5.5, we need to designate which of the two currents is the flux- or the torque-producing current. Commonly, the d-axis stator current is assigned as the flux-producing current, whereas the q-axis stator current is assigned as the torque-producing current. By obtaining these orthogonal currents, the requirement 1 can be fulfilled.

In this case it should be noted that in order for the assigned d-axis stator current to become the true flux-producing current, the d-axis should be located at the actual position of the flux vector λ in the induction motor as shown in Fig. 5.6.

Here, the flux vector λ in the induction motor is the rotating magnetic field generated by the three-phase currents as mentioned in Chapter 3. Unlike the stationary field flux of a DC motor, the field flux of an induction motor is rotating at the synchronous speed ω_e corresponding to the supply frequency. Thus we have to use the transformation into the synchronous reference frame rotating at the synchronous speed ω_e. The d-axis will then be located at the actual position of the flux vector. Through this process, the d-axis stator current i^e_{ds} in the synchronous reference frame becomes the true flux-producing current, and then can be used for controlling the magnitude of the flux. This d-axis stator current i^e_{ds} corresponds to the field current of a DC motor.

Moreover, in this case, since the field flux exists only on the d-axis, the torque equation of Eq. (4.99) for an induction motor becomes similar to that of a DC motor as

$$\begin{aligned} T_e &= \frac{3}{2}\frac{P}{2}\frac{L_m}{L_r}(\lambda^e_{dr} i^e_{qs} - \lambda^e_{qr} i^e_{ds}) \\ &= k|\lambda| i^e_{qs} \quad \left(\lambda^e_{dr} = |\lambda|,\ \lambda^e_{qr} = 0,\ k = \frac{3}{2}\frac{P}{2}\frac{L_m}{L_r}\right) \end{aligned} \tag{5.5}$$

The flux used in Eq. (5.5) is assumed to be the flux linkage of the rotor.

In this condition, if the magnitude of the flux vector is kept constant by controlling i^e_{ds}, Eq. (5.5) becomes as

$$T_e = k' i^e_{qs} \quad (k' = k|\lambda|) \tag{5.6}$$

This torque equation is similar to that of a DC motor. We can see easily that the q-axis current acts like the armature current of a DC motor. The torque can be directly controlled by the q-axis current i^e_{qs}, so the q-axis current is considered as the torque-producing current.

There is no cross-coupling between the d-axis stator current (thus, the flux vector) and the q-axis stator current because they are orthogonal to each other. In other words, i^e_{qs} is not influenced by i^e_{ds} and vice versa. Thus the flux-producing current and the torque-producing current can be controlled independently. In this case the magnitude of the flux vector can be controlled independently by the d-axis stator current i^e_{ds}, while the torque can be controlled independently by the q-axis stator current i^e_{qs}. This indicates that the requirement 2 can be fulfilled.

Finally, the torque-producing current can be controlled instantaneously by means of the current regulator as will be described in Chapter 6, and thus the requirement 3 can also be fulfilled. Likewise, through these three processes, the torque of an induction motor can be controlled instantaneously as in the separately excited DC motors.

In summary, to control the instantaneous torque of an induction motor, first, the three-phase stator currents are separated into d- and q-axes stator currents by using the synchronous reference frame transformation in which the d-axis is assigned to the position of the flux vector. Then, the flux can be controlled by

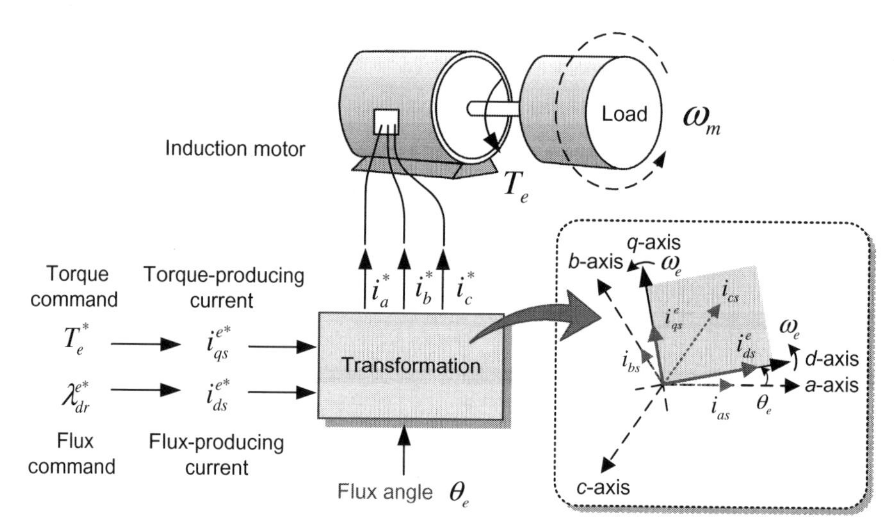

FIGURE 5.7

Instantaneous torque control method of an induction motor.

the d-axis stator current i_{ds}^e while the instantaneous torque can be controlled by the q-axis stator current i_{qs}^e independently. Finally, the instantaneous torque control can be achieved by controlling these currents instantaneously. This instantaneous torque control method for an induction motor is called the *vector control* or *field-oriented control*.

The actual implementation process of the vector control is demonstrated in Fig. 5.7.

In the vector control of an induction motor, the magnitude of the flux remains normally at a constant level of the rated value except for high-speed operations above the rated speed. Based on the rated flux command λ_{ds}^{e*}, the required flux-producing current command i_{ds}^{e*} is calculated. On the other hand, the torque-producing current command i_{qs}^{e*} is calculated based on the output torque command T_e^* required for driving a given load. These command currents i_{ds}^{e*} and i_{qs}^{e*} are given in the synchronous rotating reference frame. Thus we need to obtain the three-phase stator current commands through the inverse transformation into these $d-q$ current commands. Here, it should be noted that this transformation requires the knowledge of the angular position of the flux vector, so called the *flux angle*. Finally, by making the three-phase stator currents of the induction motor to follow these current commands fast enough, the output torque of the induction motor can be controlled to the required torque command T_e^* instantaneously.

5.2.1.1 *Important information for vector control: flux angle θ*

It was mentioned that for the vector control, we should make the d-axis to be located at the actual position of the flux vector λ. In an induction motor, this flux

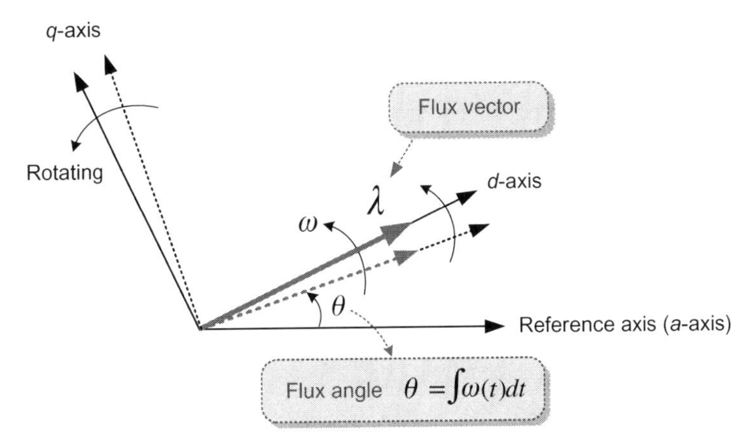

FIGURE 5.8

Rotating flux vector in the d–q axes.

vector λ is the rotating magnetic field generated by the three-phase stator currents. Thus the position of the flux vector λ varies over time as shown in Fig. 5.8.

Consider that the flux vector λ is rotating at a speed ω. In order for the d-axis to be always located at the position of the rotating flux vector λ, the d-axis should rotate along with the flux vector, i.e., the d-axis has to be moved by the flux angle $\theta\left(= \int \omega(t)dt\right)$, which is the amount of movement of the flux vector. For this purpose, it is necessary to use the synchronous d–q reference frame, which rotates at the same speed as that of the flux vector λ, i.e., the synchronous speed. Therefore, to accomplish the vector control, the flux angle θ is an inevitable information.

The vector control of an induction motor can be classified into *direct vector control* and *indirect vector control* depending on how the information about the flux angle is obtained. In the direct vector control the flux angle is obtained directly from the flux itself, which can be either measured or estimated. On the other hand, in the indirect vector control, the flux is not used to obtain the flux angle. Instead, it uses the slip angular velocity, which is needed to separate the stator currents into the flux- and the torque-producing currents at the desired values. From this slip angular velocity and the rotor angular velocity, the flux angle can be obtained.

An induction motor has three fluxes available for reference: *stator flux*, *rotor flux*, and *air-gap flux*. The vector control of an induction motor can be classified according to the reference flux as

- Rotor flux–oriented (RFO) control
- Stator flux–oriented (SFO) control
- Air-gap flux–oriented (AFO) control

Among these, the RFO control is the most popular due to its simplicity. For the AFO control, there is a coupling between the slip angular velocity and the air-gap flux, and for the SFO control, there is a coupling between the q-axis current and the stator flux. These couplings can make the associated vector control complex. Thus in this book, we will focus only on the RFO control.

5.2.2 DIRECT VECTOR CONTROL BASED ON THE ROTOR FLUX

In the vector control based on the rotor flux, the rotor flux linkage is used as the reference flux. The rotor flux linkage generated by the three-phase rotor currents can be expressed as a complex vector $\boldsymbol{\lambda}_r$ as

$$\begin{aligned}\boldsymbol{\lambda}_r &= \frac{2}{3}(\lambda_{ar} + a\lambda_{br} + a^2\lambda_{cr}) \\ &= |\lambda_r|e^{j\omega_e t} = |\lambda_r|e^{j\theta_e} \quad \left(\theta_e = \int \omega_e(t)dt + \theta(0)\right)\end{aligned} \tag{5.7}$$

The rotor flux linkage vector $\boldsymbol{\lambda}_r$ rotates at the speed ω_e corresponding to the operating frequency of the stator currents (i.e., synchronous speed), and thus the position of this flux linkage vector varies over time.

For the vector control based on the rotor flux, the position of the rotor flux linkage vector should be assigned to the d-axis. Thus we need to use the synchronous d–q reference frame rotating at the speed equal to that of the rotor flux linkage vector. By doing so, the d-axis is always locked at the position of the rotor flux linkage vector. As a result, the rotor flux linkage in the synchronous d–q reference frame has entirely the d-axis component and no q-axis component, i.e., $\boldsymbol{\lambda}_r = \lambda_{dr}^e$, and $\lambda_{qr}^e = 0$.

In such a direct vector control, the position of the rotor flux linkage vector can be identified as follows.

The rotor flux linkage vector in the stationary reference frame as shown in Fig. 5.9 has both the d–q axes components, λ_{dr}^s and λ_{qr}^s, whose magnitudes are changed according to the position of the rotor flux linkage vector.

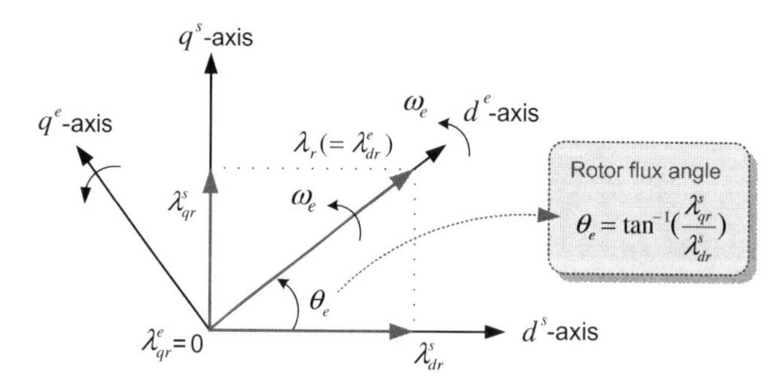

FIGURE 5.9

Rotor flux angle.

If these stationary $d-q$ rotor flux linkages are known, then we can obtain the rotor flux angle from the following equation.

$$\theta_e = \tan^{-1}\left(\frac{\lambda_{qr}^s}{\lambda_{dr}^s}\right) \tag{5.8}$$

Like this, the direct vector control uses the rotor flux angle calculated by using Eq. (5.8) from the stationary $d-q$ rotor flux linkages, which are obtained by detection or estimation.

Next, we are going to discuss the meaning of the $d-q$ axes stator currents in the synchronous reference frame oriented to the rotor flux linkages vector $\boldsymbol{\lambda}_r$.

5.2.2.1 Relation between the d-axis stator current and the rotor flux linkage

In the rotor flux–oriented vector control, there is an intimate relation between the d-axis stator current i_{ds}^e and the rotor flux linkage. The d-axis rotor voltage equation in the synchronous reference frame is

$$0 = R_r i_{dr}^e + p\lambda_{dr}^e - (\omega_e - \omega_r)\lambda_{qr}^e \tag{5.9}$$

Here, p denotes a differential operator.

The necessary and sufficient condition for the rotor flux–oriented vector control is $\lambda_{dr}^e = \boldsymbol{\lambda}_r$ and, thus $\lambda_{qr}^e = 0$.

Applying $\lambda_{qr}^e = 0$ to Eq. (5.9), we can obtain the d-axis rotor current as

$$i_{dr}^e = -\frac{p\lambda_{dr}^e}{R_r} \tag{5.10}$$

If this rotor current is substituted into the following d-axis rotor flux linkage equation,

$$\lambda_{dr}^e = L_r i_{dr}^e + L_m i_{ds}^e \tag{5.11}$$

then the following relation between the d-axis stator currents i_{ds}^e and the rotor flux linkage λ_{dr}^e can be found.

$$\lambda_{dr}^e = \frac{R_r L_m}{R_r + L_r p} i_{ds}^e = \frac{L_m}{1 + T_r p} i_{ds}^e \quad \left(T_r = \frac{L_r}{R_r}\right) \tag{5.12}$$

From this, it can be seen that the rotor flux linkage $\boldsymbol{\lambda}_r (= \lambda_{dr}^e)$ is proportional to the d-axis stator current by the first-order lag with the rotor time constant T_r. Thus the rotor flux linkage to a sudden change in the d-axis current will change

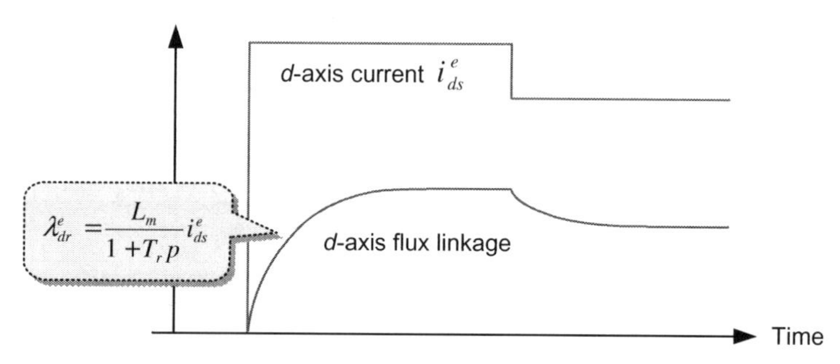

FIGURE 5.10

Relation between the *d*-axis stator current and the rotor flux.

exponentially with the time constant of T_r as shown in Fig. 5.10. This phenomenon occurs, as can be seen from Eq. (5.10), because the variation in the rotor flux linkage by the changing in the *d*-axis stator current induces a rotor current i_{dr}^e decaying with a rotor time constant T_r.

Although there is a time delay, the magnitude of the rotor flux linkage can be regulated by the *d*-axis stator current. Thus the *d*-axis stator current i_{ds}^e can be called the *flux-producing current*.

With the *d*-axis current i_{ds}^e held constant, the magnitude of the rotor flux linkage λ_{dr}^e is directly related to i_{ds}^e as

$$|\lambda_r| = \lambda_{dr}^e = L_m\, i_{ds}^e \tag{5.13}$$

Normally, the magnitude of the rotor flux linkage is kept constant for operations below the rated speed, so the *d*-axis current command is given as a constant value of $i_{ds}^{e*}(=\lambda_{dr}^{e*}/L_m)$.

Next, we will examine the relationship between the *q*-axis stator current and the output torque.

5.2.2.2 Relationship between the q-axis stator current and the output torque

By applying $\lambda_{qr}^e = 0$ to Eq. (4.99), the torque equation becomes

$$T_e = \frac{3}{2}\frac{P}{2}\frac{L_m}{L_r}(\lambda_{dr}^e i_{qs}^e - \lambda_{qr}^e i_{ds}^e) = \frac{3}{2}\frac{P}{2}\frac{L_m}{L_r}\lambda_{dr}^e i_{qs}^e \tag{5.14}$$

With the rotor flux held constant, the torque equation reduces further to

$$T_e = K_T\, i_{qs}^e \quad \left(K_T = \frac{3}{2}\frac{P}{2}\frac{L_m}{L_r}\lambda_{dr}^e\right) \tag{5.15}$$

The torque is directly related to the *q*-axis stator current i_{qs}^e, so this current can be called the *torque-producing current*. By controlling the *q*-axis stator current

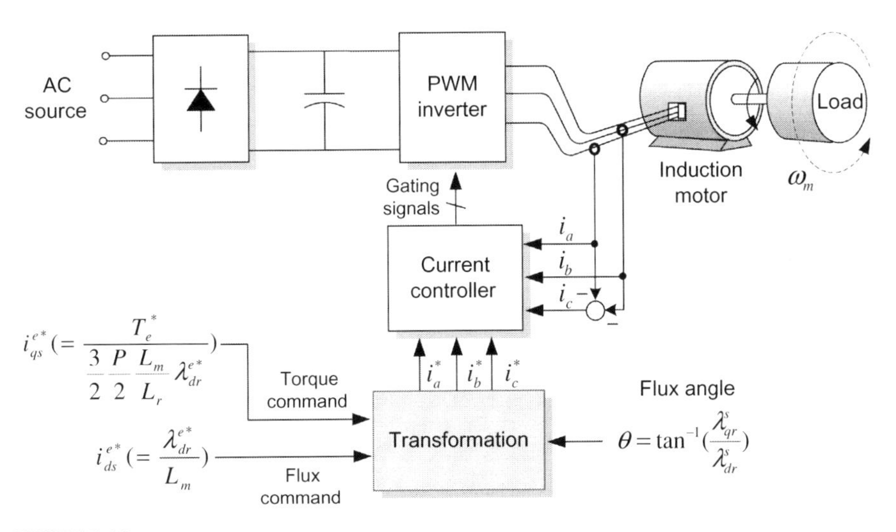

FIGURE 5.11

Induction motor drive system by the direct vector control.

instantaneously, the torque of an induction motor can be controlled instantaneously as in a DC motor.

By combining Eqs. (5.12) and (5.14), the torque can also be expressed in terms of the stator currents as

$$T_e = \frac{3}{2}\frac{P}{2}\frac{L_m^2}{L_r}\frac{1}{\left(1+\frac{L_r}{R_r}p\right)}i_{ds}^e\, i_{qs}^e \tag{5.16}$$

With the d-axis current i_{ds}^e held constant, the torque is given as

$$T_e = \frac{3}{2}\frac{P}{2}\frac{L_m^2}{L_r}i_{ds}^e\, i_{qs}^e \tag{5.17}$$

5.2.2.3 Induction motor drive system by the direct vector control

An induction motor drive system that adapts the direct vector control based on the rotor flux is shown in Fig. 5.11.

In the direct vector control of an induction motor, the rotor flux linkage is kept constant and the torque is regulated by controlling i_{qs}^e. Accordingly, the d-axis current command i_{ds}^{e*} will be given from the rotor flux command λ_{dr}^{e*} by using Eq. (5.13) as

$$i_{ds}^{e*} = \frac{\lambda_{dr}^{e*}}{L_m} \tag{5.18}$$

The rated value of the rotor flux linkage command λ_{dr}^{e*} depends on the rated voltage and frequency of a given motor. The ways to calculate the rated rotor flux linkage for a given motor will be described later in this section.

After applying the d-axis current command i_{ds}^{e*} of Eq. (5.18), as shown in Fig. 5.10, a time to build up the rotor flux linkage to its final value is required. This build-up time depends on the rotor time constant T_r and will be normally about $5T_r$.

After the rotor flux linkage arrives at the final value, the q-axis current can be applied to produce the required torque. Using Eq. (5.15), the q-axis current command i_{qs}^{e*} can be calculated from the output torque command T_e^*, which may be given from a speed or torque controller, as

$$i_{qs}^{e*} = \frac{T_e^*}{K_T} \quad \left(K_T = \frac{3}{2}\frac{P}{2}\frac{L_m}{L_r}\lambda_{dr}^{e*}\right) \tag{5.19}$$

We need to regulate the stator currents of the induction motor to follow these current commands. It should be noted that these $d-q$ current commands i_{ds}^{e*} and i_{qs}^{e*} are variables in the synchronous reference frame. We need to know the three-phase current commands i_{as}^*, i_{bs}^*, i_{cs}^* of the induction motor. To obtain these three-phase current commands, we need to transform i_{ds}^{e*} and i_{qs}^{e*} into i_{as}^*, i_{bs}^*, i_{cs}^* by using the rotor flux angle θ_e. Alternatively, after transforming these synchronous current commands into stationary current commands as

$$i_{ds}^{s*} = i_{ds}^{e*}\cos\theta_e - i_{qs}^{e*}\sin\theta_e \tag{5.20}$$

$$i_{qs}^{s*} = i_{ds}^{e*}\sin\theta_e + i_{qs}^{e*}\cos\theta_e \tag{5.21}$$

then, we can transform these stationary current commands into three-phase current commands as

$$i_{as}^* = i_{ds}^{s*} \tag{5.22}$$

$$i_{bs}^* = -\frac{1}{2}i_{ds}^{s*} + \frac{\sqrt{3}}{2}i_{qs}^{s*} \tag{5.23}$$

$$i_{cs}^* = -\frac{1}{2}i_{ds}^{s*} - \frac{\sqrt{3}}{2}i_{qs}^{s*} \tag{5.24}$$

Finally, if the stator currents of an induction motor are regulated to instantaneously follow these three-phase current commands by using a current controller, then the instantaneous torque control of an induction motor can be achieved.

On the other hand, there is a more effective control method in regulating the stator currents. In this method of using a synchronous reference frame current

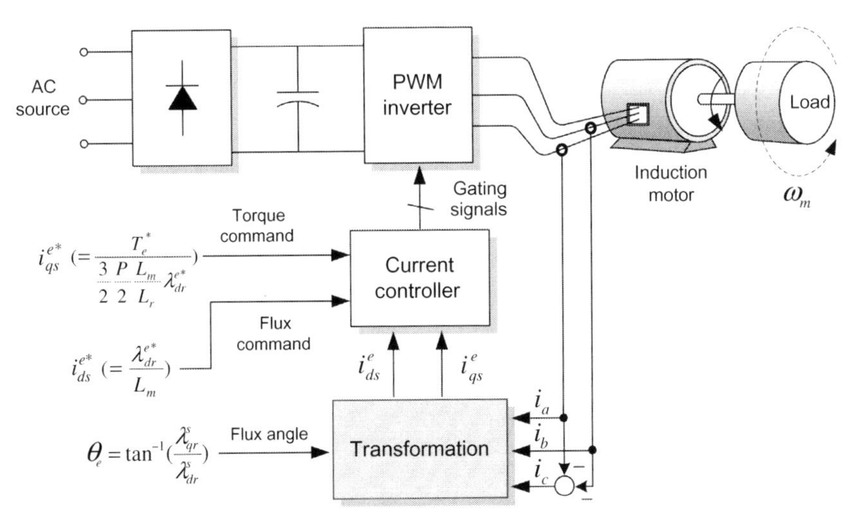

FIGURE 5.12

Direct vector control system with the synchronous reference current regulator.

regulator, the synchronous current commands, i_{ds}^{e*} and i_{qs}^{e*}, are used directly instead of the three-phase current commands as shown in Fig. 5.12. In this case, the three-phase stator currents, i_{as}, i_{bs}, i_{cs}, of the induction motor are transformed into synchronous dq currents, i_{ds}^e and i_{qs}^e. This synchronous reference frame regulator, which will be described in detail in Chapter 6, shows an excellent current control performance. Thus, most vector control systems employ this synchronous regulator.

In this control, as it is called "Vector control", we can see that both the magnitude and the phase of the stator current vector vary according to the output torque command. As an example, Fig. 5.13 describes how the stator current command vector alters when the torque command is changed. The stator current vector of the induction motor is assumed to be $\boldsymbol{i}_s^*$ initially. If a larger torque needs to be produced, the q-axis current command will be increased from i_{qs}^{e*} to i_{qs1}^{e*}, but the d-axis current command i_{ds}^{e*} will remain the same. This requires a new stator current command vector $\boldsymbol{i}_{s1}^*$, and thus both the magnitude and the phase of the stator current vector are changed. Likewise, if the flux command is changed, both the magnitude and the phase of the stator current vector will be also changed.

Fig. 5.14 shows experimental results for the direct vector control on 5-hp, 4-pole induction motor when the speed command is changed from 500 to 1500 r/min.

The most difficult aspect of implementing the direct vector control is finding the instantaneous position of the rotor flux linkage, i.e., the rotor flux angle,

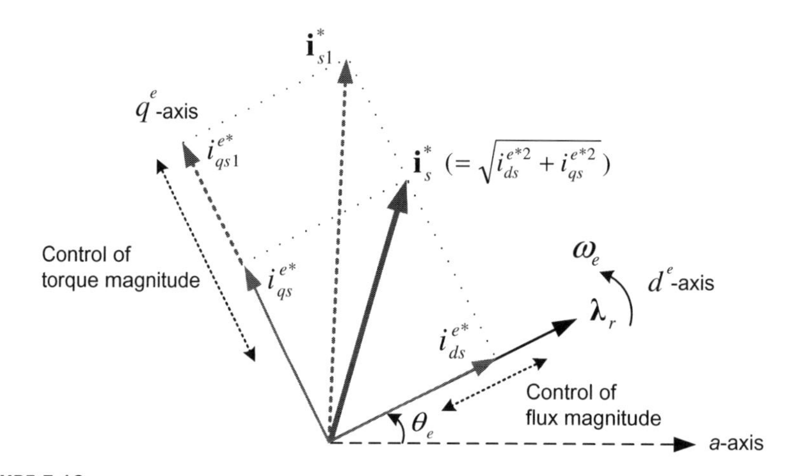

FIGURE 5.13

Stator d–q current commands.

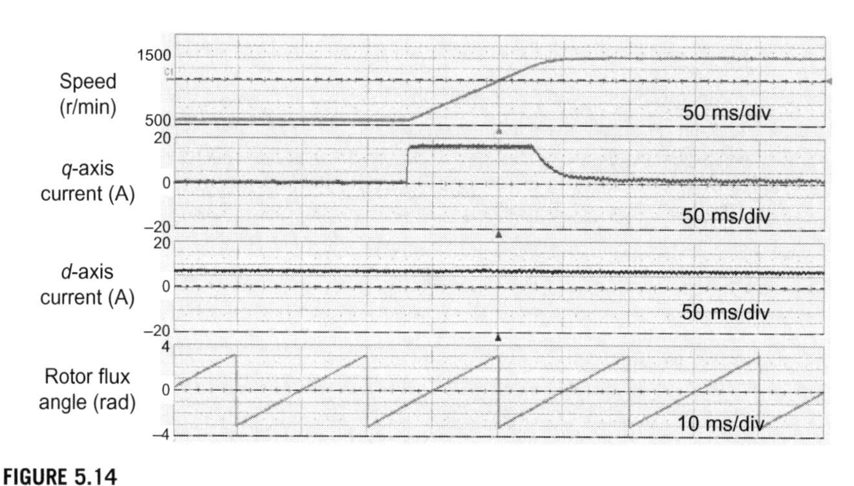

FIGURE 5.14

Experimental results for the direct vector control of 5-hp induction motor.

accurately. It is not an overstatement to say that the performance of the instantaneous torque control depends on the accuracy of the rotor flux angle. A way to obtain the rotor flux angle is to calculate the rotor flux linkage from the air-gap flux detected by Hall effect sensors or sensing coils, which are installed in the applied motor. However, this method is costly and is not practical. A widely used method is to estimate the rotor flux linkage from voltages and currents of the motor, which will be described in more detail in Chapter 6.

MATLAB/SIMULINK SIMULATION: VECTOR CONTROL OF AN INDUCTION MOTOR

The rated values and parameters of an induction motor are as follows.

- 5-hp, 220 V, 60 Hz, 4-Pole, 1750 r/min, 12.8 A rms

$$R_s = 0.295\Omega, \quad R_r = 0.379\Omega, \quad L_m = 59 \text{ mH}, \quad L_{ls} = L_{lr} = 1.794 \text{ mH}$$

The rated d- and q-axes stator currents can be obtained as follows.

- Rated torque: $T_e = (P/2)(\text{power}/\omega_s(1-s)) = 20.35$ Nm
- Rated d-axis current: $I_{ds}^e \cong \dfrac{V_{ph\_peak}}{\sqrt{R_s^2 + X_s^2}} = 7.84$ A
- Rated flux linkage: $|\lambda_r| = L_m I_{ds}^e = 0.4624$ Wb
- Rated q-axis current: $I_{qs}^e = T_e / \left(\dfrac{3}{2}\dfrac{P}{2}\dfrac{L_m^2}{L_r} I_{ds}^e\right) = 15.12$ A
- Simulation block for the direct vector control

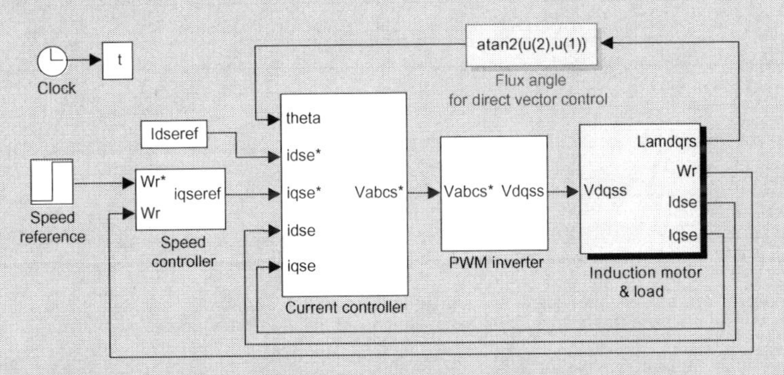

- Flux angle: $\theta_e = \tan^{-1}(\lambda_{qr}^s / \lambda_{dr}^s)$
- Load model: refer to Chapter 1.
- Speed controller: refer to Chapter 2.
- Induction motor model: refer to Chapter 4.
- Current controller: refer to Chapter 6.
- PWM inverter: refer to Chapter 7.

- Simulation results

  The q-axis current (thus torque) is applied when the rotor flux linkage builds up to the value in the steady-state condition.

(Continued)

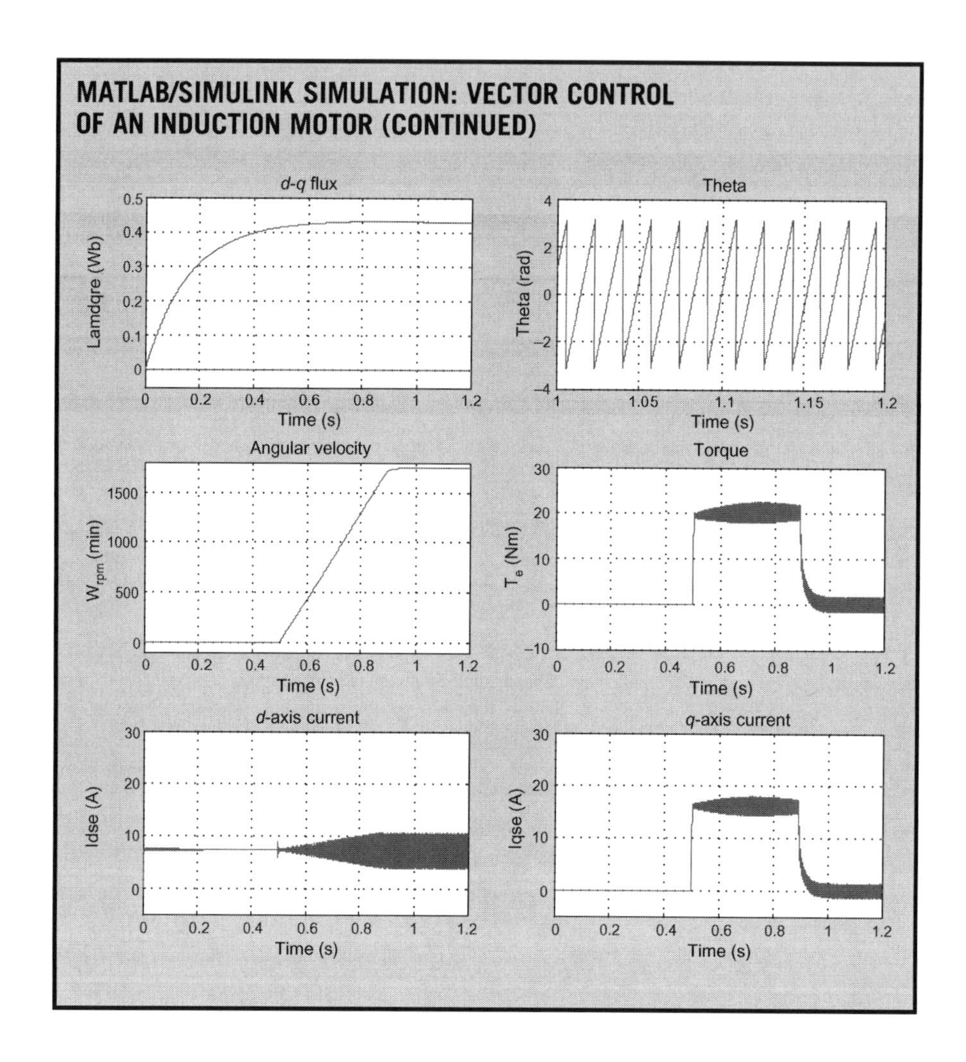

5.2.3 INDIRECT VECTOR CONTROL BASED ON THE ROTOR FLUX

As mentioned in Section 5.2.2, we need to know the accurate rotor flux linkage for obtaining the rotor flux angle in the direct vector control. However, it is not easy to obtain the rotor flux linkage accurately. The indirect vector control does not use the rotor flux linkage directly for obtaining the rotor flux angle, so it may be implemented easily.

From the steady-state equivalent circuit of an induction motor in Fig. 5.15, we can see that the slip s determines how the stator current can be divided into the flux- and torque (or rotor)-producing currents. Thus we can infer from this that the desired flux and torque can be obtained by controlling the slip. The flux- and

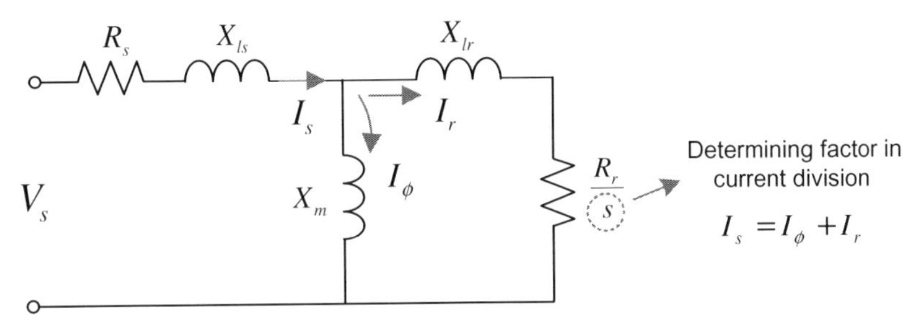

FIGURE 5.15

Steady-state equivalent circuit of an induction motor.

the torque-producing currents in the stated-state equivalent circuit are not equal to those in the vector control. However, similar to this concept, the principle of the indirect vector control is to control the slip so that the stator current can be divided into the desired flux- and the torque-producing components. Now we will take a closer look at the principle of the indirect vector control.

As stated in Section 5.2.2, in the vector control, since the d-axis is assigned to the position of the rotor flux linkage, there is no rotor flux linkage in the q-axis, i.e., $\lambda^e_{qr} = 0$.

Thus in the vector control, the rotor voltage equation becomes

$$v^e_{qr} = R_r i^e_{qr} + p\lambda^e_{qr} + (\omega_e - \omega_r)\lambda^e_{dr} = R_r i^e_{qr} + (\omega_e - \omega_r)\lambda^e_{dr} = 0 \tag{5.25}$$

From this, the slip angular frequency ω_{sl} required in the vector control can be derived as

$$\omega_e - \omega_r = \omega_{sl} = -\frac{R_r i^e_{qr}}{\lambda^e_{dr}} \tag{5.26}$$

Here, the rotor current can be obtained as Eq. (5.28) from the rotor flux linkage equation of Eq. (5.27).

$$\lambda^e_{qr} = L_r i^e_{qr} + L_m i^e_{qs} = 0 \tag{5.27}$$

$$i^e_{qr} = -\frac{L_m}{L_r} i^e_{qs} \tag{5.28}$$

Combining Eqs. (5.26) and (5.28) yields the requirement of the slip angular frequency for the vector control as

$$\omega_{sl} = \frac{L_m R_r}{L_r} \frac{i^e_{qs}}{\lambda^e_{dr}} \tag{5.29}$$

From Eq. (5.12), this can also be expressed in d–q axes currents as

$$\omega_{sl} = \left(\frac{1}{T_r} + p\right)\frac{i^e_{qs}}{i^e_{ds}} \quad \left(T_r = \frac{L_r}{R_r}\right) \tag{5.30}$$

With the rotor flux linkage held constant, we can finally obtain the important slip requirement for the indirect vector control as

$$\omega_{sl} = \frac{1}{T_r}\frac{i_{qs}^e}{i_{ds}^e} \tag{5.31}$$

This slip relation indicates that if we want to divide the stator currents into the specific values of $d-q$ axes currents, we have to set the operating slip angular frequency of an induction motor at the value of Eq. (5.31) calculated by using the specific values. After that, using the information of the rotor velocity ω_r, we can obtain the rotor flux angle as

$$\theta_e = \int \omega_e dt = \int (\omega_{sl} + \omega_r) dt \tag{5.32}$$

Like this, in the indirect vector control, instead of using the rotor flux linkage directly, the slip angular frequency is adjusted to achieve the required division of the $d-q$ axes currents.

Fig. 5.16 shows the indirect vector control system of an induction motor, where the slip angular frequency ω_{sl}^* is calculated from $d-q$ axes current commands for the required torque and flux production.

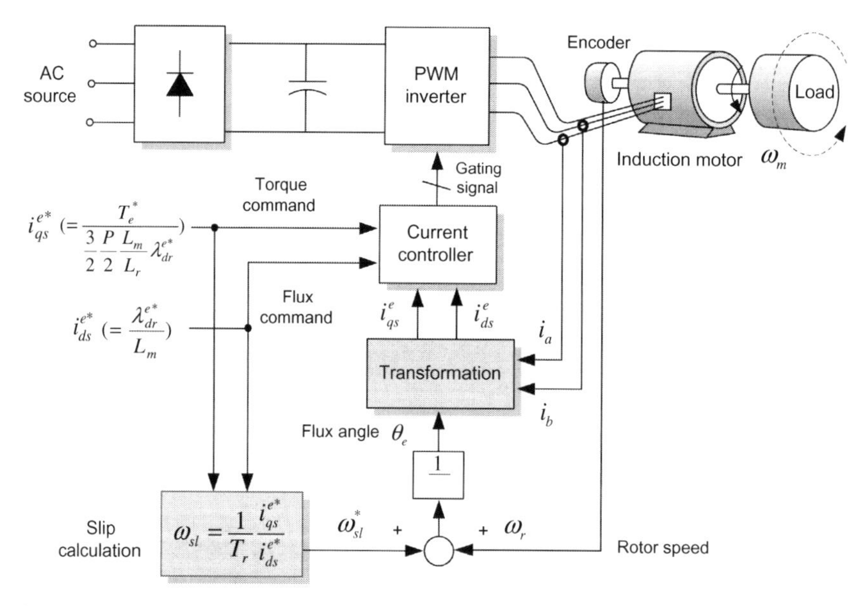

FIGURE 5.16

Induction motor drive system using the indirect vector control.

Since the indirect vector control uses the rotor flux angle obtained from adding the slip angular velocity to the rotor velocity, this is often called the *feedforward field-oriented control*. The indirect vector control that does not require the information of the rotor flux linkage has an advantage of easy implementation and a better performance in low-speed operations.

A drawback of the indirect vector control is that the rotor speed is required to obtain the rotor flux angle. The rotor speed is normally found using an encoder, a type of a position sensor, which will be discussed in Chapter 9. Recently the vector control that does not require an expensive position sensor, called the *sensorless control*, has received a lot of attention. In this case, however, we cannot use this simple indirect vector control technique.

In addition, the indirect vector control is also problematic in which the rotor open circuit time constant T_r required in the calculation of Eq. (5.31) is sensitive to both the temperature and rotor flux level. The rotor resistance R_r can easily change with temperature, and the rotor inductance L_r can vary with the magnitude of the rotor flux. These variations result in an inaccurate slip angular frequency, and thus an accurate vector control cannot be achieved. Therefore in the indirect vector control, the accuracy of the torque control depends on that of these parameters. The parameters of an induction motor can be obtained by the block test and no-load test, which were described in Chapter 3. However, these tests may lack accuracy. Thus we need to use an alternative method that can obtain these parameters more accurately. For both the direct vector control and the indirect vector control, the accuracy of these parameters is important. The parameters can be obtained by off-line auto tuning method prior to the start-up or online estimation method during a normal operation [1–3].

Now we will take a close look at how an incorrect rotor time constant (by an inaccurate R_r) influences the indirect vector control performance.

5.2.4 DETUNING IN THE INDIRECT VECTOR CONTROL

The following slip relation indicates the necessary condition for dividing the stator current into the desired values of the d- and q-axes currents correctly.

$$\omega_{sl}^{*} = \frac{1}{T_r}\frac{i_{qs}^{e*}}{i_{ds}^{e*}} = \frac{R_r}{L_r}\frac{i_{qs}^{e*}}{i_{ds}^{e*}} \tag{5.33}$$

Since the value of the rotor time constant T_r used in the slip calculator can vary easily during operations, the slip angular frequency is likely to be incorrect. The value of the rotor resistance especially varies with temperature. If the temperature increases by 10°C, the rotor resistance will increase about 4%. From the heat produced from copper losses, the resistance of a rotor in an enclosed space can increase more than that of a stator. Besides the temperature, the value of the rotor resistance can vary with the operating frequency of the rotor circuit due to the skin effect.

The rotor self-inductance varies with the value of the rotor flux linkage. The rotor flux linkage remains constant for a normal operation below the rated speed, so the rotor self-inductance may remain unchanged. However, the rotor self-inductance will change for high-speed operations, which require a reduced rotor flux level for a field-weakening operation.

The variations of the parameters due to the reasons mentioned above make the value of the rotor time constant used in the slip calculator to deviate from the correct value, which is referred to as *detuning* [4]. In this detuned operation the d-axis will be misaligned from the true rotor flux axis due to an incorrect slip angular frequency. As a result, the actual $d-q$ axes currents will be different from their commands, and in turn, the actual rotor flux linkage and torque will be different from their commands. Moreover, the decoupling of the rotor flux linkage and torque is lost, i.e., the correct vector control cannot be achieved, so steady-state and transient torque responses will be degraded. Fig. 5.17 shows the currents and output torque for a detuned operation [5].

We will take a look at the influence of the detuned rotor resistance on the rotor flux linkage and output torque. Assuming that the rotor resistance increases due to an increase in the temperature, the rotor time constant used by the slip calculator will be larger, so the produced slip frequency will be smaller than the desired value. As a result, the $d-q$ axes set by the controller will lag behind the true axis as shown in Fig. 5.18. Hence, the actual d-axis current component will increase, but the actual q-axis current will decrease. This will cause the actual rotor flux linkage to increase but not proportionally to the increase in the d-axis current due to magnetic saturation. Consequently, the output torque will be reduced, and a bigger q-axis current command will be imposed to obtain the required output torque.

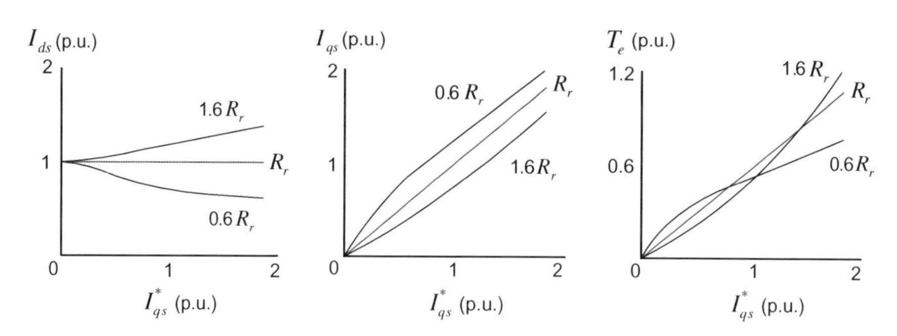

FIGURE 5.17

Detuning effects.

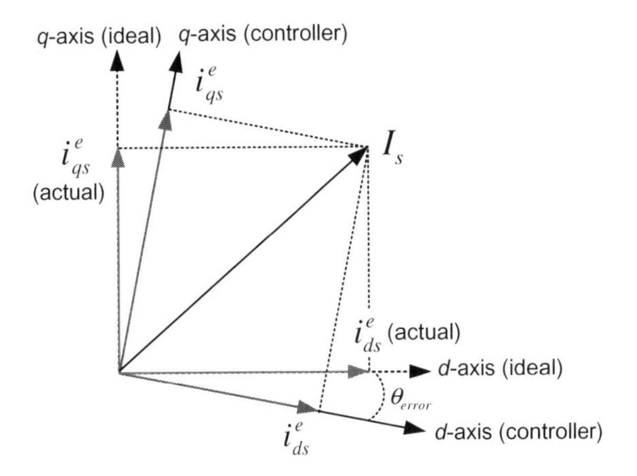

FIGURE 5.18

Reference frame error due to detuning effect.

On the contrary, when the rotor resistance decreases, the calculated slip frequency becomes larger than the desired value because the slip calculator uses a smaller time constant. Thus the actual q-axis component current increases but the actual d-axis current decreases. As a result of this, the actual rotor flux linkage is significantly reduced, and in turn, the output torque is reduced. In these two detuned cases the d-axis is misaligned from the true rotor flux axis due to the incorrect slip frequency. As a result, the distribution of the $d-q$ axes currents will be inadequate, so the induction motor will not be able to generate the desired rotor flux linkage and output torque. In addition, the detuning leads to additional motor losses and a reduction of the peak torque capability. Generally, larger high-efficiency motors with a large magnetizing inductance and a low rotor resistance are much more sensitive to detuning than small low-efficiency motors.

5.3 FLUX ESTIMATION OF AN INDUCTION MOTOR

As described in Section 5.2, the most important information for realizing the vector control is the position of the rotor flux linkage, i.e., the rotor flux angle for the $d-q$ axes transformation. The position of the rotor flux linkage is directly identified from the rotor position for permanent magnet synchronous motors (PMSMs). However, it is not possible to identify the position of the rotor flux linkage directly from the rotor position for induction motors, which operate with a slip with regard to the synchronous speed.

Now we will examine the ways to find the rotor flux linkage for the direct vector control. To find the rotor flux angle, the d–q axes rotor flux linkages in the stationary reference frame are required as we can see from Eq. (5.8) as

$$\theta_e = \tan^{-1}\left(\frac{\lambda_{qr}^s}{\lambda_{dr}^s}\right) \tag{5.34}$$

There are two methods to obtain the stationary rotor flux linkages: one is to measure the flux linkages or the physical quantity proportional to the flux linkages, and the other one is to estimate them indirectly from the mathematical motor model. In the direct method the rotor flux linkage is obtained from the air-gap flux acquired from the Hall effect sensors or the sensing coils installed inside the stator. However, since there are some difficulties in the practical usage of the direct method, the rotor flux linkages are commonly estimated from the mathematical motor model by using the motor voltage, current, and/or speed as follows.

There are two typical open-loop observers for estimating the rotor flux linkage: one is based on the stator voltage equations, which is referred to as the *voltage model*, and the other one is based on the rotor voltage equations, which is known as the *current model*.

5.3.1 ROTOR FLUX LINKAGES ESTIMATION BASED ON THE STATOR VOLTAGE EQUATIONS: VOLTAGE MODEL

In this method the stator flux linkages are obtained from the stator voltage equations and, in turn, the rotor flux linkages are estimated from these stator flux linkages. This method requires information on the stator voltages and currents.

First, the stator flux linkages can be found from the integral of stator voltages as

$$v_{ds}^s = R_s i_{ds}^s + \frac{d\lambda_{ds}^s}{dt} \tag{5.35}$$

$$v_{qs}^s = R_s i_{qs}^s + \frac{d\lambda_{qs}^s}{dt} \tag{5.36}$$

$$\lambda_{ds}^s = \int (v_{ds}^s - R_s i_{ds}^s)dt = \int e_{ds}^s dt \tag{5.37}$$

$$\lambda_{qs}^s = \int \left(v_{qs}^s - R_s i_{qs}^s\right)dt = \int e_{qs}^s dt \tag{5.38}$$

where $e_{ds}^s = v_{ds}^s - R_s i_{ds}^s$ and $e_{qs}^s = v_{qs}^s - R_s i_{qs}^s$.

The rotor flux linkages can be obtained from these stator flux linkages as follows.

The d-axis rotor flux linkage in the stationary reference frame is given by

$$\lambda_{dr}^s = L_r i_{dr}^s + L_m i_{ds}^s \tag{5.39}$$

Here, the nonmeasurable rotor current i^s_{dr} can be obtained from the d-axis stator flux linkage and the stator current as

$$i^s_{dr} = \frac{\lambda^s_{ds} - L_s i^s_{ds}}{L_m} \leftarrow \lambda^s_{ds} = L_s i^s_{ds} + L_m i^s_{dr} \tag{5.40}$$

Substituting this rotor current into Eq. (5.39), we can finally obtain the equation for the stationary d-axis rotor flux linkage as

$$\hat{\lambda}^s_{dr} = \frac{L_r}{L_m}(\lambda^s_{ds} - \sigma L_s i^s_{ds}) \tag{5.41}$$

where $\sigma = (1 - L_m^2/L_s L_r)$ is a total leakage factor.

Likewise, we can obtain the stationary q-axis rotor flux linkage as

$$\hat{\lambda}^s_{qr} = \frac{L_r}{L_m}(\lambda^s_{qs} - \sigma L_s i^s_{qs}) \tag{5.42}$$

Using the arctangent function of these estimated stationary $d-q$ axes rotor flux linkages, the flux angle can be calculated as

$$\hat{\theta}_e = \tan^{-1}\left(\frac{\hat{\lambda}^s_{qr}}{\hat{\lambda}^s_{dr}}\right) \tag{5.43}$$

The strength of this method is its simplicity and insensitivity to the parameter variations except for the stator resistance. However, since this method is fundamentally based on the estimation of the back-EMF e^s_{dqs} its performance depends on the rotor speed. Thus this method works well in the medium- to high-speed range, in which the back-EMF becomes large enough. However, in the low-speed range with a low operation voltage, it may be easily influenced by the inaccuracy of the stator resistance, the sensor noise, and the inverter-nonlinearity effect. Thus this method based on the voltage model is normally used in high-speed regions.

To estimate the rotor flux linkage accurately with this method, it should be noted that it is necessary to offer accurate stator voltages and currents. A voltage sensor may be used to obtain the stator voltages, but it is hard to measure the actual stator voltages precisely because the motor voltages are given by the inverter as a high-frequency pulse width modulator (PWM) waveform. Thus instead of measuring the actual stator voltages, the voltage commands or the voltages reconstructed by using the DC-link voltage and the commanded PWM duty are commonly used. However, the reconstructed value may differ from the actual output voltage due to inverter-nonlinearity such as the dead-time effect, turn-on/off delay time, and voltage drop across switching devices. Though such voltage errors can be reduced by a proper compensation technique, it is not easy to obtain an accurate voltage in the low-speed range of 10% below the rated speed.

Unlike the voltages, the stator currents are directly measured by current sensors. When the measured currents include a DC offset or noise, the integration of the back-EMF may be saturated. Thus instead of a pure integrator, a first-order, low-pass filter is commonly used. This low-pass filter can be considered as the combination of a pure integrator and a high-pass filter as

$$\lambda_{dqs}^{s}(s) = \left[e_{dqs}^{s}(s) \times \frac{s}{s+\alpha}\right] \times \frac{1}{s} = e_{dqs}^{s}(s) \times \frac{1}{s+\alpha} \tag{5.44}$$

However, the use of such low-pass filter can cause errors in the phase and the magnitude of the estimated flux linkages, especially in low-speed regions. The cutoff frequency of the filter in Eq. (5.44) greatly influences the errors of the flux estimator and is usually chosen as the reciprocal of the rotor time constant T_r.

A DIGITAL FIRST-ORDER, LOW-PASS FILTER FOR THE FLUX ESTIMATION

Let us find out a digital first-order, low-pass filter required for the flux estimation. Here, we discuss how to transform an analog filter designed using the Laplace transform (s-domain) into an equivalent digital filter. The first-order, low-pass filter designed in the s-domain has a transfer function of:

$$y(s) = \frac{1}{s + \frac{1}{\tau}} \cdot x(s)$$

Using the bilinear transform to replace the Laplace operator s with a z-operator, we can convert the analog first-order, low-pass filter to an equivalent digital filter as follows.

To apply the bilinear transform, we just need to replace the s by $s = 2/T((1-z^{-1})/(1+z^{-1}))$, where T is the sampling period.

$$y[n] = \frac{1}{\frac{2}{T}\left(\frac{1-z^{-1}}{1+z^{-1}}\right) + \frac{1}{\tau}} \cdot x[n] = \frac{1+z^{-1}}{(k_1+k_2)+(k_1-k_2)z^{-1}} \cdot x[n] \qquad \left(k_1 = \frac{2}{T}, k_2 = \frac{1}{\tau}\right)$$

$$\therefore \quad y[n] = \frac{(k_1-k_2)}{(k_1+k_2)} y[n-1] + \frac{1}{(k_1+k_2)}(x[n] + x[n-1])$$

Using the above equation, the digital filter for estimating the stator flux linkage can be given as

$$\therefore \quad \lambda_{dqs}^{s}[n] = \frac{(k_1-k_2)}{(k_1+k_2)} \lambda_{dqs}^{s}[n-1] + \frac{1}{(k_1+k_2)}(e_{dqs}^{s}[n] + e_{dqs}^{s}[n-1])$$

where $e_{dqs}^{s}[n] = v_{dqs}^{s}[n] - R_s i_{dqs}^{s}[n]$

5.3.2 ROTOR FLUX LINKAGES ESTIMATION BASED ON THE ROTOR VOLTAGE EQUATIONS: CURRENT MODEL

In this method the rotor flux linkages are obtained from the rotor voltage equations using the information on the stator currents and the rotor speed.

The rotor flux linkages can be directly obtained from the following rotor voltage equations as

$$v_{dr}^{\omega} = R_r i_{dr}^{\omega} + \frac{d\lambda_{dr}^{\omega}}{dt} - (\omega - \omega_r)\lambda_{qr}^{\omega} = 0 \tag{5.45}$$

$$v_{qr}^{\omega} = R_r i_{qr}^{\omega} + \frac{d\lambda_{qr}^{\omega}}{dt} + (\omega - \omega_r)\lambda_{dr}^{\omega} = 0 \tag{5.46}$$

Here, the nonmeasurable rotor currents can be replaced by the rotor flux linkage and the stator current as

$$i_{dr}^{\omega} = \frac{\lambda_{dr}^{\omega} - L_m i_{ds}^{\omega}}{L_r} \leftarrow \lambda_{dr}^{\omega} = L_r i_{dr}^{\omega} + L_m i_{ds}^{\omega} \tag{5.47}$$

$$i_{qr}^{\omega} = \frac{\lambda_{qr}^{\omega} - L_m i_{qs}^{\omega}}{L_r} \leftarrow \lambda_{qr}^{\omega} = L_r i_{qr}^{\omega} + L_m i_{qs}^{\omega} \tag{5.48}$$

Substituting these rotor currents into Eqs. (5.45) and (5.46), the equations of the rotor flux linkage can be found as

$$\frac{d\lambda_{dr}^{\omega}}{dt} = -\frac{R_r}{L_r}\lambda_{dr}^{\omega} + R_r\frac{L_m}{L_r}i_{ds}^{\omega} + (\omega - \omega_r)\lambda_{qr}^{\omega} \tag{5.49}$$

$$\frac{d\lambda_{qr}^{\omega}}{dt} = -\frac{R_r}{L_r}\lambda_{qr}^{\omega} + R_r\frac{L_m}{L_r}i_{qs}^{\omega} - (\omega - \omega_r)\lambda_{dr}^{\omega} \tag{5.50}$$

There is a difficulty in solving these equations due to the cross-coupling components between the axes. These undesirable cross-coupling components can be eliminated when the rotor flux linkages are estimated from the rotor reference frame ($\omega = \omega_r$) using the measured rotor speed ω_r as

$$\frac{d\lambda_{dr}^{r}}{dt} = -\frac{R_r}{L_r}\lambda_{dr}^{r} + \frac{R_r}{L_r}L_m i_{ds}^{r} \tag{5.51}$$

$$\frac{d\lambda_{qr}^{r}}{dt} = -\frac{R_r}{L_r}\lambda_{qr}^{r} + \frac{R_r}{L_r}L_m i_{qs}^{r} \tag{5.52}$$

The rotor flux linkages can be obtained from the integral of these two differential equations (5.51) and (5.52). Here, the knowledge of the stator currents in the rotor reference frame is necessary to solve these equations. These currents can be obtained by using the knowledge of the rotor speed ω_r as

$$i_{ds}^{r} = i_{ds}^{s}\cos\theta_r + i_{qs}^{s}\sin\theta_r \tag{5.53}$$

$$i_{qs}^{r} = -i_{ds}^{s}\sin\theta_r + i_{qs}^{s}\cos\theta_r \tag{5.54}$$

where $\theta_r\left(= \int \omega_r dt\right)$.

Finally, the required rotor flux linkage can be obtained by transforming Eqs. (5.51) and (5.52) back into the stationary reference frame as

$$\lambda_{dr}^{s} = \lambda_{dr}^{r}\cos\theta_r - \lambda_{qr}^{r}\sin\theta_r \tag{5.55}$$

$$\lambda_{qr}^{s} = \lambda_{qr}^{r}\sin\theta_r + \lambda_{qr}^{r}\cos\theta_r \tag{5.56}$$

Besides the knowledge of the rotor speed, this method requires the rotor resistance. Thus, the estimation performance is largely affected by the variation of the rotor resistance. Because of this, online identification and correction of the rotor resistance are essential for improving the estimation performance. In contrast to the voltage model that is advantageous for high-speed operations, this estimation method based on the current model is useful in low-speed regions because it may give an oscillatory response in high-speed regions.

5.3.3 COMBINED FLUX ESTIMATION METHOD

In high-speed regions where the back-EMF is large enough, the flux estimation technique based on the voltage model using an integration of the back-EMF has an advantage of accuracy and robustness to parameters variation. However, in low-speed regions where it is hard to obtain accurate information on the back-EMF, the flux estimation based on the current model using the rotor speed has an advantage over the one based on the voltage model. Therefore, it is desirable to estimate the rotor flux linkage by using the current model in low-speed regions and the voltage model in high-speed regions. For this reason, a method combining these two models was introduced to produce a good performance over the whole speed range [6]. Fig. 5.19 shows the block diagram of this combined method that estimates the rotor flux linkages gradually by using the current model in the low-speed region and the voltage model in the high-speed region.

When combining the two models, a method to alternate between these two models is important. The flux estimator in Ref. [6] uses a filter to combine the two models effectively according to the operating frequency as follows.

Fig. 5.19 can be simplified to Fig. 5.20. Here, $\lambda^s_{dqr\_cm}$ denotes the rotor flux linkage estimated by the current model, and $\lambda^s_{dqr\_vm}$ denotes the rotor flux linkage estimated by the voltage model. The gains are given by $K_p = K_1 L_r / L_m$, $K_i = K_2 L_r / L_m$.

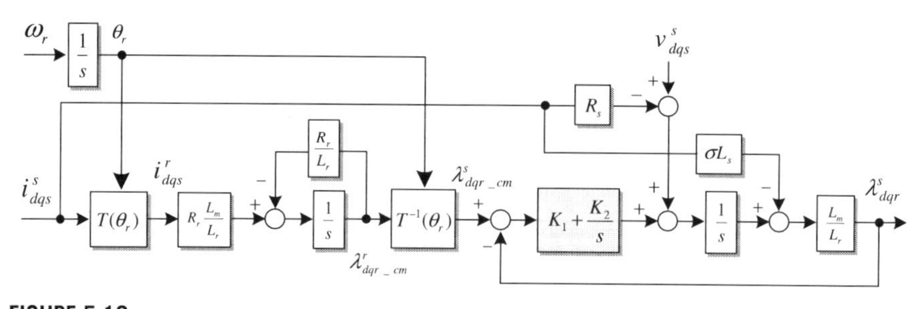

FIGURE 5.19

Block diagram of a combined method.

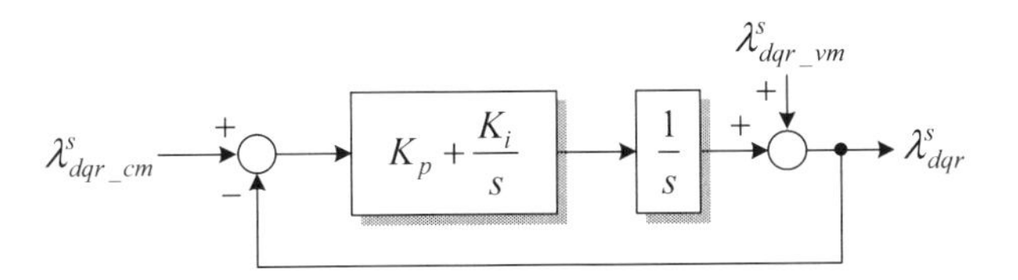

FIGURE 5.20

Simplified combined method.

The output of this simplified combined estimator can be expressed as

$$\lambda^s_{dqr} = \frac{s^2}{s^2 + K_p s + K_i}\lambda^s_{dqr\_vm} + \frac{K_p s + K_i}{s^2 + K_p s + K_i}\lambda^s_{dqr\_cm} \tag{5.57}$$

This consists of the sum of the second-order, high-pass filtered rotor flux linkage $\lambda^s_{dqr\_vm}$ based on the voltage model and the second-order, low-pass filtered rotor flux linkage $\lambda^s_{dqr\_cm}$ based on the current model.

This means that the final estimated flux linkage is automatically obtained from the current model in the low-frequency (low-speed) region and the voltage model in the high-frequency (high-speed) region.

In this estimator the transition frequency ω_c between the two models is commonly selected at 10 Hz. The transition frequency ω_c can be established through the gains of a proportional–integral (PI) controller as shown in Fig. 5.20. This utilizes the concept of designing the cutoff frequency for a second-order Butterworth filter as follows.

By comparing the transfer function of a second-order, high-pass Butterworth filter with the first term of Eq. (5.57), we can see that there exists the following relationship between gains and parameters of the filter as

$$G(s) = \frac{s^2}{s^2 + 2\zeta\omega_n s + \omega_n^2} \leftrightarrow \frac{s^2}{s^2 + K_p s + K_i} \tag{5.58}$$

$$K_p = 2\zeta\omega_n, \quad K_i = \omega_n^2 \tag{5.59}$$

TYPES OF SECOND-ORDER BUTTERWORTH FILTER

The transfer functions of second-order filters are:

- High-pass filter: $G(s) = \dfrac{s^2}{s^2 + 2\zeta\omega_n s + \omega_n^2}$
- Band-pass filter: $G(s) = \dfrac{2\zeta\omega_n s}{s^2 + 2\zeta\omega_n s + \omega_n^2}$
- Low-pass filter: $G(s) = \dfrac{\omega_n^2}{s^2 + 2\zeta\omega_n s + \omega_n^2}$

where ω_n denotes an undamped natural frequency and ζ denotes a damping ratio.

In the Butterworth filter, if $\zeta = 0.707$, then ω_n becomes the bandwidth. To set the bandwidth ω_n equal to the transition frequency ω_c, the PI gains are given as

$$K_p = \sqrt{2}\omega_c, \quad K_i = \omega_c^2 \tag{5.60}$$

Finally, K_1 and K_2 needed for the estimator in Fig. 5.19 are given as

$$K_1 = K_p \frac{L_m}{L_r} = \sqrt{2}\omega_c \frac{L_m}{L_r} \tag{5.61}$$

$$K_2 = K_i \frac{L_m}{L_r} = \omega_c^2 \frac{L_m}{L_r} \tag{5.62}$$

Besides the estimators mentioned here, there are several different estimation methods that use the modern control theory such as observers using the state equations [7]. The estimated rotor flux linkages can be also used for the sensorless vector control without using any position/speed sensors, which will be discussed in Chapter 9.

5.4 FLUX CONTROLLER OF INDUCTION MOTORS

The estimated rotor flux linkage mentioned previously is primarily used to obtain the rotor flux angle. However, this estimated rotor flux linkage may also be used for flux control to improve the stability of the torque control or to achieve the field-weakening control for high-speed operations.

From Section 5.3, we can see that the relationship between the d-axis stator currents i_{ds}^e and the rotor flux linkage λ_{dr}^e for the vector-controlled induction motor is given as

$$\lambda_{dr}^e = \frac{L_m}{1 + T_r p} i_{ds}^e \tag{5.63}$$

The magnitude of the rotor flux linkage can be controlled by the d-axis current. However, its instantaneous value is not proportional to the d-axis current when the d-axis current is changing. Thus when the magnitude of the rotor flux linkage has to be changed quickly for the field-weakening control, the rotor flux linkage needs to be controlled directly. Moreover, even when an induction motor is driven with a constant flux level, the flux control can enhance the stability of the drive system.

For the flux control, a PI controller is commonly used. Now we will take a close look at the design of such flux controller, i.e., how to select PI gains for achieving the desired control bandwidth.

5.4.1 PROPORTIONAL–INTEGRAL FLUX CONTROLLER

In the vector control based on the rotor flux, the magnitude of the rotor flux linkage can be controlled by the d-axis current i^e_{ds}, so the output of the PI flux controller has to be the d-axis current. The block diagram of the flux control system from Eq. (5.63) and a PI controller can be given as Fig. 5.21.

The gains of a PI flux controller can be determined in a similar way as those of a current/speed controller as was described in Chapter 2, as follows.

First, we assume that the dynamics of the d-axis current controller is fast enough compared to the flux change. The open-loop transfer function of the flux control system shown in Fig. 5.21 is given as

$$G^o_f(s) = \left(K_{pf} + \frac{K_{if}}{s}\right) \cdot \frac{L_m \frac{R_r}{L_r}}{s + \frac{R_r}{L_r}} = \frac{K_{pf}\left(s + \frac{K_{if}}{K_{pf}}\right)}{s} \cdot \frac{L_m \frac{R_r}{L_r}}{s + \frac{R_r}{L_r}} \tag{5.64}$$

Similar to the design of a current controller described in Section 2.6, if the zero of the PI controller is designed to cancel the pole of the system, i.e., $K_{if}/K_{pf} = R_r/L_r$, then the transfer function of Eq. (5.64) can be simplified as

$$G^o_f(s) = \frac{K_{pf} L_m \frac{R_r}{L_r}}{s} \tag{5.65}$$

The gain crossover frequency of this open-loop frequency response equals to the bandwidth of the flux control system. Hence the proportional and the integral gains to obtain the required control bandwidth ω_f can be given as

$$\text{Proportional gain:} \quad K_{pf} = \frac{L_r}{R_r L_m} \omega_f \tag{5.66}$$

$$\text{Integral gain:} \quad K_{if} = \frac{R_r}{L_r} K_{pf} = \frac{\omega_f}{L_m} \tag{5.67}$$

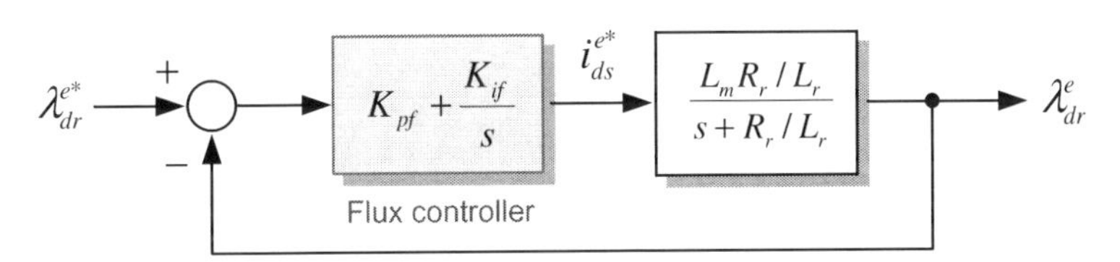

FIGURE 5.21

Flux control system including PI flux controller.

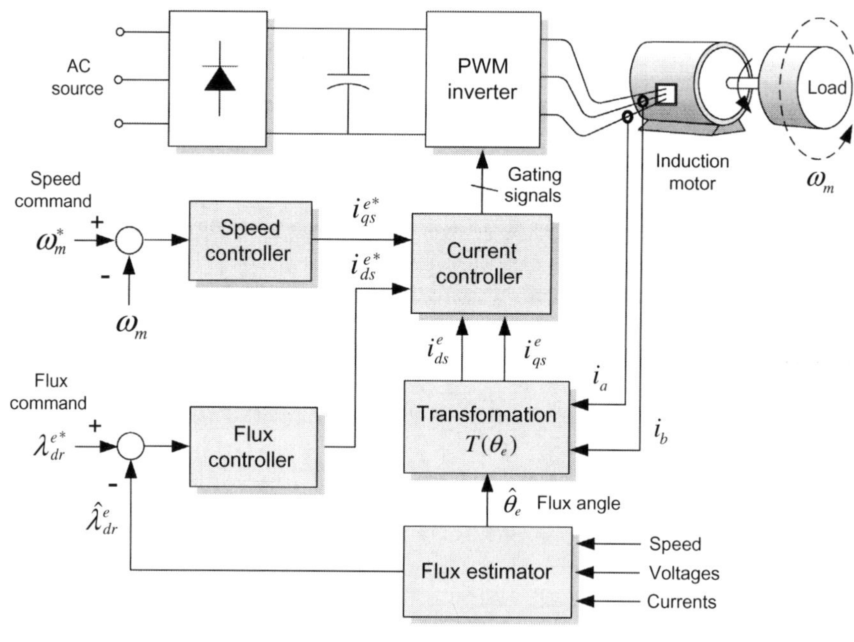

FIGURE 5.22

Vector control system including speed and flux controllers.

When the bandwidth of the current controller is several thousand radians per second, the bandwidth of the flux controller is designed generally within several ten to several hundred radians per second. This flux controller also needs an anti-windup control just like the speed/current controllers. The performance of the flux control can also be enhanced by the feedforward compensation using the flux command.

The diagram of the vector control system including speed and flux controllers for an induction motor is shown in Fig. 5.22. A PI controller as mentioned in Chapter 2 is commonly used for the speed control. The gain selection procedure of the PI speed controller was explained in detail in Section 2.7.

5.5 VECTOR CONTROL OF PERMANENT MAGNET SYNCHRONOUS MOTORS

Due to their superior performance compared to DC motors or induction motors, PMSMs have become increasingly popular in various high-performance motor drive applications. The vector control for PMSMs is relatively simple compared

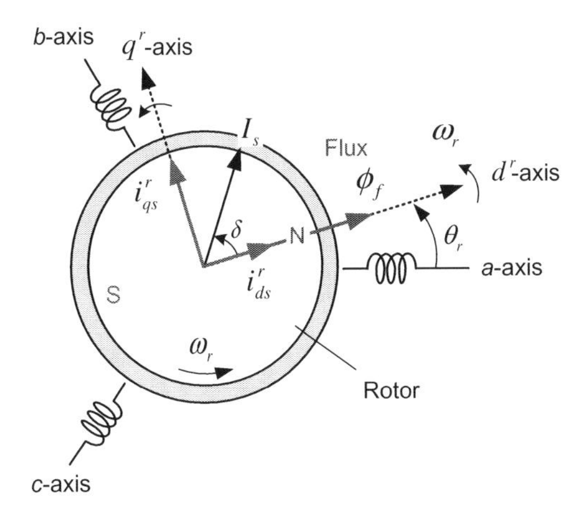

FIGURE 5.23

Rotor flux in a permanent magnet synchronous motor.

to that of induction motors. As shown in Fig. 5.23, this is because the rotor position itself is the position of the rotor flux, which is essential information for the vector control. Therefore, unlike for an induction motor, the complicated estimation algorithms are not required to obtain the rotor flux for the vector control of a PMSM.

As shown in Chapter 4, PMSMs can be classified into SPM and IPM types. These two types differ in the output torque expression and thus are slightly different in the way of implementing the vector control. First, we will begin with the vector control method for an SPMSM.

The three requirements for the instantaneous torque control of a DC motor can be rewritten to apply to a PMSM as follows:

1. The space angle between the field flux and the torque-producing current is always 90 degrees.
2. Both the field flux and the torque-producing current can be controlled independently.
3. The torque-producing current has to be controlled instantaneously.

Now we will take a look at how we can meet these three requirements for a PMSM. For a PMSM, the requirement 2 for the instantaneous torque control can be inherently fulfilled. In PMSMs the field flux of the rotor is produced by permanent magnets, while the torque-producing current is given by the current of stator windings. Thus the field flux and the torque-producing current can be controlled independently. In addition, the torque-producing current (i.e., the stator current) can be controlled instantaneously by the synchronous reference frame current regulator and the PWM inverter and thus can satisfy the requirement 3.

Likewise, the requirements 2 and 3 are relatively easy to fulfill for a PMSM. For the vector control of a PMSM, it is important to fulfill the requirement 1, i.e., to maintain the space angle between the field flux and the torque-producing current at 90 degrees. To achieve this, it is necessary to first transform the three-phase *abc* stator currents into two $d-q$ axes currents, just like in an induction motor. When transforming, the *d*-axis is aligned with the position of the rotor flux (i.e., the flux of the permanent magnet on the rotor) and then the *q*-axis current becomes the true torque-producing current, which is orthogonal to the rotor flux. In a PMSM, the permanent magnet flux of the rotor is rotating at the same speed as the rotor speed ω_r as shown in Fig. 5.23. Thus, we have to make the $d-q$ axes rotate at the rotor speed ω_r. This implies that we need to use the transformation of the stator currents into the *rotor reference frame* ($d^r - q^r$ axis). Such transformation requires the rotor position θ_r, i.e., the position of the permanent magnet. This can be obtained from the position sensors such as a resolver or an absolute encoder, which will be discussed in Chapter 9. Unlike in induction motors, it should be noted that the absolute initial position of the rotor flux is needed for the start-up of a PMSM, so an absolute type position sensor has to be used. However, instead of costly absolute type position sensors, an incremental encoder accompanied by an extra initial position detection method is often used.

5.5.1 VECTOR CONTROL OF A SURFACE-MOUNTED PERMANENT MAGNET SYNCHRONOUS MOTOR

In the vector control of an SPMSM, the stator currents are transformed into the rotor reference frame ($d^r - q^r$ axis) where the *d*-axis is assigned as the rotor flux position θ_r as shown in Fig. 5.24.

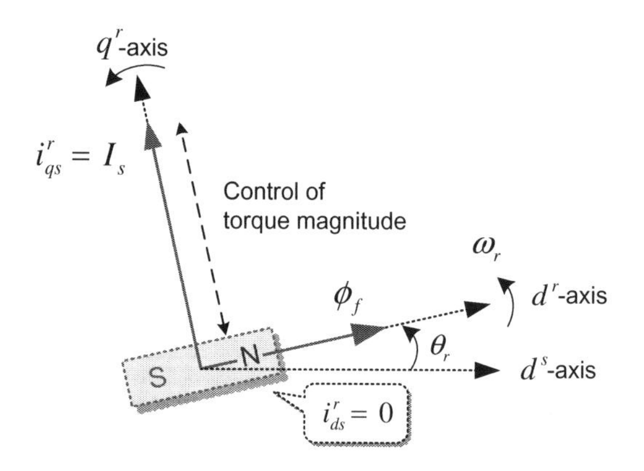

FIGURE 5.24

$d-q$ currents in the vector control of an SPMSM.

Here, it is important to note that an SPMSM does not require any d-axis current to produce the flux unlike an induction motor. Moreover, the d-axis current does not contribute to the torque production, which is proportional to only the q-axis component of the stator current. This can be seen from the following torque expression of an SPMSM.

$$T_e = \frac{P}{2}\frac{3}{2}\phi_f i_{qs}^r \tag{5.68}$$

Therefore, it is desirable to assign all the stator currents to the q-axis current to make a full use of the allowed motor current. In other words, in the vector control of an SPMSM, all the stator current I_s is used as the q-axis current i_{qs}^r, while the d-axis current is zero.

Consequently, the necessary condition for the maximum torque per ampere (MTPA) in the vector control of an SPMSM is $i_{ds}^r = 0$. In such condition, if we can control the q-axis current i_{qs}^r instantaneously, then the torque of the SPMSM can be controlled instantaneously like in the DC motors. The diagram of the vector control system for an SPMSM is shown in Fig. 5.25.

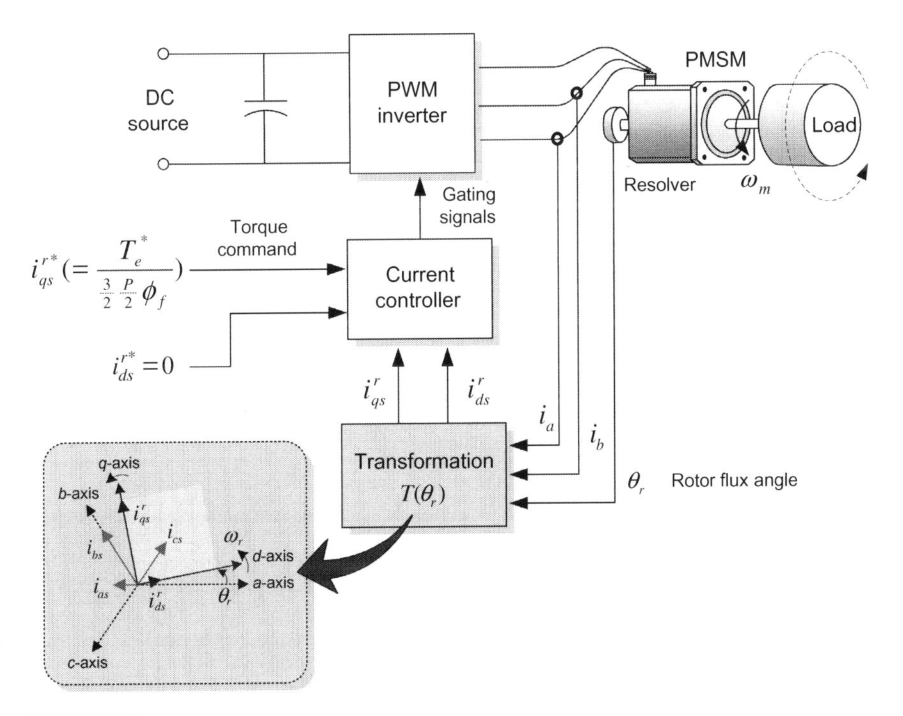

FIGURE 5.25

Block diagram of a vector control system for a PMSM.

In this system the d-axis current command $i_{ds}^{r*}=0$, while the q-axis current command is given by using Eq. (5.70) from the required torque command T_e^* as

$$i_{ds}^{r*}=0 \tag{5.69}$$

$$i_{qs}^{r*}=\frac{T_e^*}{\frac{P}{2}\frac{3}{2}\phi_f} \tag{5.70}$$

If the stator currents of the SPMSM can be regulated rapidly to follow these current commands by using a current regulator, then the instantaneous torque control can be achieved. As a current regulator, a synchronous reference frame current regulator is usually adopted due to its excellent current control performance, which will be discussed in Chapter 6, in more detail. For the synchronous current regulators, we need to transform the three-phase stator currents, i_{as}, i_{bs}, i_{cs}, of the SPMSM into the synchronous dq stator currents i_{ds}^r and i_{qs}^r by using the rotor flux angle θ_r.

5.5.2 VECTOR CONTROL OF AN INTERIOR PERMANENT MAGNET SYNCHRONOUS MOTOR

Next, we will examine the vector control strategy for an IPMSM, where $L_{ds}\neq L_{qs}$. In the vector control of an IPMSM the $d-q$ axes currents required for the MTPA are different from those of an SPMSM. This is because, in contrast to that of SPMSMs, the torque of IPMSMs consists of two terms as

$$T_e=\frac{P}{2}\frac{3}{2}\left[\phi_f i_{qs}^r+(L_{ds}-L_{qs})i_{ds}^r i_{qs}^r\right] \tag{5.71}$$

In addition to the first term of Eq. (5.71) that represents the *magnetic torque* of an SPMSM, an IPMSM has a *reluctance torque*, the second term, which is produced by the difference between the inductance values of the two axes. To produce the reluctance torque, both the d- and the q-axes components of the stator current are required. In most commonly used parallel topology IPMSMs as described in Chapter 4, the q-axis inductance L_{qs} is typically larger than the d-axis inductance L_{ds}. Thus, the d-axis current must have a negative polarity for the reluctance torque to be added to the magnet torque, i.e., $i_{ds}^r<0$. Similar to an induction motor, the stator current of an IPMSM needs to be divided into d-and q-axes components as shown in Fig. 5.26.

In this case, we need to know how the stator current is divided into the d-and q-axes stator currents properly to produce the MTPA. These optimal currents can be obtained as follows.

As seen in Eq. (5.71), there are many different combinations of d- and q-axes currents that produces the same torque. Fig. 5.27 shows the curve consisting of

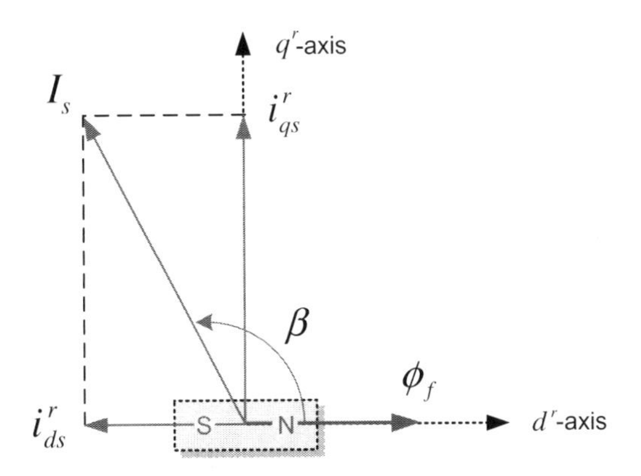

FIGURE 5.26

d–q currents in the vector control of an IPMSM.

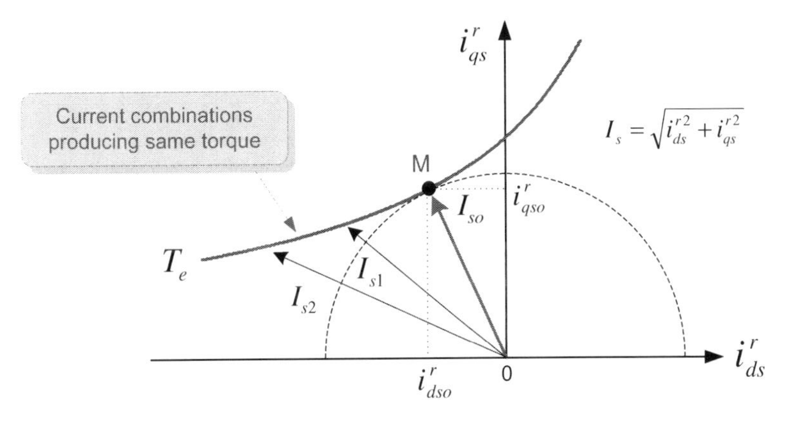

FIGURE 5.27

Curve consisting of current combinations producing the same torque.

these currents. Among these combinations, there is an optimal one that gives the smallest stator current. This optimal combination is the point M, $I_{sc}(i_{dso}^r, i_{qso}^r)$, at which the circle with a radius of I_{so} is tangent to the curve. If an IPMSM operates with the d-and q-axes currents at this optimal point, then it can produce the MTPA. Such operation method with the optimal combination of d-and q-axes currents that leads to the smallest stator current for producing the given torque command is called the *MTPA control*. The optimal combination of the d-and q-axes currents varies with the required torque.

Now we will find out the optimal combination of d-and q-axes currents for a given torque command [8]. First, the torque expression and d-and q-axes currents are normalized as

$$T_{en} = \frac{T_e}{T_b}, \quad i_{dn} = \frac{i^r_{ds}}{i_b}, \quad i_{qn} = \frac{i^r_{qs}}{i_b} \tag{5.72}$$

where $i_b(=\phi_f/(L_{ds}-L_{qs}))$ and $T_b(=\frac{P}{2}\frac{3}{2}\phi_f i_b)$.

The normalized torque can be rewritten as

$$\begin{aligned} T_{en} = \frac{T_e}{T_b} &= \frac{\left[i^r_{qs} + (L_{ds}-L_{qs})\dfrac{i^r_{ds}i^r_{qs}}{\phi_f} \right]}{i_b} \\ &= \frac{i^r_{qs}}{i_b} - \frac{i^r_{qs}i^r_{ds}}{i_b^2} = i_{qn}(1-i_{dn}) \end{aligned} \tag{5.73}$$

The process of obtaining the optimal stator current for the MTPA control is as follows. First, find the smallest d-axis current i_{dno} for producing the torque command T_{en}. Using Eq. (5.73), the stator current can be expressed as a function of i_{dno} as

$$I^2_{sno} = i^2_{dno} + i^2_{qno} = i^2_{dno} + \left(\frac{T_{en}}{1-i_{dno}} \right) \tag{5.74}$$

Differentiating Eq. (5.74) with respect to i_{dno} and making the result equal to zero gives

$$i^4_{dno} - 3i^3_{dno} + 3i^2_{dno} + i_{dno} - T^2_{en} = 0 \tag{5.75}$$

The optimal d-axis current i_{dno} can be obtained by solving Eq. (5.75) for the given torque command.

Likewise, we can find the smallest q-axis current i_{qno} for producing the torque command T_{en}. After expressing the stator current of Eq. (5.74) as a function of i_{qno}, differentiating this with respect to i_{qno} gives

$$i^4_{qno} + T_{en}i_{qno} - T^2_{en} = 0 \tag{5.76}$$

The optimal q-axis current i_{qno} can be obtained by solving Eq. (5.76) for the given torque command.

Fig. 5.28 shows the optimal current commands for the MTPA operation by solving Eqs. (5.75) and (5.76) for the torque commands. A look-up table for the optimal current commands is commonly used instead of online calculations.

In a speed controller the command to produce the output torque is often given as a stator current. For this stator current command I^*_s, we can obtain the

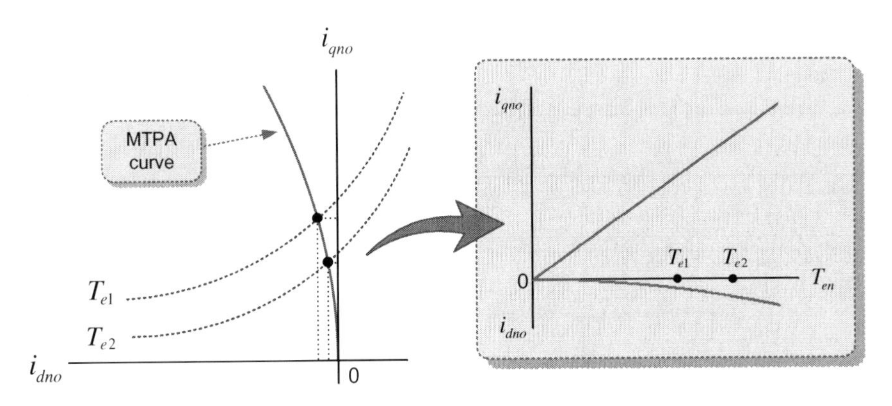

FIGURE 5.28

d–q axes current commands for MTPA operation.

d–*q* axes stator current commands for the MTPA operation as follows [9]. The torque expression of Eq. (5.71) can be rewritten as a function of the stator current I_s^* as

$$T_e = \frac{P}{2}\frac{3}{2}\left[\phi_f I_s^* \sin\beta + \frac{(L_{ds}-L_{qs})}{2} I_s^{*2} \sin 2\beta\right] \quad (5.77)$$

where β denotes the angle between the *d*-axis and the stator current vector I_s^* as shown in Fig. 5.26.

The β_{max} for producing the maximum torque can be obtained by solving $dT_e/d\beta = 0$ as

$$\frac{\partial T_e}{\partial \beta} = \frac{P}{2}\frac{3}{2}\left[\phi_f I_s^* \cos\beta + (L_{ds}-L_{qs}) I_s^{*2} \cos 2\beta\right] = 0$$
$$\rightarrow \beta_{\max} = \cos^{-1}\left(\frac{-\phi_f + \sqrt{\phi_f^2 + 8(L_{ds}-L_{qs})^2 I_s^{*2}}}{4(L_{ds}-L_{qs})I_s^*}\right) \quad (5.78)$$

From β_{max}, the optimal *d*- and *q*-axes current commands are given as

$$i_{dso}^r = I_s^* \cos\beta_{\max} = \frac{-\phi_f + \sqrt{\phi_f^2 + 8(L_{ds}-L_{qs})^2 I_s^{*2}}}{4(L_{ds}-L_{qs})} \quad (5.79)$$

$$i_{qso}^r = I_s^* \sin\beta_{\max} \quad (5.80)$$

To obtain the optimal currents for the MTPA operation by using the methods just mentioned, the exact knowledge of stator inductances and magnetic

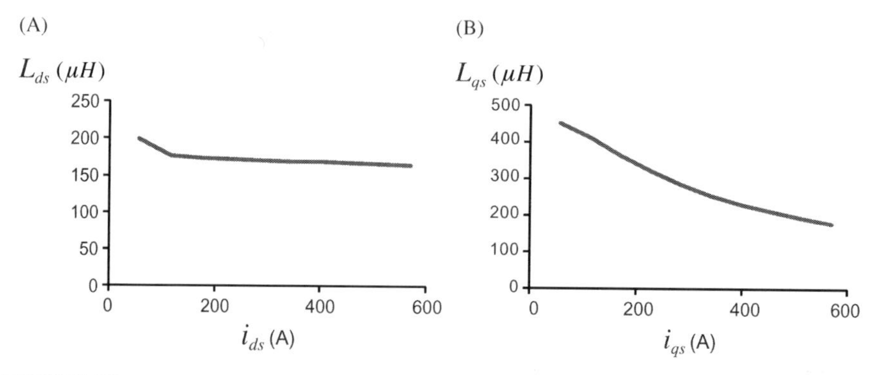

FIGURE 5.29

Variation of inductances with currents. (A) *d*-axis inductance and (B) *q*-axis inductance.

flux is necessary. The magnetic flux varies considerably with the operating temperature. In addition, each inductance value of $d-q$ axes depends on the magnitude of the current of each axis. As an example, the variation of $d-q$ axes inductances of an 80-kW IPMSM according to each current is shown in Fig. 5.29. Because of the presence of a permanent magnet, the d-axis magnetic circuit is in the saturated condition even without the current. Thus the d-axis inductance does not change much with the d-axis current as shown in Fig. 5.29A. On the other hand, the q-axis magnetic circuit consists of an iron core which is easily saturated with its current, and thus the q-axis inductance may be changed significantly with the q-axis current as shown in Fig. 5.29B.

Furthermore, it is also known that each inductance of $d-q$ axes may change with the current of the other axis due to cross-saturation effect between the $d-q$ axes [10,11]. Therefore the exact knowledge of the inductances and magnetic flux at various operating conditions is very important for a successful fulfillment of the MTPA control.

The diagram of the vector control system for an IPMSM is shown in Fig. 5.30. For the output torque command, the $d-q$ axes stator current commands for the MTPA operation are given from the calculations or a look-up table as stated earlier. If the stator currents of the IPMSM can be regulated rapidly to follow these current commands by using a synchronous reference frame current regulator, then the instantaneous torque control of the IPMSM can be achieved. When using this synchronous current regulator, we need to transform the three-phase stator currents, i_{as}, i_{bs}, i_{cs}, of the IPMSM into synchronous dq currents i^r_{ds} and i^r_{qs} by using the rotor flux angle θ_r.

Fig. 5.31 shows the experimental results for the vector control on an 800-W, 8-pole IPMSM when the speed command is changed from 200 to 2000 r/min.

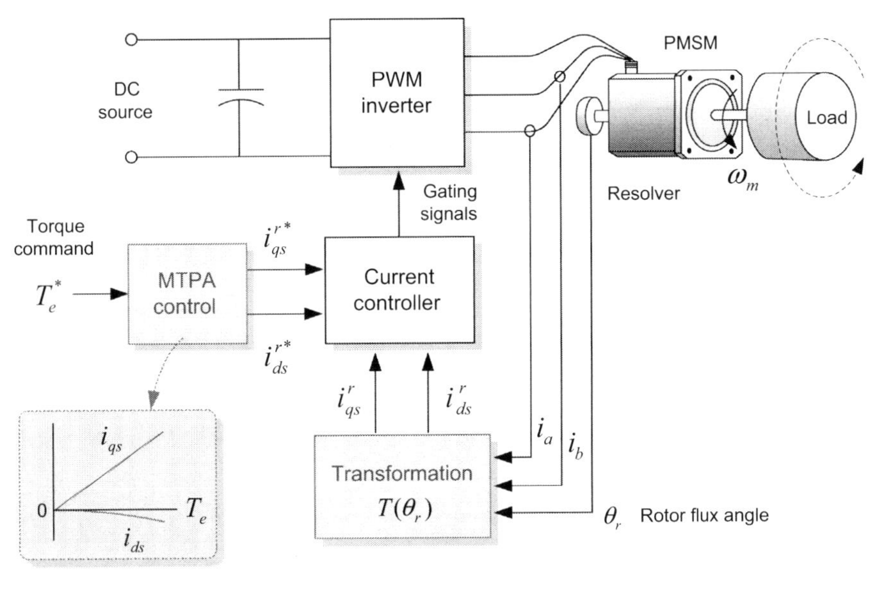

FIGURE 5.30

Block diagram of the vector control system for a PMSM.

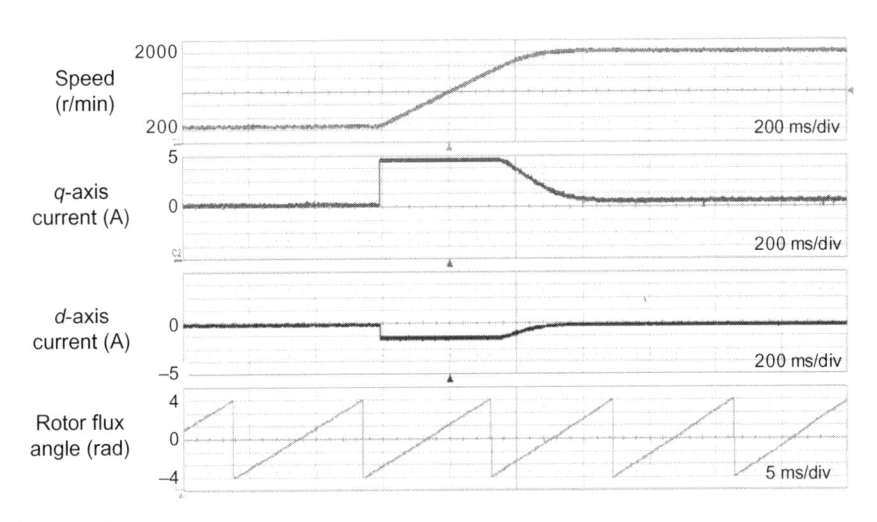

FIGURE 5.31

Experimental results for the vector control on an 800-W IPMSM.

MATLAB SIMULINK SIMULATION: VECTOR CONTROL OF A PERMANENT MAGNET SYNCHRONOUS MOTOR

- Overall simulation block (not including an inverter)

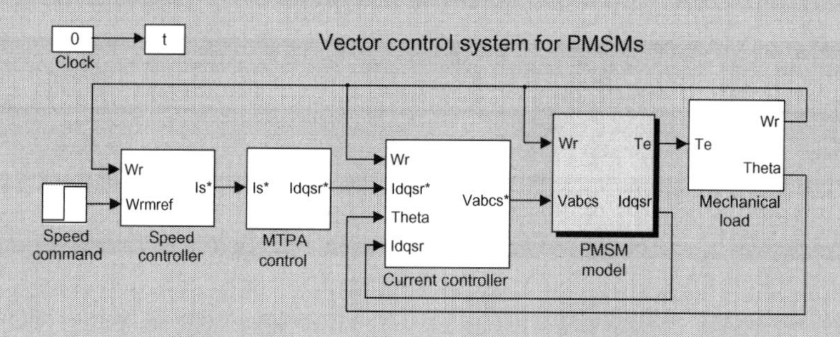

- PMSM model block
  d–q axes model in the rotor reference frame (see Chapter 4)

$$v^r_{ds} = R_s i^r_{ds} + \frac{d\lambda^r_{ds}}{dt} - \omega_r \lambda^r_{qs}$$

$$v^r_{qs} = R_s i^r_{qs} + \frac{d\lambda^r_{qs}}{dt} + \omega_r \lambda^r_{ds}$$

$$\lambda^r_{ds} = L_{ds} i^r_{ds} + \phi_f$$

$$\lambda^r_{qs} = L_{qs} i^r_{ds}$$

$$T_e = \frac{P}{2}\frac{3}{2}\left[\phi_f i^r_{qs} + (L_{ds} - L_{qs}) i^r_{ds} i^r_{qs}\right]$$

- MTPA control block

$$d\text{-axis current command:} \quad I^r_{ds} = \frac{-\phi_f + \sqrt{\phi_f^2 + 8(L_{ds} - L_{qs})^2 I_s^2}}{4(L_{ds} - L_{qs})}$$

$$q\text{-axis current command:} \quad I^r_{qs} = \sqrt{I_s^2 - I^{r2}_{ds}}$$

- Other blocks
  - Mechanical load model: refer to Chapter 1.
  - Speed controller model: refer to Chapter 2.
  - Current controller model: Chapter 6.
- Simulation conditions
  A SPMSM and an IPMSM will be simulated for acceleration characteristics when speed commands are 1200 r/min (SPMSM) and 1500 r/min (IPMSM), respectively.
- Simulation results

(*Continued*)

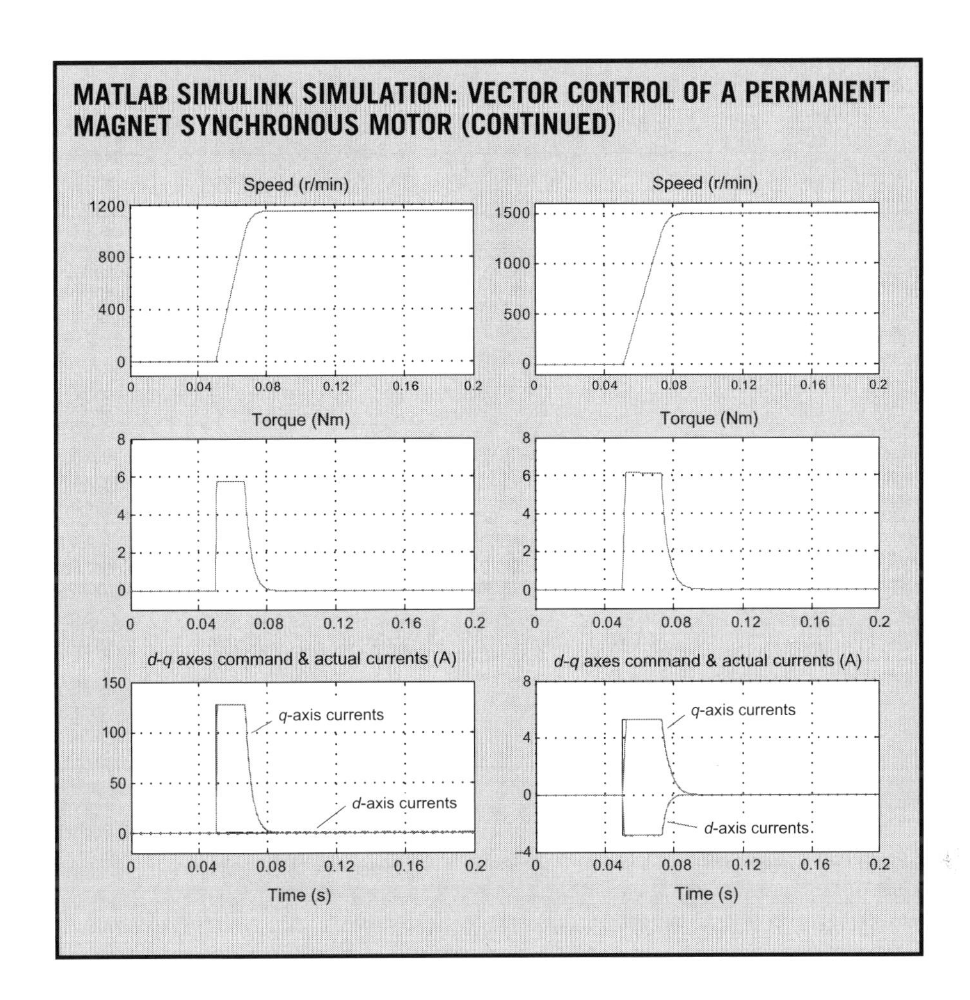

REFERENCES

[1] M. Ruff, H. Grotstollen, Off-line identification of the electrical parameters of an industrial servo drive system, in: Conf. Rec. IEEE IAS Annu. Meeting, 1996, pp. 213–220.

[2] T.M. Rowan, R.J. Kerkman, D. Leggate, A simple on-line adaptation for indirect field orientation of an induction machine, in: Conf. Rec. IEEE IAS Annu. Meeting, 1989, pp. 579–587.

[3] J.K. Seok, S.I. Moon, S.K. Sul, Induction machine parameter identification using PWM inverter at standstill, IEEE Trans. Energy Convers. 12 (2) (Jun. 1997) 127–132.

[4] K.B. Nordin, D.W. Novotny, D.S. Zinger, The influence of motor parameters deviations in feedforward field orientation drive systems, IEEE Trans. Ind. Appl. IA-21 (4) (Jul./ Aug. 1985) 1009–1015.

[5] D. Fodor, D. Diana, G. Griva, F. Profumo, IFO control performance considering parameters detuning and rotor speed error, in: Conf. Rec. IEEE IAS Annu. Meeting, 1994, pp. 719–725.

[6] P.L. Jansen, R.D. Lorenz, A physically insightful approach to the design and accuracy assessment of flux observers for field oriented induction machine drives, IEEE Trans. Ind. Appl. 30 (1) (Jan./Feb. 1994) 101–110.

[7] H. Kuboda, M. Matsuse, T. Nakano, DSP-based speed adaptive flux observer of induction motor, IEEE Trans. Ind. Appl. 29 (2) (Mar./Apr. 1993) 344–348.

[8] T.M. Jahns, G.B. Kliman, T.W. Neumann, Interior permanent-magnet synchronous motors for adjustable-speed drives, IEEE Trans. Ind. Applicat. IA-22 (4) (Jul./Aug. 1986) 738–747.

[9] S. Morimoto, T. Hirasa, Servo drive system and control characteristics of salient pole permanent magnet synchronous motor, IEEE Trans. Ind. Appl. 29 (2) (Mar./Apr. 1993) 338–343.

[10] T. Gopalarathnam, R. McCann, Saturation and armature reaction effects in surface-mount PMAC Motors, in: Proc. IEEE international Conf. IEMDC 2001, 2001, pp. 618–621.

[11] B. Stumberger, et al., Evaluation of saturation and cross-magnetization effects in interior permanent-magnet synchronous motor, IEEE Trans. Ind. Appl. 39 (5) (Sept./ Oct. 2003) 1264–1271.

CHAPTER

Current regulators of alternating current motors

6

A current regulator plays a role in making a motor current achieve its desired value, and is an essential part of the vector control system for alternating current (AC) motors. For an accurate instantaneous torque control by using the vector control, the actual motor currents must follow the current commands required to produce the flux and torque regardless of hindrances such as back-electromotive force (back-EMF), leakage inductance, and resistance of windings. Thus for the implementation of the vector control system, it is very important to design the current regulator well. In this chapter we will examine the current regulation techniques for the vector control system of AC motors.

The system configuration for the current control of a three-phase load is shown in Fig. 6.1. The current regulator plays a role in generating gating signals for the switching devices of a pulse width modulation (PWM) inverter, which can produce the output voltage to make the required current flow into the three-phase load.

In other words, the main function of the current regulator is to transform the current error into gating signals for the switching devices in the time domain.

For the case of using a voltage source type PWM inverter, there have been four different types of current regulators researched as:

- Predictive current regulator
- Hysteresis current regulator
- Ramp comparison current regulator or sine-triangle comparison regulator
- $d-q$ axes current regulator

A current regulator consists of an error compensation part and a voltage modulation part as shown in Fig. 6.2. The error compensation part produces a command voltage to reduce the error between the actual and command currents, whereas the voltage modulation part generates gating signals for switching devices to accurately produce the command voltage given by the error compensation part.

In the conventional regulators, such as a predictive current regulator, hysteresis current regulator, and ramp comparison current regulator, the error compensation part and the voltage modulation part are formed as one. In contrast, in a $d-q$ axes current regulator, these two parts are constructed separately. In this regulator a proportional integral (PI) controller is commonly used for the error compensation. Various PWM techniques are used for the voltage modulation, which will be described in Chapter 7.

Electric Motor Control. DOI: http://dx.doi.org/10.1016/B978-0-12-812138-2.00006-4

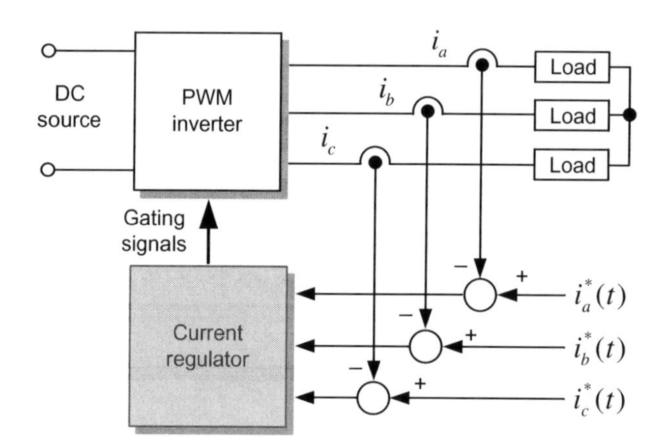

FIGURE 6.1

Current control of a three-phase load.

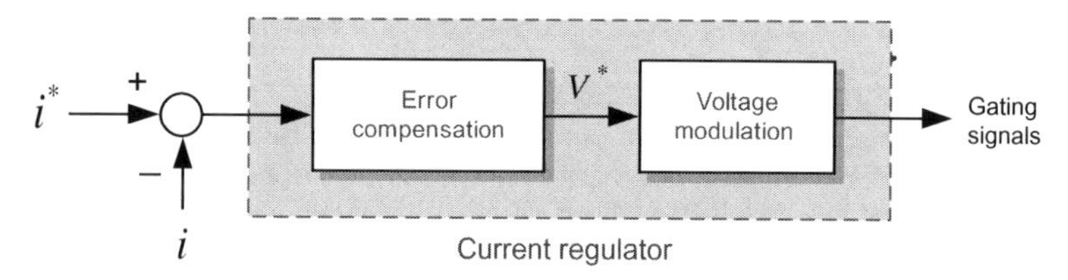

FIGURE 6.2

Configuration of a current regulator.

Now we will explore these current regulators. First, we discuss briefly the predictive current regulator, which is based on a mathematical model of the load. In the predictive current regulator the current error is calculated at every current control period, and the required voltage is determined to minimize the error using the load mathematical model. Afterward, the on/off states of the switching devices are directly chosen to produce the required voltage. The actual load current in the predictive current regulator may lag behind the command current by more than one sampling time. Furthermore, its performance will be greatly affected by the variation of parameters used in the load mathematical model. Next, we will examine the rest of the current regulators in more detail.

6.1 HYSTERESIS REGULATOR

As a type of bang–bang control, a hysteresis regulator is the simplest current controller, directly controlling the on/off states of switches according to the current error. The operation principle of a hysteresis controller is as follows [1].

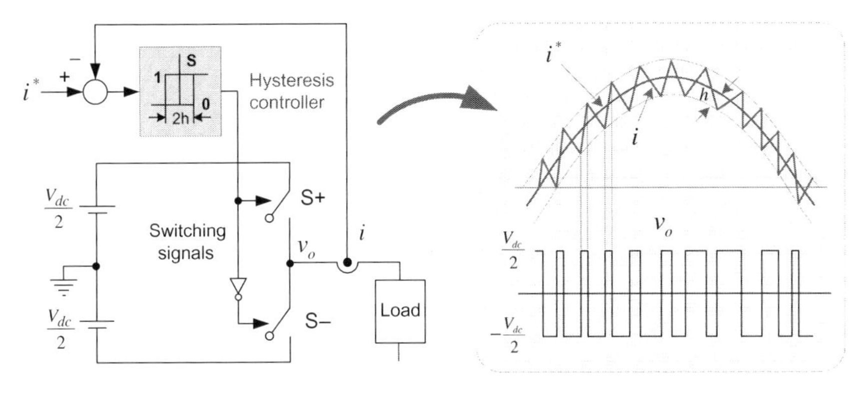

FIGURE 6.3

Operation principle of a hysteresis regulator.

In the hysteresis regulator, as shown in Fig. 6.3, if the error between the actual current and the command current is more than the preset value h (called *hysteresis band*), then the state of the switch is changed to reduce the error. In other words, the state of the switches is changed whether the actual current is greater or less than the command current by the hysteresis band h as:

- $i^* - i \leq -h$: lower switch S− is turned on to decrease the load current by producing a negative voltage $(-\frac{1}{2}V_{dc})$
- $i^* - i \geq h$: upper switch S+ is turned on to increase the load current by producing a positive voltage $(\frac{1}{2}V_{dc})$

Fig. 6.4 describes the operating range of an actual current vector **i** in the complex plane for the hysteresis control. For example, if the actual current of the phase a increases more than its command, then the actual current vector **i** moves to the positive (+) of the phase a axis. If the current error of the phase a is equal to the hysteresis band $-h$, then the current vector **i** reaches the −A line. At this moment, the lower switch (S−) of the phase a will be turned on and the output voltage will become $-V_{dc}/2$. This will cause the current to decrease. On the contrary, if the current error of the phase a becomes h, then the current vector **i** reaches the +A line. Thus the upper switch (S+) will be turned on and the output voltage will become $V_{dc}/2$. This will cause the current to increase.

Likewise, for the phases b and c, the switching actions can be done independently with their own hysteresis band. Since the error of each phase current is limited within h by the hysteresis action, the operating area of the current vector **i** can be considered to be inside the hexagon consisting of the six switching lines of the three-phase axes. However, in practice, the current errors can vary up to double the hysteresis band, $2h$, for a Y(wye)-connected three-phase load with a floating neutral point. In that load, the sum of the three-phase currents is zero (i.e., $i_{as} + i_{bs} + i_{cs} = 0$), and thus all three-phase currents cannot be controlled independently. This results in the current errors of 2h.

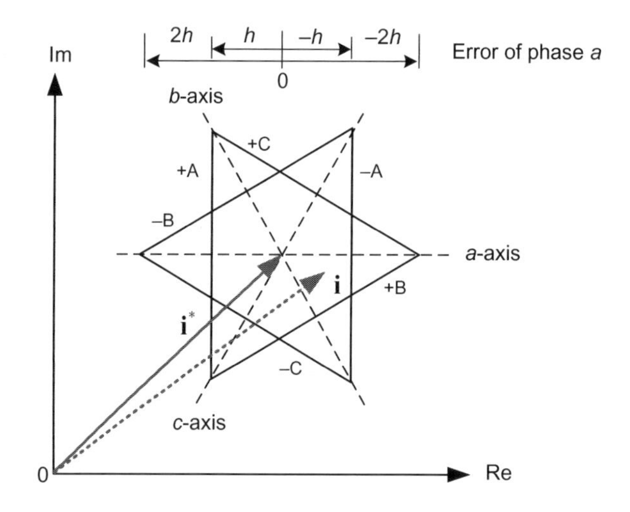

FIGURE 6.4

Operating range of current vector for hysteresis control [1].

Hysteresis current regulators have been widely used for small systems because of their simple implementation and an excellent dynamic performance. However, hysteresis controllers are inherently an analog controller. In addition, the switching frequency is not constant since the turn-on/off instants of switches can change with the back-EMF and load condition. This makes the thermal design of a switching power converter and the filter design of switching noise elimination difficult because the losses and the harmonics generated by the switching actions are a function of the switching frequency. For these reasons, its usage nowadays is limited. In addition, the switching frequency may increase sharply (called *limit cycle*) at low operating frequencies, where only the effective voltage vectors are more likely to be selected due to a small back-EMF. This problem may be improved by adding an offset to the hysteresis band.

6.2 RAMP COMPARISON CURRENT REGULATOR [1]

In a ramp comparison current regulator as shown in Fig. 6.5, the switching states are determined by comparing the current error with the triangular carrier wave based on the following principle:

- Current error i_{err} > carrier wave: upper switch S+ is turned on to produce a positive voltage ($\frac{1}{2}V_{dc}$).
- Current error i_{err} < carrier wave: lower switch S− is turned on to produce a negative voltage ($-\frac{1}{2}V_{dc}$).

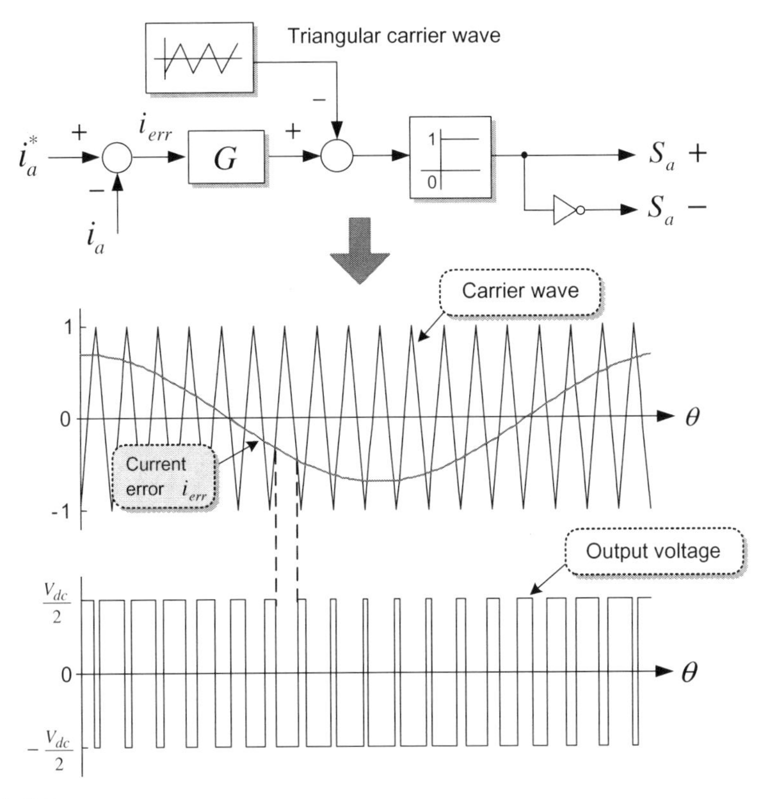

FIGURE 6.5

Operation principle of the ramp comparison current regulator.

For a proper operation of this regulator, the current error should be less than the triangle carrier wave. In this regulator it can be seen that the switching states are changed only at the intersection of the current error and the triangular carrier wave. Thus the switching frequency is equal to the triangle wave frequency and becomes constant. This is a big advantage of this regulator over the hysteresis regulators. However, one major drawback of this regulator is that it has steady-state magnitude and phase errors in the resultant current.

The performance of this regulator is shown in Fig. 6.6. We can see that the actual current has steady-state errors in amplitude and phase with regard to the reference current. These errors increase with the back-EMF, and thus it is hard to obtain a good current regulation performance at high speed operation. A compensator G such as a P (proportional) or PI controller is often used to reduce these errors as shown in Fig. 6.5. Although larger gains can reduce the errors, there is a limit to increasing the gains due to the increase in noise sensitivity. Moreover, a PI controller is unsatisfactory for alternating current (AC) regulation because it can eliminate the control error completely for only direct current (DC) regulation.

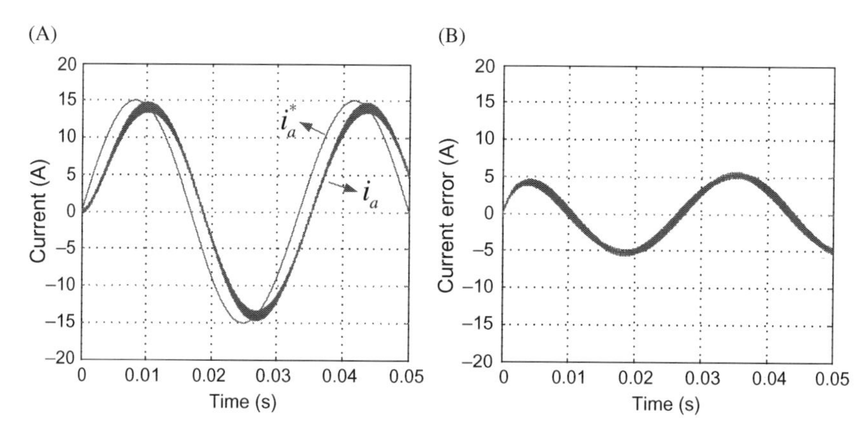

FIGURE 6.6

Performance of the ramp comparison current control. (A) Reference current and actual current and (B) current error.

6.3 *d-q* AXES CURRENT REGULATORS

Since a three-phase AC motor has three stator currents, we may consider at first that three current controllers are needed to regulate their currents individually. However, since their windings are commonly connected in wye with a floating neutral point, the sum of their currents is equal to zero and thus only two of the currents are controllable independently. In addition, the vector control uses the two currents of *d*- and *q*-axes, which the three-phase currents are transformed into. Therefore it is natural to use only two independent current controllers for the control of the three-phase currents of an AC motor.

Now we will introduce the regulator to control the two currents of *d*- and *q*-axes.

6.3.1 STATIONARY REFERENCE FRAME *d-q* CURRENT REGULATOR [2,3]

To begin with, consider the current control for a typical three-phase load as shown in Fig. 6.7. A three-phase load can be generalized as a circuit of $R-L$ and back-EMF and can be expressed as

$$\begin{aligned} v_{as} &= Ri_{as} + L\frac{di_{as}}{dt} + e_{as} \\ v_{bs} &= Ri_{bs} + L\frac{di_{bs}}{dt} + e_{bs} \\ v_{cs} &= Ri_{cs} + L\frac{di_{cs}}{dt} + e_{cs} \end{aligned} \tag{6.1}$$

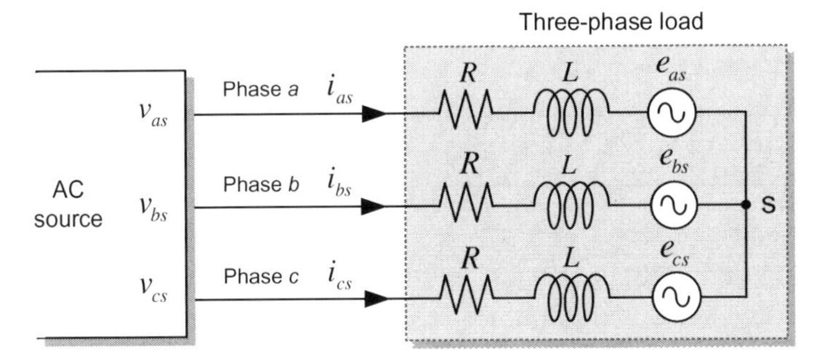

FIGURE 6.7

Three-phase load.

where v_{as}, v_{bs}, v_{cs}, i_{as}, i_{bs}, i_{cs}, and e_{as}, e_{bs}, e_{cs} are three-phase voltages, currents, and back-EMFs, respectively, and R and L are the resistance and inductance of the load, respectively.

Applying the axis transformation described in Chapter 4 we can transform the three-phase voltage equations of Eq. (6.1) into d–q forms in the stationary reference frame as

$$v_{ds}^s = Ri_{ds}^s + L\frac{di_{ds}^s}{dt} + e_{ds}^s \tag{6.2}$$

$$v_{qs}^s = Ri_{qs}^s + L\frac{di_{qs}^s}{dt} + e_{qs}^s \tag{6.3}$$

The current regulator for the d- and q-axes currents in the stationary reference frame based on Eqs. (6.2) and (6.3) is shown in Fig. 6.8.

In this stationary reference frame d–q axes current regulator, the three-phase load currents i_{as}, i_{bs}, i_{cs} should be transformed into the stationary frame d–q currents to be used as feedback currents. The PI control is normally used to regulate these currents to the desired values. The PI regulators produce d–q axes reference voltages v_{ds}^{s*}, v_{qs}^{s*} to eliminate the current errors, and these voltages are transformed into three-phase reference voltages v_{as}^*, v_{bs}^*, v_{cs}^*. A voltage source PWM inverter, which will be discussed in detail in Chapter 7.

Let us evaluate the characteristic of the stationary reference frame d–q current regulator. Fig. 6.9 shows the block diagram of this regulator.

The transfer function of this system is given as

$$I_{dqs}(s) = \frac{K_p s + K_i}{Ls^2 + (R + K_p)s + K_i} I_{dqs}^*(s) - \frac{s}{Ls^2 + (R + K_p)s + K_i} E_{dqs}(s) \tag{6.4}$$

From Eq. (6.4), it can be seen that if the current reference $I_{dqs}^*(s)$ is an AC quantity (i.e., $s \neq 0$), the actual current $I_{dqs}(s)$ never follows its reference $I_{dqs}^*(s)$ unless the current regulator has infinite gains. Thus when AC currents are regulated by the stationary regulator, the amplitude and phase errors will exist in the steady state.

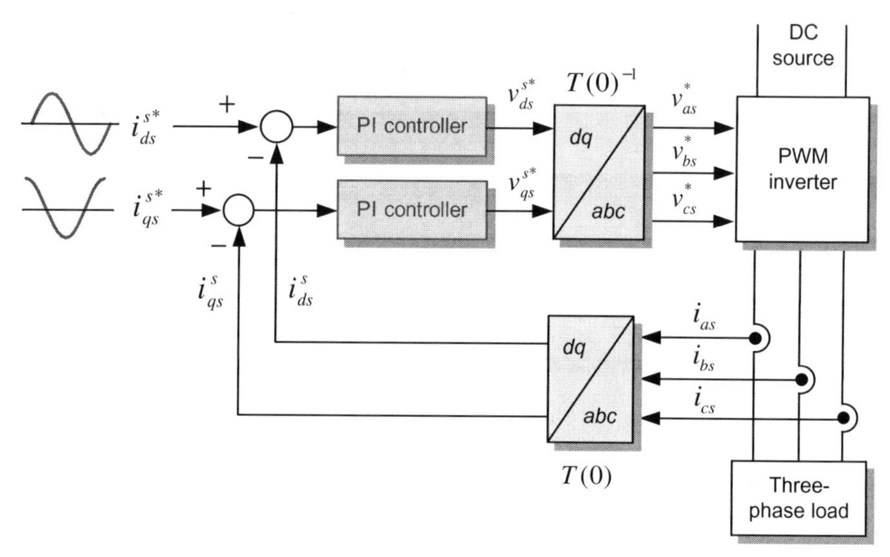

FIGURE 6.8

Stationary reference frame d–q axes current regulator.

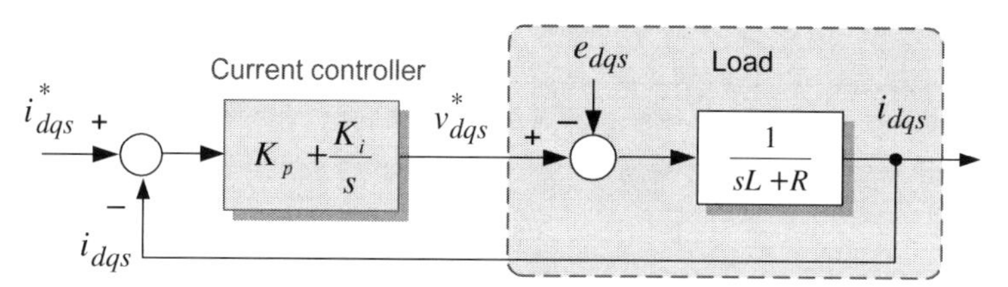

FIGURE 6.9

Block diagram for one axis of a stationary frame current regulator.

On the other hand, if the current reference $I_{dqs}^{*}(s)$ is a DC quantity (i.e., $s = 0$), Eq. (6.4) becomes

$$\frac{I_{dqs}(s)}{I_{dqs}^{*}(s)}\Big|_{s=0} = 1 \tag{6.5}$$

This indicates that the actual current $I_{dqs}(s)$ can follow its reference value $I_{dqs}^{*}(s)$ accurately, i.e., the steady-state error is zero. This fact implies that instead of controlling the load current as an AC quantity directly, it is desirable to control the load current transformed as a DC quantity. For this purpose, the current control needs to be performed in the synchronous reference frame, where the currents of the AC load can be given as DC quantities. This regulator is called the *synchronous reference frame d–q current regulator*, which is widely used for the current control of AC systems.

6.3.2 SYNCHRONOUS REFERENCE FRAME *d-q* CURRENT REGULATOR [2]

Transforming the stationary frame voltage equations of Eqs. (6.2) and (6.3) into the synchronous reference frame, which is rotating at the electrical angular frequency ω_e corresponding to the operating frequency of the three-phase currents, gives

$$v^e_{ds} = Ri^e_{ds} + L\frac{di^e_{ds}}{dt} - \omega_e Li^e_{qs} + e^e_{ds} \tag{6.6}$$

$$v^e_{qs} = Ri^e_{qs} + L\frac{di^e_{qs}}{dt} + \omega_e Li^e_{ds} + e^e_{qs} \tag{6.7}$$

Unlike the stationary reference frame, in these equations, all the electrical variables such as currents, voltages, and back-EMFs are all DC quantities in the steady state. It should be noted that, in addition to the back-EMFs e^e_{ds} and e^e_{qs}, there are $-\omega_e Li^e_{qs}$ and $\omega_e Li^e_{ds}$ (called *speed voltages*) in the synchronous reference frame expressions of Eqs. (6.6) and (6.7). These speed voltages are cross-coupled between *d*- and *q*-axes. Thus a change in the *d*-axis current may affect the control of the *q*-axis current and vice versa. These are the major disturbances on this synchronous current regulator. Thus the feedforward compensation of the speed voltages as well as the back-EMF voltages should be employed to achieve a good current control performance. This will be discussed in detail in a later section.

The block diagram for a current regulator operating on the synchronous reference frame is shown in Fig. 6.10.

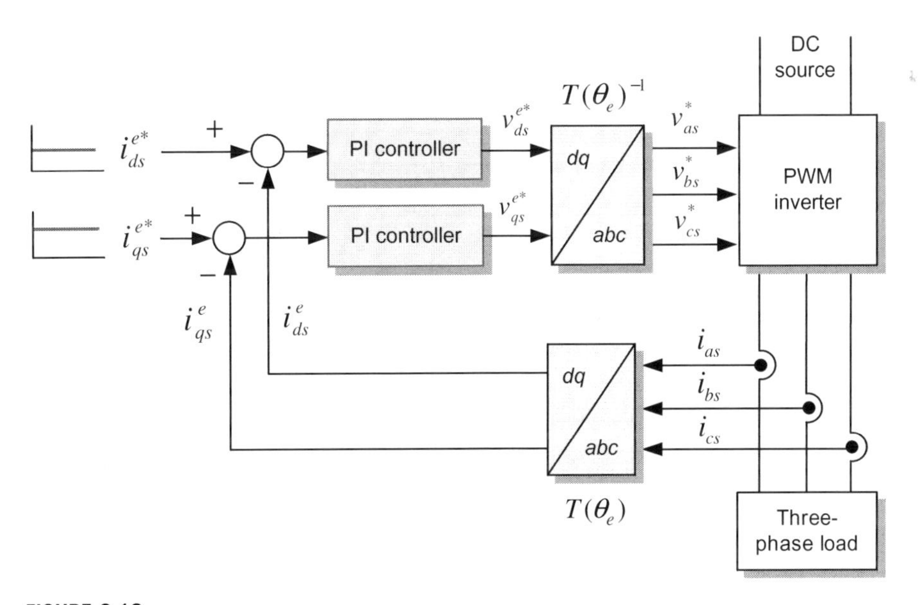

FIGURE 6.10

Block diagram of a synchronous frame current regulator.

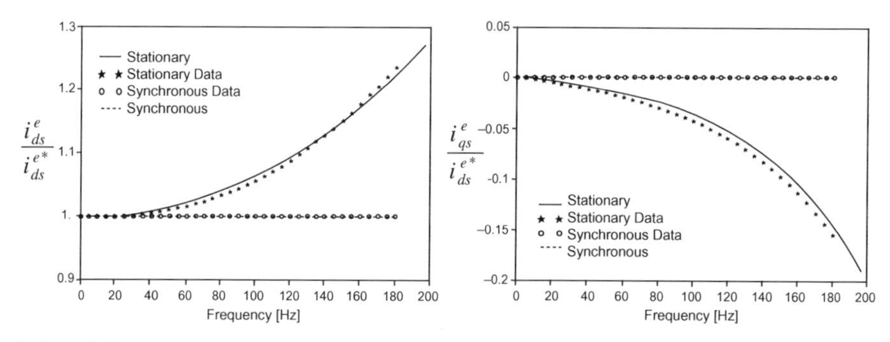

FIGURE 6.11

Comparison between the stationary and the synchronous frame regulators [2].

The structure of this regulator seems to be similar to that of a stationary frame current regulator in Fig. 6.8, but the controlled currents are DC quantities and a different transformation is used in this system. Since the synchronous reference frame uses DC quantities, the currents can be regulated well by using a PI controller.

Fig. 6.11 depicts the performance comparison between the stationary frame and the synchronous frame current regulators in the steady state for $R-L$ load [2]. From the left figure showing the performance of the d-axis current control, it can be seen that the steady-state error for the stationary frame current regulator increases with the operating frequency of the reference current. On the other hand, the synchronous frame current regulator has no steady-state error, regardless of the operating frequency. In addition, we can see that there is a cross-coupling between the axes for the stationary frame current regulator in the right figure. The control error for the q-axis current increases as the operating frequency of the d-axis reference current increases for the stationary frame current regulator.

6.3.3 GAIN SELECTION OF THE SYNCHRONOUS REFERENCE FRAME PI CURRENT REGULATOR

In Chapter 2, we described how to select the PI gains of a current controller for DC motors. Since a synchronous reference frame current regulator also controls DC quantities, we can directly use such method to select the PI gains of a synchronous reference frame current regulator. By referring to the gain selection of the current controller for a DC motor in Section 2.6, we can determine the proportional gain and the integral gain of a synchronous reference frame current regulator for the three-phase AC load as:

$$\text{Proportional gain:} \quad K_p = L \cdot \omega_c \tag{6.8}$$

$$\text{Integral gain:} \quad K_i = R \cdot \omega_c \tag{6.9}$$

where R and L denote the resistance and inductance of the load, respectively, and ω_c denotes the control bandwidth of the current regulator. Here, the proportional gain is determined by the required control bandwidth ω_c of the current regulator, and the integral gain is determined by the relationship of $K_p/K_i = L/R$ *(pole-zero cancellation)*.

On the basis of the current control for the three-phase AC load as mentioned earlier, it can readily be seen that it is better to use a synchronous frame current regulator for the current control of AC motors. For this case, the PI gains can be obtained directly from Eqs. (6.8) and (6.9). However, since the resistance and inductance values differ from system to system, we need to find the equivalent R and L values corresponding to the chosen AC motor. The values of these equivalent parameters for induction motors and synchronous motors can be obtained as follows.

6.3.3.1 Proportional–integral gains for induction motors

From Eqs. (4.86) and (4.87), the $d-q$ voltage equations of an induction motor in the synchronous reference frame are given as

$$v_{ds}^e = Ri_{ds}^e + p\lambda_{ds}^e - \omega_e\lambda_{qs}^e \tag{6.10}$$

$$v_{qs}^e = Ri_{qs}^e + p\lambda_{qs}^e + \omega_e\lambda_{ds}^e \tag{6.11}$$

The stator flux linkages in these equations can be expressed as the rotor flux linkages and the stator currents as

$$\begin{aligned}\lambda_{ds}^e &= L_s i_{ds}^e + L_m i_{dr}^e \\ &= L_s i_{ds}^e + L_m\left(\frac{\lambda_{dr}^e - L_m i_{ds}^e}{L_r}\right) = \sigma L_s i_{ds}^e + \frac{L_m}{L_r}\lambda_{dr}^e\end{aligned} \tag{6.12}$$

$$\begin{aligned}\lambda_{qs}^e &= L_s i_{qs}^e + L_m i_{qr}^e \\ &= L_s i_{qs}^e + L_m\left(\frac{\lambda_{qr}^e - L_m i_{qs}^e}{L_r}\right) = \sigma L_s i_{qs}^e + \frac{L_m}{L_r}\lambda_{qr}^e\end{aligned} \tag{6.13}$$

where $\sigma = 1 - L_m^2/L_sL_r$ and the rotor currents can be obtained from the rotor flux linkages as

$$i_{dr}^e = \frac{\lambda_{dr}^e - L_m i_{ds}^e}{L_r} \leftarrow \lambda_{dr}^e = L_r i_{dr}^e + L_m i_{ds}^e \tag{6.14}$$

$$i_{qr}^e = \frac{\lambda_{qr}^e - L_m i_{qs}^e}{L_r} \leftarrow \lambda_{qr}^e = L_r i_{qr}^e + L_m i_{qs}^e \tag{6.15}$$

Substituting Eqs. (6.12) and (6.13) into Eqs. (6.10) and (6.11) gives

$$\begin{aligned} v_{ds}^e &= R_s i_{ds}^e + p\lambda_{ds}^e - \omega_e \lambda_{qs}^e \\ &= R_s i_{ds}^e + p\left(\sigma L_s i_{ds}^e + \frac{L_m}{L_r}\lambda_{dr}^e\right) - \omega_e\left(\sigma L_s i_{qs}^e + \frac{L_m}{L_r}\lambda_{qr}^e\right) \\ &= \left(R_s + R_r\frac{L_m^2}{L_r^2}\right) i_{ds}^e + \sigma L_s \frac{di_{ds}^e}{dt} - \omega_e \sigma L_s i_{qs}^e - R_r\frac{L_m}{L_r^2}\lambda_{dr}^e - \omega_r \frac{L_m}{L_r}\lambda_{qr}^e \end{aligned} \tag{6.16}$$

$$\begin{aligned} v_{qs}^e &= R_s i_{qs}^e + p\lambda_{qs}^e + \omega_e \lambda_{ds}^e \\ &= R_s i_{qs}^e + p\left(\sigma L_s i_{qs}^e + \frac{L_m}{L_r}\lambda_{qr}^e\right) + \omega_e\left(\sigma L_s i_{ds}^e + \frac{L_m}{L_r}\lambda_{dr}^e\right) \\ &= \left(R_s + R_r\frac{L_m^2}{L_r^2}\right) i_{qs}^e + \sigma L_s \frac{di_{qs}^e}{dt} + \omega_e \sigma L_s i_{ds}^e - R_r\frac{L_m}{L_r^2}\lambda_{qr}^e + \omega_r \frac{L_m}{L_r}\lambda_{dr}^e \end{aligned} \tag{6.17}$$

In the vector control, since $\lambda_{qr}^e = 0$, the above equations will be reduced as

$$v_{ds}^e = \left(R_s + R_r\frac{L_m^2}{L_r^2}\right) i_{ds}^e + \sigma L_s \frac{di_{ds}^e}{dt} - \omega_e \sigma L_s i_{qs}^e - R_r\frac{L_m}{L_r^2}\lambda_{dr}^e \tag{6.18}$$

$$v_{qs}^e = \left(R_s + R_r\frac{L_m^2}{L_r^2}\right) i_{qs}^e + \sigma L_s \frac{di_{qs}^e}{dt} + \omega_e \sigma L_s i_{ds}^e + \omega_r \frac{L_m}{L_r}\lambda_{dr}^e \tag{6.19}$$

From these, the equivalent R and L values of induction motors are given as

$$R = R_s + R_r\left(\frac{L_m}{L_r}\right)^2 \tag{6.20}$$

$$L = \sigma L_s \tag{6.21}$$

Thus the PI gains of a synchronous reference frame current regulator for induction motors are given as

$$\text{Proportional gain:} \quad K_p = \sigma L_s \cdot \omega_c \tag{6.22}$$

$$\text{Integral gain:} \quad K_i = \left[R_s + R_r\left(\frac{L_m}{L_r}\right)^2\right] \cdot \omega_c \tag{6.23}$$

6.3.3.2 Proportional–integral gains for permanent magnet synchronous motors

From Eqs. (4.106), (4.107), (4.113), and (4.114), the d–q voltage equations of a permanent magnet synchronous motor (PMSM) in the synchronous reference frame rotating at the angular frequency ω_r of the rotor are given as

$$v_{ds}^r = R_s i_{ds}^r + L_{ds}\frac{di_{ds}^r}{dt} - \omega_r L_{qs} i_{qs}^r \tag{6.24}$$

$$v_{qs}^r = R_s i_{qs}^r + L_{qs}\frac{di_{qs}^r}{dt} + \omega_r (L_{ds} i_{ds}^r + \phi_f) \tag{6.25}$$

From these, it can be seen that the equivalent R value of a PMSM for the synchronous frame current regulator is the stator resistance R_s, and the equivalent L value is the stator inductance. For a surface-mounted PMSM, the d- and q-axes inductances have the same value. However, since they are different for an interior PMSM, the equivalent L value according to the axis is different as the following.

$$R = R_s \tag{6.26}$$

$$L = L_{ds} \quad (d\text{-axis}), \quad L = L_{qs} \quad (q\text{-axis}) \tag{6.27}$$

Thus the PI gains of the synchronous frame current regulator for a PMSM are given as

$$\text{Proportional gain:} \quad K_{pd} = L_{ds} \cdot \omega_c, \quad K_{pq} = L_{qs} \cdot \omega_c \tag{6.28}$$

$$\text{Integral gain:} \quad K_i = R_s \cdot \omega_c \tag{6.29}$$

6.4 FEEDFORWARD CONTROL

As described in the Section 6.3.2, a synchronous frame PI current regulator has been widely used for the current control for AC systems because it can achieve zero steady-state error and provide a good transient performance in spite of its simple structure. We saw in Section 6.3.2 that the voltage expressions of an AC system in the synchronous reference frame have cross-coupling components, $-\omega_e L i_{qs}^e$ and $\omega_e L i_{ds}^e$, between the axes in addition to back-EMF as

$$v_{ds}^e = R i_{ds}^e + L\frac{di_{ds}^e}{dt} - \omega L i_{qs}^e + e_{ds}^e \tag{6.30}$$

$$v_{qs}^e = R i_{qs}^e + L\frac{di_{qs}^e}{dt} + \omega L i_{ds}^e + e_{qs}^e \tag{6.31}$$

These are described as a block diagram in Fig. 6.12.

These voltage components can have a negative influence on the feedback current control although a synchronous frame current regulator is used. In particular, at high operating frequencies where they become large, these components can incur an oscillatory current response. If the bandwidth of the current controller is large enough, then their influence can be reduced. However, as explained in Section 2.6.1.1, the value of the gains is limited by the switching frequency and the current sampling frequency. Moreover, large gains make the system more sensitive to noise. To eliminate the effect of these disturbance components and improve the performance of the current control, the *feedforward control*

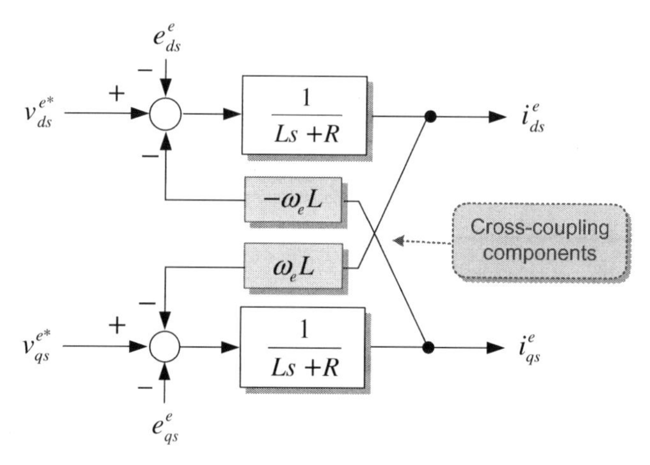

FIGURE 6.12

Voltage expressions in the synchronous frame.

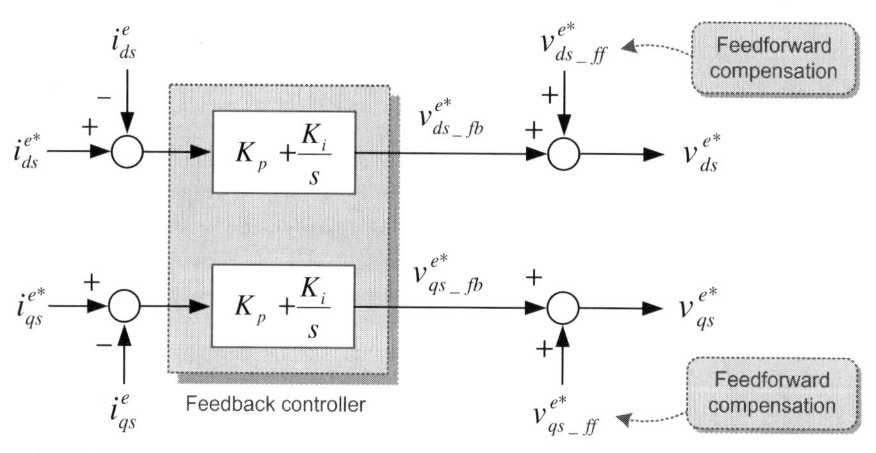

FIGURE 6.13

Current regulator with the feedforward control.

(also called the *decoupling control*), described in Section 2.5, is usually used along with the feedback control [4].

Fig. 6.13 shows a synchronous frame current regulator with the feedforward control for decoupling the cross-coupling.

When using the feedforward control, the output voltage of the synchronous frame current regulator consists of two components as follows:

$$v_{ds}^{e*} = v_{ds\_fb}^{e*} + v_{ds\_ff}^{e*} \tag{6.32}$$

$$v_{qs}^{e*} = v_{qs\_fb}^{e*} + v_{qs\_ff}^{e*} \tag{6.33}$$

where $v^{e*}_{ds\_fb}$, $v^{e*}_{qs\_fb}$ are the feedback voltage components produced by the PI current controller and $v^{e*}_{ds\_ff}$, $v^{e*}_{qs\_ff}$ are the feedforward voltage components. From Eqs. (6.30) and (6.31), the $d-q$ axes feedforward components for a three-phase $R-L$ load are given as

$$v^{e*}_{ds\_ff} = -\omega_e L i^e_{qs} + e^e_{ds} \tag{6.34}$$

$$v^{e*}_{qs\_ff} = \omega_e L i^e_{ds} + e^e_{qs} \tag{6.35}$$

In the current control of AC motors, the feedforward components vary depending on the motor. The feedforward components for an induction motor and a PMSM can be obtained as follows.

6.4.1 FEEDFORWARD CONTROL FOR INDUCTION MOTORS

To identify the feedforward voltage components for an induction motor, Eqs. (6.18) and (6.19) are rewritten as

$$v^e_{ds} = \left(R_s + R_r\frac{L_m^2}{L_r^2}\right)i^e_{ds} + \sigma L_s\frac{di^e_{ds}}{dt} \underline{- \omega_e\sigma L_s i^e_{qs} - R_r\frac{L_m}{L_r^2}\lambda^e_{dr}} \tag{6.36}$$

$$v^e_{qs} = \left(R_s + R_r\frac{L_m^2}{L_r^2}\right)i^e_{qs} + \sigma L_s\frac{di^e_{qs}}{dt} \underline{+ \omega_e\sigma L_s i^e_{ds} + \omega_r\frac{L_m}{L_r}\lambda^e_{dr}} \tag{6.37}$$

Here, the third and the fourth underlined terms on the right-hand side express the feedforward voltage components. Thus

$$v^e_{ds\_ff} = -\omega_e\sigma L_s i^e_{qs} - R_r\frac{L_m}{L_r^2}\lambda^e_{dr} \tag{6.38}$$

$$v^e_{qs\_ff} = \omega_e\sigma L_s i^e_{ds} + \omega_r\frac{L_m}{L_r}\lambda^e_{dr} \tag{6.39}$$

6.4.2 FEEDFORWARD CONTROL FOR PERMANENT MAGNET SYNCHRONOUS MOTORS

To identify the feedforward voltage components for an PMSM, Eqs. (6.24) and (6.25) are rewritten as

$$v^r_{ds} = R_s i^r_{ds} + L_{ds}\frac{di^r_{ds}}{dt} \underline{- \omega_r L_{ds} i^r_{qs}} \tag{6.40}$$

$$v^r_{qs} = R_s i^r_{qs} + L_{qs}\frac{di^r_{qs}}{dt} \underline{+ \omega_r L_{ds} i^r_{ds} + \omega_r\phi_f} \tag{6.41}$$

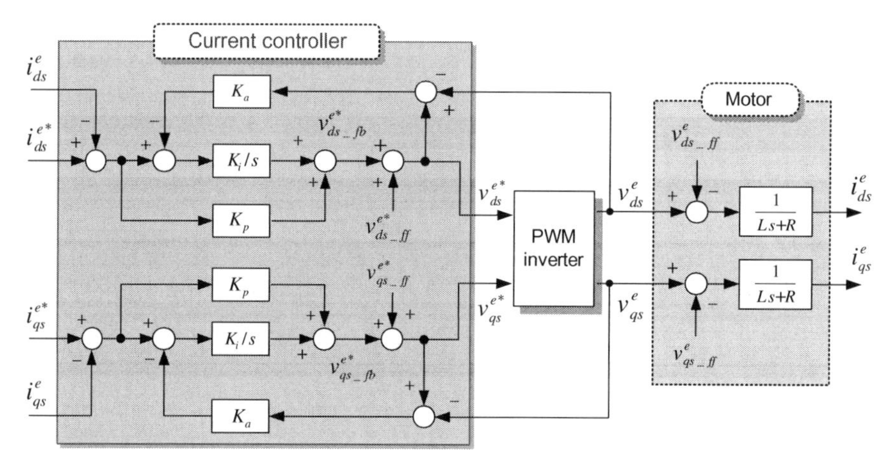

FIGURE 6.14

Synchronous frame PI current regulator with the feedforward control. *PI*, proportional–integral.

Here, the third and the fourth underlined term on the right-hand side express the feedforward voltage components. Thus

$$v^{r*}_{ds\_ff} = -\omega_r L_{qs} i^r_{qs} \tag{6.42}$$

$$v^{r*}_{qs\_ff} = \omega_r L_{ds} i^r_{ds} + \omega_r \phi_f \tag{6.43}$$

By using such feedforward control, the motor can be simplified to an $R-L$ passive circuit, so an improved current control performance can be achieved. An accurate feedforward compensation requires an accurate knowledge of inductances, flux linkage, and speed. Even when there are errors in these quantities, the feedforward control can largely reduce the effect of the disturbances by back-EMFs and cross-coupling components on the current control when compared with the feedback control alone.

In the feedforward control, it is desirable to estimate the feedforward voltage components by using the measured currents. Such feedforward control that uses measured currents for estimation is called *state feedback decoupling control*.

Fig. 6.14 shows a synchronous frame PI current regulator with the feedforward control, which has become the industry standard for the high-performance current control of AC motors. A PI current regulator needs to include an anti-windup controller to prevent integrator saturation, which was described in detail in Section 2.6.2.

6.5 COMPLEX VECTOR CURRENT REGULATOR

An ideal synchronous reference frame current regulator has a time response independent of the operating frequency. However, in reality, its transient performance degrades as the operating frequency approaches the bandwidth of the current regulator.

In the design of a PI current controller, as described in Chapter 2, the zero of the PI controller is made equal to the pole of the plant for pole-zero cancellation, i.e., $K_p/K_i = L/R$. At low operating frequencies, the controller zero can cancel the plant pole relatively completely. This allows a faster response of the system corresponding to the given control bandwidth. However, as the operating frequency increases, the plant pole and the controller zero move apart, resulting in a degradation of the system performance. The decoupling control as mentioned in Section 6.4 can make the performance of the current regulator to be independent of the operating frequency, i.e., the plant pole to be independent of the operating frequency.

A synchronous frame current regulator with the state feedback decoupling control is shown in Fig. 6.15. An accurate decoupling control requires the correct knowledge of system parameters (here, the inductance value). If the system parameters are incorrect, then the estimated value of the feedforward compensation becomes inaccurate, so the controller cannot cancel the plant pole completely. To solve this problem, a complex vector synchronous frame current regulator was proposed [5–7]. In this strategy, by using a complex vector notation, the two-input/two-output system of the synchronous frame $d-q$ current regulator is simplified to an equivalent single-input/single-output complex vector system. Because a complex vector notation is used for the analysis and design of current regulators for multiphase AC loads, this current regulator is called the *complex vector synchronous frame current regulator.*

In contrast to a current regulator with the state feedback decoupling control, a complex vector synchronous frame current regulator has decoupling control inside the PI controller without using any system parameter as shown in Fig. 6.16.

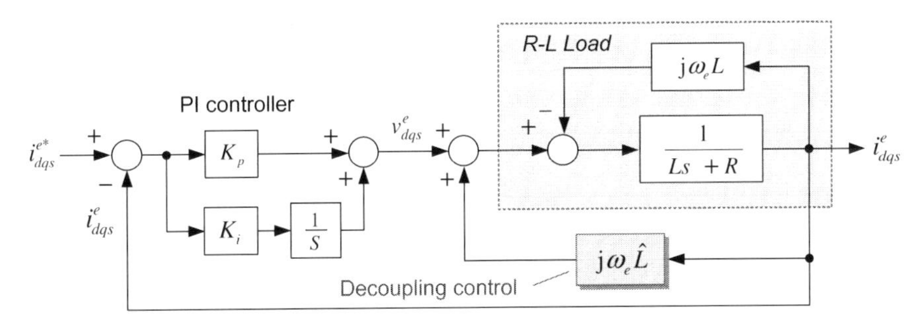

FIGURE 6.15

Synchronous current regulator with the state feedback decoupling control.

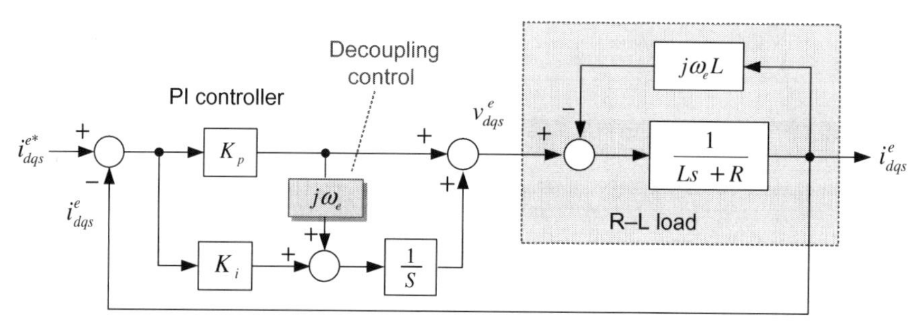

FIGURE 6.16

Complex vector synchronous frame current regulator.

Thus its performance is less sensitive to errors in system parameters, i.e., inductances, and is independent of the operating frequency. However, its response may be slightly oscillatory. If the system parameters used for decoupling control are identical to the actual system parameters, then the performances of these two current regulators are identical.

REFERENCES

[1] D.M. Brod, D.W. Novotny, Current control of VSI-PWM inverters, IEEE Trans. Ind. Appl. IA-21 (4) (1985) 562–570.

[2] T.M. Rowan, R.J. Kerkman, A new synchronous current regulator and an analysis of current-regulated PWM inverter, IEEE Trans. Ind. Appl. IA-22 (4) (1986) 678–690.

[3] M.P. Kazmierkowski, L. Malesani, Current control techniques for three-phase voltage-source PWM converters: a survey, IEEE Trans. Ind. Electron. 45 (5) (1998) 691–703.

[4] F. Briz, M.W. Degner, R.D. Lorenz, Performance of feedforward current regulators for field-oriented induction machine controllers, IEEE Trans. Ind. Appl. IA-23 (4) (1987) 597–602.

[5] F. Briz, M.W. Degner, R.D. Lorenz, Analysis and design of current regulators using complex vectors, in: Conference Record of the IEEE-IAS Annual Meeting, 1997, pp. 1504–1511.

[6] F. Briz, M.W. Degner, R.D. Lorenz, Dynamic analysis of current regulators for AC motors using complex vectors, IEEE Trans. Ind. Appl. 35 (6) (1999) 1424–1432.

[7] H. Kim, R.D. Lorenz, Analysis and design of current regulators using complex vectors, in: Conference Record of the IEEE-IAS Annual Meeting, 2004, pp. 856–863.

CHAPTER

Pulse width modulation inverters

7

In Chapter 6, it was described that the current regulators produce voltage commands for regulating the $d-q$ axes currents required in the vector control of an alternating current (AC) motor. These voltage commands are normally implemented by a pulse width modulation (PWM) inverter. As described in Chapter 3, the variable speed drives of induction motors by the *V/f* control require a PWM inverter. Inverters use the voltage modulation techniques (PWM techniques) for generating the actual voltages applied to the motor in accordance with the given voltage commands. Thus the voltage modulation technique of an inverter is a crucial factor in the AC motor drives.

In this chapter we will explore the configurations of an inverter and its voltage modulation techniques in detail.

7.1 INVERTERS

An inverter is a power electronics equipment that converts direct current (DC) power to AC power. Since an inverter can control the voltage and frequency of AC power at any value, it is essential for AC motor drives that require variable voltage and variable frequency.

Inverters can be classified into two groups according to the type of DC input source: *voltage source inverter* (VSI) and *current source inverter* (CSI). The configurations, output voltage, and current waveforms of these two inverters are shown in Fig. 7.1.

VSIs are powered by a DC voltage source. At the DC-link side of a VSI, there is usually a shunt capacitor with large capacitance to smooth out the DC input voltage. Since a VSI uses a voltage source, if the output terminals of a VSI are short-circuited, then it can be highly risky since a large current can flow. A typical output of VSI is an AC voltage of a square waveform, as shown in Fig. 7.1A. The magnitude and waveform of its output AC current depend on the connected load.

In contrast, CSIs are powered by a DC current source. Since in most cases DC current source is not available, it is normally obtained by using a large series inductor in the DC-link side powered by a DC voltage source. For CSIs, it can be dangerous when the circuit is opened because a high voltage can be produced by

Electric Motor Control. DOI: http://dx.doi.org/10.1016/B978-0-12-812138-2.00007-6

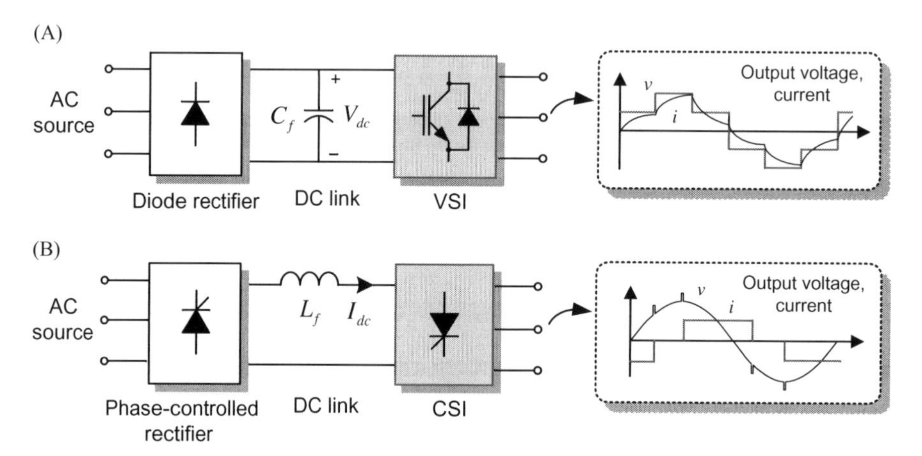

FIGURE 7.1

Two types of inverters. (A) VSI and (B) CSI.

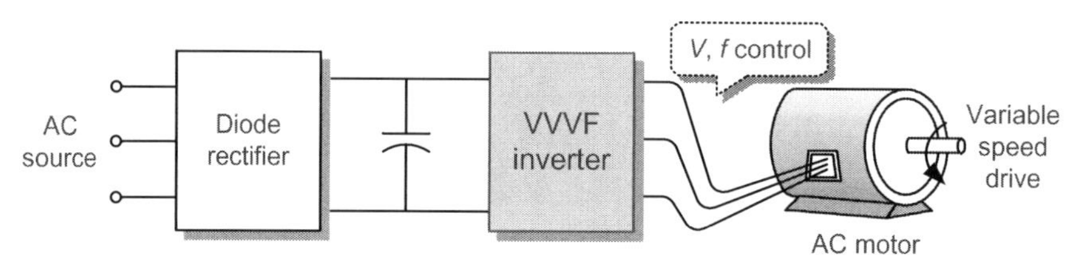

FIGURE 7.2

Inverter for AC motor drives.

Ldi/dt. A typical output of CSIs is an AC current source of a square waveform, as shown in Fig. 7.1B. The magnitude and waveform of the output voltage depend on the connected load.

During the early years of the inverter usage, CSIs were adopted in many applications. However, nowadays, VSIs have become the industry standard rather than CSIs that need a large inductor.

The inverters are used in two major applications: *AC motor drive applications* and *AC power supply applications.* In the inverter for AC motor drive applications as shown in Fig. 7.2, both the amplitude and frequency of the AC output voltages can often be varied for the variable speed drives. Thus this type of inverter is often called the *Variable Voltage Variable Frequency (VVVF)* inverter.

On the other hand, the inverter for AC power supply applications is used as a replacement of the AC voltage source supplied by the utility mains. In such an inverter both the amplitude and the frequency of the AC output voltages are fixed in accordance with the standards regarding the distribution of electricity, and thus

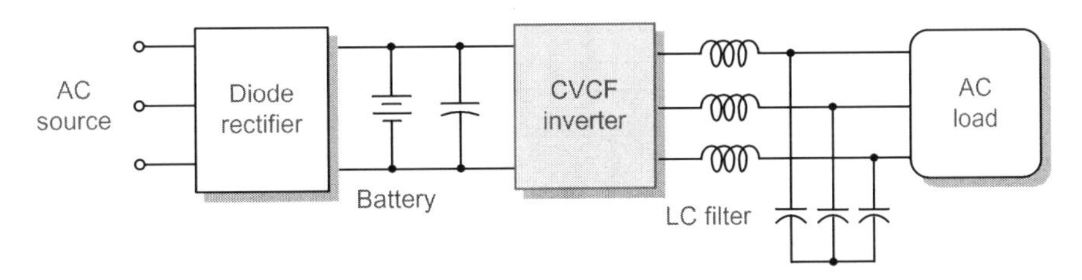

FIGURE 7.3

Example of an inverter for AC power supply.

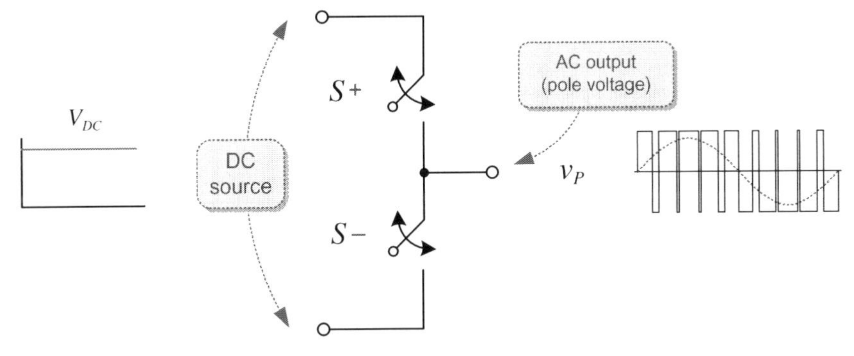

FIGURE 7.4

Basic circuit of an inverter.

this type of inverter is called the *constant voltage constant frequency (CVCF)* inverter. Typical examples of such an inverter are an *uninterruptible power supply* (UPS) that provides AC power to the load when the utility mains are not available, and a *power conditioning system* that supplies AC power from renewable energy sources like fuel cells, solar panels, or wind turbines as shown in Fig. 7.3.

Now we will describe the VSI in detail. VSIs can be classified according to whether or not the amplitude of the output voltage can be controlled. One type of VSI is the *square wave inverter*, whose frequency is controlled while the amplitude of the output voltage remains constant. Another is the *PWM inverter* in which both the amplitude and the frequency of the output voltage are controlled.

First, we will begin by looking at the basic circuit configuration of a VSI.

7.1.1 BASIC CIRCUIT CONFIGURATION OF A VOLTAGE SOURCE INVERTER

The basic circuit that forms a VSI is shown in Fig. 7.4. This basic circuit is called a *pole, arm*, or a *leg*. The basic circuit consists of a pair of switches that operate in a complementary manner to perform a rapid switching function to convert the DC power to AC power.

A DC input power source is applied to both ends of the basic circuit, and an AC output voltage is produced between the two switches. This AC output voltage, v_P, is called a *pole voltage*.

As switching devices used in the basic circuit, semiconductor power devices such as gate turn-off (GTO) thyristor, integrated gate commutated thyristor (IGCT), insulated gate bipolar transistor (IGBT), or metal oxide semiconductor field effect transistor (MOSFET) as shown in Fig. 7.5 are available on the market today. These devices are mainly selected according to the required power capacity and switching frequency. The GTO thyristor and IGCT have a high-power handling capability, but their switching frequency is very low, being below 1 kHz. The IGBT is now most widely used for medium- to high-power applications. Some of the IGBTs have switching frequencies up to 100 kHz, but most are normally below 20 kHz. The MOSFET, which can support a high switching frequency beyond 100 kHz, is the most suitable device for small-power applications. Besides these active switching devices, for driving inductive loads, it should be noted that an antiparallel diode should be connected across each switch to provide an alternate path for the load current when the switch is turned off as shown in Fig. 7.5.

Most inverter topologies have a configuration composed of basic circuits connected in parallel. However, inverters for high-power applications often employ another basic circuit with four switches connected in series as shown in Fig. 7.6.

In an inverter topology composed of the basic circuit as shown in Fig. 7.4, the waveform of the output voltage has a two-level, i.e., 0 and $+V_{dc}$, so it is called a *two-level inverter*. On the other hand, in an inverter topology composed of the basic circuit as shown in Fig. 7.6, the waveform of the output voltage has a three-level, i.e., 0, $-V_{dc}$, and $+V_{dc}$, so it is often referred to as a *three-level inverter*. A *neutral-point clamped (NPC) inverter* is a typical example of the three-level inverter [1]. An inverter circuit can even be expanded to four- or five-level topology configurations. These inverters are called a *multilevel inverter*. In this chapter, only the typical two-level inverter will be discussed.

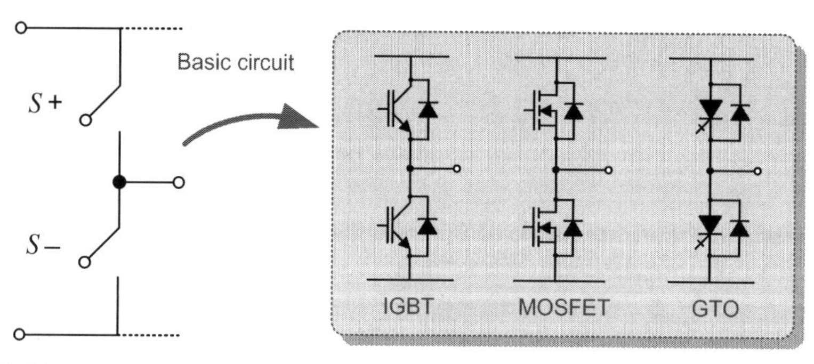

FIGURE 7.5

Switching devices for an inverter.

7.1.1.1 The output voltage of the basic circuit

The output voltage of the basic circuit, i.e., pole voltage, v_P, is determined by the switching states of two devices, $S+$ and $S-$, but is independent of the load. The pole voltage of an inverter normally is not equal to the load voltage.

There are four different situations of the basic circuit according to the on-/off-state of the two switches. When the two switches are both turned on, the DC power source is short-circuited (called *shoot-through condition*) as shown in Fig. 7.7, and this causes a current large enough to destroy the switching devices to flow. This situation should be inhibited at all times.

Hence, the two devices switch alternately, i.e., while one is on, the other one is off. Even in this case, there is still a possibility of a shoot-through when the switching devices are changing their on/off states because the turn-off time of a device is always longer than its turn-on time. Thus, to prevent a shoot-through during the switching transients, a gating signal of the turning-on device is applied after a certain delay. This is to ensure that the turning-on device is turned on after

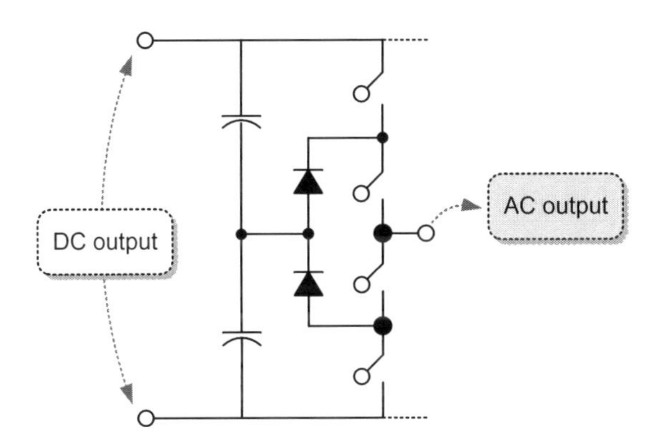

FIGURE 7.6

Basic circuit used in a three-level inverter.

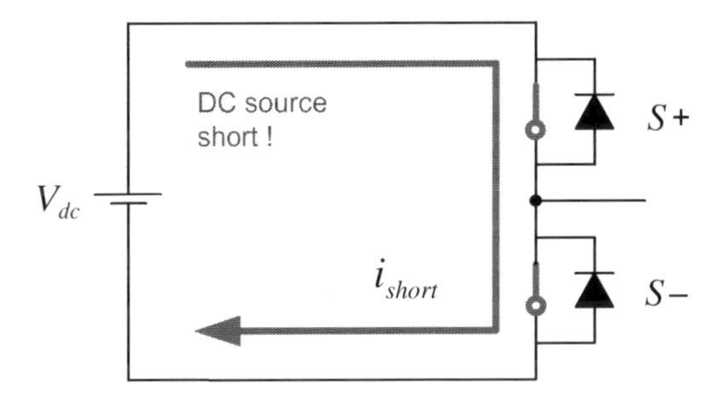

FIGURE 7.7

Shoot-through condition.

the turning-off device is completely turned off. This delay time is called *dead time*. The required value of the dead time differs depending on the type and the power rating of the device used. The dead time will be explained in more detail in Section 7.6.

Except in the case when both switches are turned on, there are three possible situations of the basic circuit. When the upper switch $S+$ is turned on and the lower switch $S-$ is turned off as shown in Fig. 7.8A, the pole voltage becomes a positive voltage as

$$v_P = \frac{V_{dc}}{2} \tag{7.1}$$

On the other hand, when $S+$ is turned off and $S-$ is turned on as shown in Fig. 7.8B, the pole voltage becomes a negative voltage as

$$v_P = -\frac{V_{dc}}{2} \tag{7.2}$$

Lastly, even if $S+$ and $S-$ are both turned off, the pole voltage can be produced if a current is flowing through the load. In this case, the pole voltage depends on the direction of the current as shown in Fig. 7.9. For example, if the

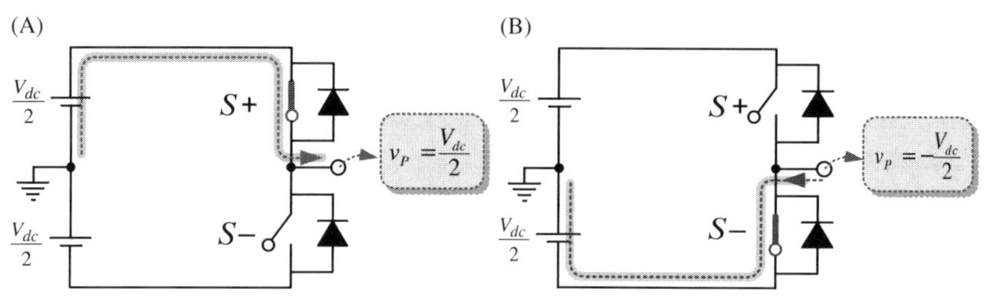

FIGURE 7.8

Circuit configurations according to the switching states. (A) $S+$ On, $S-$ Off and (B) $S+$ Off, $S-$ On.

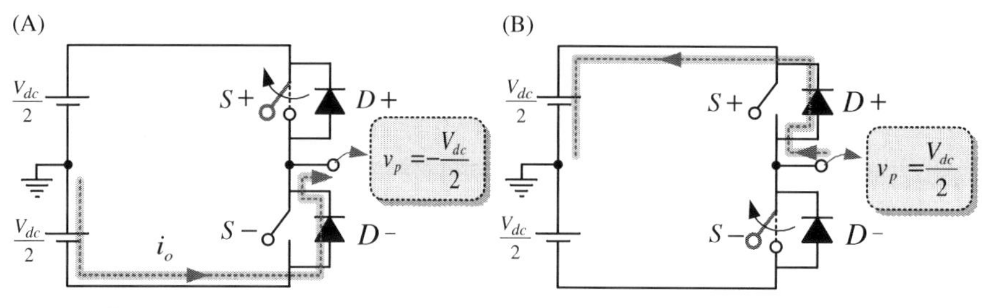

FIGURE 7.9

Circuit configurations when both switches are turned off. (A) $i_o > 0$ and (B) $i_o < 0$.

switch $S+$ is turned off when a positive current ($i_o > 0$) is flowing as shown in Fig. 7.8A, then the current flows through the lower diode so the pole voltage v_P becomes $-V_{dc}/2$ as shown in Fig. 7.9A.

On the other hand, if the switch $S-$ is turned off when a negative current ($i_o < 0$) is flowing in Fig. 7.8B, then the current flows through the upper diode so it becomes $V_{dc}/2$ as shown in Fig. 7.9B.

From the switching situations stated above, the pole voltage of the basic circuit of an inverter can be expressed using a switching function S describing the states of switch as

$$v_P = V_{dc}\left(S - \frac{1}{2}\right) \tag{7.3}$$

Here, the switching function S is defined as "1" when the upper switch $S+$ is turned on and as "0" when it is turned off.

An inverter has two operation modes depending on which device the current flows through: *powering mode* and *regeneration mode*. If the current flows through the switching device, then the inverter operates in the powering mode in which the AC power is supplied into the load from the DC power source. In contrast, if the current flows through the diode, then it operates in the regeneration mode in which the AC power is returned into the DC source from the load.

7.1.2 SINGLE-PHASE HALF-BRIDGE INVERTERS

The simplest inverter that generates single-phase AC voltage from a DC source is a *single-phase half-bridge inverter*. This inverter consists of one basic circuit as shown in Fig. 7.10.

We can simply produce a single-phase AC voltage from this inverter by turning the two switches on and off alternately every half a period $T/2$. An inverter that uses this modulation is called a *square wave inverter*. In this case the inverter produces the maximum output voltage. In this half-bridge inverter the pole voltage v_P is equal to the voltage v_o applied to the load.

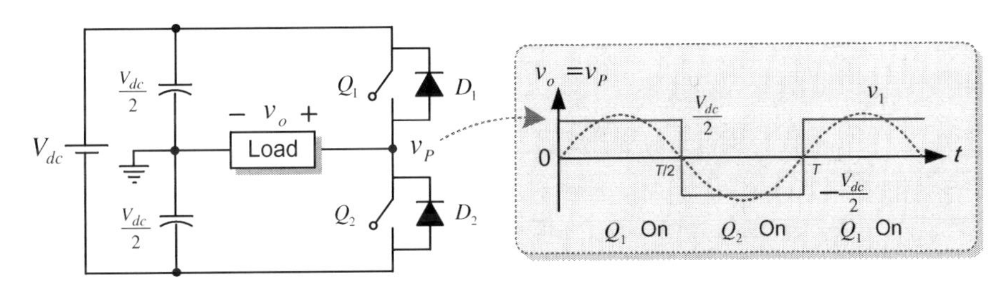

FIGURE 7.10

Single-phase half-bridge inverter and its output voltage.

The pole voltage is not a pure sinusoidal but a square wave, which contains many harmonics. To evaluate the fundamental and harmonic components included in the pole voltage, the Fourier series expansion of the pole voltage is written as

$$v_P = \frac{2V_{dc}}{\pi} \sum_{n=1,3,5...}^{\infty} \frac{\sin n\omega t}{n} \tag{7.4}$$

FOURIER SERIES

Any periodic waveform of any shape can be represented as the sum of a sine wave (so-called *fundamental component*) that has the same frequency as the original waveform and other sine waves (so-called *harmonics*) that have frequencies of integral multiples of the fundamental frequency. To find out these frequency components, we can decompose the periodic waveform into the sum of a set of sines and cosines by using a *Fourier series*. We can see the change of the shape in the waveform by combining other frequency components from the following figure.

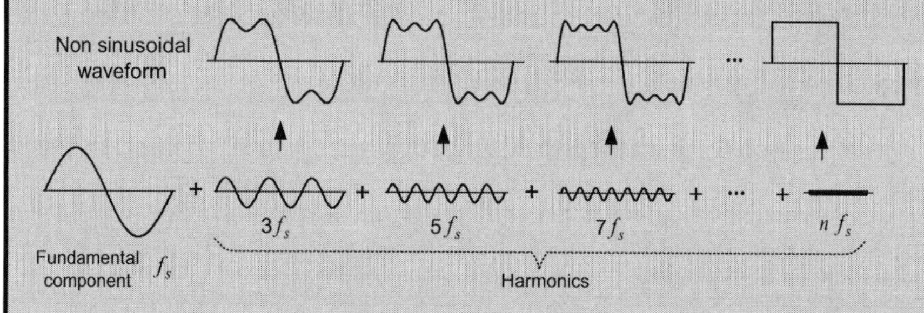

- Fourier series representation for a periodic waveform is given as

$$f(t) = a_0 + \sum_{n=1}^{\infty}(a_n \cos n\omega t + b_n \sin n\omega t)$$

$$= a_0 + \sum_{n=1}^{\infty} c_n \sin(n\omega t + \phi_n) \quad \left(c_n = \sqrt{a_n^2 + b_n^2},\ \phi_n = \tan^{-1}\frac{a_n}{b_n}\right)$$

Here, the Fourier coefficients are computed as follows:

$$a_0 = \frac{1}{T}\int_0^T f(t)dt,\ a_n = \frac{2}{T}\int_0^T f(t)\cos n\omega t d,\ b_n = \frac{2}{T}\int_0^T f(t)\sin n\omega t dt$$

- Periodic waveform is expressed as:

$$f(t) = a_0 + c_1\sin(\omega t + \phi_1) + \sum_{n=2}^{\infty} c_n \sin(n\omega t + \phi_n)$$

$$= \underset{\text{DC component}}{f_{DC}} + \underset{\text{fundamental concept}}{f_1(t)} + \underset{\text{harmonic components}}{f_h(t)}$$

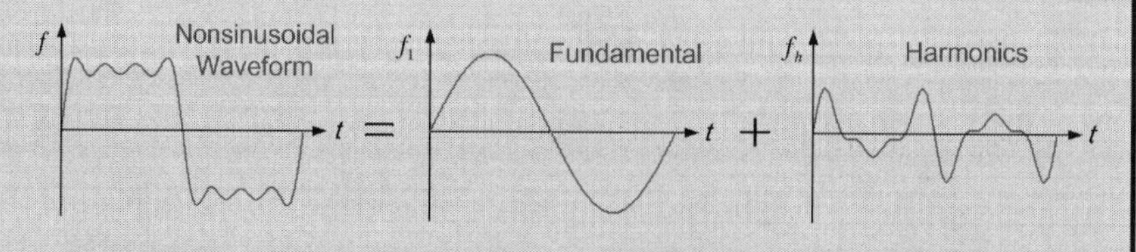

These frequency components are depicted in Fig. 7.11. Out of all the frequency components, only the fundamental component voltage can supply effective power to the load. For the square wave output of a half-bridge inverter, the fundamental rms voltage is given from Eq. (7.4) as

$$V_{P1\_rms} = \frac{1}{\sqrt{2}}\frac{2V_{dc}}{\pi} = 0.45\ V_{dc} \tag{7.5}$$

Thus this inverter can provide the load with AC rms voltage corresponding to 45% of the DC input voltage V_{dc}. Unlike the fundamental component, the harmonics will be converted into unnecessary losses such as heat or noise. Thus it is desirable for harmonics included in the output voltage to be fewer. Besides, because the harmonics cause waveform distortions, if the output AC voltage includes fewer harmonics, their shape approaches a pure sinusoidal wave.

From Eq. (7.4), we can see that the pole voltage contains only odd order harmonics, whose amplitudes are inversely proportional to the order of the harmonics. Thus low-order harmonics such as the third, fifth, and seventh components can cause a problem. This is because they have large amplitude, so they are a major source of waveform distortions. Furthermore, a large-sized, low-pass filter is required to filter them.

To measure the amount of distortion of voltage or current waveforms, we commonly use the *total harmonic distortion* (THD) defined as

$$\text{Total harmonic distortion (THD)} = \sqrt{\frac{\sum_{h>1}^{\infty} f_{h,rms}^2}{f_{1,rms}^2}} \times 100\% \tag{7.6}$$

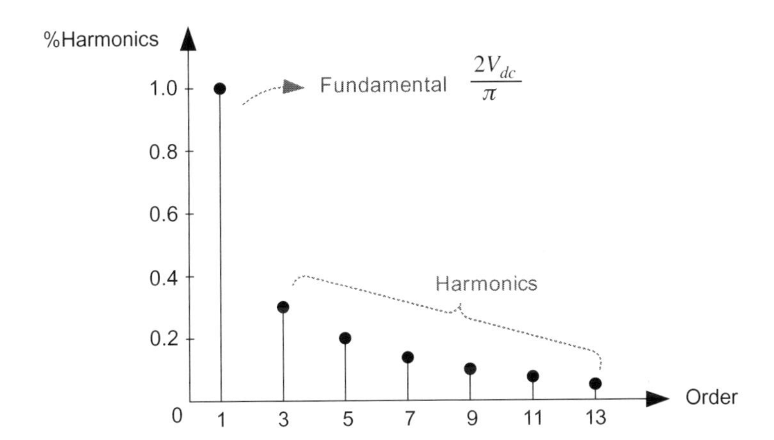

FIGURE 7.11

Harmonics in pole voltage v_P.

Here, f_{1_rms} and f_{h_rms} denote the rms values of the fundamental and harmonic components, respectively. A higher THD value indicates a more distorted waveform, i.e., the waveform deviates from the pure sinusoidal wave.

EXAMPLE 1

Calculate the THD of the output AC voltage of the single-phase half-bridge inverter.

Solution

To calculate the THD of the output voltage of the single-phase half-bridge inverter, we need to evaluate the harmonic components included in its output voltage as shown below.

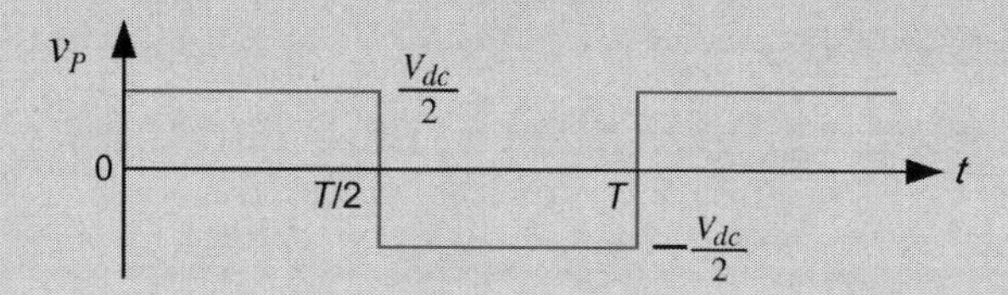

Its output voltage, i.e., pole voltage, has odd quarter-wave symmetry, and thus we need to calculate only the Fourier coefficient $b_n(n = \text{odd})$ as

$$v_P(t) = a_0 + \sum_{n=1}^{\infty}(a_n \cos n\omega t + b_n \sin n\omega t)$$

$$a_0 = 0,\ a_n = 0,\ b_{n-even} = 0$$

$$b_{n-odd} = \frac{2}{T}\int_0^T f(t)\sin n\omega t dt = \frac{4}{\pi}\int_0^{\pi/2}\frac{V_{dc}}{4}\sin n\omega t dt = \frac{2V_{dc}}{n\pi}$$

Thus Fourier series representation for the output voltage is given as

$$\therefore\ v_P(t) = \frac{2V_{dc}}{\pi}\left[\sin \omega t + \frac{1}{3}\sin 3\omega t + \frac{1}{5}\sin 5\,\omega t + \frac{1}{7}\sin 7\omega t + \cdots\right]$$

- Fundamental component:

$$v_{P1}(t) = \frac{2V_{dc}}{\pi}\sin \omega t$$

- Harmonic components:

$$v_h(t) = \sum_{n=3,5,7\cdots}^{\infty}\frac{2V_{dc}}{n\pi}\sin n\omega t$$

The THD of the output AC voltage can be calculated as

$$\text{THDv} = \sqrt{\frac{\sum_{h>1}^{\infty} V_{h,rms}^2}{V_{1,rms}^2}} \times 100\% = \sqrt{\frac{V_3^2 + V_5^2 + V_7^2 + \cdots}{V_1^2}} \times 100\%$$

$$= \sqrt{\left(\frac{1}{3}\right)^2 + \left(\frac{1}{5}\right)^2 + \left(\frac{1}{7}\right)^2 + \cdots} = 46.2\%$$

7.1.3 SINGLE-PHASE FULL-BRIDGE INVERTERS

As can be seen in the Section 7.1.2, a half-bridge inverter has a poor DC voltage utilization of 0.45 V_{dc}. Therefore a commonly used inverter topology for producing a single-phase AC voltage is the *full-bridge inverter* as shown in Fig. 7.12. This inverter consisting of two basic circuits can produce an output voltage twice as much as that of the half-bridge inverter using the same DC voltage.

In the single-phase full-bridge inverter, the load voltage v_o is the difference between two pole voltages, v_H and v_L, as

$$v_o = v_H - v_L \tag{7.6}$$

Here, each pole switches independently. The available load voltages of the full-bridge inverter according to the state of four switches $Q_1 \sim Q_4$ are shown in Table 7.1.

To produce the effective load voltage in this inverter, the phase difference between two pole voltages needs to be 180 electrical degrees as shown in Fig. 7.13. In this case, the voltage of the full-bridge inverter is double the pole voltage.

For inductive loads, the operation of the inverter is shown in Fig. 7.14. The inverter operates alternately in the powering mode and regeneration mode.

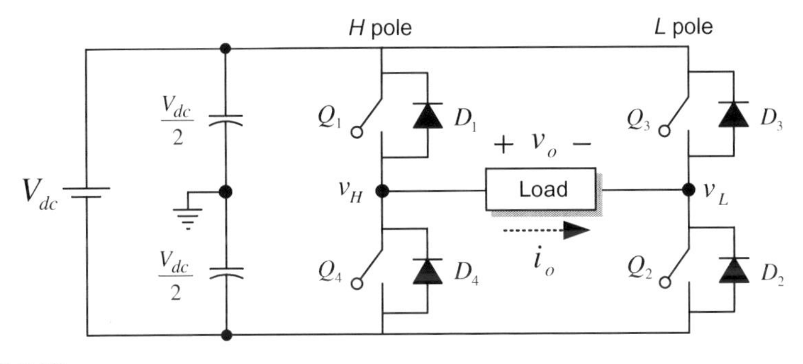

FIGURE 7.12

Single-phase full-bridge inverter.

Table 7.1 Pole Voltage and Load Voltage According to the State of Four Switches

Switch State	Pole Voltage		Load Voltage
	v_H	v_L	$v_o = v_H - v_L$
Q_1, Q_2 On	$\frac{V_{dc}}{2}$	$-\frac{V_{dc}}{2}$	V_{dc}
Q_4, Q_3 On	$-\frac{V_{dc}}{2}$	$\frac{V_{dc}}{2}$	$-V_{dc}$
Q_1, Q_3 On	$\frac{V_{dc}}{2}$	$\frac{V_{dc}}{2}$	0
Q_4, Q_2 On	$-\frac{V_{dc}}{2}$	$-\frac{V_{dc}}{2}$	0

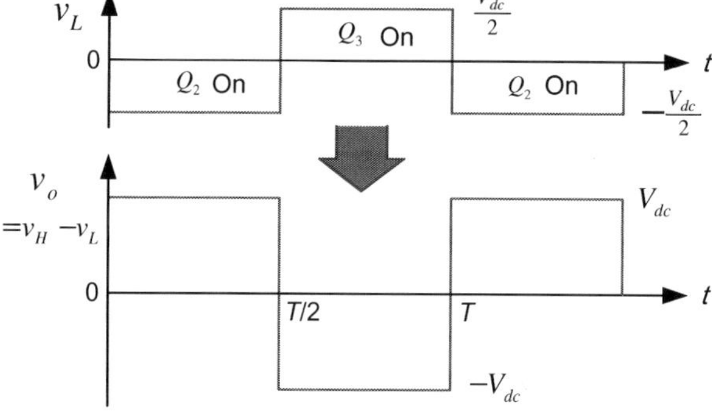

FIGURE 7.13

Load voltage of single-phase full-bridge inverter.

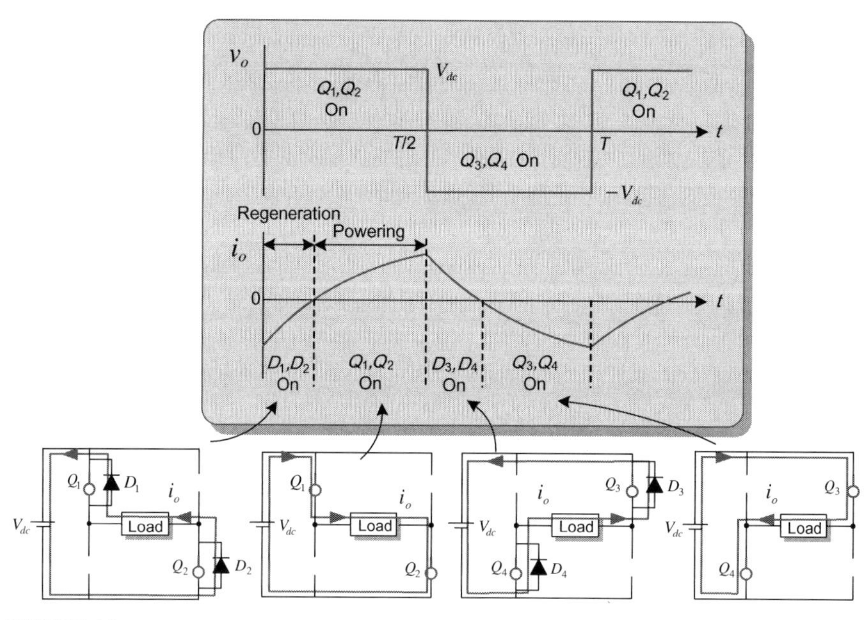

FIGURE 7.14

Operation of single-phase full-bridge inverter for inductive load.

The load voltage in a full-bridge inverter is a square waveform like the pole voltage, so it contains a lot of harmonics. Its harmonic orders are the same as those of the pole voltage. Using Fourier series expansion, the load voltage can be presented as

$$v_o = \frac{4V_{dc}}{\pi} \sum_{n=1,3,5\cdots}^{\infty} \frac{\sin n\omega t}{n} \quad (n = 1, 3, 5, \ldots) \tag{7.7}$$

This inverter can provide the load with AC rms voltage corresponding to 90% of the DC input voltage V_{dc} as

$$V_{o1\_rms} = \frac{1}{\sqrt{2}} \frac{4V_{dc}}{\pi} = 0.9V_{dc} \tag{7.8}$$

This voltage is double the voltage of the half-bridge inverter under an equal DC voltage.

We can express the output voltage v_o of this inverter using the switching function S. Assume that the switching functions for H and L poles are S_H and S_L, respectively. As described previously, the switching function S is defined as "1" when the upper switch is turned on and as "0" when it is turned off. As in Eq. (7.3), the pole voltages v_H and v_L can be expressed using switching functions as

$$v_H = V_{dc}\left(S_H - \frac{1}{2}\right) \tag{7.9}$$

$$v_L = V_{dc}\left(S_L - \frac{1}{2}\right) \tag{7.10}$$

From these pole voltage expressions, the output voltage v_o is given as

$$v_o = v_H - v_L = V_{dc}(S_H - S_L) \tag{7.11}$$

The output voltage can be easily obtained by using this expression from the states of four switches.

7.1.4 THREE-PHASE SQUARE WAVE INVERTERS (SIX-STEP INVERTER)

The circuit configuration of a three-phase inverter as shown in Fig. 7.15 consists of three poles. Each pole is switched independently of each other, and the two switches in each pole are alternately switched.

In a three-phase inverter, each pole is in charge of the production of one-phase voltage. To do so, the switches of each pole will be turned on and off alternately every half period (180° conduction interval) in the same way as the single-phase inverters.

However, to achieve three-phase voltages, each pole voltage should be displaced from each other by 120 electrical degrees. In such switching, three of

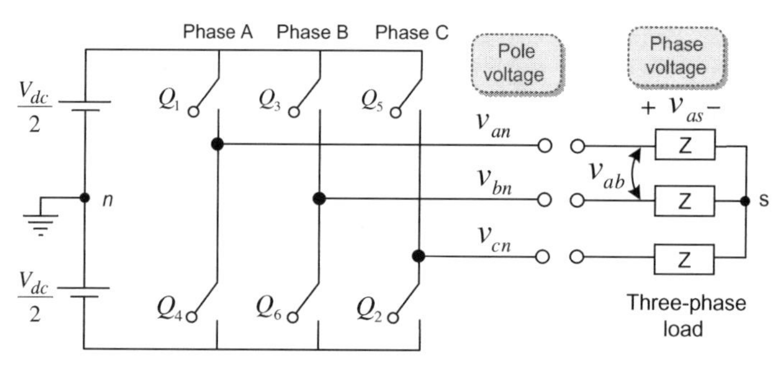

FIGURE 7.15

Configuration of the three-phase inverter circuit.

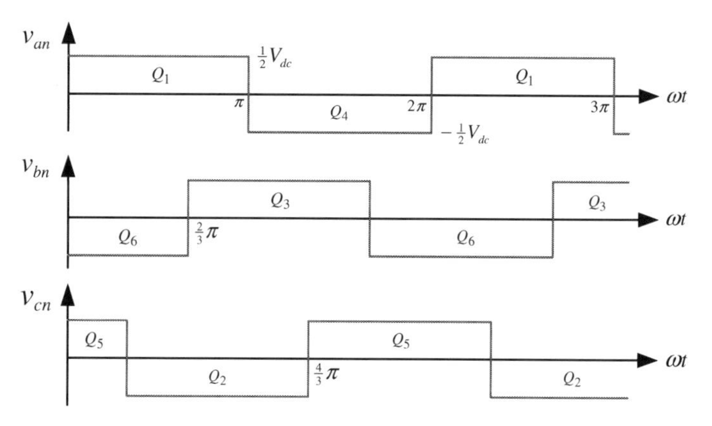

FIGURE 7.16

Pole voltages of the three-phase inverter.

six switches are always turned on, and their switching sequence goes on in the order of switch numbers shown in Fig. 7.16 such as ...123, 234, 345, 456, 561, 612 and back to 123. In this case, there are six different sections in one fundamental period of the output voltage. The inverter operated by this switching scheme is called *square wave inverter*, in which the amplitude of output voltages remains constant while only their frequency can be controlled.

Line-to-line voltage is the difference between two pole voltages (e.g., $v_{ab} = v_{an} - v_{bn}$) as shown in Fig. 7.17. It is a quasi-square wave and has 120° conduction interval.

Next we will look at the phase voltages applied to the load when a three-phase inverter drives a load. We need to know the phase voltages to solve the voltage equations of AC motors as described in Chapter 4.

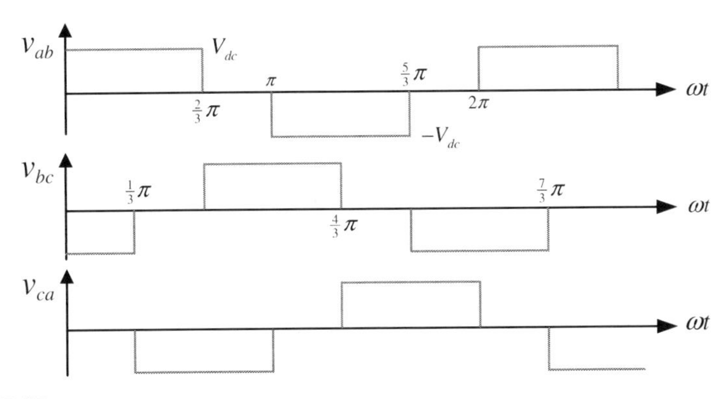

FIGURE 7.17

Line-to-line voltages.

Assume a balanced Y-connected load as shown in Fig. 7.15. In this case the line to neutral voltage, or the phase voltage, is the load voltage. From line-to-line voltages, the phase voltages for the three-phase loads can be obtained as

$$v_{as} = \frac{1}{3}(v_{ab} - v_{ca}) \tag{7.12}$$

$$v_{bs} = \frac{1}{3}(v_{bc} - v_{ab}) \tag{7.13}$$

$$v_{cs} = \frac{1}{3}(v_{ca} - v_{bc}) \tag{7.14}$$

We can see that the pole voltage of the inverter depends on the states of switches in only its own pole, but the phase voltage depends on the states of switches of other poles as well as the corresponding pole. Thus the phase voltage is determined by the states of all switches of an inverter.

The phase voltage can be also obtained directly from the states of all the six switches of an inverter as follows. First, consider the first section in Fig. 7.16, where three switches of $Q5$, $Q6$, and $Q1$, are turned on. In this section, the system can be described as the circuit shown in Fig. 7.18A. In this case, we can readily see that the phase voltages are given as

$$v_{as} = \frac{1}{3}V_{dc}, \quad v_{bs} = -\frac{2}{3}V_{dc}, \quad v_{cs} = \frac{1}{3}V_{dc} \tag{7.15}$$

Next, consider the second section, where $Q6$, $Q1$, and $Q2$ are turned on. In this section, the inverter circuit can be described as shown in Fig. 7.18B, and thus the phase voltages can be obtained as

$$v_{as} = \frac{2}{3}V_{dc}, \quad v_{bs} = -\frac{1}{3}V_{dc}, \quad v_{cs} = \frac{1}{3}V_{dc} \tag{7.16}$$

In the same way we can obtain the phase voltages for all the sections as shown in Fig. 7.19. Since there are six steps in one cycle of the phase voltage waveform,

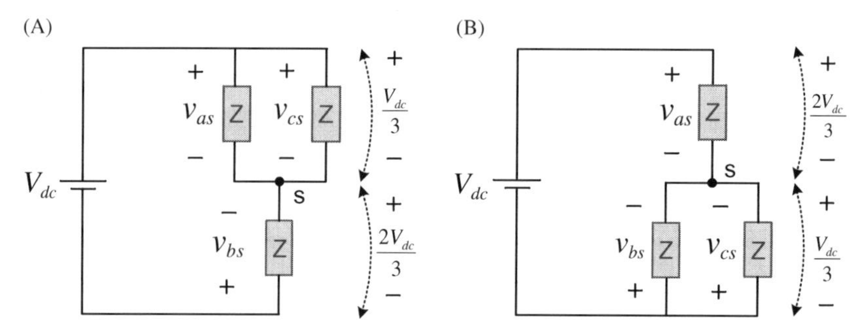

FIGURE 7.18

Phase voltages for the two different sections. (A) Q6, Q5, Q1 On and (B) Q6, Q1, Q2 On.

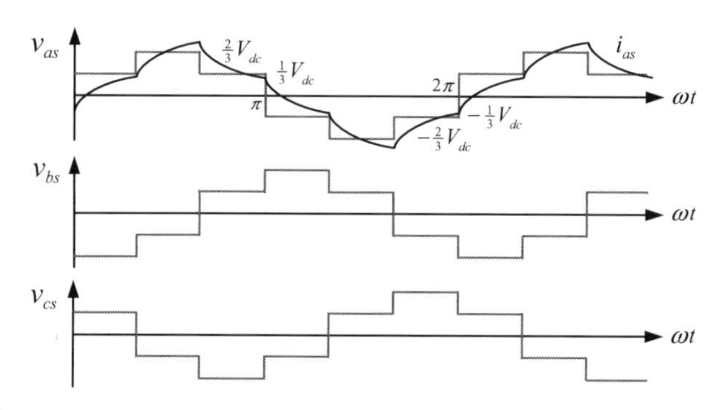

FIGURE 7.19

Phase voltages and currents.

this type of inverter is often referred to as the *six-step inverter*. When these square waveform voltages are applied to inductive loads such as motors, the load currents are not a square waveform, because the harmonics of the voltage is filtered by the inductive component of the load.

Table 7.2 summarizes the pole voltage and the phase voltage (load voltage) in a three-phase inverter according to the states of the switches. Here, the switching state is defined as "1" when the upper switch is in on-state and as "0" when it is in off-state. For the six switches of a three-phase inverter, there are only eight possible switch combinations, i.e., eight different switching states.

As mentioned earlier, the output voltages of a three-phase inverter have the shape of a square wave not a pure sinusoidal wave, so they include many harmonics. Now we will evaluate the fundamental and harmonic components included in the output voltages.

Table 7.2 Pole Voltages and Phase Voltages According to the States of Switches

Switch States			**Pole Voltages**			**Phase Voltages**		
S_a	S_b	S_c	v_{an}	v_{bn}	v_{cn}	v_{as}	v_{bs}	v_{cs}
0	0	0	$-\frac{1}{2}V_{dc}$	$-\frac{1}{2}V_{dc}$	$-\frac{1}{2}V_{dc}$	0	0	0
0	0	1	$-\frac{1}{2}V_{dc}$	$-\frac{1}{2}V_{dc}$	$\frac{1}{2}V_{dc}$	$-\frac{1}{3}V_{dc}$	$-\frac{1}{3}V_{dc}$	$\frac{2}{3}V_{dc}$
0	1	0	$-\frac{1}{2}V_{dc}$	$\frac{1}{2}V_{dc}$	$-\frac{1}{2}V_{dc}$	$-\frac{1}{3}V_{dc}$	$\frac{2}{3}V_{dc}$	$-\frac{1}{3}V_{dc}$
0	1	1	$-\frac{1}{2}V_{dc}$	$\frac{1}{2}V_{dc}$	$\frac{1}{2}V_{dc}$	$-\frac{2}{3}V_{dc}$	$\frac{1}{3}V_{dc}$	$\frac{1}{3}V_{dc}$
1	0	0	$\frac{1}{2}V_{dc}$	$-\frac{1}{2}V_{dc}$	$-\frac{1}{2}V_{dc}$	$\frac{2}{3}V_{dc}$	$-\frac{1}{3}V_{dc}$	$-\frac{1}{3}V_{dc}$
1	0	1	$\frac{1}{2}V_{dc}$	$-\frac{1}{2}V_{dc}$	$\frac{1}{2}V_{dc}$	$\frac{1}{3}V_{dc}$	$-\frac{2}{3}V_{dc}$	$\frac{1}{3}V_{dc}$
1	1	0	$\frac{1}{2}V_{dc}$	$\frac{1}{2}V_{dc}$	$-\frac{1}{2}V_{dc}$	$\frac{1}{3}V_{dc}$	$\frac{1}{3}V_{dc}$	$-\frac{2}{3}V_{dc}$
1	1	1	$\frac{1}{2}V_{dc}$	$\frac{1}{2}V_{dc}$	$\frac{1}{2}V_{dc}$	0	0	0

The pole voltage is always a square wave regardless of the number of the phase. Hence, Fourier series for three-phase pole voltages is the same as Eq. (7.4), but there is a phase difference as

$$v_{an} = \frac{2V_{dc}}{\pi} \sum_{n=1,3,5,\ldots}^{\infty} \frac{\sin n\omega t}{n} \tag{7.17}$$

$$v_{bn} = \frac{2V_{dc}}{\pi} \sum_{n=1,3,5,\ldots}^{\infty} \frac{\sin[n(\omega t - 120^\circ)]}{n} \tag{7.18}$$

$$v_{cn} = \frac{2V_{dc}}{\pi} \sum_{n=1,3,5,\ldots}^{\infty} \frac{\sin[n(\omega t + 120^\circ)]}{n} \tag{7.19}$$

Fourier series expansion for the line-to-line voltage, v_{ab}, is expressed as

$$\begin{aligned} v_{ab} &= v_{an} - v_{bn} \\ &= \frac{4V_{dc}}{n\pi} \sum_{n=1,3,5,\ldots}^{\infty} \cos\frac{n\pi}{6} \sin n\left(\omega t + \frac{\pi}{6}\right) \\ &= \frac{2\sqrt{3}V_{dc}}{\pi}\left[\sin\left(\omega t + \frac{\pi}{6}\right) - \frac{1}{5}\sin 5\left(\omega t + \frac{\pi}{6}\right) - \frac{1}{7}\sin 7\left(\omega t + \frac{\pi}{6}\right)\right. \\ &\quad \left. + \frac{1}{11}\sin 11\left(\omega t + \frac{\pi}{6}\right) + \frac{1}{13}\sin 13\left(\omega t + \frac{\pi}{6}\right) + \cdots\right. \end{aligned} \tag{7.20}$$

Because the line-to-line voltage is given as the difference between two pole voltages, it does not have any harmonics that are multiples of three, which exist in the pole voltages. Since the pole voltages have a phase difference of 120° with each other, the harmonics of multiples of three included in these will have a phase difference that are multiples of 360° with each other and thus they are actually in phase with each other. Therefore the harmonics of multiples of three are absent in the line-to-line voltages. Accordingly, the line-to-line voltage contains harmonics of order $6n \pm 1$ ($n =$ integer).

The phase voltages also have harmonics of order $6n \pm 1$ like line-to-line voltages. From Eq. (7.20), Fourier series expansion for phase *as* voltage can be written as

$$v_{as} = \frac{2V_{dc}}{\pi}\left[\sin \omega t - \frac{1}{5}\sin 5\omega t - \frac{1}{7}\sin 7\omega t + \frac{1}{11}\sin 11\omega t + \cdots\right] \tag{7.21}$$

Phase *bs* and *cs* voltages have harmonics identical to phase *as* voltage, but their phases are different. Fig. 7.20 shows the frequency spectrum of the line-to-line voltages and the phase voltages. Among harmonics, low-order ones such as the fifth and seventh harmonic are the major cause of waveform distortions.

In a three-phase inverter with a balanced Y-connected three-phase load, the neutral voltage v_{sn} of the load is not zero but fluctuates at a frequency which is three times the output frequency as shown in Fig. 7.21. This is because the inverter output voltages are not pure sinusoidal.

Using Fourier series, the neutral voltage can be presented as

$$v_{sn} = \frac{4}{\pi}\frac{V_{dc}}{6}\sum_{n=1,3,5,\ldots}^{\infty}\frac{1}{n}\sin 3n\omega t \tag{7.22}$$

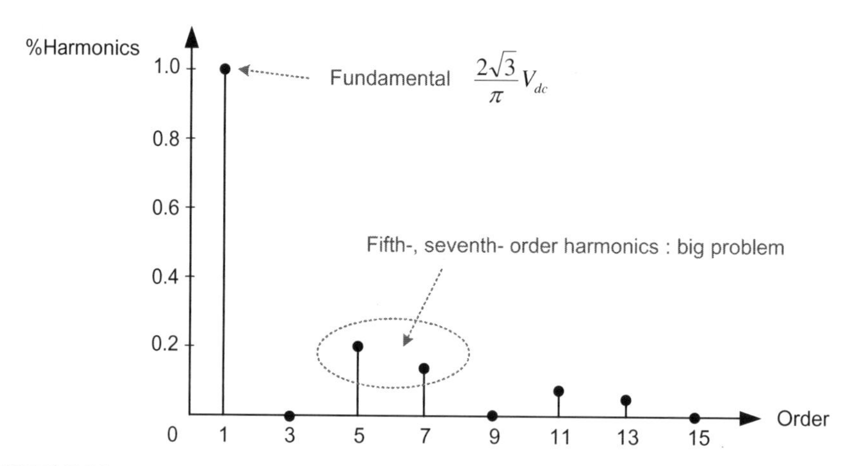

FIGURE 7.20

Frequency spectrum of line-to-line voltage and phase voltage.

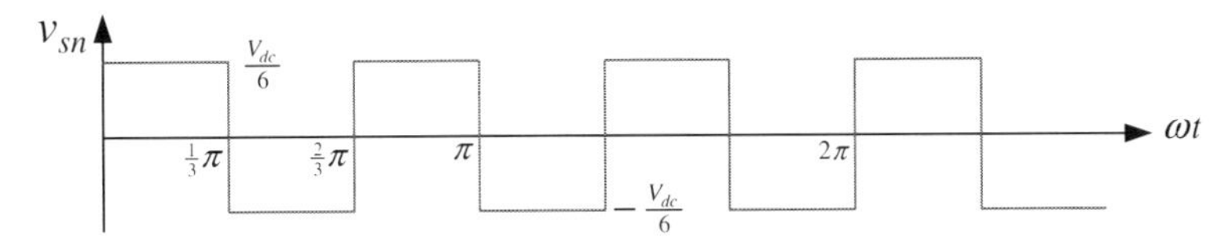

FIGURE 7.21

Neutral voltage in a Y-connected three-phase load.

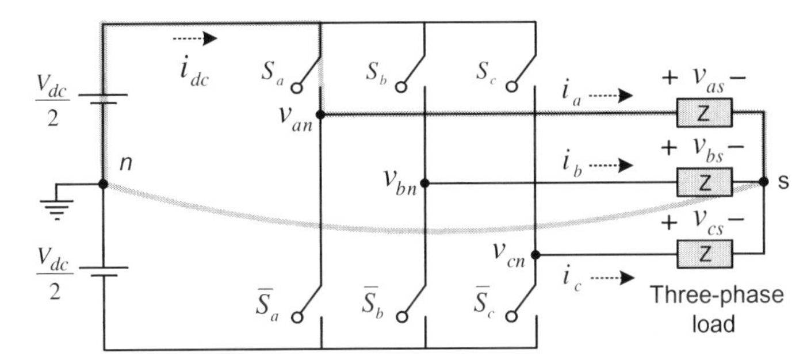

FIGURE 7.22

Three-phase inverter.

EXAMPLE 2

Calculate the maximum output voltage available by a three-phase inverter.

Solution

From Eq. (7.20), the maximum fundamental component of the line-to-line voltage, which a three-phase inverter can apply to the load, is

$$V_{ll-1rms} = \sqrt{3}\frac{2V_{dc}}{\pi}\frac{1}{\sqrt{2}} = \frac{\sqrt{6}}{\pi}V_{dc} \cong 0.78\ V_{dc}$$

To support a 220-V_{rms} load, a three-phase inverter requires at least 282 V DC voltage.

7.1.5 MODELING OF A THREE-PHASE INVERTER USING SWITCHING FUNCTIONS

As mentioned earlier, the phase voltages of a three-phase inverter depends on the states of all six switches. These phase voltages can be expressed by using the switching function describing the states of switches in a pole. This switching function expression for the phase voltages are useful for analysis and simulation of a three-phase inverter.

Let us find the expressions of the phase voltages by the switching function. As shown in Fig. 7.22, assume a balanced Y-connected load with a floating neutral point S.

Let S_a, S_b and S_c be the switching functions for a, b, and c poles, respectively. Again, the switching function is defined as "1" when the upper switch is turned on and as "0" when it is turned off.

When Kirchhoff's voltage law is applied to the closed-loop in the circuit shown in Fig. 7.22, the voltage equation is written as

$$v_{an} = v_{as} + v_{sn} \tag{7.23}$$

For two other phase circuits, we can obtain the voltage equations as

$$v_{bn} = v_{bs} + v_{sn} \tag{7.24}$$

$$v_{cn} = v_{cs} + v_{sn} \tag{7.25}$$

Using the corresponding switch function, each pole voltage is given as

$$v_{an} = V_{dc}\left(S_a - \frac{1}{2}\right) \tag{7.26}$$

$$v_{bn} = V_{dc}\left(S_b - \frac{1}{2}\right) \tag{7.27}$$

$$v_{cn} = V_{dc}\left(S_c - \frac{1}{2}\right) \tag{7.28}$$

The sum of these three equations is

$$v_{an} + v_{bn} + v_{cn} = v_{as} + v_{bs} + v_{cs} + 3v_{sn} \tag{7.29}$$

Here, the sum of three-phase voltages is zero for a balanced Y-connected load with a floating neutral point, i.e.,

$$v_{as} + v_{bs} + v_{cs} = Z(i_{as} + i_{bs} + i_{cs}) = 0 \tag{7.30}$$

Hence the neutral voltage v_{sn} is given as the average of three pole voltages, i.e.,

$$v_{sn} = \frac{1}{3}(v_{an} + v_{bn} + v_{cn}) = \frac{V_{dc}}{3}\left(S_a + S_b + S_c - \frac{3}{2}\right) \tag{7.31}$$

By substituting this neutral voltage into Eqs. (7.23)–(7.25), we can obtain the phase voltages. For example, by substituting Eq. (7.31) into Eq. (7.23), the voltage of phase as is given as

$$\begin{aligned} v_{as} = v_{an} - v_{sn} &= V_{dc}\left(S_a - \frac{1}{2}\right) - \frac{V_{dc}}{3}\left(S_a + S_b + S_c - \frac{3}{2}\right) \\ &= \frac{V_{dc}}{3}(2S_a - S_b - S_c) \end{aligned} \tag{7.32}$$

Likewise, the expressions of phase bs and cs voltages can be obtained.

The expressions of phase voltages in a three-phase inverter using switching functions are as follows:

$$v_{as} = \frac{V_{dc}}{3}(2S_a - S_b - S_c) \quad (7.33)$$

$$v_{bs} = \frac{V_{dc}}{3}(2S_b - S_c - S_a) \quad (7.34)$$

$$v_{cs} = \frac{V_{dc}}{3}(2S_c - S_a - S_b) \quad (7.35)$$

These expressions are also obtained directly by combining Eqs. (7.12)–(7.14) with Eqs. (7.26)–(7.28). Fig. 7.23 shows the voltage of phase *as* obtained by using Eq. (7.33) and the switching functions.

The input DC current in a three-phase inverter can also be expressed by using switching functions. The input DC current i_{dc} can be expressed as one-phase current for each section as follows:

$$\begin{aligned} i_{dc} = & -i_b\left(0 \le \omega t \le \frac{\pi}{3}\right) \\ & i_a\left(\frac{\pi}{3} \le \omega t \le \frac{2\pi}{3}\right) \\ & -i_c\left(\frac{2\pi}{3} \le \omega t \le \pi\right) \\ & i_b\left(\pi \le \omega t \le \frac{4\pi}{3}\right) \\ & -i_a\left(\frac{4\pi}{3} \le \omega t \le \frac{5\pi}{3}\right) \\ & i_c\left(\frac{5\pi}{3} \le \omega t \le \frac{2\pi}{3}\right) \end{aligned} \quad (7.36)$$

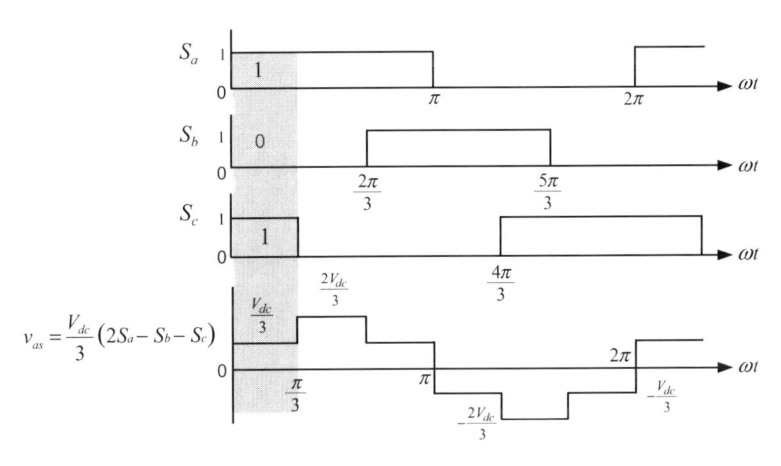

FIGURE 7.23

Phase *as* voltage obtained by using switching functions.

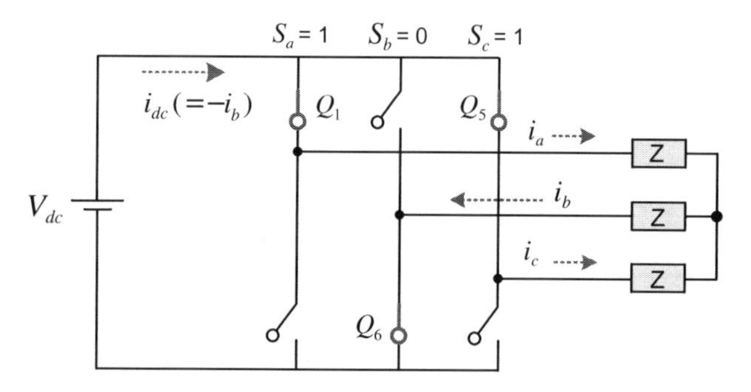

FIGURE 7.24

Input DC current and phase currents.

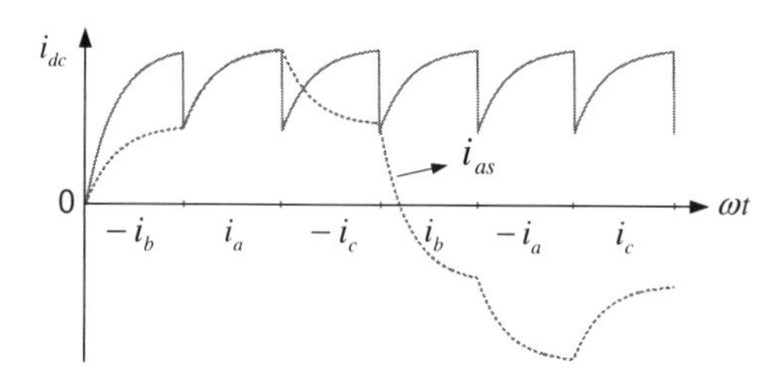

FIGURE 7.25

Input DC current of a three-phase inverter.

As one example, Fig. 7.24 shows that the input current i_{dc} is $-i_b(=i_a+i_c)$ in the first section where $Q5$, $Q6$, and $Q1$ are turned on. From Eq. (7.36), the input current of a three-phase inverter can be expressed as switching functions as

$$i_{dc} = S_a i_a + S_b i_b + S_c i_c \tag{7.37}$$

The input current for the six-step operation is shown in Fig. 7.25. We can see that it has a ripple component six times the output frequency. We need to consider this component in the filter design for reducing the DC voltage ripple.

MATLAB/SIMULINK SIMULATION: SIX-STEP INVERTER

- Simulation conditions: $V_{dc} = 300$ V, $R-L$ load, output frequency 60 Hz
- Overall simulation block diagram

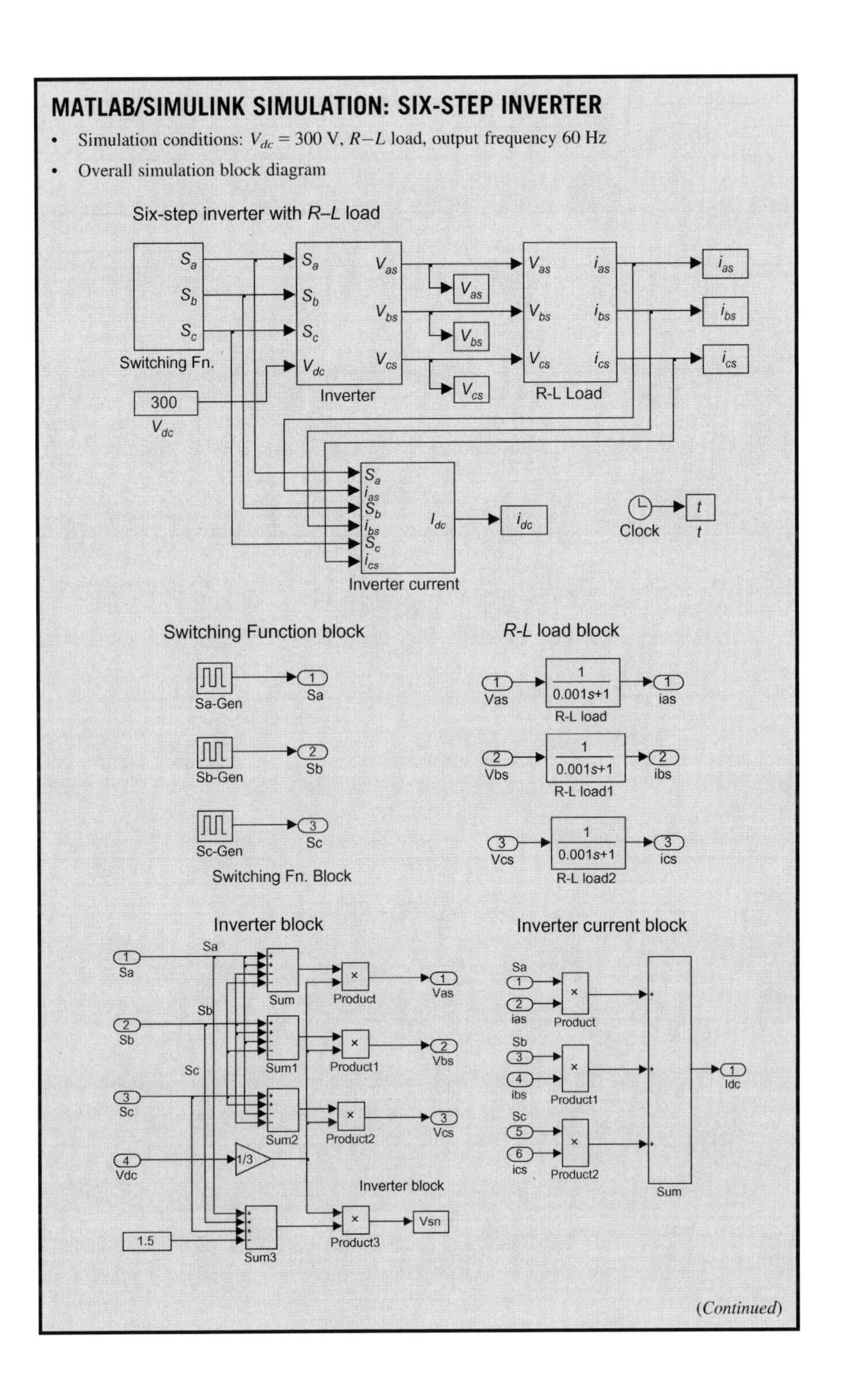

(Continued)

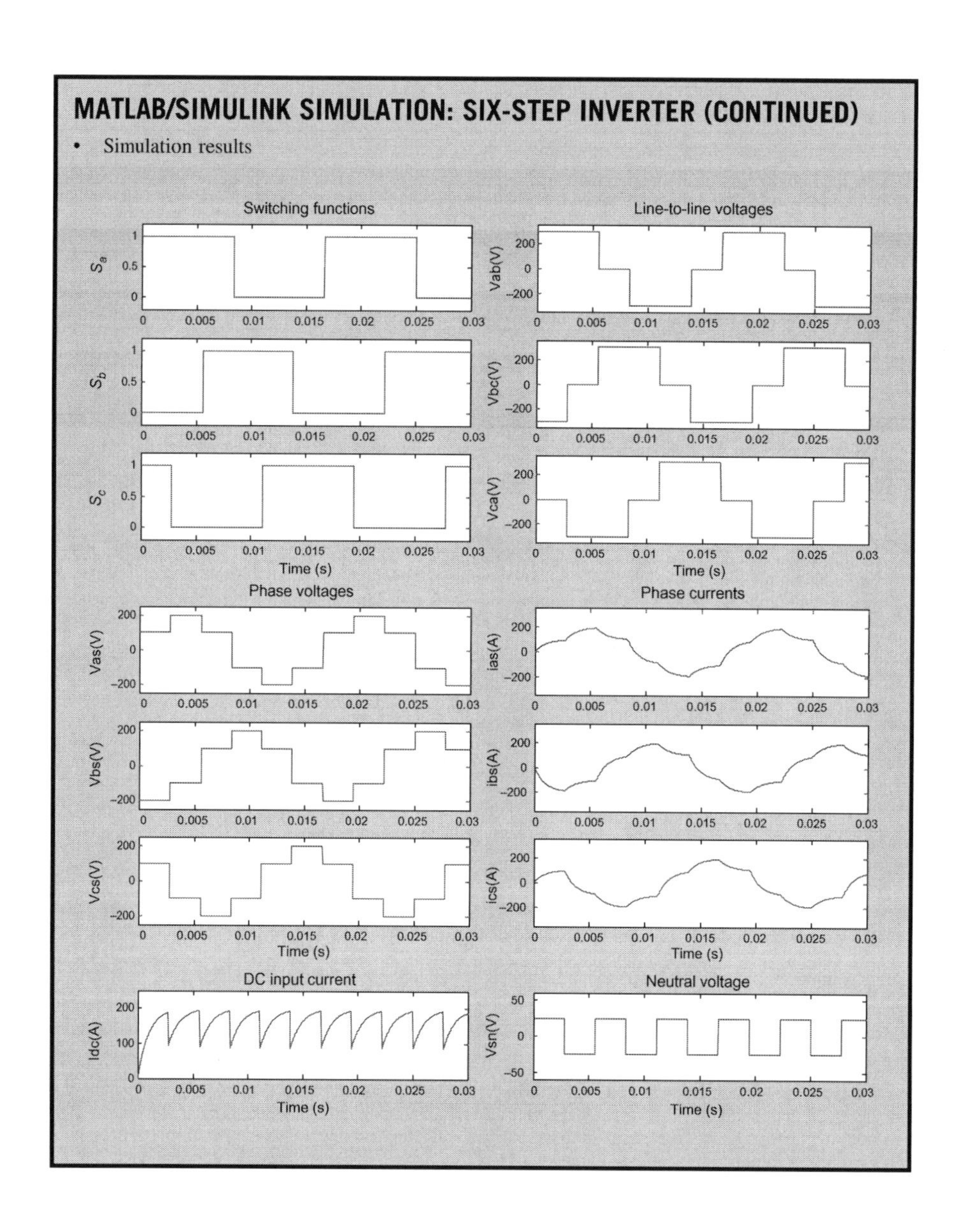

7.2 PULSE WIDTH MODULATION INVERTERS

In the Section 7.1, we illustrated a square wave inverter where only the frequency of the output voltage could be controlled, not its amplitude. We can use this square wave operation for producing the maximum output voltage of an inverter, i.e., for the maximum DC voltage utilization of an inverter. To control the amplitude and the frequency of the output voltage of a VSI, the PWM technique is commonly used as shown in Fig. 7.26.

For such PWM inverters, the following control is possible:

- Linear control of the fundamental output voltage
- Control of frequency of the fundamental output voltage
- Control of harmonics included in the output voltage

The ultimate goal of the PWM is to generate gating pulses of the switches for the inverter to produce an output voltage with the desired fundamental amplitude and frequency. In this modulation process, the switching patterns to eliminate unnecessary harmonics and/or to minimize switching losses can be also developed. To improve these control performances, a variety of PWM techniques have been proposed ever since the sinusoidal pulse width modulation (SPWM) was first introduced in 1964 [2].

There are several performance criteria to evaluate the PWM techniques. First, the range of linearly controllable fundamental output voltage under a given DC voltage is assessed. It is obvious that a larger linear range means better DC voltage utilization. Next, harmonics included in the output voltage is assessed. If the output voltage contains fewer harmonics, it approaches a pure sinusoidal AC. Thus this item is important in assessing the quality of the inverter output voltage. Commonly a higher switching frequency leads to fewer harmonics, and thus this item should be evaluated under the same switching frequency. For the inverter utilized in AC motor drives, since the developed torque of the motor depends on its current, the harmonics included in the current instead of the output voltage is often evaluated. Finally the losses of the switching devices and the losses of the motor generated by adopting PWM are assessed.

Over the past 40 years, many PWM techniques, different in concept and performance, have been developed to improve these criteria. These techniques can be classified into three groups as follows:

- Programmed/optimal PWM
- Carrier-based PWM
- Space vector PWM

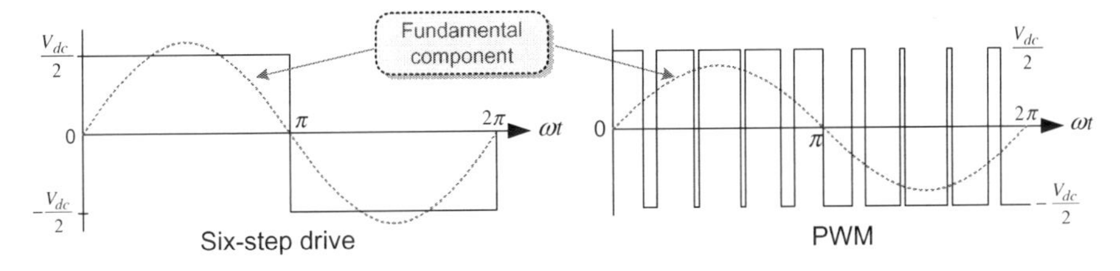

FIGURE 7.26

Output voltage control of an inverter.

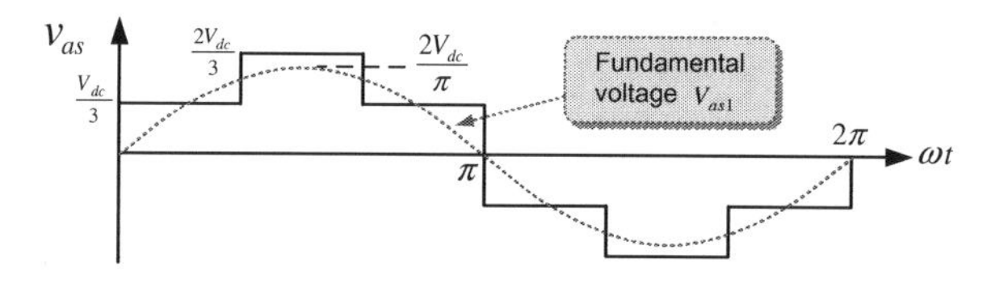

FIGURE 7.27

Fundamental phase voltage of a six-step inverter.

Among these, the carrier-based PWM techniques have been the most widely used method ever since the early development of PWM technology. However, the space vector pulse width modulation (SVPWM) is known to be the most superior in terms of the voltage linearity range and harmonic content.

Before describing these PWM techniques, it is helpful to define the *modulation index* (*MI*) term, which indicates the voltage utilization level as

$$\text{Modulation index:} \quad MI = \frac{V_{1peak}}{\left(\frac{V_{dc}}{2}\right)} \tag{7.38}$$

The modulation index denotes the ratio of the fundamental component magnitude, V_{1peak}, of the inverter phase voltage to the half of DC-link voltage, $V_{dc}/2$. For example, the *MI* in the six-step inverter shown in Fig. 7.27 becomes

$$MI = \frac{V_{1\,peak}}{\left(\frac{V_{dc}}{2}\right)} = \frac{\left(\frac{2V_{dc}}{\pi}\right)}{\left(\frac{V_{dc}}{2}\right)} = \frac{4}{\pi} = 1.273 \tag{7.39}$$

This modulation index value is the maximum *MI* attainable from a three-phase inverter.

Now we will look at a detailed description of the typical PWM techniques of the three groups.

7.2.1 PROGRAMMED PWM TECHNIQUE [3]

In this technique, the PWM switching patterns for generating an output voltage with the desired amplitude of the fundamental component are precalculated off-line and then stored in memory for online access. Inverters are operated by using the PWM switching patterns read in accordance with the operating condition from the memory. Thus, this technique is called programmed PWM.

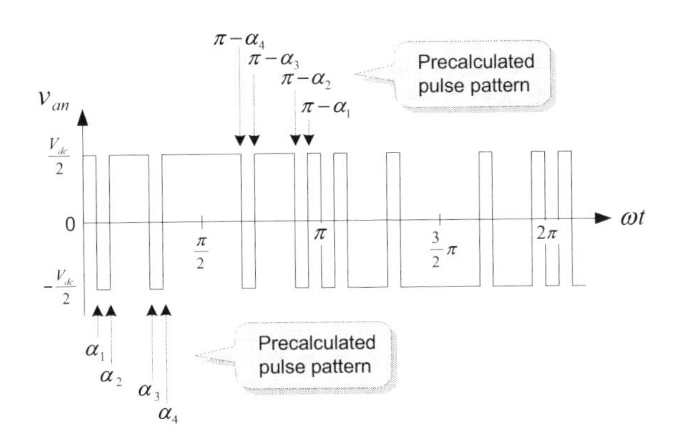

FIGURE 7.28

Programmed PWM technique.

This PWM technique not only controls the fundamental component of the inverter output voltage but also optimizes the desired performance criteria such as eliminating unnecessary low-order harmonics, reducing total harmonics, reducing harmonic torque, or minimizing the switching frequency. Thus this technique is also called *optimal PWM*.

Fourier series are used to obtain the switching instants in this PWM technique. For example, consider eliminating unnecessary low-order harmonics, while producing the desired fundamental component. Suppose that the number of switching angles α_k in 1/4 period for the quarter-wave symmetry square wave is k (e.g., $k=4$ in Fig. 7.28). In this case, k equations, which can control k frequency components of output voltage, can be obtained from Fourier series representation of the waveform with k switching angles. By solving these k equations, the switching angles α_k can be obtained for producing the desired fundamental voltage while eliminating $k-1$ harmonics.

Besides eliminating unnecessary harmonics, the switching patterns can be obtained to optimize specific performance criteria such as harmonic losses or harmonic torque. This technique has advantages of better DC voltage utilization and optimization of switching frequency. However, a considerable computational effort is required to solve transcendental equations when obtaining switching patterns. Moreover, these precalculated steady-state switching patterns are difficult to produce the desired output voltage correctly in the transient state or DC-link voltage variation. Thus this technique cannot be used for systems requiring a fast dynamic.

7.2.2 SINUSOIDAL PWM TECHNIQUE [2]

Sinusoidal PWM is a typical PWM technique. In this PWM technique, the sinusoidal AC voltage reference v_{ref} is compared with the high-frequency triangular carrier wave v_c in real time to determine switching states for each pole in the

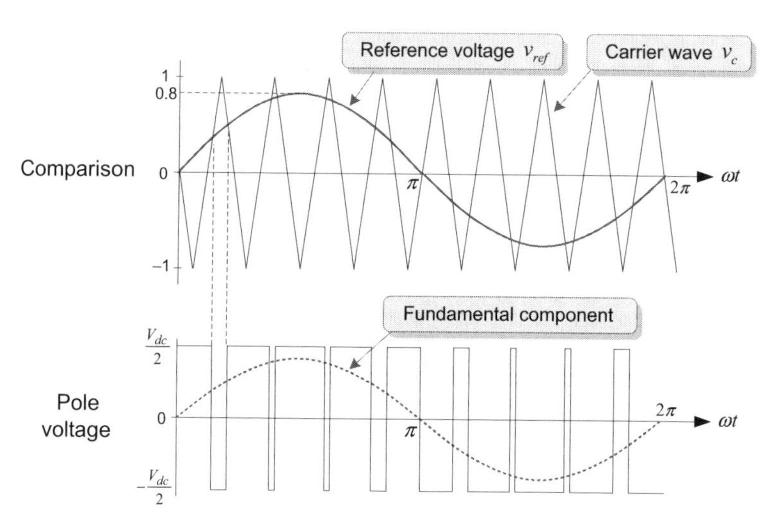

FIGURE 7.29

Sinusoidal PWM technique.

inverter. After comparing, the switching states for each pole can be determined based on the following rule:

- Voltage reference $v_{ref} >$ Triangular carrier v_c: upper switch is turned on (pole voltage $= V_{dc}/2$)
- Voltage reference $v_{ref} <$ Triangular carrier v_c: lower switch is turned on (pole voltage $= -V_{dc}/2$)

Here, the peak-to-peak value of the triangular carrier wave is given as the DC-link voltage V_{dc}. In this PWM technique, the necessary condition for linear modulation is that the amplitude of the voltage reference v_{ref} must remain below the peak of the triangular carrier v_c, i.e., $v_{ref} \leq V_{dc}/2$. Since this PWM technique utilizes a high-frequency carrier wave for voltage modulation, this kind of PWM technique is called a carrier-based PWM technique. Especially, this carrier-based technique is called SPWM since the reference is given as the shape of a sine wave. This is also called the triangle-comparison PWM technique since this uses the carrier of a triangular wave. Fig. 7.29 depicts the sinusoidal PWM technique for one phase.

MODULATING WAVE AND CARRIER WAVE

In the carrier-based PWM techniques, the desired voltage reference waveform is referred to as *modulating wave*. In addition, a wave which is modulated with the modulating wave is referred to as *carrier wave* or *carrier*. The carrier wave usually has a much higher frequency than the modulating wave. The triangular waveform is the most commonly used carrier in the PWM technique for modulating AC voltage. On the other hand, different forms of modulating wave can be used according to the PWM technique. Typical SPWM technique uses the sinusoidal modulating waveform.

Difference Between Pole Voltage and Phase Voltage References
An inverter output determined by comparing a voltage reference with the triangular carrier wave is the pole voltage. Thus the voltage reference that is compared with the triangular carrier wave is considered as the pole voltage reference. Typical SPWM technique uses a phase voltage reference as the pole voltage reference. On the other hand, different pole voltage reference can be used according to the PWM techniques.

In this PWM based on comparison with the triangular wave, if the ratio of carrier frequency to fundamental frequency is large enough (greater than 21), then the fundamental component of the output voltage varies linearly with the reference voltage v_{ref} for a constant DC-link voltage as

$$v_{o1} = v_{ref} \sin \omega t \tag{7.40}$$

In addition, the fundamental frequency of the output voltage is identical to that of the reference voltage.

The output voltage of Eq. (7.40) can be rewritten in terms of the modulation index *MI* as

$$v_{01} = \frac{V_{dc}}{2} MI \sin\omega t \tag{7.41}$$

Here, since $v_{ref} \leq V_{dc}/2$, so $0 \leq MI \leq 1$.

The range of $0 \leq MI \leq 1$ is called the *linear modulation range* because, in this range, the inverter can generate an output voltage linearly proportional to the reference voltage as shown in Fig. 7.30. In this case, the PWM inverter is considered to be simply a voltage amplifier with a unit gain.

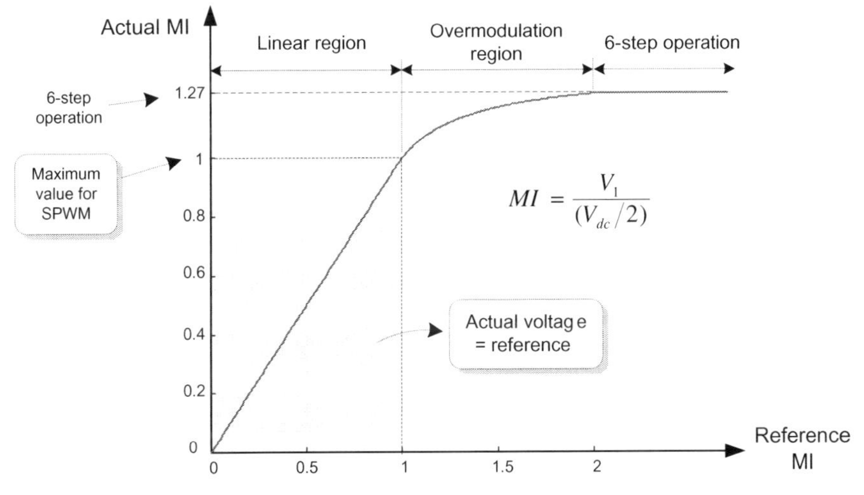

FIGURE 7.30

Voltage modulation range for SPWM.

However, when the reference exceeds the peak of the triangular carrier (i.e., $MI > 1$), the inverter cannot produce an output voltage linearly proportional to the voltage reference. The range of $MI > 1$ is called *overmodulation region*, where the linearity of the modulation is lost. We will discuss the overmodulation techniques in Section 7.5.

The maximum linear output voltage, $V_{dc}/2$, attainable by the SPWM technique corresponds to 78.5% of the maximum output voltage, $2V_{dc}/\pi$, by the six-step inverter. Therefore, when using the PWM technique, the attainable maximum limit of the linear modulation range is inevitably less than the maximum output voltage of an inverter.

Fig. 7.31 shows the SPWM technique for a three-phase inverter.

In the SPWM technique, the switching frequency of an inverter is equal to that of a carrier wave. From Figs. 7.29 and 7.31, we can see that the switch is turned on/off once every period of the triangular carrier wave. Thus the SPWM technique has an advantage of having a constant switching frequency. A constant switching frequency makes it possible to calculate the losses of switching devices, so the thermal design for them becomes easier. In addition, since the harmonic characteristics will be well-defined, the design of a low-pass filter to eliminate the harmonics will become easier.

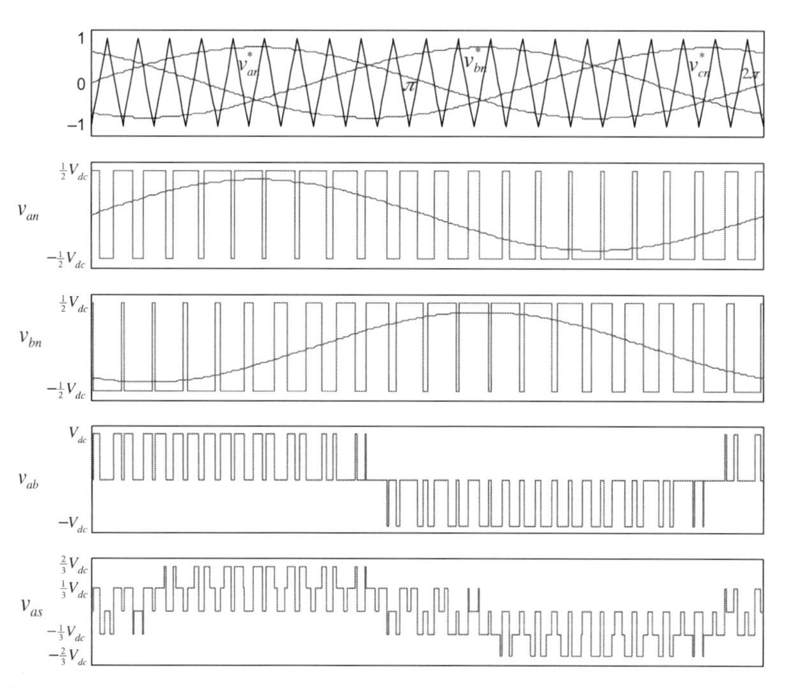

FIGURE 7.31

SPWM technique for a three-phase inverter.

Now we will evaluate which harmonics are contained in the output voltage generated by the SPWM technique. First, we will investigate the harmonic components of the pole voltage v_P as shown in Fig. 7.29. It is widely known that the pole voltage contains harmonics at the carrier frequency f_c and frequencies of its integer multiples (M), and the sidebands (N) of all these frequencies [4]. Thus these harmonics, which are known as switching frequency harmonics, can be expressed as

$$\begin{aligned} v_{o-h} &= V_h \sin[2\pi(Mf_c \pm Nf_o)t + \phi_h] \\ &= V_h \sin[2\pi f_o(Mm_f \pm N)t + \phi_h] \end{aligned} \tag{7.42}$$

Here, f_o is the fundamental frequency of the output voltage and m_f is the *frequency modulation index*, which denotes the ratio of the carrier frequency to the fundamental frequency, i.e., $m_f = f_c/f_o$. M and N are integers, and $M+N$ is odd. ϕ_h denotes the phase of harmonic component. From Eq. (7.42), the orders of harmonics are given as

$$\begin{aligned} &m_f,\ m_f \pm 2,\ m_f \pm 4,\ m_f \pm 6, \ldots \\ &2m_f \pm 1,\ 2m_f \pm 3,\ 2m_f \pm 5,\ 2m_f \pm 7, \ldots \\ &3m_f,\ 3m_f \pm 2,\ 3m_f \pm 4,\ 3m_f \pm 6, \ldots \\ &4m_f \pm 1,\ 4m_f \pm 3,\ 4m_f \pm 5,\ 4m_f \pm 7, \ldots \end{aligned} \tag{7.43}$$

Among the harmonics, the component of order m_f has the largest magnitude. This means that the harmonic with the frequency equal to the switching frequency f_c is the largest one.

As an example, Fig. 7.32 shows the frequency spectrum for the pole voltage of $f_o = 50$ Hz and $m_f = 21$. In this case, the harmonic of 1050 Hz($=21 \times 50$ Hz), i.e., the switching frequency is the largest component.

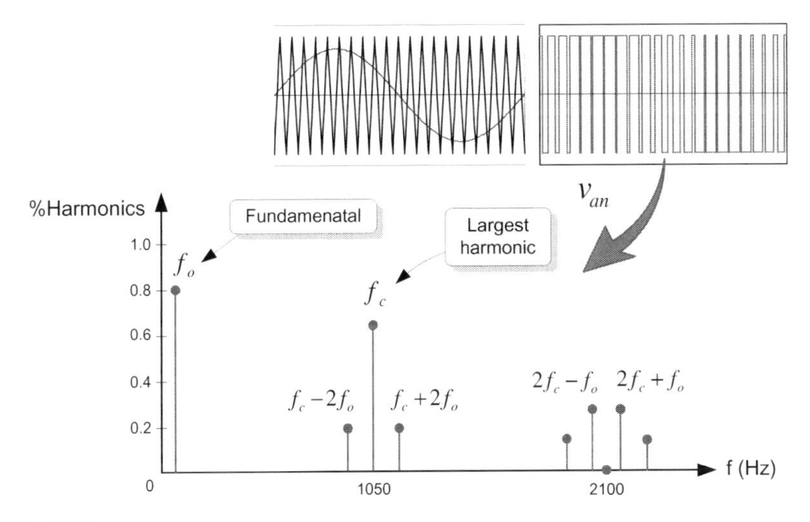

FIGURE 7.32

Frequency spectrum of the pole voltage for the SPWM.

The higher the switching frequency is, the higher the order of the major harmonic is. Thus, when a higher switching frequency is used, the quality of the voltage waveform can be improved and filtering can be made easier. However, this leads to greater switching losses. Therefore it is important to consider the overall performance of the system when selecting the switching frequency.

Next we will examine the harmonic components for the line-to-line and phase voltages. Since the line-to-line voltage is the difference between the two pole voltages, they do not have any harmonic at multiples of three, which exist in the pole voltages. As mentioned earlier, this is because the harmonics at multiples of three included in the pole voltages will have no phase difference with each other. Hence, if we select the value of m_f as multiples of three, then the total harmonics will be reduced in the line-to-line voltage due to the elimination of the harmonics at multiples of three. For this reason, the value of m_f is usually selected as multiples of three. Furthermore, among these values, only the odd values can eliminate the even harmonics for the symmetry of three-phase PWM patterns. In that case, the harmonic of order $2m_f \pm 1$ becomes the largest component for the range of $MI < 0.9$, while $m_f \pm 2$ around $MI = 1$. For example, Fig. 7.33 depicts the harmonic spectrum for the line-to-line voltage in the case of $m_f = 21$ and $MI = 0.8$. In this case, unlike that of the pole voltage, the largest harmonic component becomes the order of $2m_f \pm 1$. The phase voltages have harmonic components identical to those of the line-to-line voltages, but their magnitudes are different.

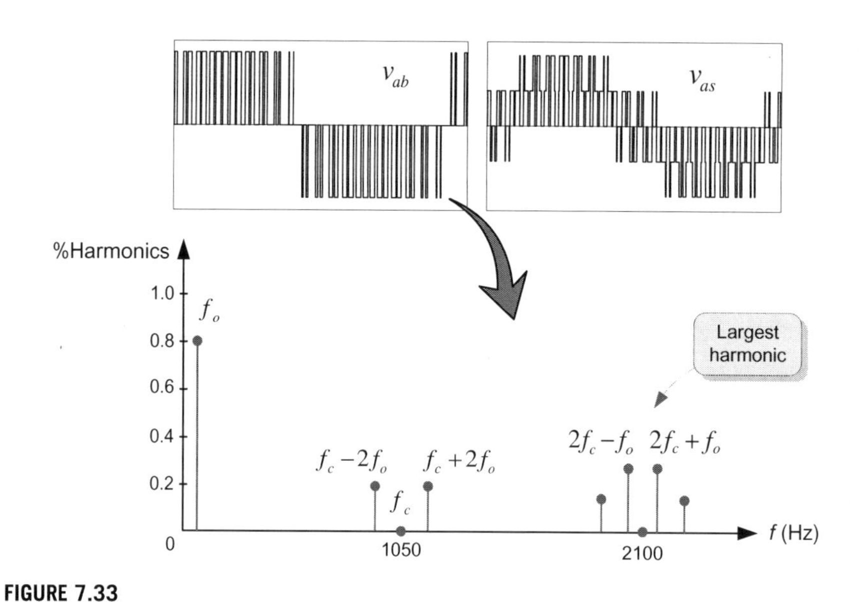

FIGURE 7.33

Frequency spectrum of the line-to-line voltage for the SPWM ($MI = 0.8$, $m_f = 21$).

The SPWM technique has been widely popular due to the simplicity of its principle and analog implementation. In the analogue implementation of the SPWM (referred to as *naturally sampled PWM*), an analog integrator is used to generate a triangular carrier wave, and an analog comparator is used to determine the intersection instants of the triangular carrier wave and modulating signal.

In contrast, its software-based implementation using a digital technique or microprocessor is not easy because this requires solving the transcendental equation, which defines points of intersection used to determine the switching instants. Instead, as shown in Fig. 7.34, the so-called *regular-sampled PWM* is used in which the sinusoidal reference is held at a constant sampled value for the carrier interval, and the sampled value is compared with the carrier wave to determine the switching instants [5]. In the regular-sampled PWM, there are two types of sampling, *symmetric* and *asymmetric*. In the symmetrical sampling of Fig. 7.34A, the sinusoidal reference is sampled once at the peak of the triangular carrier wave, whereas in the asymmetrical sampling of Fig. 7.34B, it is sampled twice at both the positive and negative peaks of the triangular carrier wave. Nowadays, its digital implementation can be easily done by using microcontrollers supporting the dedicated module for the PWM signal generation.

Since the SPWM technique can perform voltage modulation every sampling interval with a fixed switching frequency, it exhibits a better dynamic performance than the programmed PWM. However, this technique has a limited voltage linearity range (only 78.5% of six-step operation) and a poor waveform quality in the high modulation range. To overcome these problems, many improved PWM techniques have been developed. Improvements to extend the voltage linearity range have been mainly done through the modification of the modulating signal, resulting in nonsinusoidal modulating signals. As a typical example of the improvement, the *third harmonic injection PWM* makes it possible to increase the fundamental component of the output voltages by 15.5% more than the conventional SPWM technique. Now we will discuss the third harmonic injection PWM.

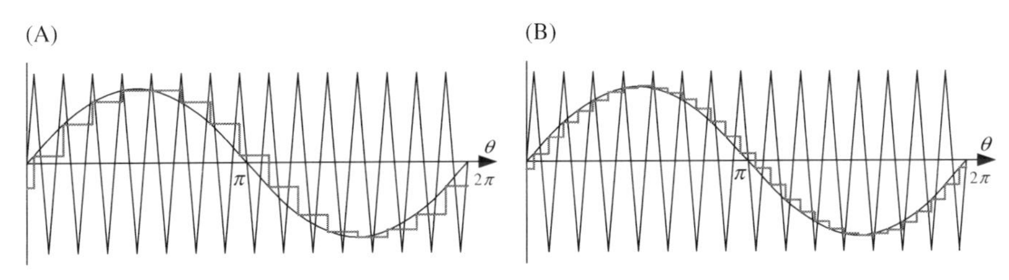

FIGURE 7.34

Regular-sampled PWM technique (A) Symmetrically sampling and (B) asymmetrically sampling.

7.2.3 THIRD HARMONIC INJECTION PWM TECHNIQUE [6]

The SPWM technique operates properly when the sinusoidal voltage reference v^* remains below the peak of the triangular carrier. This limits the range of linear modulation in the SPWM technique. When the peak of the voltage reference v^* exceeds the peak of the triangular carrier (i.e., $MI > 1$) as shown in Fig. 7.35, a pulse dropping, which indicates no intersection between the voltage reference and the triangular carrier, happens. As a result, the linear relationship between the voltage reference and the output voltage cannot be maintained.

Recall that the only effective voltage to a load is the fundamental component contained in the output voltage. Thus, if we select the voltage reference of whose fundamental component exceeds the peak of the triangular carrier but its own peak does not, then it can be expected that the linear modulation range can be extended. By adding a third harmonic to the voltage reference waveform, this improvement can be attainable. This is because when a third harmonic is added to the voltage reference waveform, the peak of the resultant waveform becomes less than that of the original waveform as shown in Fig. 7.36. The technique that adopts this principle is the *third harmonic injection pulse width modulation* (THIPWM). By using the THIPWM, the fundamental of the output voltage can be increased by 15.5% more than the conventional SPWM technique.

This deliberate third harmonic voltage is not present on the line-to-line and phase voltages for a three-phase load with a floating neutral point such as AC

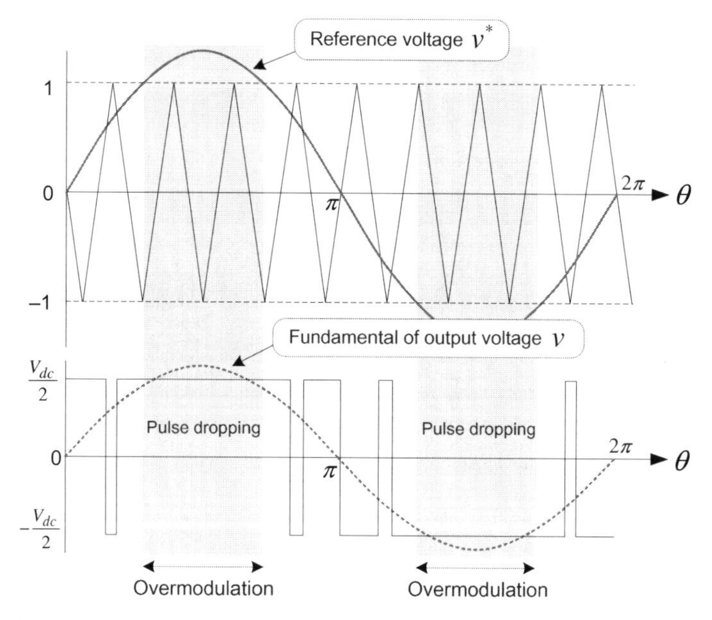

FIGURE 7.35

Overmodulation in the SPWM technique.

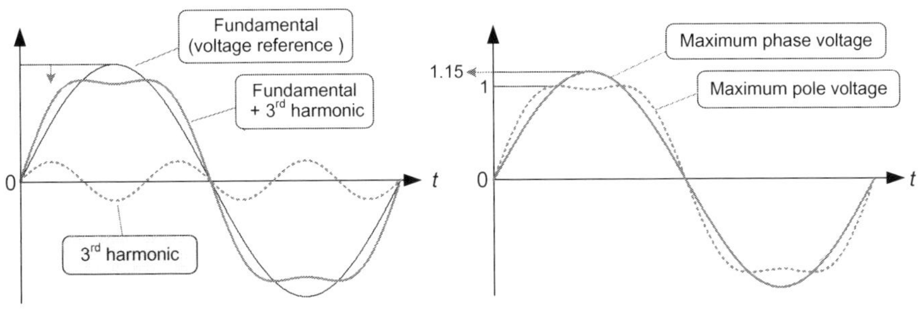

FIGURE 7.36

Principle of THIPWM.

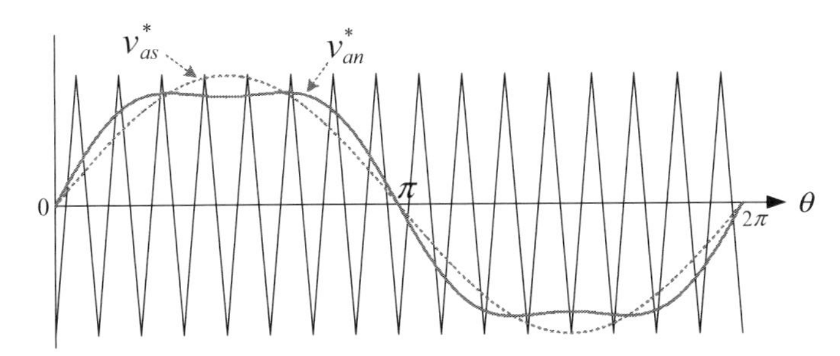

FIGURE 7.37

Third harmonic injection pulse width modulation.

motors. Thus the injected third harmonic voltage does not cause any distortion on the line-to-line and phase voltages.

Let us find the third harmonic voltage that gives an optimal performance in the THIPWM. Assume that the voltage reference of phase *as* is $v_{as}^* = V_1 \sin\omega t$ as shown in Fig. 7.37.

By adding the third harmonic to this voltage reference, the resultant voltage reference becomes

$$v_{an}^* = V_1 \sin \omega t + V_3 \sin 3\omega t \tag{7.44}$$

For Eq. (7.44), the optimum value of V_3 that maximizes the fundamental of the phase voltage is $V_1/6$ [6]. The addition of a third harmonic with an amplitude of 1/6 of the voltage reference can reduce the peak value of the voltage reference by a factor of 0.866 without changing the amplitude of the fundamental component. Accordingly, the fundamental component of the phase voltage can be increased by 15.5% (=1/0.866). This value corresponds to 90.7% of the output voltage for a six-step inverter.

Instead of the factor of 1/6, it is known that a third harmonic with an amplitude of 1/4 of the voltage reference can lead to a minimum harmonic distortion on the output voltage. However, this amplitude incurs a slight decrease in the maximum linear modulation value to $MI = 1.12$. The THIPWM technique has a disadvantage of implementation complexity of the third harmonic and steady-state current harmonic characteristics inferior to the SVPWM method. In addition to the third harmonic, there is another THIPWM technique that uses a higher order triple harmonic such as the ninth harmonic.

7.2.4 SPACE VECTOR PWM TECHNIQUE [7]

In the PWM techniques mentioned previously, the three-phase voltage references are modulated individually. Unlike these, there is a PWM technique that uses a different approach based on the concept of a space vector, called SVPWM. In the SVPWM technique, the three-phase voltage references are represented as a space vector $\boldsymbol{v}_{abc}$ in the complex plane, and this voltage reference vector is modulated by output voltage vectors available from an inverter. The SVPWM technique is now widely used in many three-phase inverter applications because it produces fundamental output voltage 15.5% more than the one produced by the SPWM technique and gives less harmonic distortion of the load current, lower torque ripple in AC motors, and lower switching losses.

The concept of the space vector required in this PWM technique was discussed in Section 4.3. For this PWM technique, since the three-phase voltage references are given as a voltage space vector, the possible output voltages of an inverter also need to be expressed as a space vector. In the Section 7.1.4, we saw that there are eight possible switching states in a three-phase inverter. The output voltage vectors, $V_o - V_7$, corresponding to the eight possible switching states, are shown in Table 7.3. Refer to Example 3, which shows how to obtain these vectors. Fig. 7.38 depicts these output voltage vectors of a three-phase inverter in the complex plane.

Table 7.3 Output Voltage Vectors Corresponding to Switching States

Switch States			Phase Voltages			Space Voltage Vector
S_a	S_b	S_c	v_{as}	v_{bs}	v_{cs}	$V_n(n=1-7)$
0	0	0	0	0	0	$V_0 = 0\angle 0°$
1	0	0	$\frac{2}{3}V_{dc}$	$-\frac{1}{3}V_{dc}$	$-\frac{1}{3}V_{dc}$	$V_1 = \frac{2}{3}V_{dc}\angle 0°$
1	1	0	$\frac{1}{3}V_{dc}$	$\frac{1}{3}V_{dc}$	$-\frac{2}{3}V_{dc}$	$V_2 = \frac{2}{3}V_{dc}\angle 60°$
0	1	0	$-\frac{1}{3}V_{dc}$	$\frac{2}{3}V_{dc}$	$-\frac{1}{3}V_{dc}$	$V_3 = \frac{2}{3}V_{dc}\angle 120°$
0	1	1	$-\frac{2}{3}V_{dc}$	$\frac{1}{3}V_{dc}$	$\frac{1}{3}V_{dc}$	$V_4 = \frac{2}{3}V_{dc}\angle 180°$
0	0	1	$-\frac{1}{3}V_{dc}$	$-\frac{1}{3}V_{dc}$	$\frac{2}{3}V_{dc}$	$V_5 = \frac{2}{3}V_{dc}\angle 240°$
1	0	1	$\frac{1}{3}V_{dc}$	$-\frac{2}{3}V_{dc}$	$\frac{1}{3}V_{dc}$	$V_6 = \frac{2}{3}V_{dc}\angle 300°$
1	1	1	0	0	0	$V_7 = 0\angle 0°$

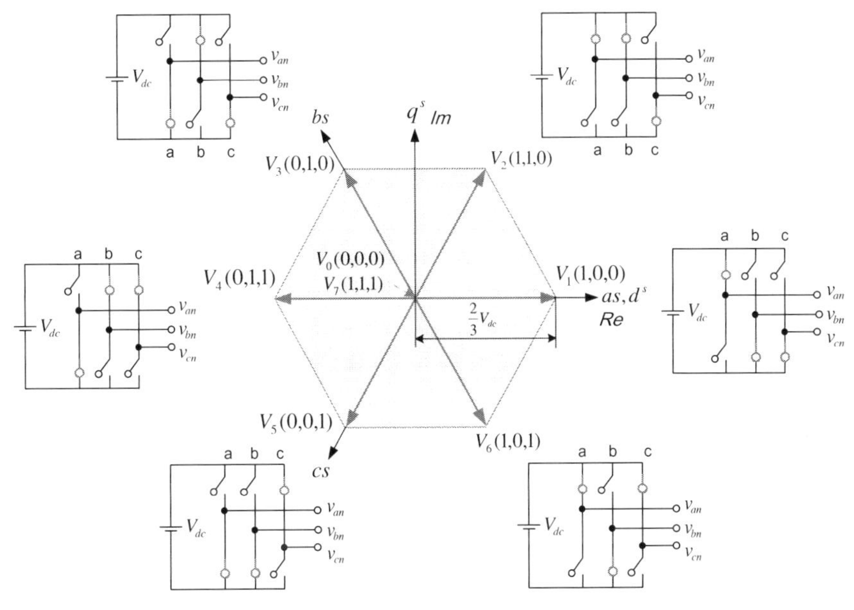

FIGURE 7.38

Output voltage vectors in the complex plane (or d–q axes stationary frame).

Six of these vectors, $V_1 - V_6$, which are called *active voltage vector*, offer an effective voltage to the load. The magnitude of all active vectors is equal to $2V_{dc}/3$. However, they are 60° out of phase with each other. By contrast, the two vectors, V_0 and V_7, are called *zero voltage vector*, which cannot yield an effective voltage to the load.

EXAMPLE 3

Express the output phase voltage of a three-phase inverter as a space vector.

Solution

As an example, we will find the voltage vector V_2 for the switching state of $S_a = 1$, $S_b = 1$, $S_c = 0$.

- Voltage space vector definition:

$$V = \frac{2}{3}(v_{as} + a v_{bs} + a^2 v_{cs}) = v_{as} + \mathrm{j}\frac{1}{\sqrt{3}}(v_{bs} - v_{cs})$$

(Here, $a = e^{\mathrm{j}2\pi/3} = -\frac{1}{2} + \mathrm{j}\frac{\sqrt{3}}{2},\ \ a^2 = e^{\mathrm{j}4\pi/3} = -\frac{1}{2} - \mathrm{j}\frac{\sqrt{3}}{2}$)

- Phase voltages: $v_{as} = \frac{1}{3}V_{dc},\ \ v_{bs} = \frac{1}{3}V_{dc},\ \ v_{cs} = -\frac{2}{3}V_{dc}$
- Voltage vector $V_2 = \left(\frac{1}{3} + \mathrm{j}\frac{1}{\sqrt{3}}\right)V_{dc} = \frac{2}{3}V_{dc}\ \angle 60^\circ$

We can obtain voltage vectors for the remaining switching states in a similar manner as shown in Table 7.3.

7.2.4.1 Principle of space vector pulse width modulation technique

As the three-phase voltage references vary with time, the voltage reference vector V^* rotates in the counterclockwise direction in the complex plane as illustrated in Fig. 7.39. This vector completes one revolution per electrical period of the reference voltage.

In the SVPWM technique, a voltage reference is given as a space vector of V^* and this voltage reference vector V^* is generated by using the output voltage vectors of a three-phase inverter. By using the two active voltage vectors adjacent to V^* and the zero vectors among the available eight voltage vectors, the SVPWM technique produces a voltage that has the same fundamental volt-second average as the given voltage reference vector V^* over a modulation period T_s.

Now we will describe how to generate a voltage reference vector. The voltage reference vector V^* is assumed to be inside a hexagon, which is formed by six output voltage vectors of a three-phase inverter. Only when this condition is satisfied, the voltage reference vector can be modulated properly.

For example, consider a voltage reference vector V^* given in sector ① of the six segments in the hexagon shown in Fig. 7.39. In this case, the inverter cannot generate the required voltage reference vector directly because there is no inverter output vector that has the magnitude and the phase equal to those of the voltage reference vector. Thus as an alternative, of the six active vectors, the two voltage vectors adjacent to the voltage reference vector and the zero vectors are used to generate a voltage that has the same fundamental volt-second average as the given voltage reference vector V^*. This modulation repeats each modulation period T_s depending on the switching frequency.

Let us now examine this modulation process in more detail. The voltage reference vector V^* is assumed to remain constant over the modulation period T_s. The process to synthesize the voltage vectors for generating the reference vector V^* consists of three steps as shown in Fig. 7.40.

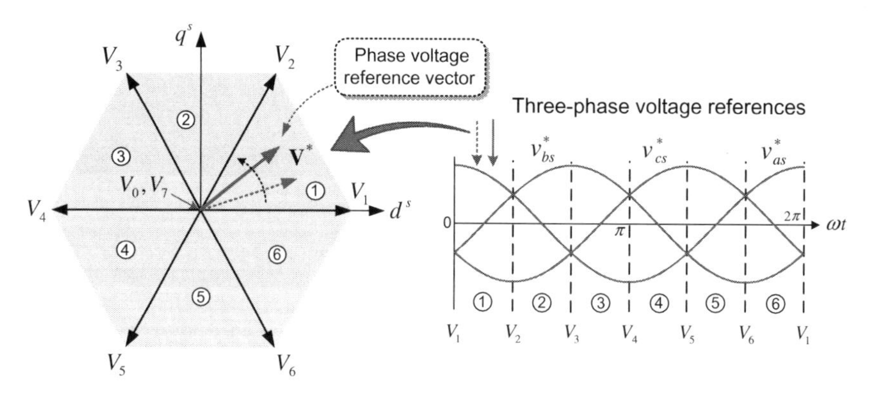

FIGURE 7.39

Rotation of voltage vector.

In the first step, one of the two adjacent active voltage vectors, V_1, is applied first during time T_1. As a result, an output voltage with the magnitude of $V_1 \cdot (T_1/T_s)$ in the direction of the vector V_1 is generated. Next, another vector V_2 is applied during time T_2 to meet the magnitude and the phase of the voltage reference vector V^*. Through these two steps, it is possible to generate the same output voltage as the voltage reference vector over the modulation period T_s. Lastly, if $T_1 + T_2 < T_s$, then one of the zero vectors, V_0 or V_7, is applied during the remaining time $T_0(=T_s - T_1 - T_2)$.

The duration time (T_1, T_2, and T_0) of each voltage vector for generating a given reference vector V^* can be calculated as follows. The above modulation process can be expressed mathematically as

$$\int_0^{T_s} V^* dt = \int_0^{T_s} V_n dt + \int_{T_1}^{T_1+T_2} V_{n+1} dt + \int_{T_1+T_2}^{T_s} V_{0,7} dt \tag{7.45}$$

Assuming a constant DC-link voltage during T_s, Eq. (7.45) can be rewritten as

$$V^* \cdot T_s = V_n \cdot T_1 + V_{n+1} \cdot T_2 \tag{7.46}$$

As an example, if the voltage reference vector V^* is given in the first sector ① $(0 \le \theta \le 60^\circ)$, Eq. (7.46) can be decomposed into two components as

$$\begin{cases} T_s \cdot |V^*| \cos\theta = T_1 \cdot \left(\dfrac{2}{3} V_{dc}\right) + T_2 \cdot \left(\dfrac{2}{3} V_{dc}\right) \cos 60^\circ \\ T_s \cdot |V^*| \sin\theta = T_2 \cdot \left(\dfrac{2}{3} V_{dc}\right) \sin 60^\circ \end{cases} \tag{7.47}$$

Solving Eq. (7.47) yields the duration times as

$$T_1 = T_s \cdot a \cdot \frac{\sin(60^\circ - \theta)}{\sin 60^\circ} \tag{7.48}$$

$$T_2 = T_s \cdot a \cdot \frac{\sin\theta}{\sin 60^\circ} \tag{7.49}$$

$$T_0 = T_s - (T_1 + T_2) \tag{7.50}$$

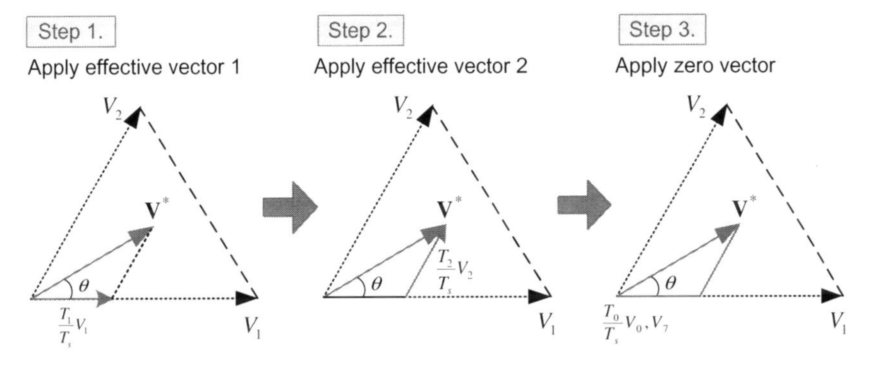

FIGURE 7.40

Modulation process for voltage generation.

Here, $a = |V^*|/\frac{2}{3}V_{dc}$. The duration times for the voltage reference vector in the other sectors ②–⑥ can be calculated in a similar manner.

Let us examine the achievable range of the output voltage by using the SVPWM technique. In the SVPWM technique, the sum of the duration times for the two active voltage vectors should not exceed the modulation period, i.e., $T_1 + T_2 \leq T_s$. Using Eqs. (7.48) and (7.49), the magnitude of the voltage reference to meet this requirement can be obtained as

$$T_1 + T_2 \leq T_s \;\rightarrow\; \mathrm{V}^* \leq \frac{V_{dc}}{\sqrt{3}} \frac{1}{\sin(60^\circ + \theta)} \tag{7.51}$$

This equation indicates that the possible range of the voltage reference vector V^* is inside the hexagon formed by joining the extremities of the six active vectors as shown in Fig. 7.41. However, for a voltage reference vector over one electrical period, the range of the voltage reference vector should be inside the inscribed circle of the hexagon to obtain the equal magnitude. Therefore the radius of the inscribed circle, $V_{dc}/\sqrt{3}$, is the maximum fundamental phase voltage in the SVPWM technique. This value is about 15.5% larger than that of the SPWM technique and is equal to that of the THIPWM technique. The value corresponds to 90.7% of the output voltage in the six-step operation.

7.2.4.2 Symmetrical space vector pulse width modulation technique

In the Section 7.2.4.1, we described how to obtain the duration times of the active and the zero voltage vectors to produce a given voltage reference vector. For given selected vectors and their duration times, there are many ways to place them within the modulation interval T_s.

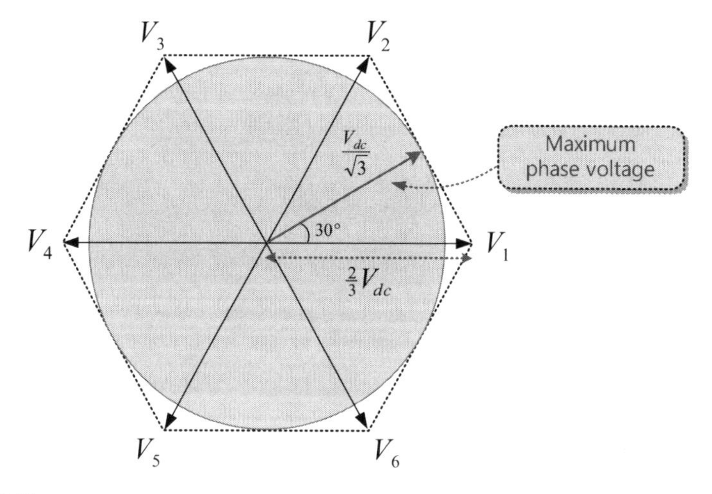

FIGURE 7.41

Possible range of the voltage reference vector in the SVPWM.

The placement of these vectors does not affect the average value of the output voltage over the interval T_s, but it significantly influences the linear modulation range and harmonic characteristics of the output voltage. In particular, the placement of the effective voltage vectors is important. Fig. 7.42 compares how the magnitude and frequency of the current ripple of the load can vary with the placement of the effective voltage vectors. Fig. 7.42A shows a case where the effective voltage vector is placed in the middle of the modulation interval. Compared with Fig. 7.42B, this placement results in a smaller current ripple and a higher ripple frequency. Thus the placement of Fig. 7.42A can give a better harmonic performance than that of Fig. 7.42B.

It was known that placing the effective voltage vectors at the center of the modulation interval shows superior harmonic characteristics as shown in Fig. 7.42A [8,9]. The SVPWM technique of this placement scheme is called *symmetrical SVPWM* technique. Moreover, this centering can increase the pulse width further to modulate the effective voltage, resulting in an improved voltage modulation range. The placement of the zero voltage vectors determines the position of the effective voltage vectors within the modulation interval T_s. Thus, in the *symmetrical SVPWM* technique, the two zero vectors, V_0 and V_7, during an equal time of $T_0/2$ are distributed at the beginning and the end of the modulation interval T_s as shown in Fig. 7.43. Furthermore, in this case, to obtain the minimum switching frequency, it is necessary to arrange the switching sequence in the order of $V_0(000) \rightarrow V_1(100) \rightarrow V_2(110) \rightarrow V_7(111)$.

For this purpose, the two zero vectors of V_0 and V_7 are used alternately in the interval T_s. In such a switching sequence, the transition from one vector to another vector can be performed by switching only one switch. In the next interval, the switching sequence is reversed, i.e., $V_7 \rightarrow V_2 \rightarrow V_1 \rightarrow V_0$ as shown in Fig. 7.44 [9]. This interval also requires only one switching for the transition.

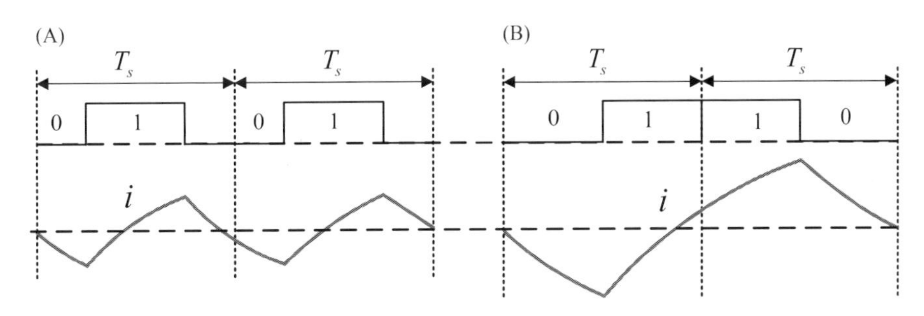

FIGURE 7.42

Current characteristics according to the placement of active voltage pulses. (A) Best case and (B) worst case.

D.G. Holmes, The significance of zero space vector placement for carrier-based PWM schemes, IEEE Trans. Ind. Appl., 32 (5) (1996) 1122–1129.

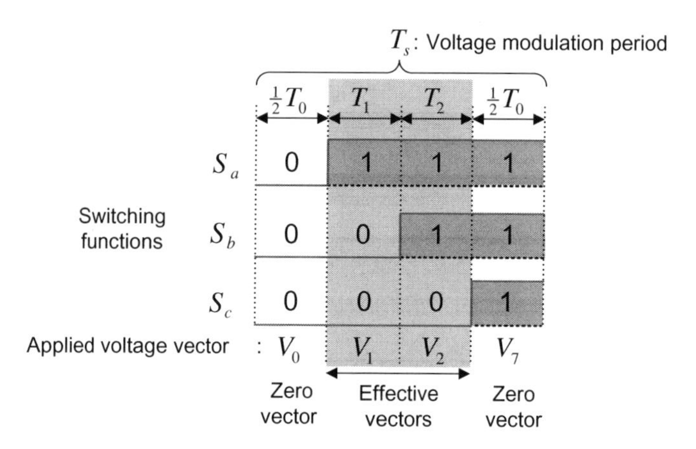

FIGURE 7.43

Switching sequence.

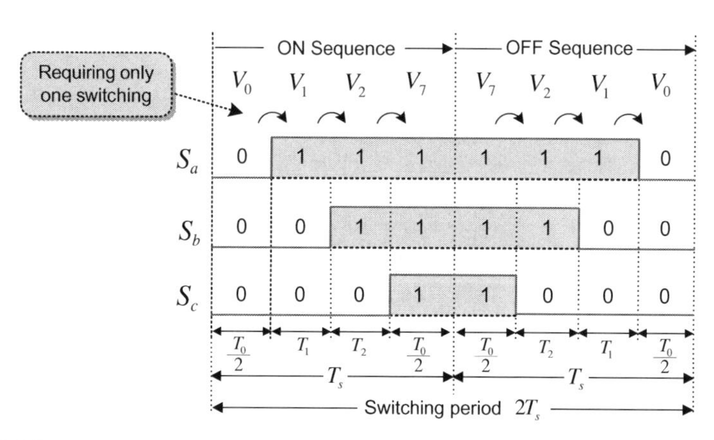

FIGURE 7.44

Alternate switching sequence.

Such alternating switching sequences at every modulation interval allow the switching frequency to be reduced.

Fig. 7.45 shows the switching sequences in all six sectors for the symmetrical *SVPWM* technique.

For this alternate switching sequence, the two modulation intervals $2T_s$ become one switching period. For example, the modulation interval T_s of 100 μs indicates a switching frequency of 5 kHz. Commonly, the modulation interval T_s refers to the current control period. This is because a new voltage reference is given at every current control period.

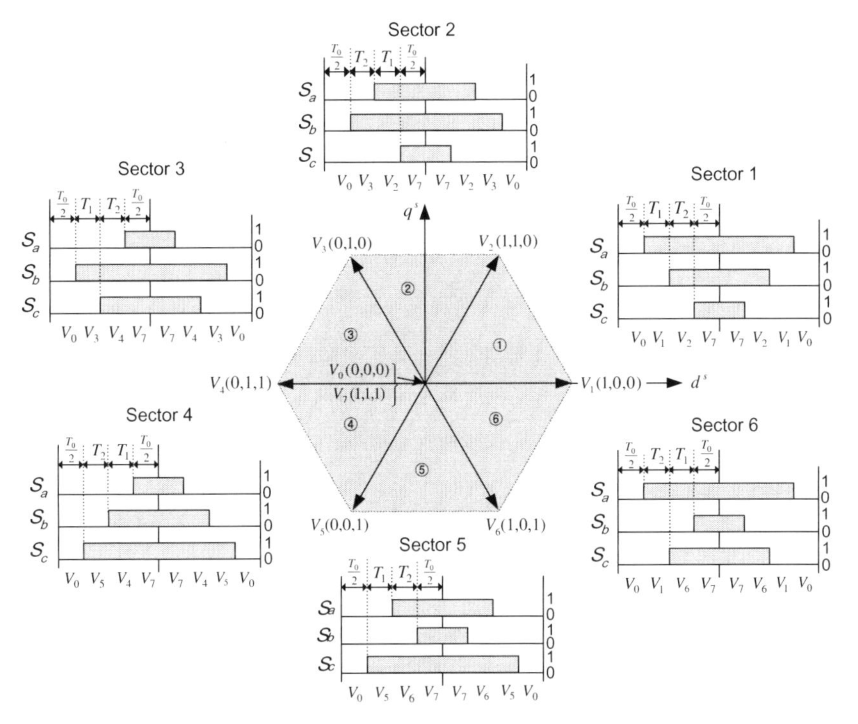

FIGURE 7.45

Switching sequences in six sectors.

EXAMPLE 4

Describe the pole voltage for the symmetrical SVPWM technique.

Solution

From the vector duration times calculated for each sector as shown in Fig. 7.45, we can obtain the pole *an* voltage for the SVPWM technique. The pole *an* voltage for a given command vector V^* in the first sector ① ($0° \le \theta \le 60°$) from Eqs. (7.48) and (7.49) is

$$v_{an1} = \frac{T_1 + T_2}{T_s} \cdot \frac{V_{dc}}{2} = \frac{\sqrt{3}}{2}|V^*|\cos\left(\frac{\pi}{6} - \theta\right)$$

In a similar manner, the pole voltages for the remaining sectors can be obtained as:

Sector ② ($60° \le \theta \le 120°$):

$$v_{an2} = \frac{T_1 - T_2}{T_s} \cdot \frac{V_{dc}}{2} = \frac{3}{2}|V^*|\cos\left(\frac{\pi}{6} - \theta\right)$$

Sector ③ ($120° \le \theta \le 180°$):

$$v_{an3} = \frac{-T_1 - T_2}{T_s} \cdot \frac{V_{dc}}{2} = -\frac{3}{2}|V^*|\cos\left(\frac{\pi}{6} - \theta\right)$$

Sector ④ ($180° \le \theta \le 240°$):

$$v_{an4} = \frac{-T_1 - T_2}{T_s} \cdot \frac{V_{dc}}{2} = -\frac{3}{2}|V^*|\cos\left(\frac{\pi}{6} - \theta\right)$$

Sector ⑤ ($240° \le \theta \le 300°$):

$$v_{an5} = \frac{-T_1 + T_2}{T_s} \cdot \frac{V_{dc}}{2} = \frac{3}{2}|V^*|\cos\left(\frac{\pi}{6} - \theta\right)$$

Sector ⑥ ($300° \le \theta \le 360°$):

$$v_{an6} = \frac{T_1 + T_2}{T_s} \cdot \frac{V_{dc}}{2} = \frac{\sqrt{3}}{2}|V^*|\cos\left(\frac{\pi}{6} - \theta\right)$$

The following figure shows the pole *an* voltage obtained in all sectors for $MI = 1$ (i.e., $|V^*| = V_{dc}/\sqrt{3}$).

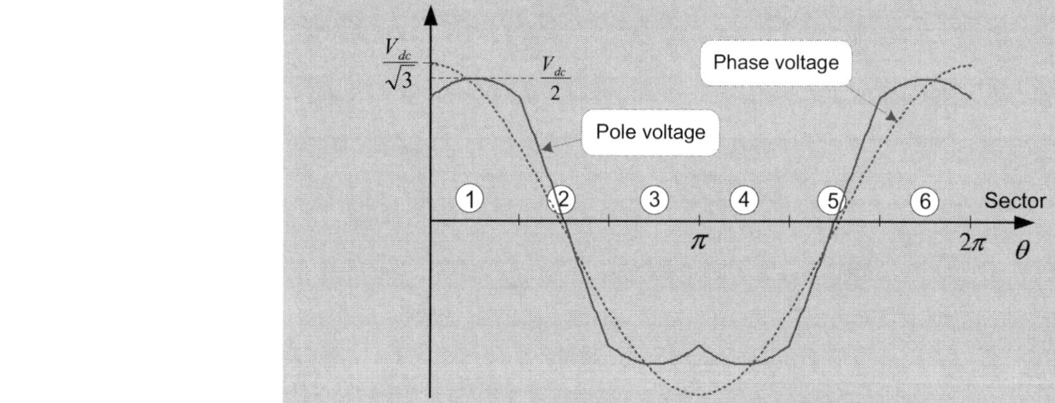

As can be seen in Example 4, it is interesting to note that the pole voltage is not a pure sinusoidal wave but a waveform that contains a third harmonic component similar to the modulation waveform of the THIPWM technique. In the THIPWM technique, we can see that the addition of a third harmonic to the voltage reference can increase the fundamental of the phase voltage. Since the SVPWM technique also exploits the third harmonic component, it can also be considered to extend the linear modulation range more than the SPWM technique.

Fig. 7.46 compares the output torque and the harmonic distortion of the load current for the SPWM and the SVPWM techniques [7]. It can be readily seen that the SVPWM technique allows a value of *MI* up to 1.15 and can produce a harmonic distortion lower than that of the SPWM technique in the high modulation range.

Contrary to the symmetrical SVPWM technique, the effective voltages in the SPWM technique are not placed at the center, resulting in harmonic characteristics inferior to those of the symmetrical SVPWM technique.

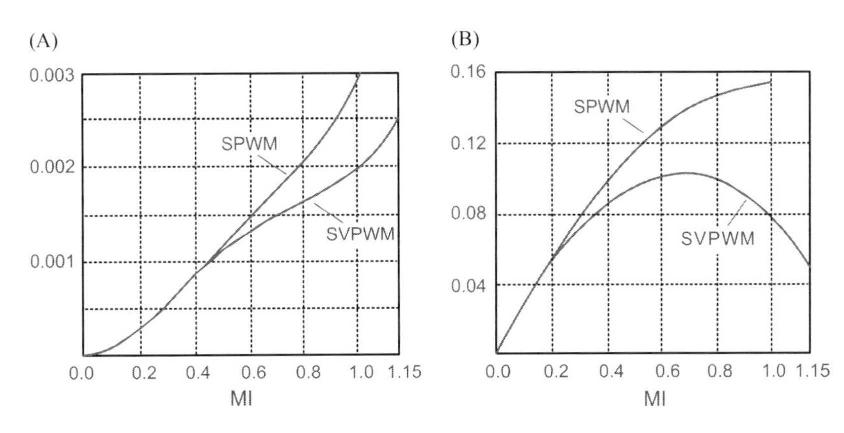

FIGURE 7.46

Comparison between the SPWM and the SVPWM techniques. (A) Square of rms harmonic current and (B) torque harmonic.

H.W. Van Der Broeck, et al., Analysis and realization of a pulse width modulator based on voltage space vectors, IEEE Trans. Ind. Appl., 24 (1) (1988) 142–150.

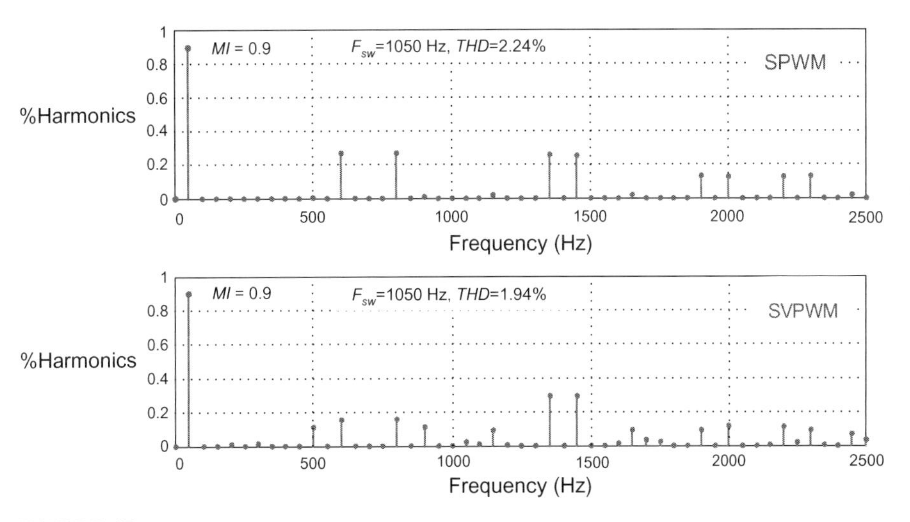

FIGURE 7.47

Frequency spectrum of the phase voltages for the SPWM and SVPWM.

The harmonic orders contained in the SVPWM technique are similar to those of the SPWM technique expressed in Eq. (7.42). However, from Fig. 7.47, it can be seen that the relative magnitude of the switching frequency harmonics for the SVPWM technique is smaller than that of the SPWM technique, and the SVPWM technique spreads its switching harmonic energy effectively into the sidebands. Consequently, the THD of the SVPWM technique is less than that of the SPWM technique, even though the harmonic of the double switching frequencies for the SVPWM technique increases.

The disadvantage of the SVPWM technique is that it requires trigonometric calculations and a greater computational effort to calculate the switching times of active voltage vectors. However, nowadays, this method can be simply implemented by the carrier-based PWM technique using an offset voltage. This technique will be discussed in more detail in Section 7.4.

The SVPWM technique described above uses the symmetrical modulation method in which the two zero vectors are distributed during an equal duration time. In this case, all switches in three poles operate at every modulation period. Thus this is called *three-phase modulation* or *continuous modulation*. By contrast, we can select different duration times for two zero vectors. Depending on the selection, we can obtain various modulation methods, which differ in modulation performances such as harmonic characteristics, voltage linearity, and switching losses. Thus the placement of the zero vectors is a degree of freedom for modulation in the SVPWM technique. A typical example of having unequally distributed zero vectors is a modulation technique in which switches in only two of the three poles operate to reduce the switching frequency. This is called *two-phase modulation* or *discontinuous modulation* in which the switches of one leg are kept inactive in the modulation interval as shown in Fig. 7.48. In this discontinuous modulation technique, only one zero vector, either T_0 or T_7, is used. Accordingly, this causes a loss of symmetry in the placement of effective voltage vectors. However, due to switching in only two of the three poles, the overall switching frequency can be reduced by 1/3 compared to the three-phase modulation. In the higher modulation range, the discontinuous modulation produces a lower harmonic distortion because of its higher net switching frequency. However, in the low modulation range, due to a loss of symmetry, the discontinuous modulation has a harmonic distortion higher than the continuous modulation.

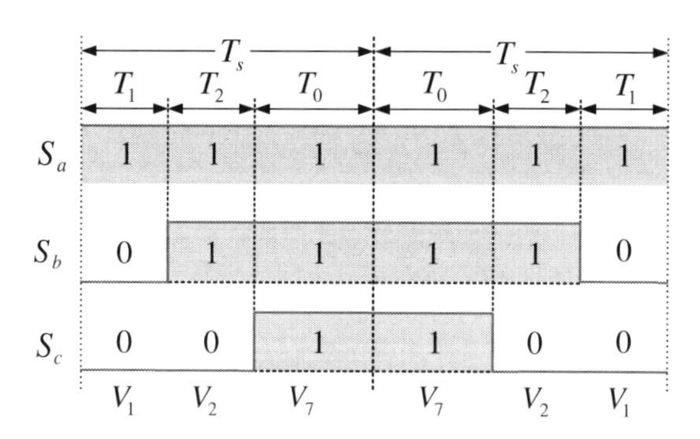

FIGURE 7.48

Switching sequence in two-phase modulation.

7.3 DISCONTINUOUS PWM TECHNIQUES [10]

The PWM techniques described previously use a continuous modulation in which the switching actions occur in all the three poles of an inverter. By contrast, in a discontinuous modulation (or two-phase modulation) that reduce the switching frequency, the switching actions occur in only two of the three poles. Though the aim of the discontinuous modulation is to reduce the switching frequency, some characteristics such as switching losses, harmonics, and voltage linearity vary according to the location of the interval of the switches being kept inactive (i.e., unmodulated section). Thus there could be various discontinuous PWM (DPWM) techniques according to the location of the unmodulated section. Here, six popular discontinuous PWM techniques will be described.

For the symmetry in a three-phase system, the scope where the modulation ceases can be a total of 120° section at most per fundamental cycle as shown in Fig. 7.49. This indicates that, by using the unmodulated section fully, the effective switching frequency can be reduced to 66.7% of the frequency of the continuous PWM (CPWM) methods. Though the unmodulated section can be placed at any position, it is normally placed around the peak of the phase current to reduce the switching losses effectively. This is because the switching losses are proportional to the magnitude of the current. Thus it is necessary to place unmodulated sections differently according to the power factor of a load.

7.3.1 60° DISCONTINUOUS PWM TECHNIQUE

This 60° discontinuous pulse width modulation (DPWM) technique ceases to switch each switching device for two 60° sections around both the positive and the negative

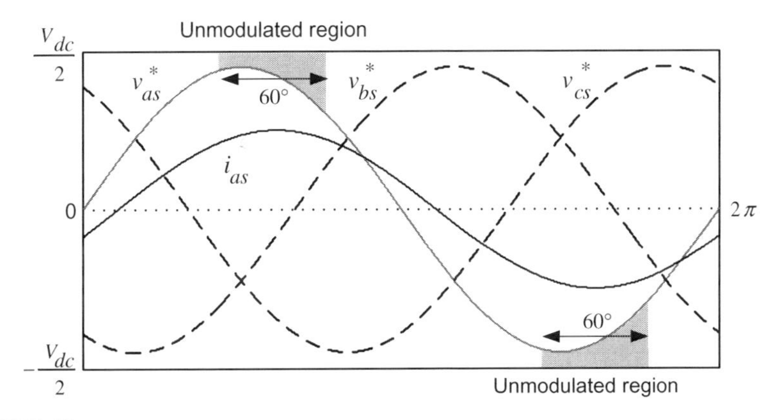

FIGURE 7.49

Unmodulated sections.

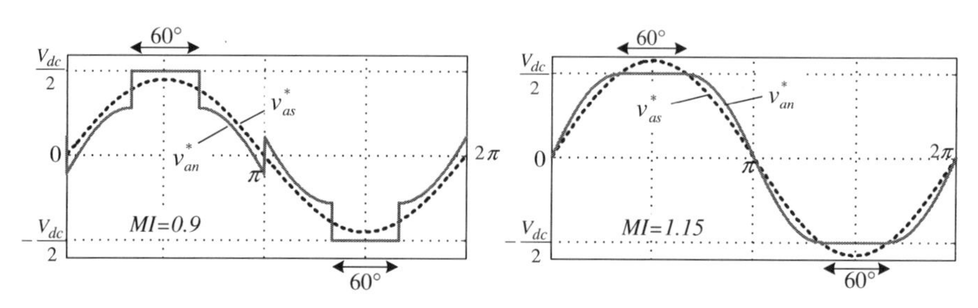

FIGURE 7.50

60° discontinuous PWM technique: pole voltage and phase voltage.

peaks of each phase voltage waveform. This DPWM technique shows the minimum switching loss characteristic for near unity power factor loads such as utility interface, UPS applications, and permanent magnet synchronous motors.

Fig. 7.50 shows the waveforms of the pole and phase voltages for $MI = 0.9$ and 1.15, respectively, by using the 60° DPWM technique. In the 60° DPWM technique, the pole voltage reference v_{an}^* is clamped to the positive DC rail ($V_{dc}/2$) for the 60° section around the positive peak of the phase voltage waveform v_{as}^*. So, the upper switch is kept on the "on" state for that section as shown in Fig. 7.50. In addition, the pole voltage reference is clamped to the negative DC rail ($-V_{dc}/2$) for the 60° section around the negative peak of the phase voltage waveform. So, the lower switch is kept on the "on" state for that section. Otherwise, the magnitude of the pole voltage reference v_{an}^* is adjusted to produce the desired fundamental voltage.

We will discuss how to obtain the pole voltage reference for the DPWM technique in the next section.

7.3.2 60° (±30°) DISCONTINUOUS PWM TECHNIQUE

To achieve the minimum switching losses for lagging power factor loads such as induction motor drives, it is desirable that the modulation cease for the 60° section shifted by +30° phase with respect to the peaks of the phase voltage as shown in Fig. 7.51A. On the other hand, the minimum switching losses for leading power factor loads such as induction generators will need the 60° unmodulated section shifted by -30° with respect to the peaks of the phase voltage waveform as shown in Fig. 7.51B.

7.3.3 ±120° DISCONTINUOUS PWM TECHNIQUE

For simplicity, there is a 120° DPWM technique in which the modulation ceases for a total of 120° section. The unmodulated section can be either the largest or the lowest 120° section of the phase voltage waveform as shown in Fig. 7.52.

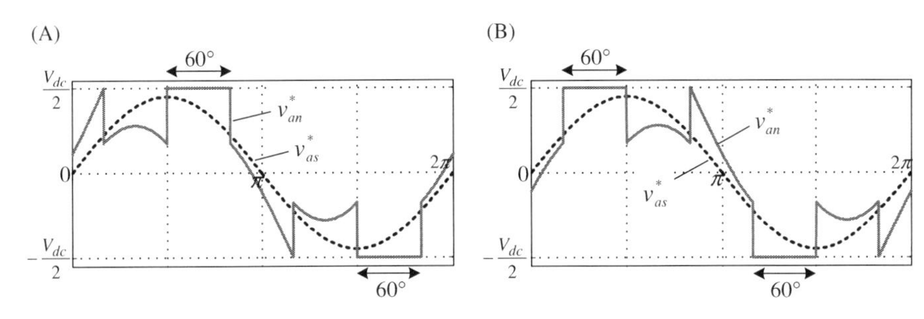

FIGURE 7.51

60° (±30°) DPWM technique: pole and phase voltages ($MI = 0.9$) (A) 60° (+30°) DPWM and (B) 60° (−30°) DPWM.

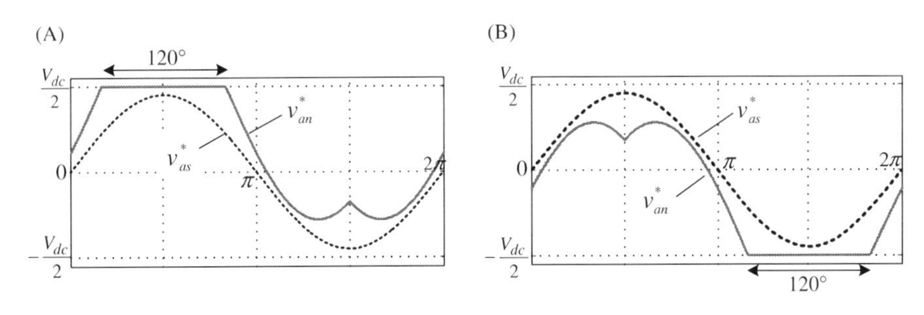

FIGURE 7.52

120° DPWM technique: pole and phase voltages ($MI = 0.9$). (A) +120° DPWM and (B) −120° DPWM.

However, these techniques have nonuniform thermal stress on the switching devices.

7.3.4 30° DISCONTINUOUS PWM TECHNIQUE

This 30° DPWM technique has less harmonic distortion than other DPWM techniques. As shown in Fig. 7.53, the 30° DPWM technique has four unmodulated sections, each equal to 30° section.

It should be noted that, as shown in Fig. 7.50–7.53, the pole voltage references in the DPWM techniques become a nonsinusoidal waveform whose shape depends on the position of the unmodulated section.

The DPWM techniques can lead to a reduction in the switching losses and an improved waveform quality in the high modulation range. However, a poor low modulation range performance and its implementation complexity have limited the application of these DPWM techniques. In addition, they exhibit a nonlinear

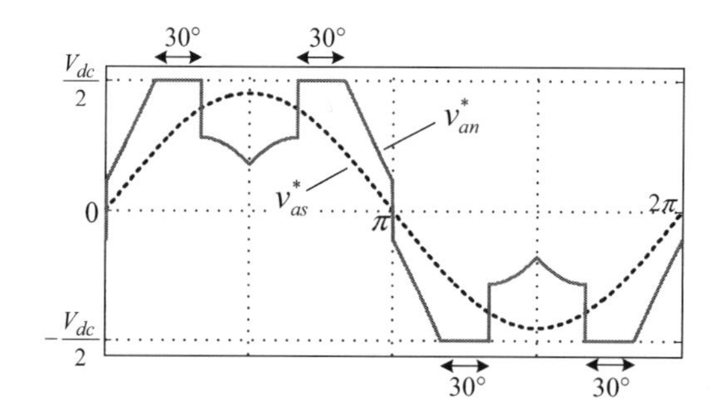

FIGURE 7.53

30° DPWM technique: pole and phase voltages ($MI = 0.9$).

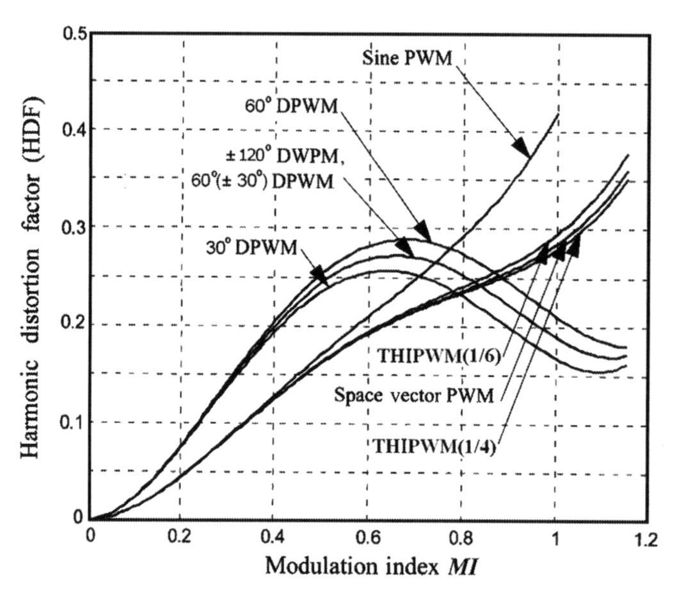

FIGURE 7.54

HDF comparison with different voltage modulation techniques.

A.M. Hava, R.J. Kerkman, T.A. Lipo, A high-performance generalized discontinuous PWM algorithm, IEEE Trans. Ind. Appl., 34 (5) (1998) 1059–1071.

relation between the reference and output voltage in very low modulation range. Fig. 7.54 shows harmonic distortion factor (HDF) curves of all the discussed PWM techniques under an equal inverter average switching frequency (thus the HDF of the DPWM methods is multiplied by $(2/3)^2$) [10]. We can see from the

HDF curves that the CPWM methods have a better HDF in the low modulation range, while the DPWM techniques are superior in the high modulation range of $MI > 0.8$.

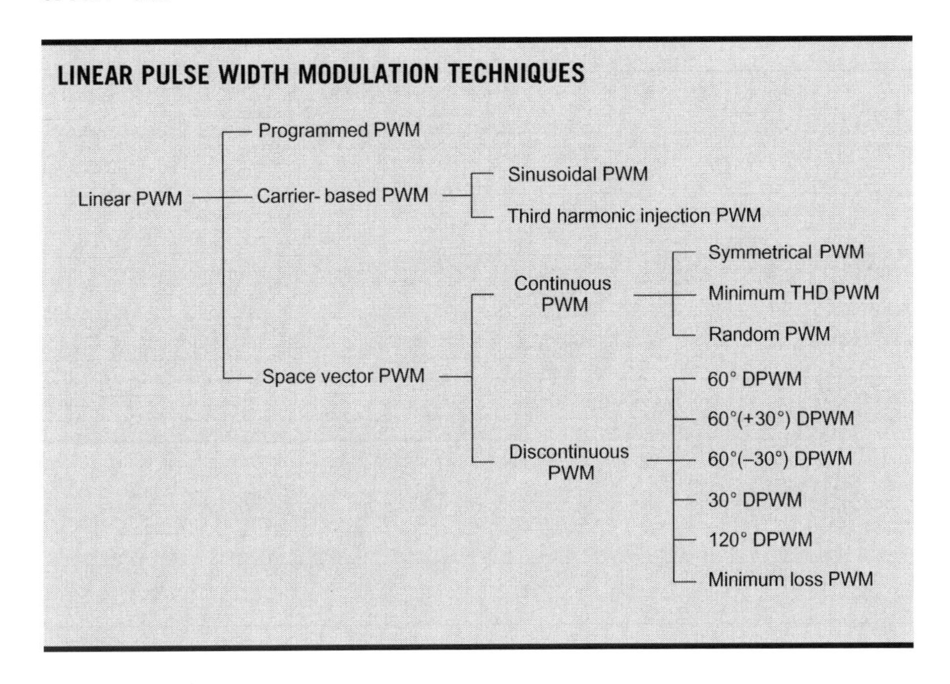

The DPWM techniques explained above can be equivalently implemented by changing the duration time of zero vectors, V_0 and V_7, in the SVPWM technique. This implies that modifying the pole voltage in the triangle-comparison PWM (or carrier-based PWM) technique is considered to be equivalent to changing the duration time of zero vectors in the SVPWM technique. Like this, we can see that there is a correlation between the SVPWM and the triangle-comparison PWM techniques.

Fig. 7.55 shows the six discontinuous PWM techniques and their distribution of zero vectors [11,12]. Here, μ denotes the distribution factor (or apportioning factor), which indicates the distribution of zero vectors inside the modulation interval as

$$\mu = \frac{t_{0\_v7}}{t_{0\_v0} + t_{0\_v7}} \tag{7.46}$$

where t_{0_v0} and t_{0_v7} are the duration times of zero vectors, V_0 and V_7, respectively. The distribution factor has a range of $0 \leq \mu \leq 1$. The range of $0 < \mu < 1$ results in a continuous modulation, while $\mu = 0$ or 1 results in a discontinuous modulation. The best choice is $\mu = 0.5$ (i.e., $t_{0\_v0} = t_{0\_v7} = T_0/2$), which results in the conventional symmetrical SVPWM technique. The use of $\mu = 1$ (i.e., $t_{0\_v0} = 0$ and $t_{0\_v7} = T_0$) in which only the zero vector of V_7 is employed, results in $+120°$ DPWM technique. On the other hand, the use of $\mu = 0$ (i.e.,

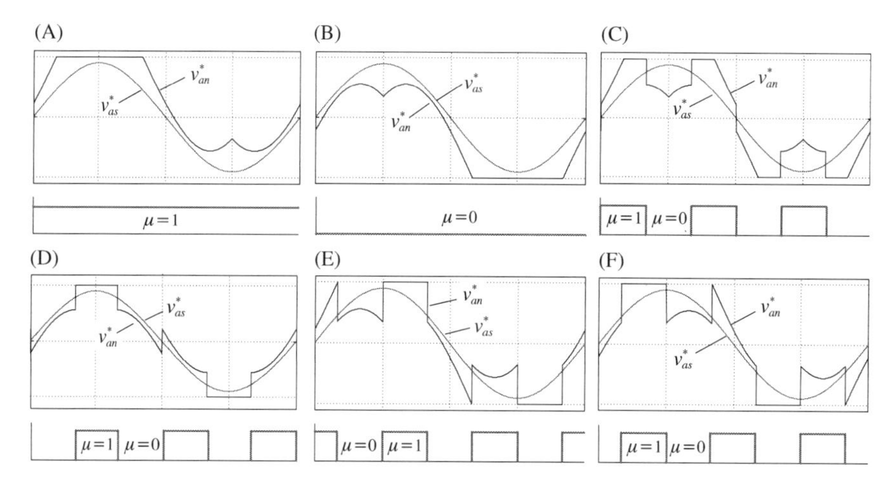

FIGURE 7.55

Several DPWM techniques and their distribution of zero vectors. (A) +120° DPWM, (B) −120° DPWM, (C) 30° DPWM, (D) 60° DPWM, (E) 60°(+30°) DPWM, (F) +60°(−30°) DPWM.

E. Silva, E. Santos Jr., C. Jacobina, Pulsewidth modulation strategies nonsinusoidal carrier-based PWM and space vector modulation techniques, IEEE Trans. Ind. Electron. Mag. (2011) 37–45.

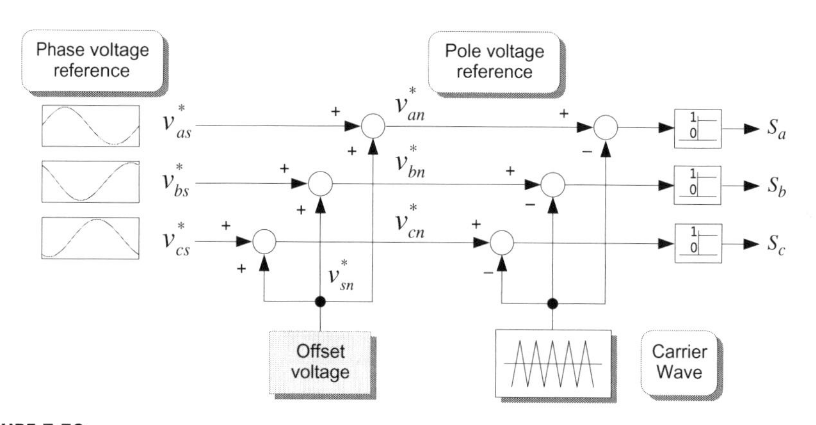

FIGURE 7.56

PWM technique employing the zero-sequence injection principle.

$t_{0\_v0} = T_0$ and $t_{0\_v7} = 0$) in which only the zero vector of V_0 is employed, results in −120° DPWM technique. The alternating use of $\mu = 0$ or 1 for each 60° section can realize other four DPWM techniques as shown in Fig. 7.55C–F. In all the PWM techniques discussed above, it is interesting to note that all the phase voltage references are a pure sinusoidal waveform. However, they are different in the pole voltage references, i.e., the modulating signals, resulting in a different modulation performance.

7.4 PWM TECHNIQUE BASED ON OFFSET VOLTAGE [13,14]

In the Section 7.3.4, we saw that there is a correlation between the SVPWM and the triangle-comparison PWM techniques. In this section, we will discuss a new PWM technique as shown in Fig. 7.56, which allows the SVPWM technique to be implemented easily based on the triangle-comparison PWM technique.

In addition to the SVPWM technique, this generalized PWM technique can actually allow almost all the PWM techniques to be implemented based on the triangle-comparison PWM technique. This technique can reproduce the existing PWM techniques by modifying the pole voltage with the *offset voltage* (or the *zero-sequence voltage*).

Prior to explaining this technique, it should be noted that the output voltage of an inverter is the pole voltage. Thus, in the triangle-comparison PWM techniques, the modulating wave, which is compared with the triangular carrier wave, is the pole voltage reference. The SPWM technique use the phase voltage reference as the modulating wave as shown in Fig. 7.57. Thus, in the SPWM technique, the pole voltage reference is equal to the phase voltage reference.

On the other hand, in the THIPWM technique, the voltage reference, which is produced by adding a third harmonic to the phase voltage reference, is compared with the triangular carrier wave as shown in Fig. 7.37. Thus the pole voltage reference is not equal to the phase voltage reference. In addition, we can see from Section 7.3 that the DPWM techniques use pole voltage references that are different from the phase voltage reference.

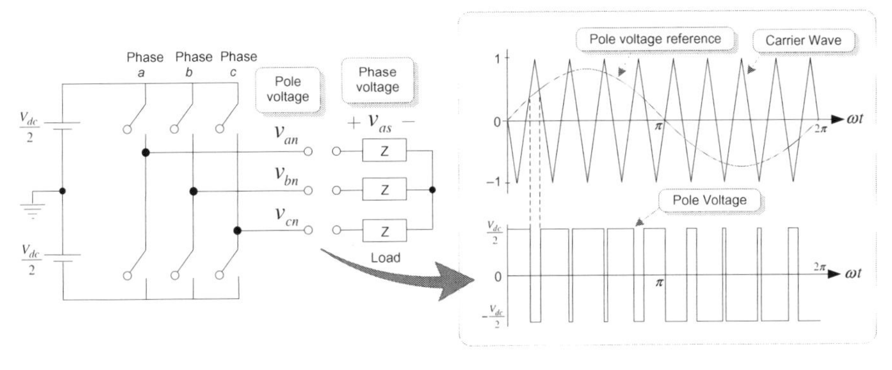

FIGURE 7.57

SPWM technique.

Based on this knowledge, we can generalize that the pole voltage reference in the PWM techniques can be given by adding an offset voltage v_{sn} to the phase voltage reference as

$$\begin{aligned} v_{an}^* &= v_{as}^* + v_{sn} \\ v_{bn}^* &= v_{bs}^* + v_{sn} \\ v_{cn}^* &= v_{cs}^* + v_{sn} \end{aligned} \tag{7.52}$$

This is based on the fact that the added offset voltage disappears in the line-to-line voltages and the phase voltages for the three-phase inverter. Since the line-to-line voltage is given as the difference between two pole voltages, this offset voltage is canceled in both the line-to-line voltages and the phase voltages as

$$v_{ab}^* = v_{an}^* - v_{bn}^* \tag{7.53}$$

$$v_{as}^* = \frac{1}{3}\left(v_{ab}^* - v_{ca}^*\right) \tag{7.54}$$

Therefore, although the phase voltage reference in an inverter is always given as a pure sinusoidal waveform, the pole voltage reference can be modified by the applied offset voltage. According to the applied offset voltage, the modulation performance such as voltage linearity, harmonics, and switching losses can be changed.

The choice of this offset voltage v_{sn} is completely arbitrary. However, there is a limitation on the value of the offset voltage. For the linear modulation, since the reference voltage should not exceed the peak of the triangular carrier wave, the value of the offset voltage should be chosen so that the resultant pole voltage references $v_{an}^*, v_{bn}^*, v_{cn}^*$ satisfy the following condition.

$$-\frac{V_{dc}}{2} \le v_{an}^*, v_{bn}^*, v_{cn}^* \le \frac{V_{dc}}{2} \tag{7.55}$$

By adding the arbitrary offset voltage within the allowed range, various pole voltage references can be obtained. Therefore the offset voltage is considered to be the degree of freedom in choosing various modulation techniques.

In the SPWM technique shown in Fig. 7.58A, the output voltage is determined by comparing the phase voltage reference with the triangular carrier wave.

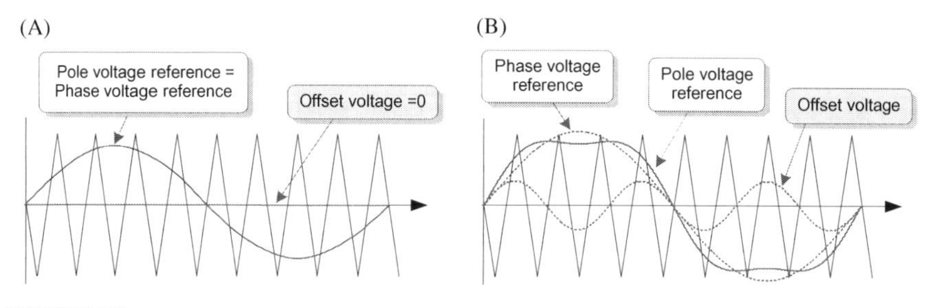

FIGURE 7.58

Offset voltage for (A) SPWM and (B) THIPWM techniques.

Thus the phase voltage reference is equal to the pole voltage reference (e.g., $v^*_{an} = v^*_{as}$). In this case, the offset voltage $v_{sn} = 0$.

On the other hand, in the THIPWM technique shown in Fig. 7.58B, since the pole voltage reference generated by adding a third harmonic to the phase voltage reference is compared with the triangular carrier wave, the offset voltage v_{sn} is the third harmonic voltage.

Now we will discuss how the symmetrical SVPWM technique can be equivalently implemented as the triangle-comparison PWM technique employing the offset voltage.

7.4.1 SPACE VECTOR PWM TECHNIQUE IMPLEMENTATION BASED ON OFFSET VOLTAGE

In the symmetrical SVPWM technique, the effective voltage vectors are always placed at the center of the modulation interval T_s. In the triangle-comparison PWM technique, if the absolute values of the maximum V_{max} and the minimum V_{min} in the three-phase voltage references are equal to each other, then the resultant effective voltage generated from the comparison with a triangular carrier wave is also placed at the center of the modulation interval as shown in Fig. 7.59.

If this requirement is met by adding the offset voltage, then the original SVPWM can be equivalently implemented as the triangle-comparison PWM

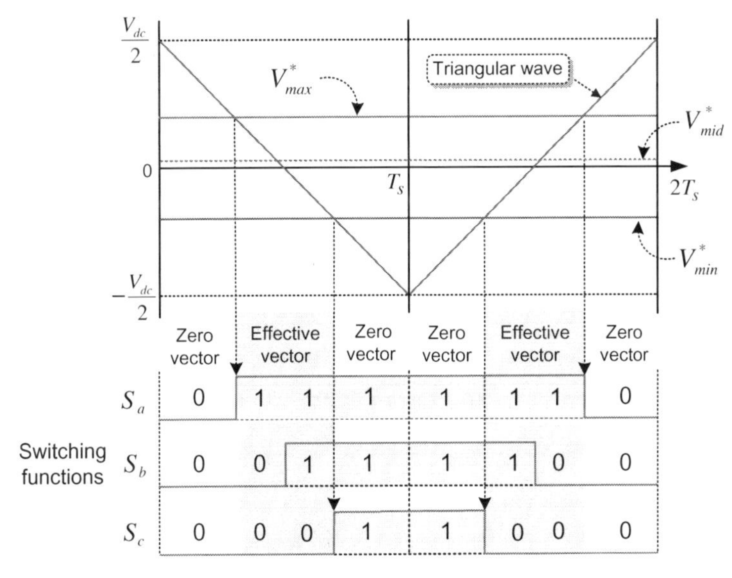

FIGURE 7.59

Symmetrical SVPWM technique.

technique. For this purpose, we establish the offset voltage to obtain the same absolute values of the maximum and the minimum voltage references as

$$V_{max} + v_{sn} = -(V_{min} + v_{sn}) \rightarrow v_{sn} = -\frac{V_{max} + V_{max}}{2} \tag{7.56}$$

Using the pole voltage references obtained by adding the offset voltage of Eq. (7.56) to the phase voltage references, an equivalent SVPWM technique can be readily implemented without a complicated calculation as shown in Fig. 7.60.

Fig. 7.61 shows the pole voltage reference, the phase voltage reference, and the offset voltage in this equivalent SVPWM technique for $MI = 0.9$. It can be seen that the SVPWM technique employs the pole voltage obtained by adding the offset voltage, which is a third harmonic in the shape of a triangular wave. This modified pole voltage is equal to the pole voltage as shown in Example 4.

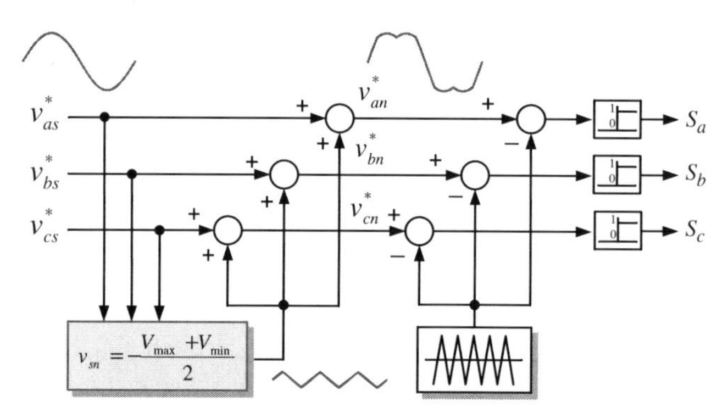

FIGURE 7.60

Equivalent SVPWM technique by using the offset voltage.

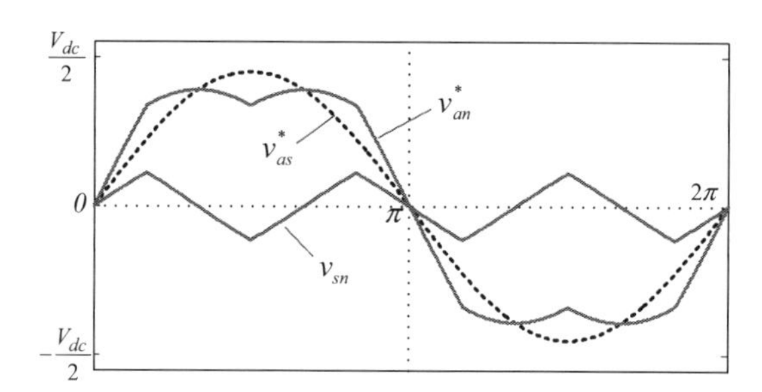

FIGURE 7.61

Voltages for equivalent SVPWM by using the offset voltage ($MI = 0.9$).

Next gives a description on the THIPWM technique implementation by using the triangle-comparison PWM technique. As noted in the earlier section, the THIPWM technique uses the pole voltage obtained by adding a third harmonic with the amplitude of 1/6 of the phase voltage reference to the phase voltage reference. Thus this third harmonic is considered to be an offset voltage. From the three-phase voltage references, we can find that the offset voltage is given as [12]

$$v_{sn} = -\frac{v_{as}^{*} v_{bs}^{*} v_{cs}^{*}}{v_{as}^{*2} + v_{bs}^{*2} + v_{cs}^{*2}} \tag{7.57}$$

Fig. 7.62 shows the pole voltage reference, the phase voltage reference, and the offset voltage for this case.

Likewise, other DPWM techniques can be also implemented by choosing an appropriate offset voltage as shown in Fig. 7.63. As one example, the offset voltage for the +120° DPWM technique is given as [13]

$$v_{sn} = \frac{V_{ref}}{2} - V_{max} \tag{7.58}$$

In this manner, various PWM techniques can be effectively implemented based on the modified pole voltage reference by using an offset voltage. This modification results in the change of the distribution time of zero vectors. As shown in Fig. 7.64, when the three pole voltage references are changed by the offset voltage, the position of the effective voltages is changed without any change in its duration time. Thus the SVPWM and the triangle-comparison PWM techniques have different approaches for the modulation, but there is a close correlation between the two techniques.

The pole voltage references used in each PWM technique ultimately defines the switching instants by their comparison with the triangular carrier wave. However, without such comparison, the switching instant of the gating pulse for each switch can be easily calculated as follows. For example, the

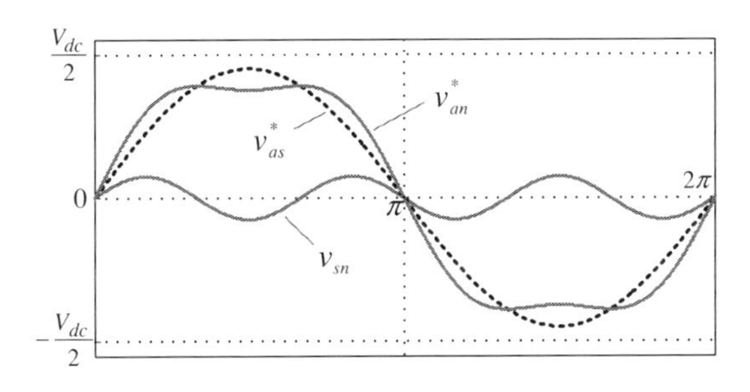

FIGURE 7.62

THIPWM technique implementation by using the offset voltage ($MI = 0.9$).

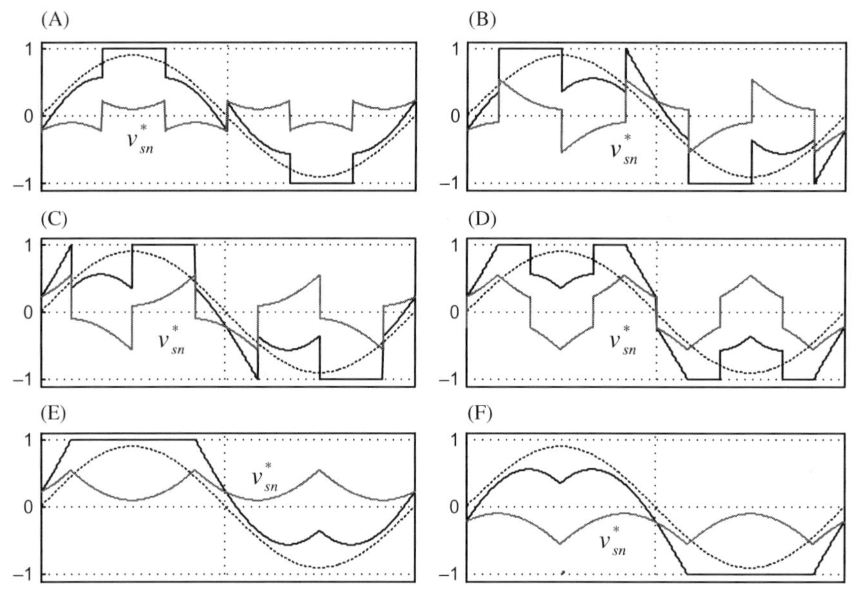

FIGURE 7.63

Implementation of different DPWM techniques by using the offset voltage. (A) +60° DPWM, (B) 60° (−30°) DPWM, (C) 60°(+30°) DPWM, (D) 30° DPWM, (E) +120° DPWM, (F) −120° DPWM.

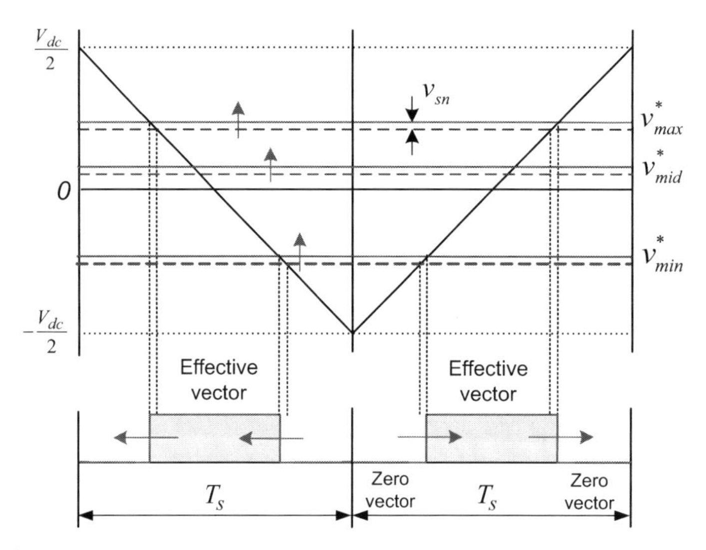

FIGURE 7.64

Position change of the effective voltage according to the offset voltage.

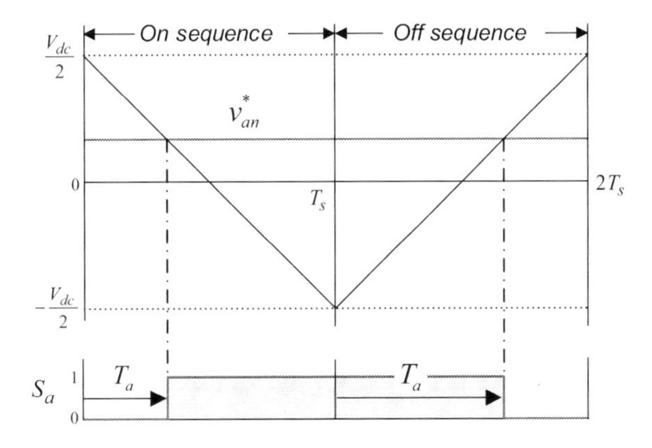

FIGURE 7.65

Switching instant of the gating pulse for each switch.

switching instant of the gating pulse for the upper switch of phase *as* in Fig. 7.65 can be calculated as:

- Off gating sequence:

$$\frac{T_a}{T_s} = \frac{\frac{V_{dc}}{2} + v_{an}^*}{V_{dc}} = \frac{1}{2} + \frac{v_{an}^*}{V_{dc}} \rightarrow T_a = \frac{T_s}{2} + \frac{v_{an}^*}{V_{dc}} T_s \tag{7.59}$$

- On gating sequence:

$$T_a = \frac{T_s}{2} - \frac{v_{an}^*}{V_{dc}} T_s \tag{7.60}$$

7.5 OVERMODULATION

A PWM inverter can correctly generate the required voltage reference by adopting any PWM technique previously explained. However, the linear modulation range for its output voltage is limited and varies with PWM techniques. In the SPWM technique, when the amplitude of the voltage reference wave becomes larger than the peak of the triangular carrier wave, or in the SVPWM technique when the voltage reference vector lies outside the hexagon boundaries as shown in Fig. 7.66, the PWM inverter cannot correctly produce the desired voltage reference. These conditions are called *overmodulation*.

The beginning of the overmodulation region depends on the type of PWM technique used as shown in Fig. 7.67. For the SPWM technique, the overmodulation starts at 78.5% ($V_{dc}/2$) in reference to the output voltage of six-step operation ($2V_{dc}/\pi$). On the other hand, for the conventional SVPWM, the overmodulation starts at 90.7% ($V_{dc}/\sqrt{3}$).

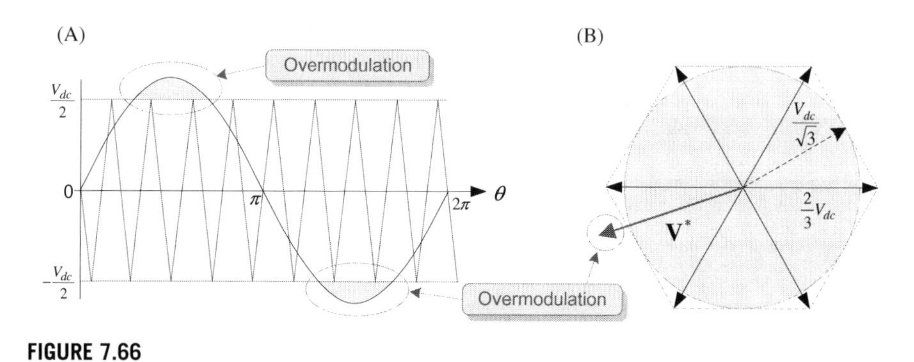

FIGURE 7.66

Overmodulation. (A) SPWM and (B) SVPWM.

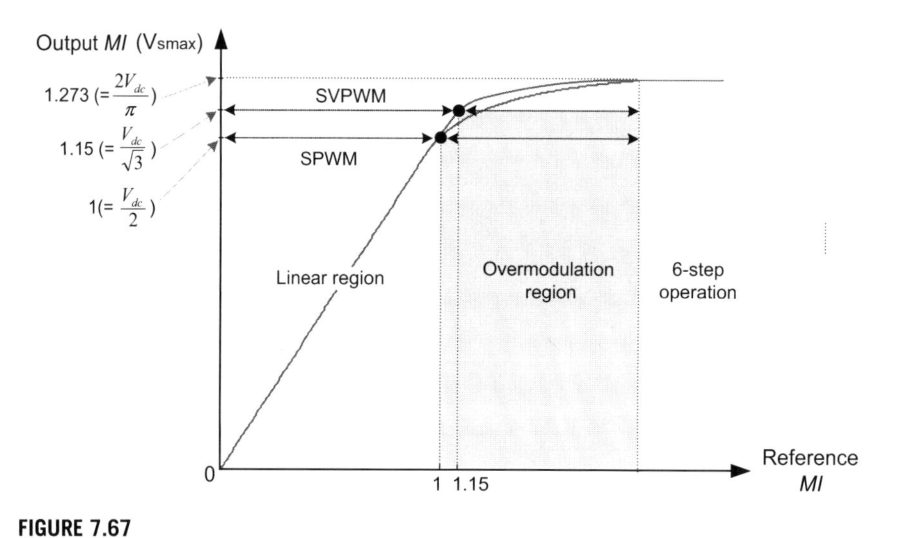

FIGURE 7.67

Overmodulation region.

In the overmodulation region, the voltage linearity is lost because the output voltage fundamental component becomes less than its command value. Furthermore, the output voltage contains a lot of harmonics. As a result, the performance of the current regulator will be degraded, and the vector control may not be performed correctly. When a voltage reference exceeds the linear modulation range of an inverter output voltage, an appropriate overmodulation method is necessary to modify the voltage reference within the linear modulation range or to produce the output voltage equal to its reference on average.

The overmodulation method can be classified into two groups: *dynamic overmodulation method* for the transient operation and *steady-state overmodulation method* for the steady-state operation.

- *Dynamic overmodulation methods* [15,16]
  The dynamic overmodulation methods handle the excessive voltage during transients such as the change of speed command or load torque. This is necessary for improving the dynamic response of a closed-loop current regulator in the high-performance AC motor drives such as vector control.
- *Steady-state overmodulation methods* [17,18]
  The steady-state overmodulation methods handle the overmodulation range voltage required in the steady-state operation. This method is important for the AC motor drives that require full inverter voltage utilization such as traction drives. In this case, the overmodulation methods should support a smooth transition from linear PWM operation to six-step mode operation.

Several overmodulation methods for the two popular SPWM and SVPWM techniques have been developed. However, we will describe overmodulation methods for only the SVPWM technique.

7.5.1 DYNAMIC OVERMODULATION METHODS

In the SVPWM technique, if the voltage reference vector lies outside the hexagon, an inverter cannot produce that vector, and thus a new voltage reference vector on the hexagon boundary must be selected instead. There are three choices for modification of a voltage reference vector as follows.

7.5.1.1 Minimum-phase-error pulse width modulation method

This method modifies the voltage reference vector $\boldsymbol{V}^*$ so that its magnitude is limited to the hexagon boundary while the phase is maintained as shown in Fig. 7.68.

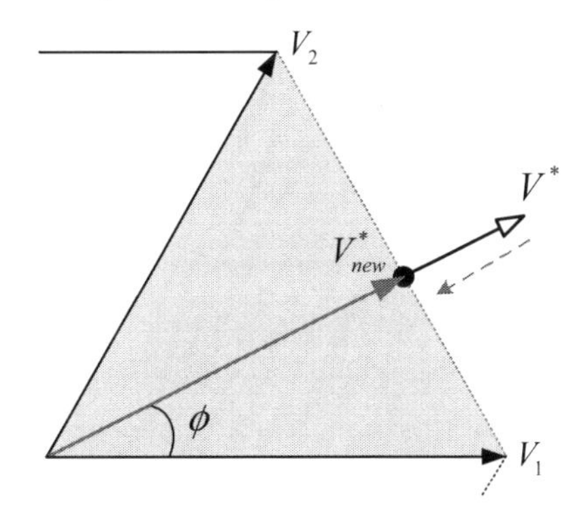

FIGURE 7.68

Minimum-phase-error PWM method.

This method is widely used due to its simplicity in implementation. However, since the new reference vector, $\boldsymbol{V}^*_{\text{new}}$, is selected irrespective of the magnitude of the original reference vector as shown in Fig. 7.68, this method has sluggish transient dynamics. In addition, it is impossible to carry out six-step operation with this method, and thus full DC input voltage utilization of an inverter cannot be achieved. For implementation of this method, the switching times of active vectors for generating a voltage reference vector can be simply modified as

$$T_1' = \frac{T_s}{T_1 + T_2} T_1 \tag{7.61}$$

$$T_2' = \frac{T_s}{T_1 + T_2} T_2 \tag{7.62}$$

7.5.1.2 Minimum-magnitude-error pulse width modulation method

This method focuses on maintaining the magnitude of the voltage reference vector $\boldsymbol{V}^*$ rather than its phase. To minimize the magnitude error between the modified and the original voltage reference vector, the modified voltage vector, $\boldsymbol{V}^*_{\text{new}}$, which is perpendicular to the hexagon boundary, is selected as shown in Fig. 7.69.

For implementation of this method, the switching times of active vectors for generating a voltage reference vector can be modified as

$$T_1' = T_1 - \frac{T_1 + T_2 - T_s}{2} \tag{7.63}$$

$$T_2' = T_2 - \frac{T_1 + T_2 - T_s}{2} \tag{7.64}$$

This overmodulation strategy has a better dynamic performance than the minimum-phase-error pulse width modulation method (MPEPWM) and also enables a smooth transition to six-step operation.

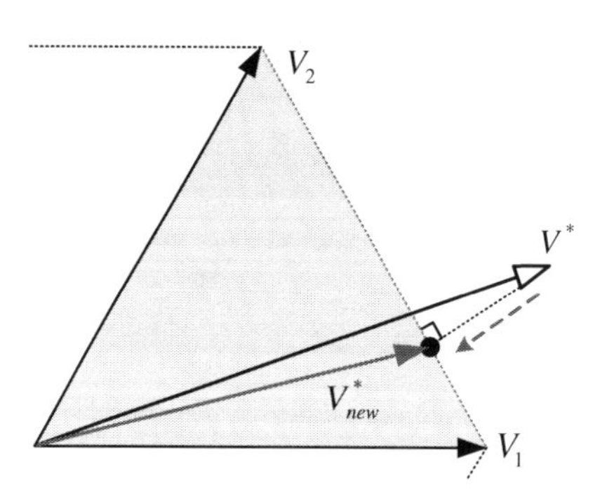

FIGURE 7.69

Minimum-magnitude-error pulse width modulation method.

7.5.1.3 Overmodulation method considering the direction of current [16]

This is an overmodulation strategy that is focused on the transient performance of the current controller in the high-performance AC motor drives. The voltage reference vector $\boldsymbol{V}^*$ for regulating motor currents consists of the PI output voltage vector $\boldsymbol{V}^*_{PI}$ and the back-EMF vector as was described in Chapter 6. The direction of the PI output voltage vector $\boldsymbol{V}^*_{PI}$ determines the transient response of the current. Thus this method selects a modified vector $\boldsymbol{V}^*_{\text{new}}$ that can maintain the direction of the PI output voltage vector as shown in Fig. 7.70. Even though this overmodulation strategy shows a better dynamic performance, its implementation is complex and needs information on back-EMF.

7.5.2 STEADY-STATE OVERMODULATION METHODS

Here, we will introduce two techniques to handle the overmodulation range voltage required in the steady-state operation.

First, we begin with the steady-state overmodulation method by using the compensated modulation technique [17]. An inverter can be considered as a voltage amplifier with unit gain, which can produce a fundamental component of the output voltage equal to its reference value. However, as the modulation index increases in the overmodulation region, the voltage gain reduces sharply in a nonlinear manner. In other words, the output voltage becomes less than its reference value. The reduction of the voltage gain G of the inverter depends on the modulation technique as shown in Fig. 7.71. Here, since the modulation index MI in the overmodulation region becomes greater than 1, we redefine the modulation index $M = v_{ref}/(2V_{dc}/\pi)$ as the ratio of the fundamental output voltage to the fundamental voltage of six-step operation. Thus, the output voltage of six-step operation, $2V_{dc}/\pi$, corresponds to $M = 1$.

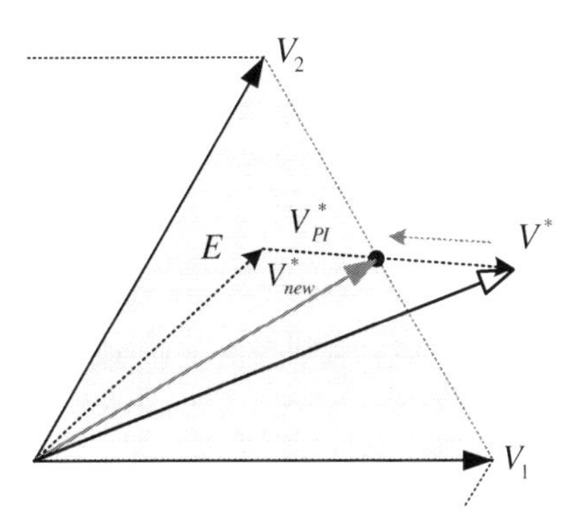

FIGURE 7.70

Method considering the direction of current.

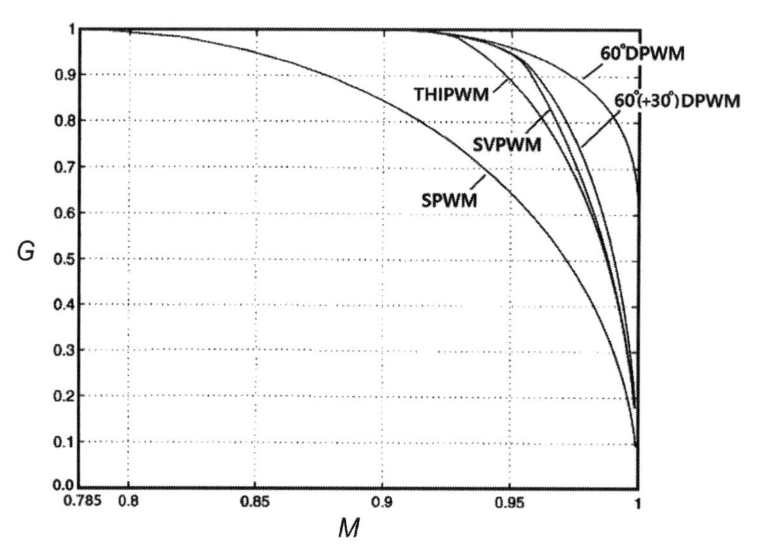

FIGURE 7.71

Voltage gain G for various PWM techniques [17].

A.M. Hava, R.J. Kerkman, T.A. Lipo, Carrier-based PWM-VSI overmodulation strategies: analysis, comparison, and design, IEEE Trans. Power Electron., 13 (4) (1998) 674–689.

For the SPWM method, the overmodulation begins at $M = 0.785$, and its nonlinear voltage gain G in the overmodulation region can be expressed in terms of the modulation index as

$$G = \frac{2}{\pi}\left[\sin^{-1}\left(\frac{1}{M}\right) + \frac{1}{M}\sqrt{1 - \frac{1}{M^2}}\right] \tag{7.65}$$

The compensated modulation technique compensates for the reduced gain in the overmodulation region. In this method, after calculating the voltage gain for each PWM technique, the voltage reference is multiplied by its inverse function so that the nonlinearity is canceled as shown in Fig. 7.72. However, the drawback of this method is that the online computation of the gain function and its inverse are very difficult.

Next we will discuss another steady-state overmodulation method, which modifies the voltage reference vector in the SVPWM technique. In the SVPWM technique, when the voltage reference vector $\boldsymbol{V}^*$ lies outside the hexagon boundaries, a PWM inverter cannot correctly produce the required voltage. Thus this method modifies the magnitude and/or the phase of the voltage reference vector to produce an output voltage equal to its reference on average per fundamental cycle [18]. As shown in Fig. 7.73, the phase and the magnitude of the voltage

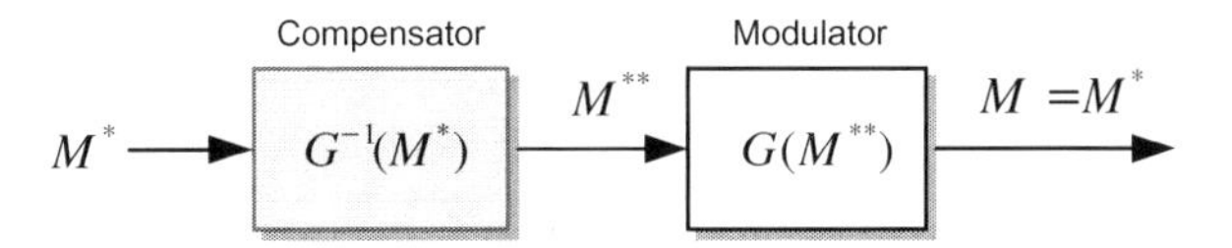

FIGURE 7.72

Steady-state overmodulation by using compensated modulation technique.

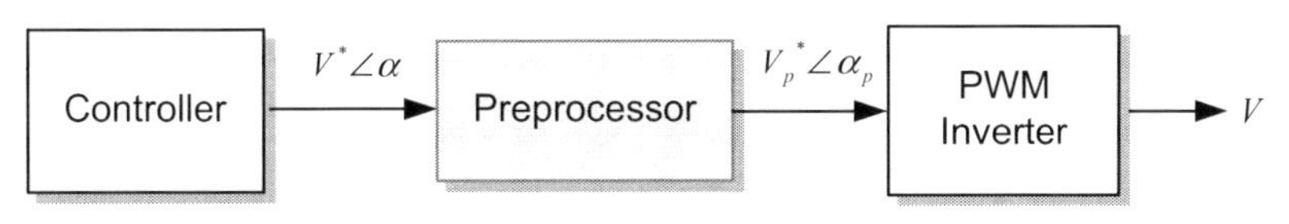

FIGURE 7.73

Steady-state overmodulaion method for the SVPWM technique.

reference vector is changed by a preprocessor. The modified reference vector is then fed to the modulator.

In this method, the overmodulation region is subdivided into two different modes: *mode* I ($0.907 \leq M \leq 0.952$) and *mode* II ($0.952 < M \leq 1.0$).

7.5.2.1 Overmodulation mode I ($0.907 \leq M \leq 0.9523$)

In this range, the preprocessor modifies the magnitude of the voltage reference vector $\boldsymbol{V}^*$ while its phase is transmitted without any change. Although the SVPWM technique utilizes a linear modulation range limited up to $V_{dc}/\sqrt{3}$, it can modulate more voltage according to the position of the voltage vector. For example, a voltage of $2V_{dc}/3$ can be achieved at the vertex of the hexagon. Exploiting this fact, the reduced fundamental component around the center portion of the hexagon sides can be compensated by producing more fundamental around the vertices as shown in Fig. 7.74.

Accordingly, it is possible to produce an output voltage equal to its reference on average per fundamental cycle. In Fig. 7.74, the *dashed circle* represents the trajectory of the the required reference vector $\boldsymbol{V}^*$ and the *solid line* represents the trajectory of the new reference vector $\boldsymbol{V}^*_{new}$ modified by the preprocessor. The limit of the overmodulation mode I is reached when the modified trajectory fully coincides with the hexagon. The maximum modulation index of this mode is $M = 0.9523$. This value can be calculated from the radius of a circle with an area equal to that of the hexagon as

$$V_{max} = \sqrt{\frac{2}{\pi\sqrt{3}}} V_{dc} = 0.606\, V_{dc} \tag{7.66}$$

Fig. 7.75 depicts the actual phase voltage reference according to the modulation index M in the overmodulation mode I. As the modulation index M increases,

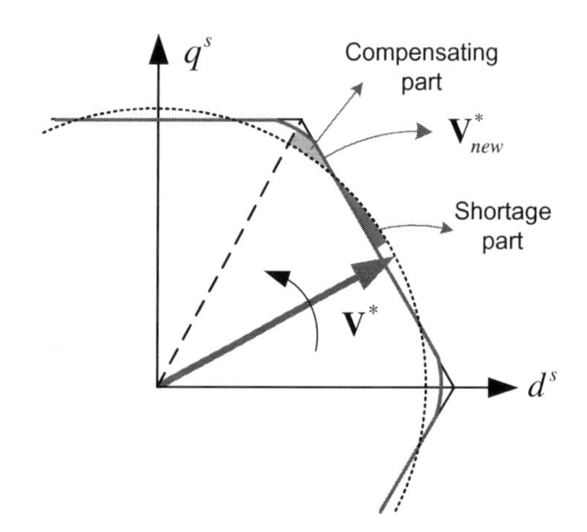

FIGURE 7.74

Overmodulation mode I.

J. Holtz, W. Lotzkat, A. Khambadkone, On continuous control of PWM inverter in the overmodulation range including the six-step mode, in Conf. Rec. IEEE IECON, 1992, pp. 307–312.

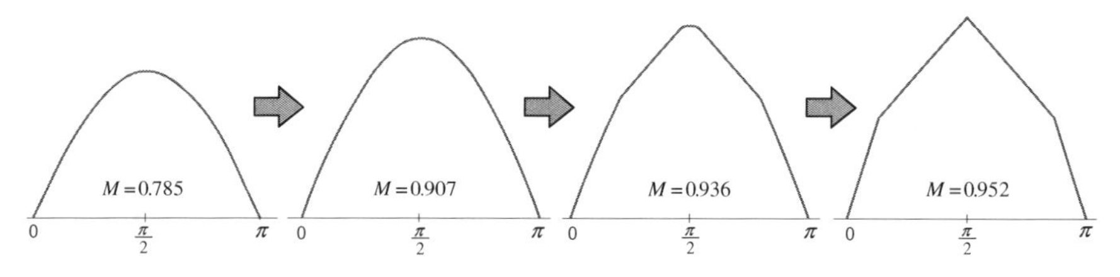

FIGURE 7.75

Phase voltage reference in the overmodulation mode I.

the actual phase voltage reference is turned into a nonsinusoidal waveform. This implies that the harmonics are increased along with the fundamental content of the voltage reference.

7.5.2.2 Overmodulation mode II ($0.9523 < M \leq 1$)

In this range from $M > 0.9523$ to six-step operation mode, the preprocessor changes both the magnitude and the phase of the voltage reference vector V^*. In the six-step operation producing the maximum output voltage, the six output vectors placed at the vertices of the hexagon are held sequentially within each discrete 60° interval. Based on this fact, in this mode, an output voltage equal to its reference on average can be achieved by adjusting the duration time for which the six vectors are held. Fig. 7.76 illustrates this more in detail. Although the voltage reference vector V^* moves along a circle with the fundamental angular frequency, the modified reference vector V_p^* remains fixed at the vertex for a

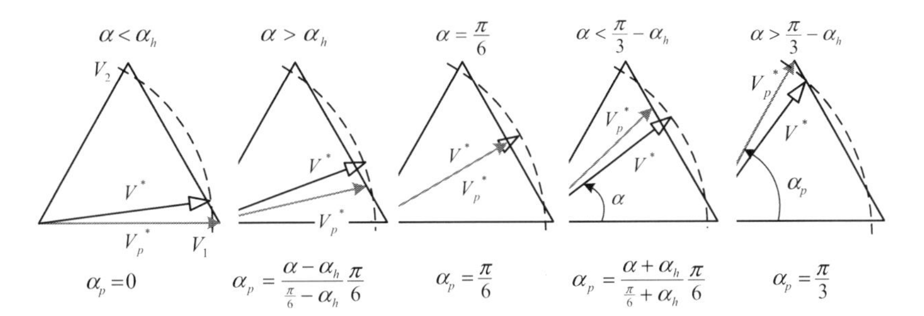

FIGURE 7.76

Overmodulation mode II.

J. Holtz, W. Lotzkat, A. Khambadkone, On continuous control of PWM inverter in the overmodulation range including the six-step mode, in Conf. Rec. IEEE IECON, 1992, pp. 307–312.

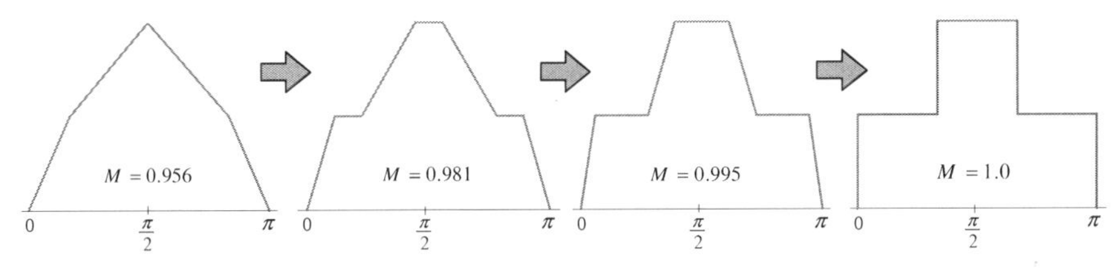

FIGURE 7.77

Actual phase voltage reference in the overmodulation mode II.

particular interval (is called a hold angle α_h) and moves along the side of the hexagon for the rest of the fundamental period. Here, α and α_P are the angles of the voltage reference vector and the modified reference vector, respectively.

The hold angle α_h controls the fundamental voltage and is given as a nonlinear function of the modulation index M [18]. Fig. 7.77 depicts the actual phase voltage reference according to the modulation index M in the overmodulation mode II. As the modulation index M increases, the actual phase voltage reference turns into a typical square waveform of the six-step operation. In this method, the waveform of the phase voltage is adjusted according to the modulation index M, as can be seen in Figs. 7.75 and 7.77, while Ref. [19] proposed a overmodulation method by adjusting the waveform of the pole voltage directly according to M.

7.6 DEAD TIME [20–23]

When both switches in an inverter leg are turned on simultaneously, the DC source is short-circuited, resulting in a current dangerous enough to destroy switches. Hence, this switching state should be inhibited at all times. Therefore, the pair of switches in a leg always changes their switching states in a complementary manner. Nevertheless, there is a possibility of an unexpected short-circuit of the leg. The main reason is that, due to the storage time in the turn-off process, the turn-off time of a switching device is always longer than the turn-on time. For example, Fig. 7.78 clearly shows a difference between the turn-on and the turn-off times from device characteristics for an IGBT [24].

Therefore, when the gating signals for changing the switching states are given at the same time, a momentary short-circuit is likely to happen. This is because one of the devices is being driven "on" while the other one may be still conducting. To prevent such undesirable short-circuit of the DC-link, the incoming switch is turned on after a sufficient time has passed to allow the turn-off of the outgoing switch as shown in Fig. 7.79.

Such small time delay, which is inserted prior to the gate signal of the turning-on device, is called *dead time* or *blanking time*. The dead time depends on the type and the size of a semiconductor device. A value of 1−3 μs is typical for a medium-power IGBT.

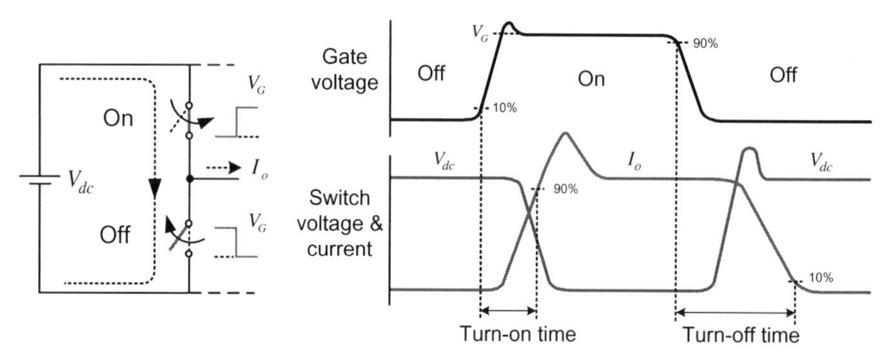

FIGURE 7.78

Switching characteristics of an IGBT.

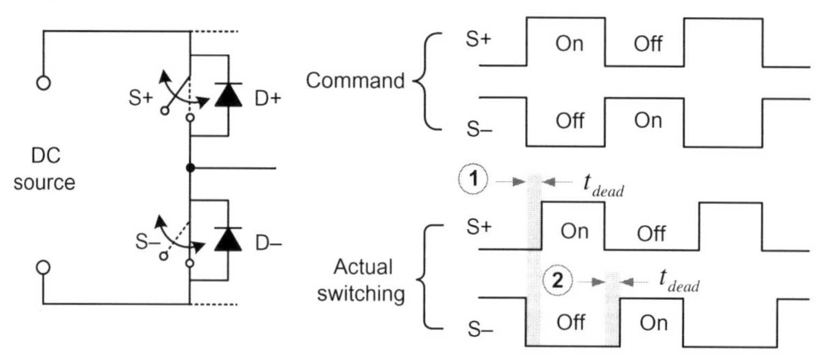

FIGURE 7.79

Dead time injection.

7.6.1 DEAD TIME EFFECT

During the dead time, there is a problem of which the output voltage of the inverter may be different from the voltage command. Thus we need to examine the effects of the dead time on the output voltage. During the dead time, since the load current does not conduct through a switching device but through a diode, the output voltage of the inverter is different according to the direction of the current. For example, examine the output voltage for section ① of the dead time as shown in Fig. 7.79. In this section, the lower switch $S-$ changes from on-state to off-state, and the upper switch $S+$ changes from off-state to on-state. In addition, the turn-on signal of $S+$ is delayed by the dead time. During the dead time, if the load current is negative, then the current will flow through the upper diode $D+$. This results in an output voltage of $V_{dc}/2$ equal to the desired voltage command. On the other hand, if the load current is positive as shown in Fig. 7.80, then it flows through the lower diode $D-$. In this case, the output voltage

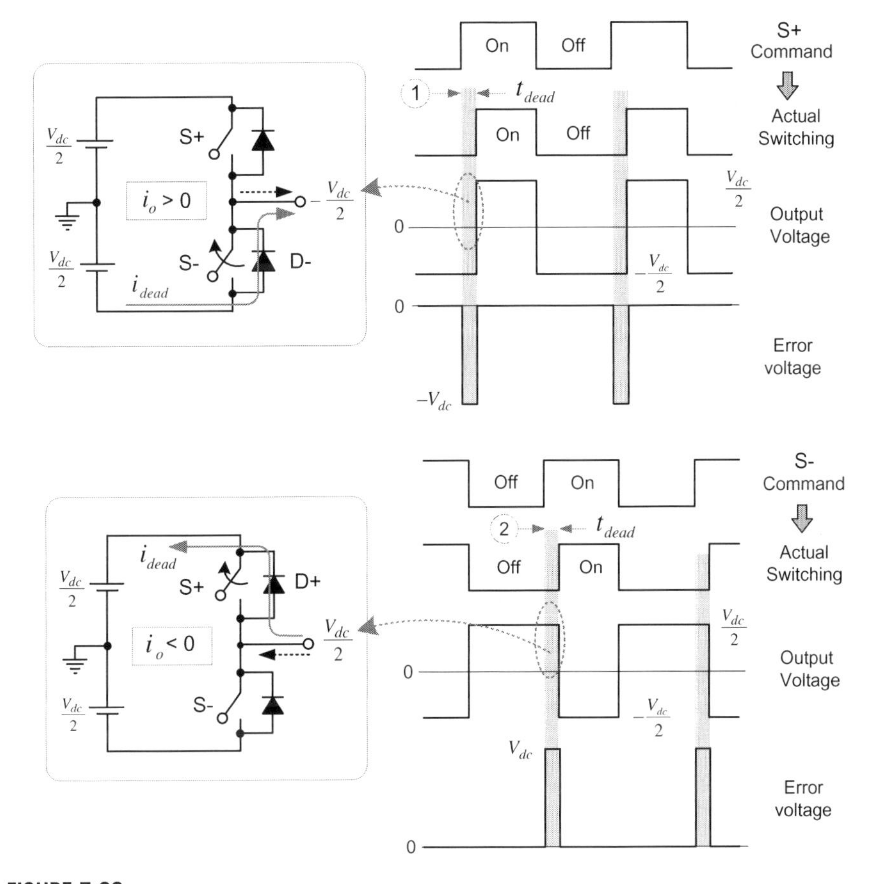

FIGURE 7.80

Voltage error due to the dead time.

becomes $-V_{dc}/2$, and thus there is an error value of $-V_{dc}$ when compared with the voltage command. On the other hand, in Section 7.2, the lower switch *S*- changes from off-state to on-state in order to produce the output voltage of $-V_{dc}/2$ and the turn-on signal of *S*- is delayed by the dead time. In this case, if the load current is negative as shown in Figure 7.80, then the current will flow through the upper diode $D+$ during the dead time. The output voltage becomes $V_{dc}/2$, and thus, there is an error value of V_{dc} when compared to the voltage command.

Fig. 7.81 shows the voltage errors due to the dead time over one cycle of the voltage reference. The magnitude of each error is V_{dc} and its width is the dead time t_{dead}, and its polarity depends on the direction of the current. Although each error is small, the accumulation of these errors over a half cycle is often sufficient to distort the output voltage. The average magnitude of the voltage error V_{err} is expressed as [20]

$$V_{err} = t_{dead} V_{dc} f_{sw} \tag{7.67}$$

where f_{sw} is the switching frequency.

This voltage deviation causes fundamental voltage drop and voltage waveform distortion. The distorted voltage induces the low-order harmonics of fifth and seventh in motor currents, causing current waveform distortion. This also results in torque pulsations as well as additional losses. Moreover, due to the dead-time effect, the estimated flux angle for the vector control of an induction motor includes the sixth-order harmonic. Fig. 7.82 shows that the phase currents and the estimated flux angle are distorted due to the dead-time effect. In particular, the dead-time effect becomes more significant at low speeds. Since the motor voltage

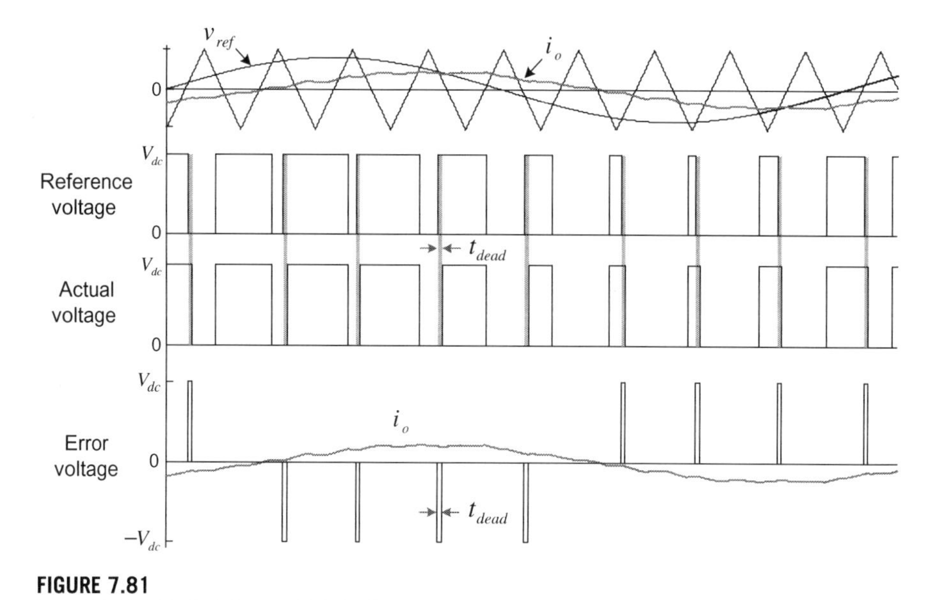

FIGURE 7.81

Actual output voltage and voltage error due to dead time.

in the low-speed region is low, it is susceptible to the distortion voltage. In the open-loop AC motor drives, the instability problem may occur at a light load or no load (implying low current) [21]. Thus it is necessary to compensate the voltage deviation due to the dead time.

In addition to the dead time, the nonlinearity caused by the turn-on/off delay and the voltage drop across power switches and diodes as shown in Fig. 7.83 can distort

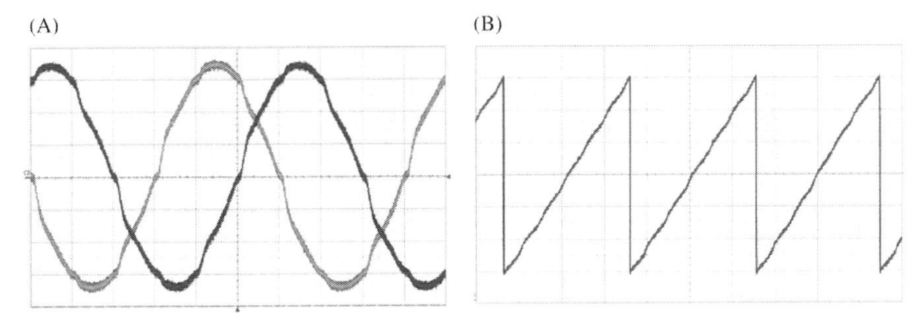

FIGURE 7.82

Dead-time effects. (A) Phase currents and (B) flux angle.

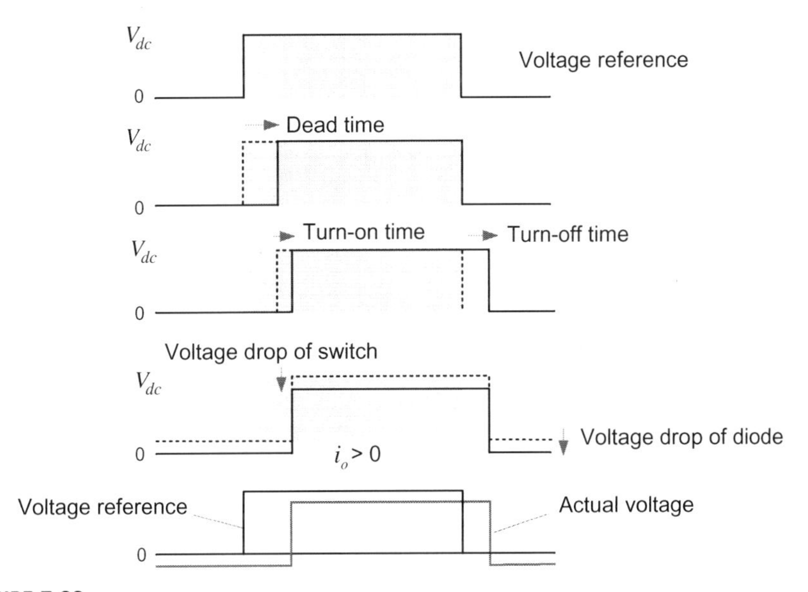

FIGURE 7.83

Nonlinearity of an inverter.

N. Urasaki, T. Senjyu, T. Funabashi, A novel method utilizing trapezoidal voltage to compensate for inverter nonlinearity adaptive dead-time compensation strategy for permanent magnet synchronous motor drive, IEEE Trans. Energy Conv., 22 (2) (2007), 271–280.

the output voltage of an inverter [22]. Thus it is necessary to compensate the voltage distortion due to the nonlinearity of an inverter in addition to the dead-time effect.

There are two types of dead-time compensation techniques. One is *feedforward compensation technique* that adds the error voltage to the voltage reference and the other is the *pulse-based compensation technique* that corrects the pulse errors due to the dead-time effect on a pulse-by-pulse basis. The feedforward compensation technique has a better performance than the pulse-based compensation technique. However, the pulse-based compensation technique can be easily implemented. Now we will discuss the pulse-based dead time compensation technique in detail.

7.6.2 DEAD TIME COMPENSATION [23]

In the pulse-based dead time compensation technique, the voltage distortion is corrected by adjusting the switching instants. Since the voltage error due to the dead time is different depending on the direction of the current, the voltage error needs to be compensated according to the sign of the operating current as follows. Now we will examine the dead time compensation for a switching sequence with a dead time added to the turning-on pulse.

7.6.2.1 Positive current: $i_o > 0$

If $i_o > 0$ during the dead time t_{dead}, then the current conducts through the lower diode, and thus the output voltage always becomes $-V_{dc}/2$. Hence, as shown in Fig. 7.84, when $S+$ is turning on and $S-$ is turning off, the actual voltage

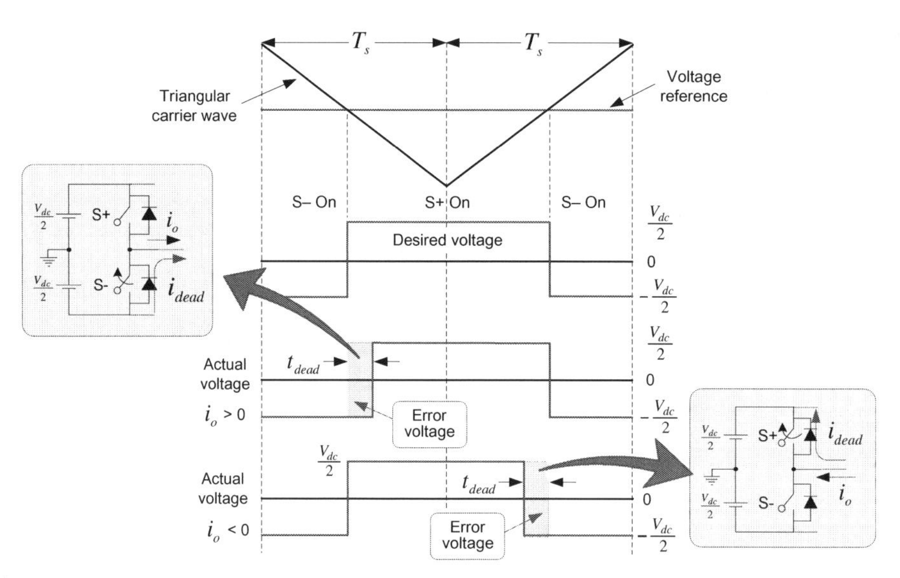

FIGURE 7.84

Dead-time effect.

of $-V_{dc}/2$ is less than the voltage reference of $V_{dc}/2$ during the dead time. The resultant error is $-V_{dc}$. On the other hand, if $S-$ is turning on and $S+$ is turning off, then the actual voltage during the dead time is equal to the voltage reference.

The left side of Fig. 7.85 shows the compensation for the voltage error by the dead time when $S+$ is turning on and $S-$ is turning off. In this case, the pulse transition time is advanced by the dead time t_{dead} to obtain the actual voltage identical to the voltage reference.

7.6.2.2 Negative current: $i_o < 0$

If $i_o < 0$ during the dead time t_{dead}, then the current conducts through the upper diode, and thus the output voltage always becomes $V_{dc}/2$. Hence, when $S+$ is turning on and $S-$ is turning off, the actual voltage during the dead time becomes equal to the voltage reference of $V_{dc}/2$. On the other hand, when $S-$ is turning on and $S+$ is turning off, the actual voltage of $V_{dc}/2$ during the dead time becomes greater than the voltage reference of $-V_{dc}/2$. The resultant error is V_{dc}. The right side of Fig. 7.85 shows the compensation for the voltage error by the dead time when $S-$ is turning on and $S+$ is turning off. In this case the

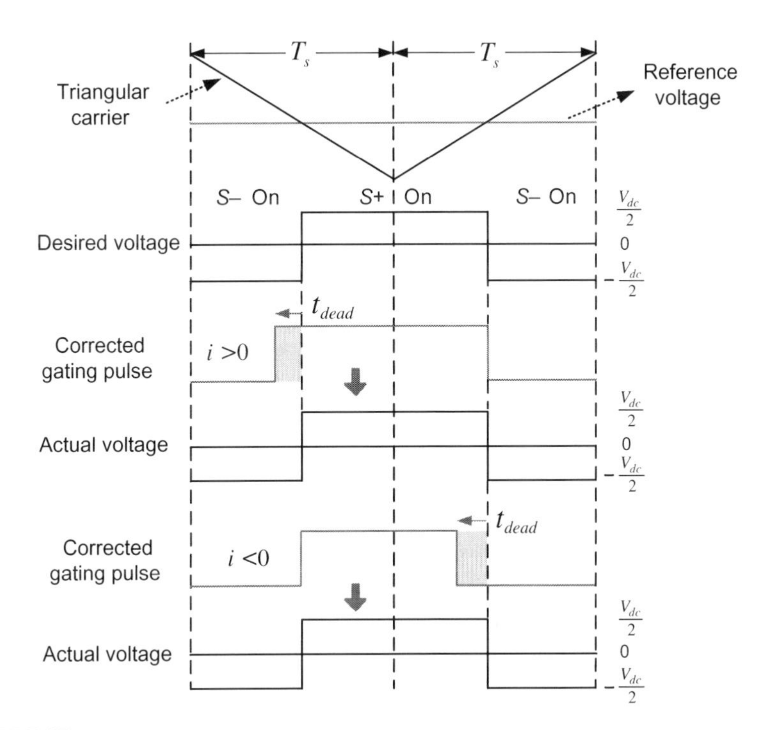

FIGURE 7.85

Dead-time compensation.

pulse transition time is advanced by the dead time t_{dead} to obtain the actual voltage identical to the voltage reference.

As explained above, the dead-time compensation technique is based on information of the polarity of the phase current. Hence, an accurate measurement of the polarity of the current is needed to correctly compensate for the dead time. In particular, it is important to detect the current exactly around the zero crossings, where the detection of the current polarity is susceptible to noise and switching ripple.

7.7 CURRENT MEASUREMENT [25,26]

The current regulator is essential for accomplishing the torque control of AC motors. The torque control performance greatly depends on the performance of the current regulator. As described in Chapter 6, since the current regulation is performed using a feedback current, an accurate current measurement is the most important. In addition, the attainable current control bandwidth depends on the current sampling method. The motor current driven by a PWM inverter contains not only the fundamental component but also ripple components due to the switching action. We need to detect only the fundamental component, which produces effective torque and flux.

For measuring the fundamental component, the instantaneous sampling method has gained a wide acceptance [25]. This method demonstrates that it is possible to detect the fundamental component of current by sampling the current at the midpoints of zero vectors in the pulse patterns of the SVPWM technique. The typical waveforms of the voltage and phase current in the symmetrical SVPWM technique are shown in Fig. 7.86. It can be seen that the instantaneous value sampled at the midpoints of zero vectors in the PWM pulse patterns is equal to the fundamental current component. As we discussed in Section 7.4, the midpoints of the zero vectors in the SVPWM technique correspond to the peaks of the triangular carrier wave. Thus an instantaneous current sampling synchronized with the triangular carrier wave is normally used. The current can be sampled at the two peaks of the triangular carrier wave, i.e., top and bottom points (i.e., twice per switching period) or only at one peak (once per switching period). The attainable bandwidth of the current control depends directly on the number of sampling [26]. When sampling the currents twice per switching period, which is called the *double sampling*, the maximum attainable bandwidth of the synchronous reference frame current regulator is about 1/10 of the switching frequency.

Even though the current is sampled at the midpoints of the zero vectors, it may deviate from the fundamental component due to a delay in the feedback loop. When the current is being measured from a sensor, the signal of the sensor is first filtered to eliminate the noise on the signal and then converted into a digital value. Due to the phase delay caused by the low-pass filter, the sampled current through the filter may be different from the actual current at the desired

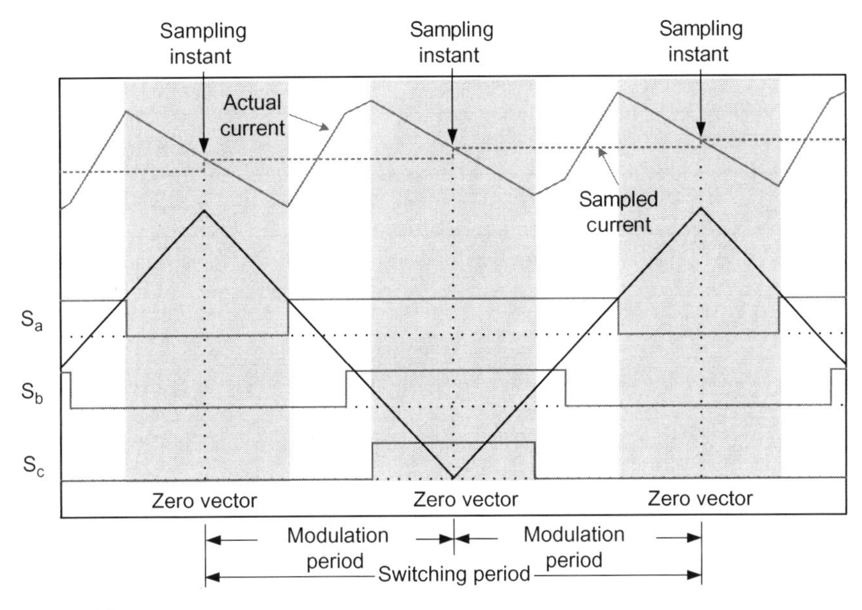

FIGURE 7.86

Typical waveforms of voltage and phase current in the symmetrical SVPWM.

instant. Moreover, a ripple component rather than the fundamental component of the current may be sampled. In Ref. [25], the delayed sampling method was used to reduce the current sampling error caused by the filter delay.

REFERENCES

[1] A. Nabae, I. Takahashi, H. Akagi, A new neutral-point-clamped PWM inverter, IEEE Trans. Ind. Appl. IA-17 (5) (Sep./Oct., 1981) 518–523.

[2] A. Schnung, H. Stemmler, Static frequency changers with 'subharmonic' control in conjunction with reversible variable-speed A.C. drives, Brown Boveri Rev. (Aug./Sep., 1964) 555–577.

[3] H.S. Patel, R.G. Hoft, Generalized techniques of harmonic elimination and voltage control in thyristor inverters: part I-harmonic elimination, IEEE Trans. Ind. Appl. IA-9 (3) (May/Jun., 1973) 310–317.

[4] V.G. Agelidis, P.D. Ziogas, G. Joos, 'Dead-Band' PWM switching patterns, IEEE Trans. Power Electron. 11 (4) (July, 1996) 522–531.

[5] S.R. Bowes, R.R. Clements, Computer-aided design of PWM inverter systems, IEE Proc. 129, Pt. B (1) (Jan. 1982) 1–17.

[6] J.A. Houldsworth, D.A. Grant, The use of harmonic distortion to increase the output voltage of a three-phase PWM inverter, IEEE Trans. Ind. Appl. IA-20 (5) (Sep./Oct. 1984) 1224–1228.

[7] H.W. Van Der Broeck, et al., Analysis and realization of a pulse width modulator based on voltage space vectors, IEEE Trans. Ind. Appl. 24 (1) (Jan./Feb. 1988) 142–150.
[8] D.G. Holmes, The significance of zero space vector placement for carrier-based PWM schemes, IEEE Trans. Ind. Appl. 32 (5) (Sep./Oct. 1996) 1122–1129.
[9] J. Sun, H. Grotstollen, Optimized space vector modulation and regular-sampled PWM: a reexamination, in: Conf. Rec. IEEE IAS Annu. Meeting, 1996, pp. 956–963.
[10] A.M. Hava, R.J. Kerkman, T.A. Lipo, A high-performance generalized discontinuous PWM algorithm, IEEE Trans. Ind. Appl. 34 (5) (Sep./Oct. 1998) 1059–1071.
[11] E. Silva, E. Santos Jr., C. Jacobina, Pulsewidth modulation strategies nonsinusoidal carrier-based pwm and space vector modulation techniques, IEEE Trans. Ind. Electron. Mag. (2011) 37–45.
[12] V. Blasko, A hybrid PWM strategy combining modified space vector and triangle comparison methods," in: Proc. IEEE Power Electronics Specialists Conference (PESC'96), 1996, pp. 1872–1878.
[13] D.W. Chung, J.S. Kim, S.K. Sul, Unified voltage modulation technique for real-time three-phase power conversion, IEEE Trans. Ind. Appl. 34 (2) (Mar./Apr. 1998) 374–380.
[14] A.M. Hava, R.J. Kerkman, T.A. Lipo, Simple analytical and graphical methods for carrier-based PWM-VSI drives, IEEE Trans. Power Electron. 14 (1) (Jan. 1999) 49–61.
[15] A.M. Hava, S.K. Sul, R.J. Kerkman, T.A. Lipo, Dynamic overmodulation characteristics of triangle intersection PWM methods, IEEE Trans. Ind. Appl. 35 (4) (Jul./Aug. 1999) 896–907.
[16] J.K. Seok, J.S. Kim, S.K. Sul, Overmodulation strategy for high-performance torque control, IEEE Trans. Power Electron. 13 (4) (Jul. 1998) 786–792.
[17] A.M. Hava, R.J. Kerkman, T.A. Lipo, Carrier-based PWM-VSI overmodulation strategies: analysis, comparison, and design, IEEE Trans. Power Electron. 13 (4) (Jul. 1998) 674–689.
[18] J. Holtz, W. Lotzkat, A. Khambadkone, On continuous control of PWM inverter in the overmodulation range including the six-step mode, in: Conf. Rec. IEEE IECON, 1992, pp. 307–312.
[19] D.-W. Han, S.-H. Kim, An overmodulation strategy for SVPWM Inverter using pole voltage, Trans. Korean Inst. Power Electron. 7 (2) (Apr. 2002) 149–157.
[20] Y. Murai, T. Watanabe, H. Iwasaki, Waveform distortion and correction circuit for PWM inverters with switching lag-times, IEEE Trans. Ind. Appl. IA-23 (5) (Sep./Oct. 1987) 881–886.
[21] S.-G. Jeong, M.-H. Park, The analysis and compensation of dead- time effects in PWM inverters, IEEE Trans. Ind. Electron. 38 (2) (Apr. 1991) 108–114.
[22] Y.S. Park, S.K. Sul, A novel method utilizing trapezoidal voltage to compensate for inverter nonlinearity, IEEE Trans. Power Electron. 27 (12) (Dec. 2012) 4837–4846.
[23] D. Leggate, J. Kerkman, Pulse-based dead-time compensator for pwm voltage inverters, IEEE Trans. Ind. Electron. 44 (2) (Apr. 1997) 191–197.
[24] IGBT Basics 1, Application Note 9016, Fairchild Semiconductor, 2001.
[25] S.H. Song, J.W. Choi, S.K. Sul, Current measurements in digitally controlled AC drives, IEEE Ind. Appl. Mag. (Jul./Aug. 2000) 51–62.
[26] V. Blasko, V. Kaura, W. Niewiadomski, Sampling of discontinuous voltage and current signals in electrical drives: a system approach, IEEE Trans. Ind. Appl. 35 (5) (Sep./Oct. 1987) 1123–1130.

CHAPTER

High-speed operation of alternating current motors

8

In variable speed drive applications such as alternating current (AC) servo and traction drive systems, a high-speed operation capability as well as a fast torque response of motors are needed. For example, for spindle drives in machine tools, a high-speed operation of more than several 10,000 r/min is required for enhancing cutting efficiency and accuracy. Traction drives such as electric cars, electric trains, electric forklifts, and dehydration of washing machines also require a high-speed operation of several times the rated speed. In addition, motors for driving compressors or motors for driving a torpedo and a missile need to be operated at a high-speed region of several 10,000 r/min. In this chapter, we will discuss the high-speed operation of AC motors. We introduce the field-weakening methods, which achieve the maximum torque capability of motors as well as a high-speed operation.

In the variable speed drives, the voltage applied to the motor should be increased with the operating speed. This is to maintain the current needed for the torque production of the motor. Since the back-EMF increases with the operating speed, the motor voltage should also be increased to control the current properly. As the speed increases, the motor reaches a speed (commonly called *base speed*) at which the applied motor voltage is equal to its rated value. From this speed, since the motor voltage is fixed at the rated voltage, we need to limit the back-EMF to a suitable level despite the increase in the speed.

For this purpose, a field-weakening control to reduce the field flux of the motor is necessary. As can be seen from Chapter 2 and Chapter 3, the back-EMF of direct current (DC) motors and AC motors is directly proportional to the operating speed ω_m (or frequency f) and the field flux as

$$E_{DC} \propto \phi \cdot \omega_m \quad \text{or} \quad E_{AC} \propto \phi \cdot f \tag{8.1}$$

If the field flux of a motor decreases with the increased speed, then the back-EMF is limited to a suitable level. Thus the voltage margin to control the current correctly can be maintained.

As an example, consider a high-speed operation in the DC motor drives. First, we will assume that the field flux ϕ is constant and the torque production to drive

Electric Motor Control. DOI: http://dx.doi.org/10.1016/B978-0-12-812138-2.00008-8

a load requires an armature current I_a. From the voltage equation of Eq. (2.1), the armature current in the steady-state is given as

$$I_a = \frac{V_a - E_a}{R_a} = \frac{V_a - k_e\phi\omega_m}{R_a} \tag{8.2}$$

Since the back-EMF increases as the operating speed increases, the armature voltage V_a applied to the motor should be increased with the increase in the speed to achieve the required current I_a. However, there is a limitation to the voltage applied to the motor because the motor voltage should not be greater than the rated voltage. Therefore the motor voltage required to regulate the command current may be insufficient in the high-speed region. As shown in Fig. 8.1, when the back-EMF is close to the full voltage of the motor as the speed increases, the current (thus output torque) of the motor is decreased rapidly. As a result, the speed no longer increases.

However, as shown in Fig. 8.2, if we limit the back-EMF to a constant value over the high-speed region by using the field-weakening control, which reduces the field flux in proportion to the inverse of the speed, then the voltage margin that regulates the current can be maintained and the output torque can be developed adequately. As a result, the speed of the motor can increase more than the rated speed.

For DC motors, the back-EMF can easily be maintained at a constant by reducing the field flux in proportion to the inverse of the speed so that the high-speed operation can function properly. However, the high-speed operation of AC motors cannot be achieved by such a simple method, and the high-speed

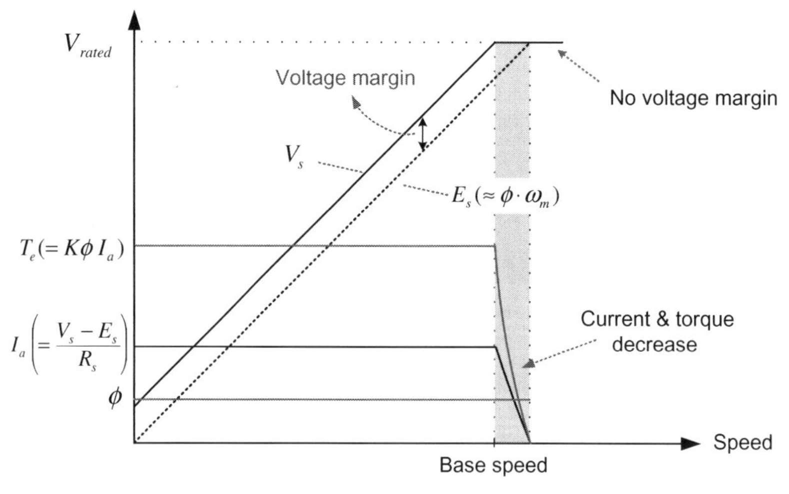

FIGURE 8.1

Motor characteristics according to speed (without field-weakening control).

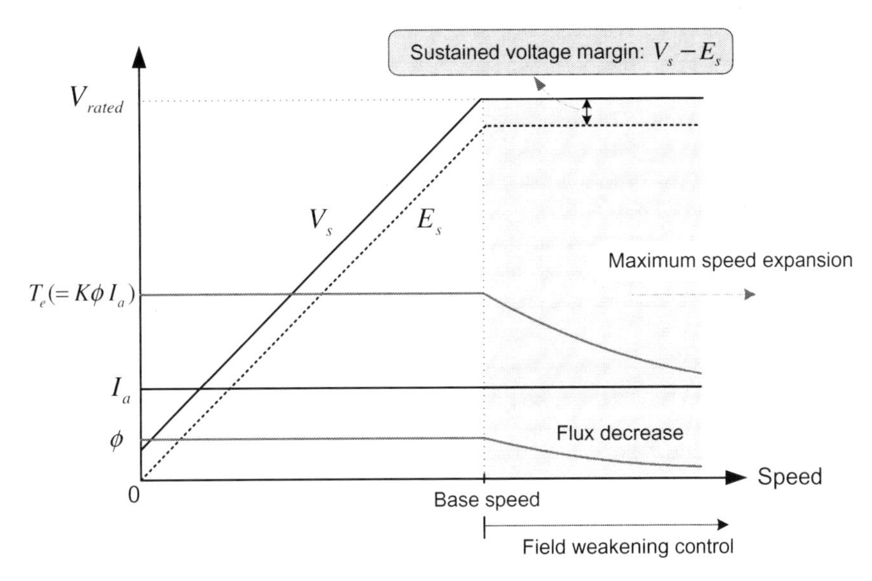

FIGURE 8.2

Motor characteristics according to speed (with field-weakening control).

operation method is different depending on the type of AC motor. Now we will discuss the field-weakening method for the high-speed operation of vector-controlled AC motors. We will start with an induction motor.

8.1 FIELD-WEAKENING CONTROL FOR INDUCTION MOTORS [1,2]

Induction motors are beneficial for high-speed operations due to their mechanical robustness. In addition, the output torque characteristic of induction motors in the constant power region matches well with a load (i.e., load torque inversely proportional to speed) requiring high-speed operation. Moreover, since the field flux of the vector-controlled induction motor can be easily weakened by reducing the d-axis current as the rotor speed increases, high-speed operations can be done easily. However, even though the same voltage and current are used, the developed torque capability of the motor at high speeds can vary according to the applied field-weakening strategy.

The field-weakening control method for induction motors can be classified into two methods: the *feedforward method* [1], which derives the flux command required for high-speed operation from the steady-state voltage equations of an induction motor, and the *feedback method* [2], which obtains the flux command from the voltage feedback. Now we will discuss the feedforward field-weakening control method in the vector control based on the rotor flux of induction motors.

8.1.1 CLASSIC FIELD-WEAKENING CONTROL METHOD

The method that was previously used for the field-weakening operation of induction motors was to simply vary the flux-producing current, i.e., d-axis stator current i_{ds}^{e} in proportion to the inverse of the rotor speed ω_r in the same way for DC motors as

$$i_{ds}^{e*} = \frac{\omega_{r\text{-}base}}{\omega_r} \cdot I_{d\text{-}rated} \tag{8.3}$$

where $I_{d\text{-}rated}$ is the rated d-axis current and $\omega_{r\text{-}base}$ is the base speed.

From this d-axis current command, the rotor flux command according to the speed in the field-weakening region is given as

$$\lambda_{dr}^{e*} = L_m i_{ds}^{e*} = L_m \left(\frac{\omega_{r\text{-}base}}{\omega_r} \right) \cdot I_{d\text{-}rated} \tag{8.4}$$

However, the flux command λ_{dr}^{e*} given by Eq. (8.4) cannot properly retain the voltage margin required for current control in the high-speed region. This is because the command flux level is too high to reduce the back-EMF appropriately. As a result, it is impossible to regulate the current commands required to perform the vector control properly, and thus the output torque capability of the induction motor is degraded. Moreover, the drive system may even fail to operate.

In Ref. [1], the authors introduced the optimal field-weakening method, which achieve the maximum torque capability of the induction motor and the high-speed operation. Now we will discuss this feedforward field-weakening method in more detail.

Since the developed torque of a motor depends on the available voltage and current, we need to begin with investigating the range of voltage and current available for an induction motor. The voltage applied to the motor is never greater than the rated value. On the other hand, the current often has a short-time rating, being two to three times greater than the rated current for a high-acceleration/high-deceleration torque production. Therefore we need to consider the available motor voltage and current for deriving the effective field-weakening method.

8.1.2 VOLTAGE- AND CURRENT-LIMIT CONDITIONS

In this section, we will examine the boundary of the current available for the torque production of an induction motor. This boundary can be obtained from the available voltage and current as follows.

8.1.2.1 Voltage-limit condition

For an inverter-driven motor, the motor voltage is supplied by an inverter. The maximum output voltage $V_{s\,max}$ of an inverter is determined by the available DC input voltage V_{dc}. Even under the same DC voltage, the maximum output voltage depends on the pulse width modulation (PWM) technique used. For example, as stated in Chapter 7, the space vector pulse width modulation (SVPWM) technique can produce a maximum voltage of $V_{dc}/\sqrt{3}$, while the sinusoidal pulse width modulation (SPWM) technique can produce that of $V_{dc}/2$.

For the maximum voltage $V_{s\,max}$ of a motor, the $d-q$ axes stator voltages should always satisfy the following relation.

$$v_{ds}^2 + v_{qs}^2 \leq V_{s\,max}^2 \tag{8.5}$$

This is called *voltage-limit condition*. Here, the selection of the value of the maximum phase voltage $V_{s\,max}$ has a strong influence on the field-weakening operation performance. Considering the voltage drops of the switching devices, dead-time effects, and control stability, it is desirable to set this value at 90–95% of the maximum attainable voltage by a PWM technique in an inverter. If the value of $V_{s\,max}$ is inappropriately selected, then it can cause degradation or even a failure of high-speed operations.

The synchronous $d-q$ axes voltage equations of a vector-controlled induction motor based on the rotor flux are rewritten from Eqs. (6.16) and (6.17) as

$$v_{ds}^e = R_s i_{ds}^e + p\left(\sigma L_s i_{ds}^e + \frac{L_m}{L_r}\lambda_{dr}^e\right) - \omega_e \sigma L_s i_{qs}^e \tag{8.6}$$

$$v_{qs}^e = R_s i_{qs}^e + p\sigma L_s i_{qs}^e + \omega_e\left(\sigma L_s i_{ds}^e + \frac{L_m}{L_r}\lambda_{dr}^e\right) \tag{8.7}$$

When considering high-speed operations, the voltage drops on the stator resistance is negligible compared to the back-EMFs, so the above equations can be simplified in the steady-state as

$$v_{ds}^e \approx -\omega_e \sigma L_s i_{qs}^e \tag{8.8}$$

$$v_{qs}^e \approx \omega_e\left(\sigma L_s i_{ds}^e + \frac{L_m}{L_r}\lambda_{dr}^e\right) \approx \omega_e L_s i_{ds}^e \tag{8.9}$$

By substituting these equations into Eq. (8.5), the voltage-limit condition can be expressed in terms of $d-q$ axes stator currents as

$$\left(\omega_e \sigma L_s i_{qs}^e\right)^2 + \left(\omega_e L_s i_{ds}^e\right)^2 \leq V_{s\,max}^2 \tag{8.10}$$

This expression illustrates the boundary of the controllable $d-q$ axes currents by using the maximum available voltage $V_{s\,max}$. This voltage-limit boundary is expressed as an ellipse equation, which is a function of operating frequency ω_e as shown in Fig. 8.3. The inside of the ellipse indicates the controllable current region. Any $d-q$ current command inside the ellipse is controllable by using the given voltage $V_{s\,max}$. Thus the current commands should remain inside the voltage-limit ellipse given at each operating frequency. The radius of the ellipse becomes smaller as the operating frequency increases. This means that, under a given voltage, the boundary of controllable currents becomes smaller. This is because the back-EMF increases as the operating frequency increases. Unlike DC

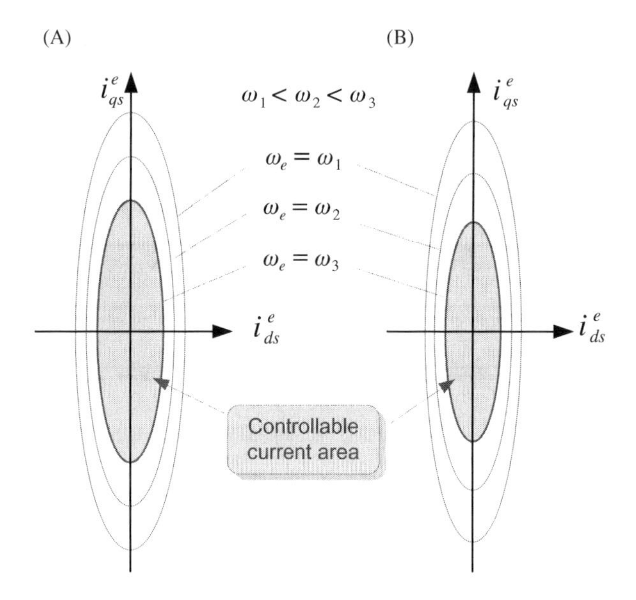

FIGURE 8.3

Voltage-limit condition. (A) Small leakage factor and (B) large leakage factor.

motors, the back-EMF of induction motors increases with the operating frequency rather than the rotor speed.

Fig. 8.3A and B shows several ellipses for different operating frequencies. We can see from Eq. (8.10) that the leakage inductance (or leakage factor σ) has a significant influence on the shape of the ellipse at a specific operation frequency. The area of the ellipse is smaller for an induction motor with a large leakage factor as in Fig. 8.2B than for one with a small leakage factor as in Fig. 8.2A at the same operating frequency. This implies that for an induction motor with a large leakage factor, more voltage is necessary for regulating an equal value of current commands. As stated in Chapter 3, an induction motor with a large leakage factor has a narrow range of the constant power region. This is because the leakage factor has an influence on the maximum values of the slip. Fig. 8.3 explains this phenomenon. Hence, the leakage factor is an important factor in the design of an induction motor.

Next, we will discuss the current-limit condition for a motor.

8.1.2.2 Current-limit condition

The motor current is usually limited by the rated current. A 150–300% rated current, however, is often allowed for a short period of time to produce a high-acceleration/high-deceleration torque, given that it does not exceed the motor thermal rating. The current rating of an inverter should be selected so that it can provide the maximum motor current.

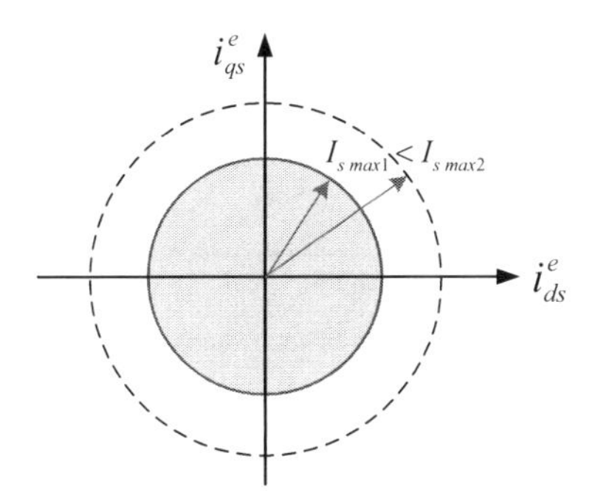

FIGURE 8.4

Current-limit condition.

Under the maximum available stator current $I_{s\,max}$, the $d-q$ axes stator currents of an induction motor are restricted to the following condition.

$$i_{ds}^2 + i_{qs}^2 \le I_{s\,max}^2 \tag{8.11}$$

This is referred to as *current-limit condition*. This current-limit boundary can be expressed as a circle in the $d-q$ axes current plane as shown in Fig. 8.4, whose radius is the maximum stator current $I_{s\,max}$. To satisfy this current-limit condition, the $d-q$ axes stator current commands should always be inside this circle. Unlike the voltage-limit condition, the boundary of the current-limit condition remains constant regardless of the operating frequency.

When we drive an induction motor, the two conditions of the voltage- and current-limit should be always satisfied. Considering both the voltage- and the current-limit conditions, the region of the controllable currents is the common area between the current-limit circle and the voltage-limit ellipse for a given operating frequency, which is the *dark area* shown in Fig. 8.5.

Thus the $d-q$ axes stator current commands in the induction motor drive must be inside this area for a given operating frequency.

Now we will find out the optimal flux under the voltage- and the current-limit conditions for the field-weakening control, which can obtain the maximum output torque capability of an induction motor over the whole high-speed region.

8.1.3 FIELD-WEAKENING CONTROL FOR PRODUCING THE MAXIMUM TORQUE

In a vector-controlled induction motor, the output torque can be expressed as a function of $d-q$ axes stator currents as

$$T_e = \frac{3}{2}\frac{P}{2}\frac{L_m^2}{L_r} i_{ds}^e i_{qs}^e = k i_{ds}^e i_{qs}^e \quad \left(k = \frac{3}{2}\frac{P}{2}\frac{L_m^2}{L_r}\right) \tag{8.12}$$

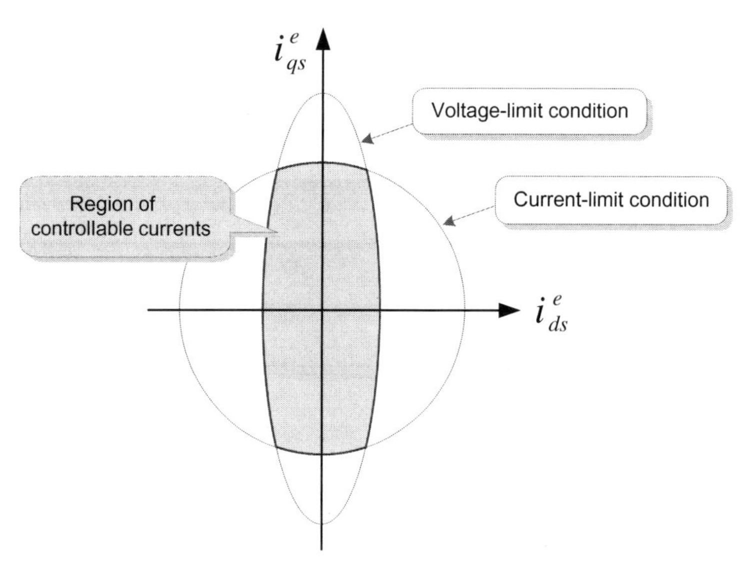

FIGURE 8.5

Region of controllable currents for voltage- and current-limit conditions.

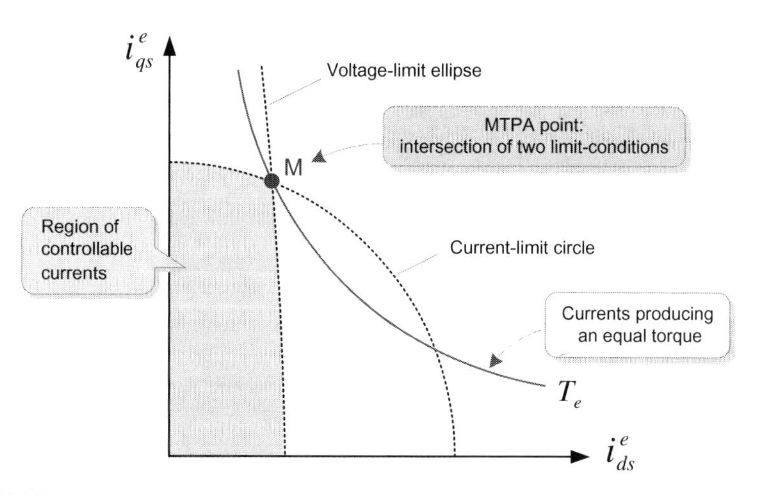

FIGURE 8.6

Voltage- and current-limit conditions and MTPA point.

The value of the output torque varies according to the chosen combinations of $d-q$ axes stator currents, i_{ds}^{e} and i_{qs}^{e}. Fig. 8.6 illustrates numerous combinations of the stator currents to produce an equal output torque, along with the voltage- and the current-limit conditions. Here, we will consider only Quadrant 1, which indicates the motoring operation mode for the forward driving.

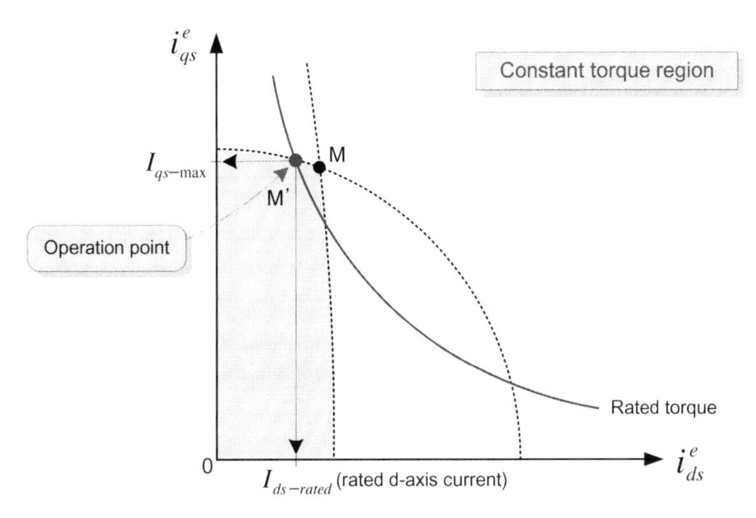

FIGURE 8.7

Optimal operating point in the constant torque region.

In the available region under both the voltage- and current-limit conditions, the optimal current combination maximizing the output torque becomes the value at the intersection (M point) of the circle and the ellipse at a given operating frequency. This operating point indicates the minimum stator current for producing the required torque. Thus this is the *maximum torque per ampere* (MTPA) operating point. This optimal operating point varies according to the operating speed, and it can be obtained as follows.

8.1.3.1 Constant torque region ($\omega_e \leq \omega_{base}$)

In the low- and mid-speed range, the d-axis stator current at the optimal operating point M is usually greater than rated d-axis current $I_{ds\text{-}rated}$ for producing the rated rotor flux linkage $\lambda^e_{dr\text{-}rated}$ as shown in Fig. 8.7. Even if the d-axis current is increased above the rated value, the flux level does not increase accordingly because of the saturation of the iron core. For this reason, in this speed range, the d-axis stator current command (thus, rotor flux command) is maintained at the rated value as

$$i^{e*}_{ds} = I_{ds\text{-}rated} \tag{8.13}$$

$$\lambda^{e*}_{dr} = \lambda^e_{dr\text{-}rated} = L_m i^{e*}_{ds} = L_m I_{ds\text{-}rated} \tag{8.14}$$

In this case, the available output torque depends only on the available q-axis stator current. The available q-axis stator current is limited by the d-axis current command as

$$i^{e*}_{qs\text{-}max} = \sqrt{I^2_{s\ max} - i^{e*2}_{ds}} = \sqrt{I^2_{s\ max} - I^2_{ds\text{-}rated}} \tag{8.15}$$

Thus, in this speed range, the operating point is given as M′ point as shown in Fig. 8.7. This condition is still held at operating speeds below the rated speed (more accurately, the base speed).

Next, we will examine the optimal flux for producing the maximum torque in the high-speed range. The high-speed region, which is called *field-weakening region*, has two subregions as follows.

8.1.3.2 Constant power region ($\omega_{base} \leq \omega_e < \omega_{BT}$): Field-weakening region I

In the constant torque region as stated earlier, when the operating frequency is increased, the boundary of the voltage-limit ellipse will be reduced. This indicates that, due to a limited $V_{s\,max}$, the controllable current region is reduced as the back-EMF increases with the operating frequency.

Even though the voltage-limit ellipse is reduced by the increase in operating frequency, the rated *d*-axis current can still be used, provided that the controllable current region includes the rated *d*-axis current. As the operating frequency is further increased, the *d*-axis stator current at the intersection of the circle and ellipse will coincide with the rated *d*-axis current as shown in Fig. 8.8. The operating frequency at this moment is the onset frequency at which the constant power region begins. The speed at this onset frequency is called the *base speed* ω_{base}. For the operation above the base speed, the *d*-axis stator current should be lowered below the rated *d*-axis current. In other words, the field-weakening operation should start reducing the flux level.

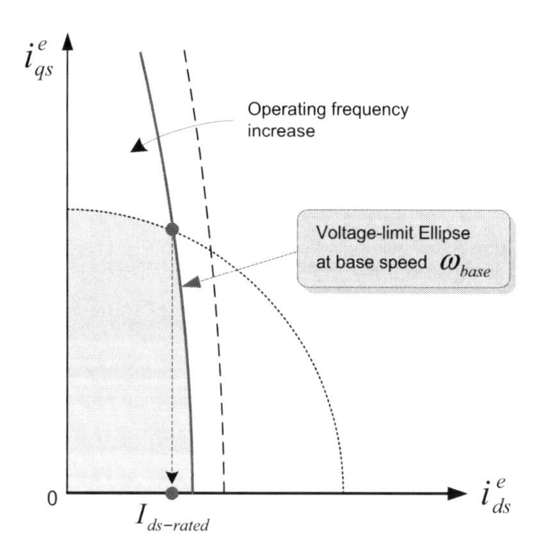

FIGURE 8.8

Onset of field-weakening operation.

From Eqs. (8.10) and (8.11), the base speed is given as

$$\omega_{base} = \frac{\sqrt{V_{s\,max}^2[(L_sI_d)^2+(\sigma L_sI_q)^2]^2 + [R_sI_dI_q(L_s-\sigma L_s)]^2} - R_sI_dI_q(L_s-\sigma L_s)}{(L_sI_d)^2+(\sigma L_sI_q)^2} \tag{8.16}$$

here, $I_d = I_{ds\text{-}rated}$ and $I_q = \sqrt{I_{s\,max}^2 - I_d^2}$ are the $d-q$ axes stator currents used in the constant torque region, respectively.

The base speed depends heavily on the motor parameters, and thus, varies with motors. Even for the same motor, the base speed varies according to the flux level used in the constant torque region and the available stator voltage and current. A higher flux level, low $V_{s\,max}$, or high $I_{s\,max}$ results in a lower base speed, and thus in such cases, the field-weakening operation should start at a lower speed.

From the base speed and the slip frequency, the base rotor speed is obtained by

$$\omega_{r\text{-}base} = \omega_e - \omega_{sl} \tag{8.17}$$

This base rotor speed may be different from the rated rotor speed according to operating conditions.

In this field-weakening region, the optimal current combination (i.e., optimal current vector) producing the maximum output torque is always given as the value at the intersection (M point) of the circle and ellipse. The optimal current vector moves along the current-limit boundary as shown in Fig. 8.9 because the voltage-limit ellipse dwindles as the operating frequency increases. This means that the d-axis stator current, and thus the rotor flux linkage λ_{dr}^e should be reduced as the operating frequency increases.

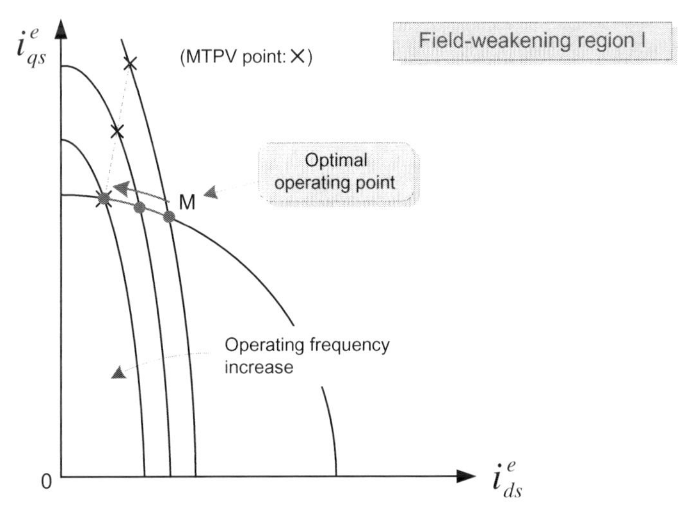

FIGURE 8.9

Optimal current trajectory in the field-weakening region I.

The d-axis stator current at this optimal point can be derived from the voltage- and current- limit conditions as

$$i_{ds}^{e}=\sqrt{\frac{\left(\frac{V_{s\,max}}{\omega_e}\right)^2-(\sigma L_s I_{s\,max})^2}{L_s^2(1-\sigma)}} \tag{8.18}$$

In this case, the available maximum q-axis stator current can be increased by the decreasing amount of the d-axis current to use the allowable stator current $I_{s\,max}$ fully as

$$i_{qs\text{-}max}^{e}=\sqrt{I_{s\,max}^2-i_{ds}^{e*2}} \tag{8.19}$$

From the optimal d-axis stator current, the optimal rotor flux command is given as

$$\lambda_{dr}^{e*}=L_m i_{ds}^{e*}=L_m\sqrt{\frac{\left(\frac{V_{s\,max}}{\omega_e}\right)^2-(\sigma L_s I_{s\,max})^2}{L_s^2(1-\sigma)}} \tag{8.20}$$

It is important to note that the magnetizing inductance L_m varies due to the flux reduction in this region. Thus the optimal flux should be obtained by considering this inductance variation. Also, as stated in Chapter 5, since the relation between the d-axis stator currents i_{ds}^{e} and the rotor flux linkage λ_{dr}^{e} is a first-order lag, a flux controller should be used to control the flux accurately for an effective field-weakening operation.

In this region the stator voltage and current are maintained at a constant as $V_{s\,max}$ and $I_{s\,max}$, respectively. Thus this region is referred to as *constant power (constant VA) region*. In this region, as we can see from the slip equation of Eq. (5.32), the slip frequency increases as the operating frequency increases because the d-axis stator current decreases and the q-axis stator current increases. As the operating frequency is further increased, the slip frequency reaches its maximum value, and then the field-weakening region II begins.

8.1.3.3 Breakdown torque region ($\omega_e > \omega_{BT}$): Field-weakening region II

As the operating frequency is further increased in the field-weakening region I, the voltage-limit ellipse will be reduced further, and a large portion will be included in the current-limit circle as shown in Fig. 8.10. This implies that the voltage-limit condition is included in the current-limit condition.

Furthermore, if the optimal current for producing the maximum torque subject to only the voltage-limit condition (point X on the voltage-limit ellipse) is included inside the current-limit circle, we can consider only the voltage-limit condition for obtaining the optimal flux. This situation corresponds to *field-weakening region II*. This optimal point subject to only the voltage-limit condition is called the *maximum torque per voltage* (MTPV) operating point.

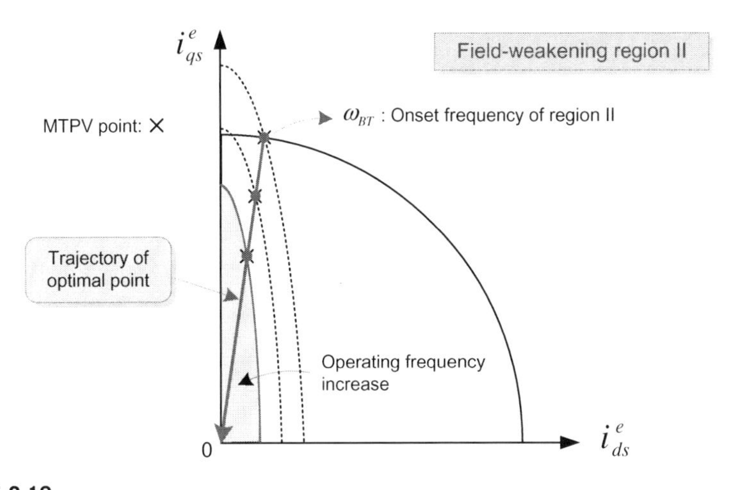

FIGURE 8.10

Maximum torque operation point for the field-weakening region II.

The onset speed of the field-weakening region II is the frequency ω_{BT}, at which the optimal current vector for the MTPV operation is just on the circumference of the current-limit circle, and can be given as

$$\omega_{BT} = \sqrt{\frac{1+\sigma^2}{2(\sigma L_s)^2}} \times \left(\frac{V_{s\,max}}{I_{s\,max}}\right) \tag{8.21}$$

This onset frequency ω_{BT} depends on the leakage inductance as well as the maximum values of the voltage and current. Induction motors with a large leakage factor will begin the operation of the field-weakening region II at a lower speed. Thus they have a narrow range of the constant power region.

In this field-weakening region II, the optimal stator currents for producing the maximum torque are given as

$$i_{ds}^e = \frac{V_{s\,max}}{\sqrt{2}\omega_e L_s} \tag{8.22}$$

$$i_{qs}^e = \frac{V_{s\,max}}{\sqrt{2}\omega_e \sigma L_s} \tag{8.23}$$

From the optimal d-axis stator current, the optimal rotor flux command is given as

$$\lambda_{dr}^{e*} = L_m i_{ds}^{e*} = \frac{V_{s\,max} L_m}{\sqrt{2}\omega_e L_s} \tag{8.24}$$

In this region, the optimal stator vector moves into the origin as the frequency increases as shown in Fig. 8.10. Thus, unlike the field-weakening region I, i_{qs}^{e*} is

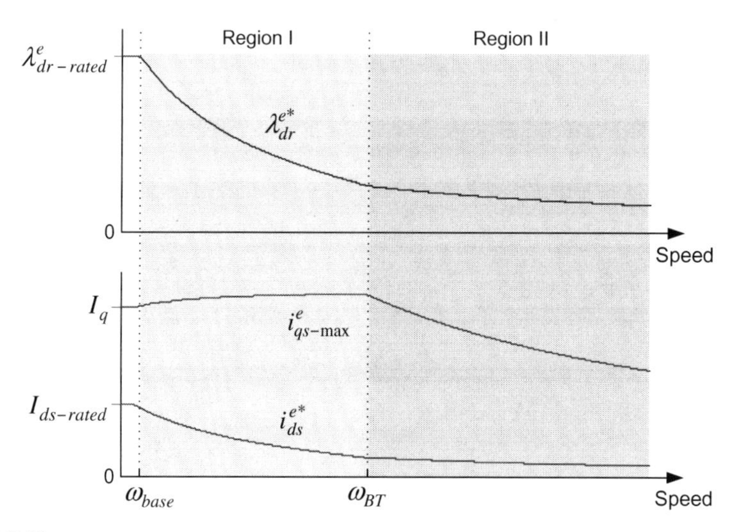

FIGURE 8.11

Optimal currents and flux in the field-weakening regions I and II.

also decreased along with the reduction of i^{e*}_{ds}. As a result, the total stator current will be reduced, and the output power will also be reduced. Fig. 8.11 shows the optimal d–q currents and the rotor flux command according to the speed in the field-weakening regions I and II.

From the optimal d–q axes currents of Eqs. (8.22) and (8.23), we can readily see that the slip frequency remains constant over this region as its maximum value, which is given as

$$\omega_{sl\text{-}max} = \frac{1}{T_r}\frac{i^e_{qs}}{i^e_{ds}} = \frac{1}{T_r\sigma} \tag{8.25}$$

The maximum torque developed by an induction motor in this region is quickly reduced by $1/\omega_e^2$ as

$$T_{e\text{-}max} = \frac{3}{2}\frac{P}{2}\frac{1}{2\sigma L_s^2}\left(\frac{V_{s\,max}}{\omega_e}\right)^2 \tag{8.26}$$

Fig. 8.12 compares the high-speed performance of the optimal field-weakening operation with that of the "$1/\omega_r$" method.

Since the level of the rotor flux command given at each operating frequency is inappropriate in the "$1/\omega_r$" method, the drive system does not exploit the output torque capability of the motor fully. Moreover, it will be hard to regulate the q-axis stator current command as the speed increases, so the vector control cannot be carried out in the high-speed region. As a result, the drive system will lose its controllability.

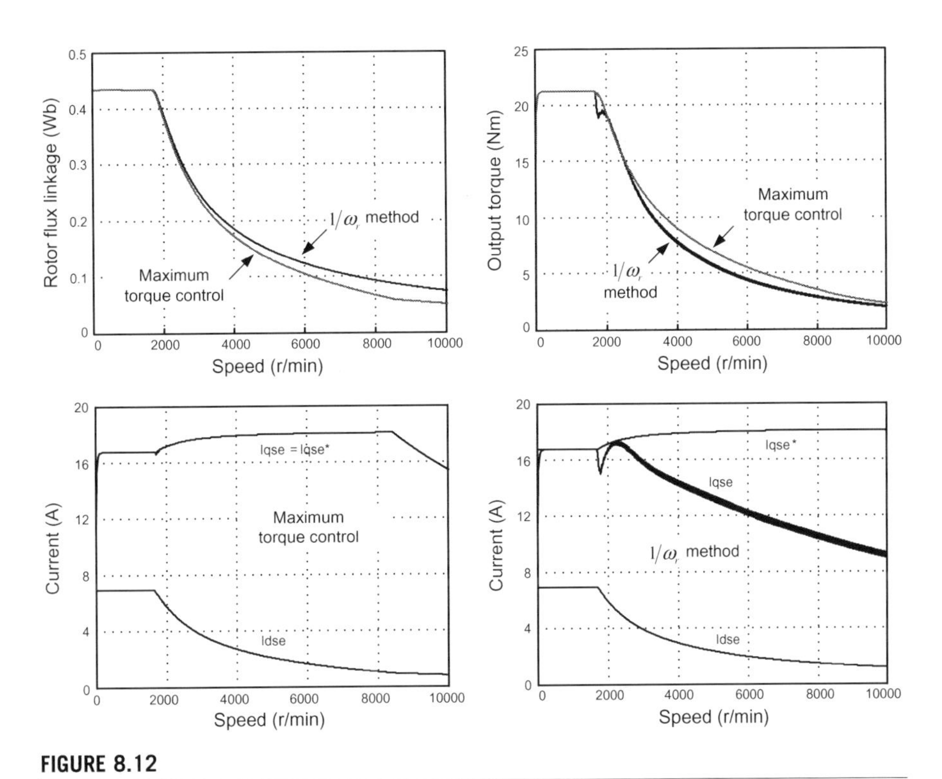

FIGURE 8.12

Comparison of the performance of the field-weakening methods (5-hp, 4-pole).

Fig. 8.13 shows a vector-controlled induction motor drive system with the field-weakening control for high-speed operations. The optimal flux reference can be given from Eq. (8.20) or (8.24) according to the operating speed.

Fig. 8.14 shows the experimental results on a 6-pole, 2.2-kW induction motor for the vector control system as shown in Fig. 8.13 [1]. The induction motor was accelerated up to 3000 r/min in the field-weakening region II by using the optimal strategy just described. The d-axis stator current is regulated to reduce the rotor flux according to the operating speed. Here, since the flux linkage is estimated by the voltage model, we can see that the estimated flux is inaccurate below the mid-speed range. While the q-axis current is increased in the field-weakening region I, it is decreased in the field-weakening region II.

To successfully carry out the optimal field-weakening operation, the optimal rotor flux reference at each operating frequency needs to be calculated correctly. In this feedforward field-weakening control method, the accuracy of the calculated optimal flux reference depends on the motor parameters such as the magnetizing inductance and leakage inductance. The magnetizing inductance will especially vary significantly due to the variation of the rotor flux level in the field-weakening operation. Therefore the variation of the magnetizing

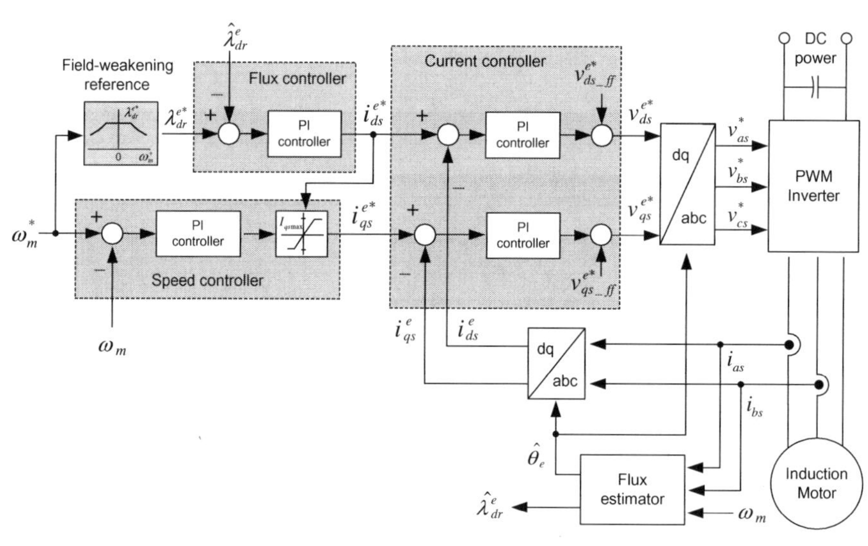

FIGURE 8.13

Vector control system of an induction motor with the field-weakening control.

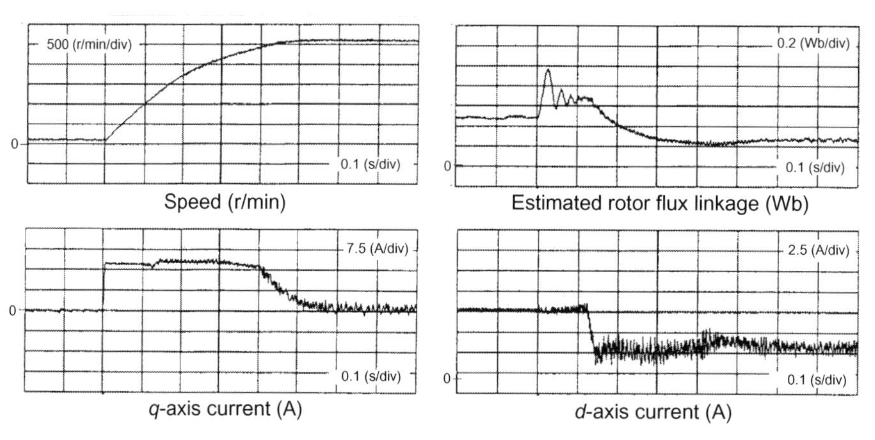

FIGURE 8.14

Optimal field-weakening operation (6-pole, 2.2-kW IM).

inductance should be considered. On the other hand, the leakage inductance is not dependent on the variation of the rotor flux level but on the magnitude of the stator current.

Unlike the feedforward field-weakening method sensitive to motor parameters, Kim and Sul [2] proposed another field-weakening control strategy that uses the feedback of the motor voltage without any motor parameter as shown

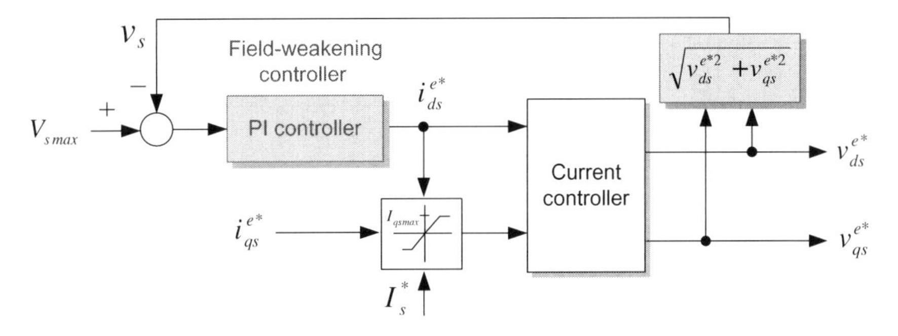

FIGURE 8.15

Field-weakening control strategy using the voltage feedback.

in Fig. 8.15. The feedback field-weakening control strategy is insensitive to motor parameters but cannot give dynamic response as fast as the feedforward strategy can.

The concept of the optimal field-weakening control methods of induction motors can be also used for high-speed operation of permanent magnet synchronous motors (PMSMs) as follows.

8.2 FLUX-WEAKENING CONTROL FOR PERMANENT MAGNET SYNCHRONOUS MOTORS [3–6]

As stated in Section 4.6, surface-mounted permanent magnet synchronous motors (SPMSMs) with magnets mounted on the surface of the rotor are unsuitable for high-speed operations with a higher centrifugal force. Furthermore, a reduction in stator inductance due to the magnets on the surface of the rotor hinders the flux-weakening operation, which reduces the magnet flux. On the contrary, interior permanent magnet synchronous motor (IPMSMs) are favorable for the high-speed operation since their magnets are inserted inside the rotor core and have a large inductance. For these reasons, IPMSMs are widely used for applications requiring a high-speed operation. Now we will discuss the flux-weakening control for the high-speed operation of PMSMs in more detail.

For motors such as an induction motor, separately excited DC motor, and synchronous motor with a field winding on the rotor, the field flux can be reduced directly by controlling the field current for high-speed operation. This technique is referred to as *field-weakening control*. By contrast, for PMSMs such as IPMSM or SPMSM, the field flux cannot be controlled directly because it is generated by a permanent magnet. As can be seen from Eq. (4.113), the stator flux linkage of PMSMs is given by

$$\lambda_{ds}^r = L_{ds} i_{ds}^r + \phi_f \tag{8.27}$$

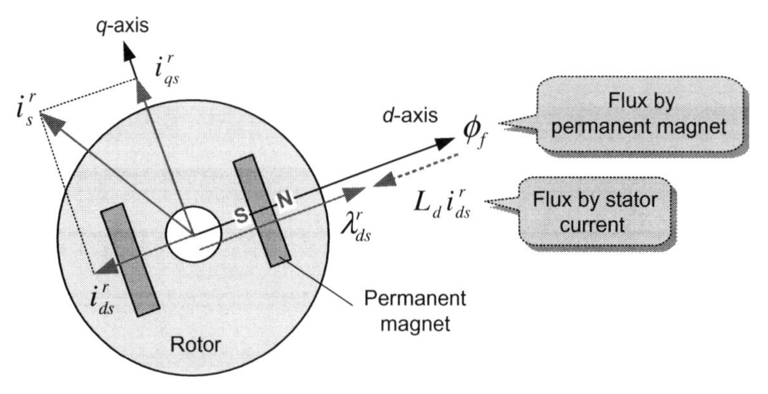

FIGURE 8.16

Concept of flux weakening.

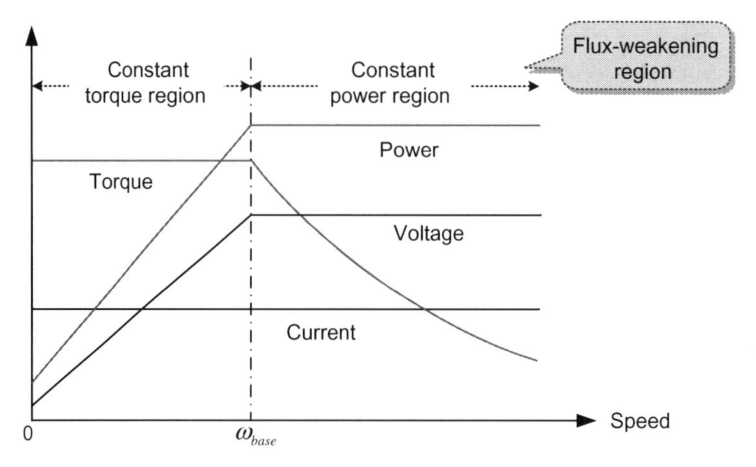

FIGURE 8.17

Operation region of PMSMs.

For PMSMs, if we produce the flux in the direction opposite of the magnet flux by using the negative d-axis stator current as shown in Fig. 8.16, we can reduce the effective stator flux linkage. This technique is referred to as *flux-weakening control*. From Eq. (8.27), we can see that a large d-axis inductance is desirable for an effective flux-weakening control.

The operation region of PMSMs can be normally divided into the following two speed regions.

- *Constant torque region*: Speed range below base speed
- *Constant power region*: Speed range above base speed (flux-weakening region)

Fig. 8.17 shows the characteristics in two speed regions.

The speed regions of PMSMs are basically similar to those of separately excited DC motors. However, PMSMs may have an additional flux-weakening region similar to the breakdown torque region of induction motors depending on their magnetic system design. This is determined by the magnitude relation between the magnet flux ϕ_f and the maximum d-axis stator flux linkage (i.e., $L_{ds}I_{s\,max}$). More practically, Fig. 8.18 shows the output power characteristics of the PMSMs according to the design. In the case of $\phi_f = L_{ds}I_{s\,max}$, PMSMs have a characteristic nearly similar to the ideal characteristic as shown in Fig. 8.17. The characteristics according to the magnetic system design will be explained in more detail in the Section 8.2.1.

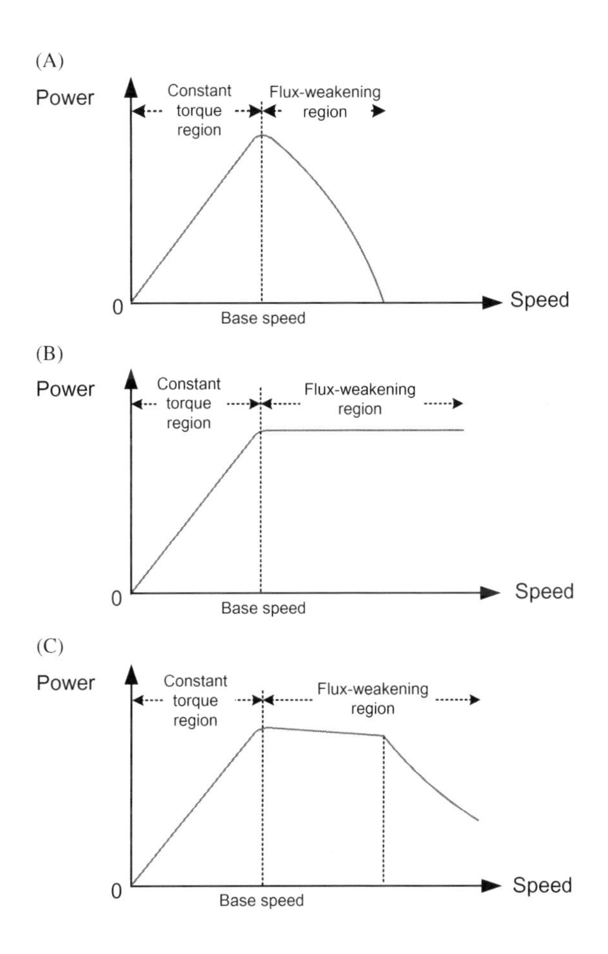

FIGURE 8.18

Output power according to speed regions of PMSM. (A) $\phi_f > L_{ds}I_{s\,max}$, (B) $\phi_f = L_{ds}I_{s\,max}$, and (C) $\phi_f < L_{ds}I_{s\,max}$.

The flux-weakening control techniques for the high-speed operation of PMSMs can be divided into three methods: *feedforward method*, which obtains the current commands required for the high-speed operation from steady-state voltage equations, *feedback method*, which obtains the current commands from the voltage feedback, and a *combined method* of these two techniques [3–5]. Here, we will explore the feedforward flux-weakening control method for IPMSMs and SPMSMs, separately.

8.2.1 HIGH-SPEED OPERATION OF AN INTERIOR PERMANENT MAGNET SYNCHRONOUS MOTOR

Before examining the high-speed operation of an IPMSM, we will first review the operation of an IPMSM in the constant torque region.

8.2.1.1 Constant torque region ($\omega_e \leq \omega_{base}$)

For the operation of an IPMSM in the constant torque region, which indicates an operating range below the base speed, the MTPA control is used as shown in Fig. 8.19.

In Section 5.5, we discussed how to obtain the optimal currents for MTPA operation. In the MTPA operation, the output torque of an IPMSM is mainly limited by only the maximum available stator current $I_{s\,max}$. However, for the high-speed range, where the back-EMF voltage becomes large, the voltage margin is insufficient to control the current commands for the MTPA operation. Thus the output torque of an IPMSM is limited by the available voltage rather than the available current.

Now we will examine the speed attainable under the maximum available stator voltage $V_{s\,max}$. In the vicinity of rated speed with a large back-EMF, since the

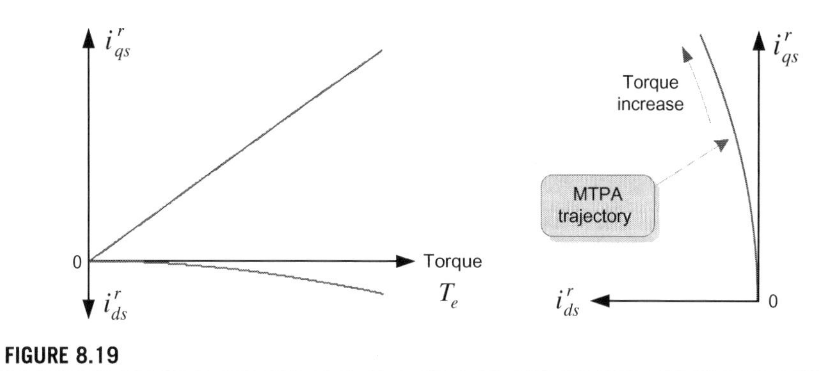

FIGURE 8.19

d–q axes current commands for MTPA operation.

voltage drops on the stator resistance is negligible, the $d-q$ axes stator voltages of an IPMSM in the steady-state are given from Eqs. (6.24) and (6.25) as

$$v_{ds}^r = -\omega_r L_{qs} i_{qs}^r \tag{8.28}$$

$$v_{qs}^r = \omega_r (L_{ds} i_{ds}^r + \phi_f) \tag{8.29}$$

These stator voltages should satisfy the following relation, which is the *voltage-limit condition* stated in Section 8.1.

$$v_{ds}^{r2} + v_{qs}^{r2} \le V_{s\,max}^2 \tag{8.30}$$

From Eqs. (8.28)–(8.30), the maximum speed $\omega_{r\text{-}base}$ of the constant torque operation region can be obtained as

$$\omega_{r\text{-}base} = \frac{V_{s\,max}}{\sqrt{(L_{ds} I_{ds}^r + \phi_f)^2 + (L_{qs} I_{qs}^r)^2}} \tag{8.31}$$

where I_{ds}^r and I_{qs}^r are the $d-q$ axes currents used for MTPA operation, respectively.

The speed $\omega_{r\text{-}base}$ is the base speed at which the flux-weakening control should start for the constant power operation. The base speed varies with the required current (or torque) and the available voltage.

Next, we will discuss the flux-weakening control method in the constant power region.

8.2.1.2 Constant power region ($\omega_e > \omega_{base}$)

Similar to induction motors, the IPMSM operation is also confined to the voltage- and the current-limit conditions, which was explained in Section 8.1.2. Thus we need to find out the optimal flux-weakening control method that ensures a maximum torque production in the high-speed region under the voltage- and the current-limit conditions.

To begin with, we will express the voltage- and the current-limit boundaries for an IPMSM in the $d-q$ axes current plane. Similar to an induction motor, under the maximum available stator current $I_{s\,max}$, the $d-q$ axes stator currents of an IPMSM are restricted to the following condition.

$$i_{ds}^{r2} + i_{qs}^{r2} \le I_{s\,max}^2 \tag{8.32}$$

The current-limit boundary in the $d-q$ axes currents plane is shown in Fig. 8.20. This current-limit boundary is expressed as a circle, whose radius is $I_{s\,max}$. On the other hand, the voltage-limit boundary can be expressed from Eqs. (8.28)–(8.30) as

$$(\omega_r L_{qs} i_{qs}^r)^2 + (\omega_r L_{ds} i_{ds}^r + \omega_r \phi_f)^2 \le V_{s\,max}^2 \tag{8.33}$$

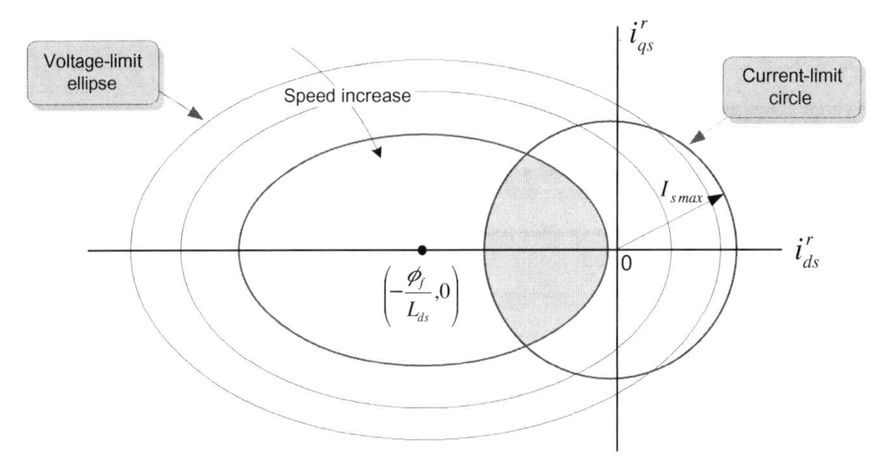

FIGURE 8.20

Voltage- and current-limit boundaries.

This voltage-limit boundary as shown in Fig. 8.20 is an ellipse equation, whose center is given as

$$(i^r_{ds0}, i^r_{qs0}) = \left(-\frac{\phi_f}{L_{ds}}, 0\right) \tag{8.34}$$

This expression illustrates the boundary of controllable $d-q$ axes currents by using the given maximum stator voltage $V_{s\,max}$. The inner region of the ellipse reduces as the operating speed ω_r increases. This indicates that the boundary of controllable currents becomes smaller as the operating speed increases. This is because the back-EMF increases as the operating speed increases.

The IPMSM drives should always satisfy both the voltage- and the current-limit conditions. Considering these two limit conditions, the region of controllable currents is the common area between the current-limit circle and the voltage-limit ellipse for a given operating speed, which is the *dark area* shown in Fig. 8.20. Thus the $d-q$ axes stator current commands must be inside this area for a given operating speed.

Now let us investigate the optimal operating point for producing the maximum output torque under these voltage- and current-limit conditions. For an IPMSM, as can be seen in Fig. 8.18, the maximum operating speed and the output power characteristic vary according to the magnitude relationship between the magnet flux ϕ_f and the maximum d-axis stator flux linkage (i.e., $L_{ds}I_{s\,max}$). This relationship can also be expressed by whether or not the center of the voltage-limit ellipse is inside the current-limit circle. According to this relationship, a different flux-weakening control strategy is required [6].

The maximum attainable operating speed of an IPMSM can be given by letting $i^r_{ds} = -I_{s\,max}$, $i^r_{qs} = 0$ in Eq. (8.31) as

$$\omega_{r\text{-}max} = \frac{V_{s\,max}}{\phi_f - L_{ds}I_{s\,max}} \tag{8.35}$$

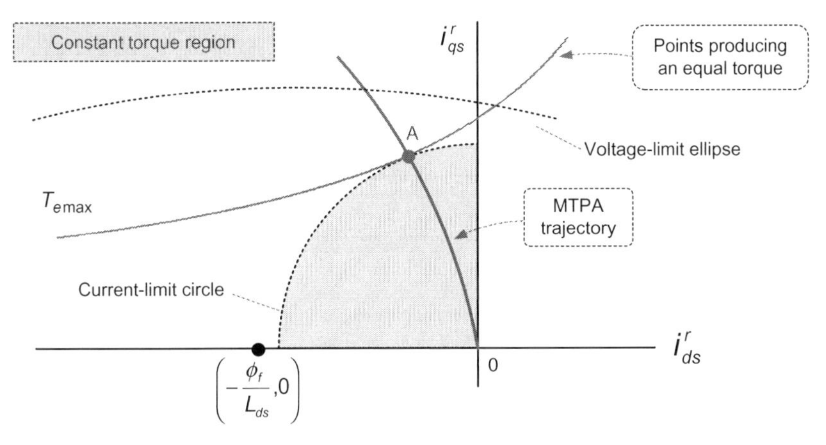

FIGURE 8.21

Current-limit circle and MTPA trajectory.

In the case of $\phi_f > L_{ds}I_{s\ max}$ it can be seen from Eq. (8.34) that there is a limitation on the operating speed at which the magnetic flux can be reduced by using the stator current, i.e., this type of motor has a *finite maximum speed*. On the other hand, in the case of $\phi_f < L_{ds}I_{s\ max}$, the maximum operating speed is theoretically infinite because the magnetic flux can be reduced fully by using the stator current, i.e., this type of motor has an *infinite maximum speed*. In this case, the operating speed is limited only by the mechanical strength. For these two cases, we need to use different flux-weakening control methods. First, let us take a look at the flux-weakening operation in the case of $\phi_f > L_{ds}I_{s\ max}$.

Fig. 8.21 illustrates numerous combinations of the current commands to produce an equal output torque, along with the voltage- and current-limit boundaries in the $d-q$ axes current plane.

In the low- and mid-speed range, the voltage-limit ellipse is large enough to encompass the current-limit circle. Thus we do not have to be concerned about the voltage-limit constraint, and the output torque depends only on the available current. In this case the optimal operating point is on the MTPA trajectory according to the given torque command. The maximum torque $T_{e\text{-}max}$ is produced at the intersection A of the current-limit circle and the MTPA trajectory. As the rotor speed increases, the boundary of the voltage-limit ellipse will be reduced. The boundary of the voltage-limit ellipse will encounter the MTPA operation point A at a specific speed as shown in Fig. 8.22. This specific speed is the base speed $\omega_{r\text{-}base}$ at which the flux-weakening control begins. When the operating speed is further increased above the base speed $\omega_{r\text{-}base}$, the voltage-limit ellipse shrinks more, so the operation point A will be outside the ellipse. As shown in Fig. 8.22, at speed $\omega_B(>\omega_{r\text{-}base})$, it can be readily seen that the currents at point A can be no longer regulated because it deviates from the region of controllable currents, i.e., the voltage is insufficient to regulate those currents.

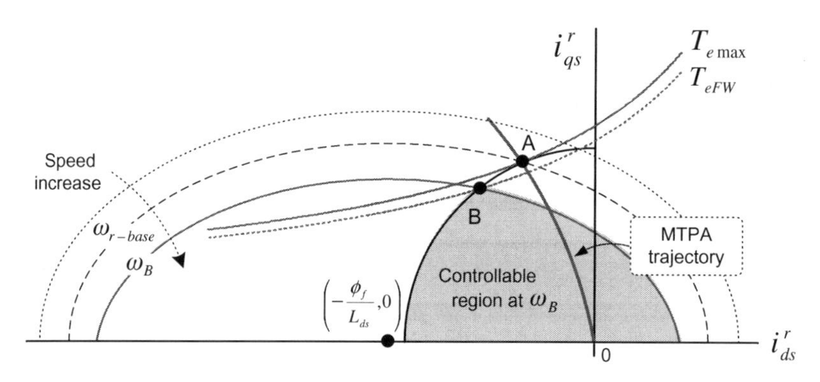

FIGURE 8.22

Onset of flux-weakening operation.

Thus the current command should be moved to the controllable operating point. Considering both the voltage-limit and the current-limit conditions, the optimal point for producing the maximum output torque is the intersection (point B) of the circle and the ellipse. The developed torque T_{eFW} at this point is less than the maximum torque $T_{e\text{-}max}$ in the constant torque region.

In such operation, the d-axis current increases in the negative direction. This is the flux-weakening control that reduces the effective stator flux linkage. The optimal d–q axes currents for the stator current command I_s given as a torque command are expressed as

$$i_{ds}^{r} = \frac{L_{ds}\phi_f - \sqrt{(L_{ds}\phi_f)^2 + (L_{qs}^2 - L_{ds}^2)\left(\phi_f^2 + (L_{qs}I_s)^2 - \left(\frac{V_{s\,max}}{\omega_r}\right)^2\right)}}{(L_{qs}^2 - L_{ds}^2)} \tag{8.36}$$

$$i_{qs}^{r} = \sqrt{I_{s\,max}^2 - i_{ds}^{r*2}} \tag{8.37}$$

The trajectory of the optimal point moves along the current-limit boundary in a counterclockwise direction as the operating speed increases as shown in Fig. 8.23. Accordingly, the d-axis stator current increases in the negative direction while the q-axis stator current decreases.

This operation continues until $i_{ds}^{r*} = I_{s\,max}$ and $i_{qs}^{r*} = 0$, at which the operating speed reaches its maximum. Thus we can identify that this type of motor has a *finite maximum speed.* In such a flux-weakening operation, the stator voltage and current of an IPMSM remain constant, resulting in constant power operation. Fig. 8.24 shows the optimal currents in the flux-weakening operation of an IPMSM.

Next, let us take a look at the flux-weakening operation in the case of $\phi_f < L_{ds}I_{s\,max}$. In this case the center of the voltage-limit ellipse is inside the current-limit circle as shown in Fig. 8.25, and the trajectory of the optimal current is somewhat different from the case of $\phi_f > L_{ds}I_{s\,max}$. In the beginning of the flux-weakening operation, the trajectory of the optimal point moves along the

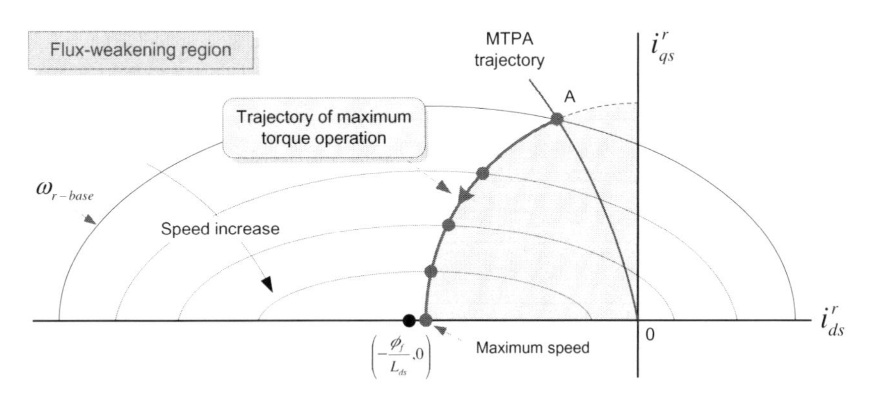

FIGURE 8.23

Optimal current vector in the flux-weakening operation ($\phi_f \geq L_{ds} I_{s\ max}$).

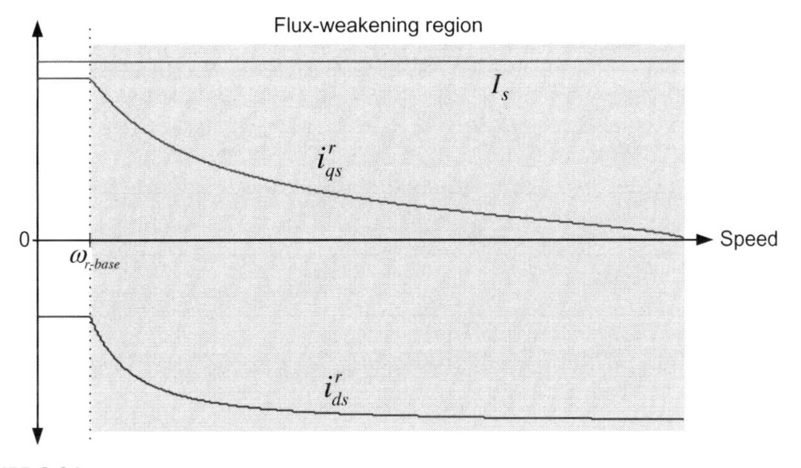

FIGURE 8.24

Optimal currents in the flux-weakening operation ($\phi_f > L_{ds} I_{s\ max}$).

current-limit boundary as the speed increases as in the case of $\phi_f > L_{ds} I_{s\ max}$. However, above a certain speed, the voltage-limit ellipse is gradually included in the current-limit circle. This means that the voltage-limit condition is included in the current-limit condition. If the optimal currents for the maximum torque subject to only the voltage-limit condition are included inside the current-limit circle (e.g., point C), then only the voltage-limit is the constraint that should be considered for obtaining the optimal currents. In this case the optimal point should move into the center of the ellipse as the speed increases instead of moving along the current-limit circle as shown in Fig. 8.25. Thus the optimal point is not point D, but point C. This flux-weakening operation is called as the MTPV (the maximum torque per voltage) operation. This operation region corresponds to the field-weakening region II of an induction motor.

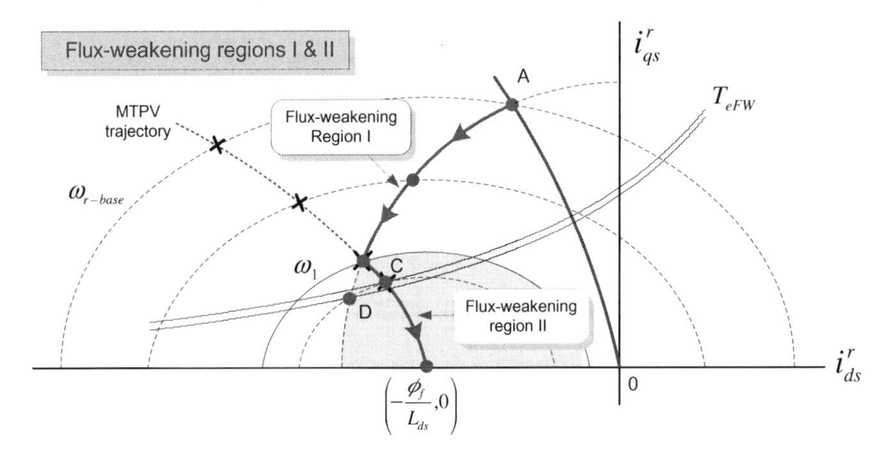

FIGURE 8.25

Optimal current trajectory in the flux-weakening region ($\phi_f < L_{ds} I_{s\ max}$)

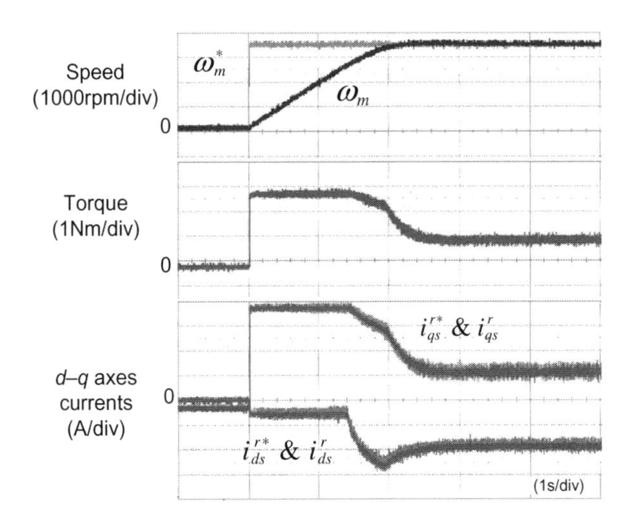

FIGURE 8.26

Flux-weakening operation of an 800-W IPMSM.

Fig. 8.26 shows the experimental results on the flux-weakening control for a 800-W, 8-pole IPMSM when the speed command is given as 3500 r/min. For this IPMSM, $\phi_f > L_{ds} I_{s\ max}$ and the base speed is 2500 r/min.

8.2.2 HIGH-SPEED OPERATION OF A SURFACE-MOUNTED PERMANENT MAGNET SYNCHRONOUS MOTOR

Unlike IPMSMs, the typical operating range of SPMSMs is the constant torque region. However, the high-speed operation of SPMSMs is often required in

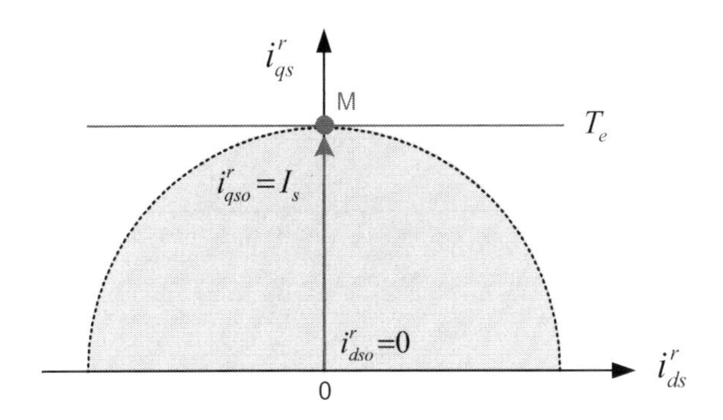

FIGURE 8.27

Optimal currents under the current-limit condition for an SPMSM.

several applications such as washing machines. The flux-weakening operation of SPMSMs is almost similar to that of IPMSMs. Now we will study the flux-weakening operation of SPMSMs. Similar to other motors, we will find out the optimal flux-weakening control method that ensures the maximum torque operation under the voltage- and the current-limit conditions.

The current-limit condition of an SPMSM is given as

$$i^{r2}_{ds} + i^{r2}_{qs} \leq I^2_{s\,max} \tag{8.38}$$

The current-limit boundary in the $d-q$ axes currents plane can be expressed as a circle whose radius is the maximum stator current $I_{s\,max}$ as shown in Fig. 8.27.

As stated in Section 5.5.1, for the MTPA operation of SPMSMs in the constant torque region, all the available current should be assigned to the q-axis stator current i^r_{qs} while making the d-axis stator current i^r_{ds} zero. Thus the optimal currents for the MTPA operation of the SPMSM under a given available stator current $I_{s\,max}$ become

$$i^{r*}_{dso} = 0 \tag{8.39}$$

$$i^{r*}_{qso} = I_{s\,max} \tag{8.40}$$

Fig. 8.27 depicts point M of the optimal currents under the current-limit condition. Similar to other motors, the possibility of the operation at this optimal point M depends on the following voltage-limit condition.

$$v^{r2}_{ds} + v^{r2}_{qs} \leq V^2_{s\,max} \tag{8.41}$$

For an SPMSM, neglecting the voltage drop of the stator resistance, from Eqs. (6.24) and (6.25), the steady-state $d-q$ axes voltage equations are expressed as

$$v^r_{ds} = -\omega_r L_s i^r_{qs} \tag{8.42}$$

$$v^r_{qs} = \omega_r (L_s i^r_{ds} + \phi_f) \tag{8.43}$$

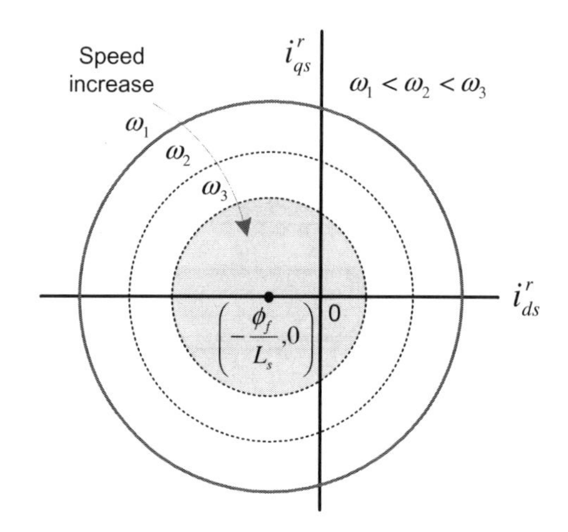

FIGURE 8.28

Voltage-limit condition for an SPMSM.

Thus the voltage-limit condition of Eq. (8.41) can be expressed from Eqs. (8.42) and (8.43) as

$$(\omega_r L_s i_{qs}^r)^2 + (\omega_r L_s i_{ds}^r + \omega_r \phi_f)^2 \leq V_{s\,max}^2 \tag{8.44}$$

Unlike an IPMSM, the voltage-limit condition is expressed as a circle with the center $(-\phi_f/L_s, 0)$ in the $d-q$ axes currents plane as shown in Fig. 8.28. Under a given maximum stator voltage $V_{s\,max}$, the boundary of this circle becomes smaller as the operating speed increases.

The voltage- and current-limit conditions for a certain speed below the base speed are shown in Fig. 8.29. In this case, the optimal currents for producing the maximum torque (i.e., point M) are inside the voltage-limit circle, so they are controllable.

8.2.2.1 Constant power region ($\omega_e > \omega_{base}$)

The voltage-limit circle dwindles with the increase in the speed. When point M begins to move out of the voltage-limit circle, the operating point should be changed. This indicates that the flux-weakening operation starts and the speed becomes the base speed.

In that case, the optimal point for producing the maximum output torque is the value at the intersection of two circles. From Eqs. (8.38) and (8.44), the optimal currents at this point are given as

$$i_{ds}^{r*} = \frac{\left(\frac{V_{s\,max}}{\omega_r}\right)^2 - \phi_f^2 - (L_s I_{s\,max})^2}{2L_s\phi_f} \tag{8.46}$$

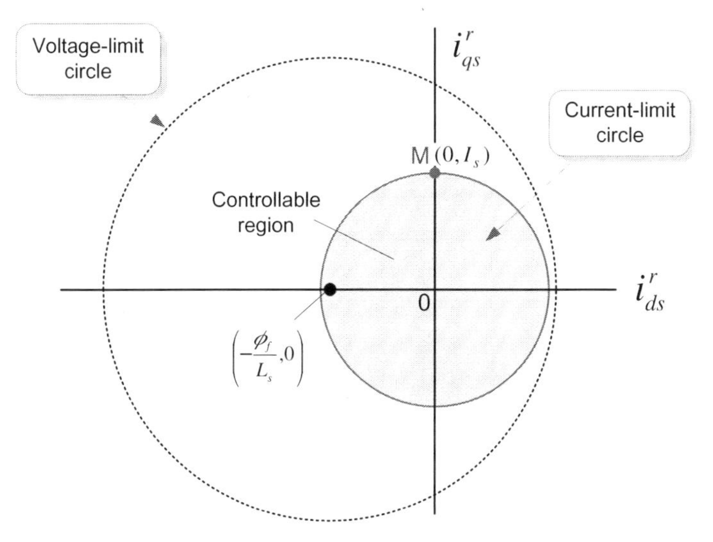

FIGURE 8.29

Voltage- and current-limit conditions for an SPMSM.

$$i_{qs}^{r*} = \sqrt{I_{s\,max}^2 - i_{ds}^{r*2}} \tag{8.47}$$

The trajectory of the optimal point moves along the current-limit boundary in a counterclockwise direction as the operating speed increases as shown in Fig. 8.30. Accordingly, the *d*-axis stator current increases in the negative direction while the *q*-axis stator current decreases.

Similar to an IPMSM, the maximum operating speed and the output power characteristic of an SPMSM vary according to the magnitude relationship between the magnet flux ϕ_f and the maximum *d*-axis stator flux linkage (i.e., $L_sI_{s\,max}$). In the case of $\phi_f > L_sI_{s\,max}$, the center of the voltage-limit circle is outside the current-limit circle, and there exists a limit on the operating speed (*finite maximum speed motor*). On the other hand, in the case of $\phi_f < L_sI_{s\,max}$, there is no limit on the maximum operating speed theoretically (*infinite maximum speed motor*) and also, there exists a speed range, which corresponds to the field-weakening region II of an induction motor. The optimal *d*-axis current of Eq. (8.46) corresponds to an SPMSM with $\phi_f > L_sI_{s\,max}$. This optimal *d*-axis current can be applied up to the maximum speed. The maximum operating speed is obtained when the *d*-axis current is equal to the stator maximum current, i.e., $i_{ds}^{r*} = I_{s\,max}$. However, in this case, the *q*-axis current becomes zero, so the maximum attainable speed will actually be less than that speed.

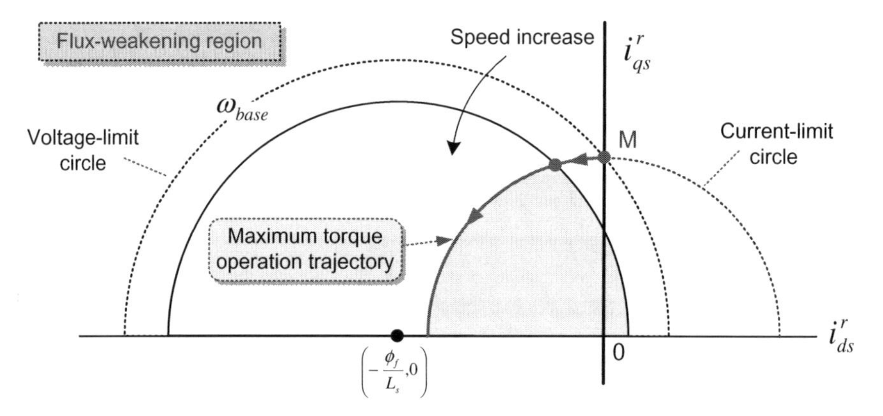

FIGURE 8.30

Trajectory of the optimal current vector (in case of $\phi_f > L_s I_{s\ max}$).

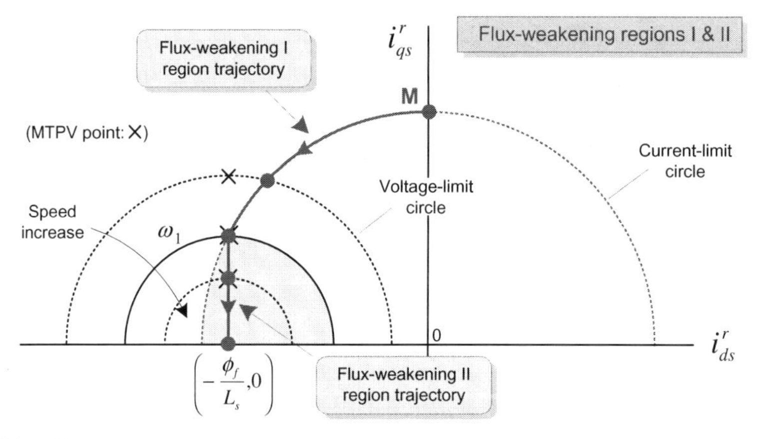

FIGURE 8.31

Trajectory of the optimal current vector (in case of $\phi_f < L_s I_{s\ max}$).

Next, we will discuss the case of $\phi_f < L_s I_{s\ max}$. In this case, the trajectory of the optimal current is identical to that in the case of $\phi_f > L_s I_{s\ max}$ in the beginning of the flux-weakening operation. However, if the optimal current for producing the maximum torque subject to only the voltage-limit constraint (the MTPV operation point X) is included inside the current-limit circle as shown in Fig. 8.31, then the MTPV operation needs to be carried out. In this operation, the d-axis stator current remains constant, while the q-axis stator current reduces as the speed increases.

REFERENCES

[1] S.-H. Kim, S.K. Sul, Maximum torque control of an induction machine in the field weakening region, IEEE Trans. Ind. Appl. 31 (4) (Jul./Aug. 1995) 787–794.

[2] S.-H. Kim, S.K. Sul, Voltage control strategy for maximum torque operation of an induction machine in the field-weakening region, IEEE Trans. Ind. Electron. 44 (4) (Aug. 1997) 512–518.

[3] J.M. Kim, S.K. Sul, Speed control of interior permanent magnet synchronous motor drive for the flux weakening operation, IEEE Trans. Ind. Appl. 33 (1) (Jan./Feb. 1997) 787–794.

[4] S.R. Macminn, T.M. Jahns, Control techniques for improved high-speed performance of interior PM synchronous motor drives, IEEE Trans. Ind. Appl. 27 (5) (Sep./Oct. 1991) 997–1004.

[5] T.S. Kwon, G.Y. Choi, M.S. Kwak, S.K. Sul, Novel flux-weakening control of an IPMSM for quasi-six-step operation, IEEE Trans. Ind. Appl. 27 (5) (Nov./Dec. 2008) 1722–1731.

[6] W.L. Soong, T.J.E. Miller, Field-weakening performance of brushless synchronous AC motor drives, IEE Proc. Elec. Power Appl. 141 (6) (Nov., 1994) 331–340.

CHAPTER

Speed estimation and sensorless control of alternating current motors

9

For the position/speed control of a motor, it is necessary for its speed/position information to be used as a feedback signal for the speed/position control loop. As it can be seen in Chapter 5, for the vector control of alternating current (AC) motors, the information on the rotor position is also necessary. To measure the position/speed of the motor, an analog or a digital position sensor is used. Resolver, synchro, and tacho-generator are well known as analog position sensors. A rotary encoder is the most widely used as the digital position sensor.

In this chapter, we will introduce position sensors used in the motor drive systems and examine the speed estimation method from the position sensor signal. Finally, we will introduce the sensorless control methods of AC motors, which do not use position sensors.

9.1 POSITION SENSORS

Besides the position/speed control of a motor, for the vector control of AC motors, a position sensor is needed. In the vector control, a different type of position sensor is used according to the motor used. As discussed in Chapter 5, the absolute position of the rotor (i.e., permanent magnet) is required for the vector control of permanent magnet synchronous motors (PMSMs). Among analog position sensors, a resolver, which can provide information on the absolute position of the rotor, is commonly used for PMSM drives. On the other hand, since the vector control of induction motors does not require the absolute position of the rotor, an incremental encoder is usually used among digital devices.

Now we will explore the resolver and rotary encoder.

9.1.1 RESOLVER

A resolver is a type of rotary electrical transformer connected to the rotating shaft. The resolver produces signals that vary sinusoidally as the shaft rotates. It is commonly used for measuring the absolute position of the rotor. Fig. 9.1

Electric Motor Control. DOI: http://dx.doi.org/10.1016/B978-0-12-812138-2.00009-X

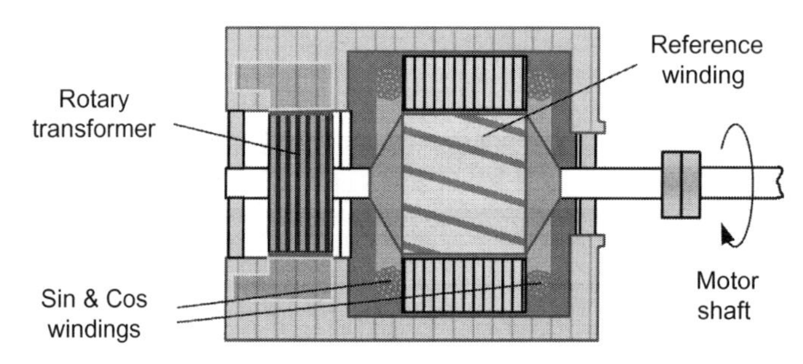

FIGURE 9.1

Resolver.

Source: http://www.amci.com/tutorials/tutorials-what-is-resolver.asp

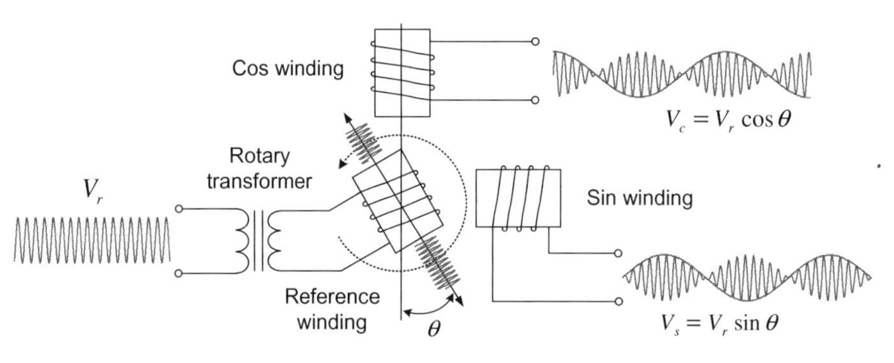

FIGURE 9.2

Resolver windings and their signals.

Source: http://www.amci.com/tutorials/tutorials-what-is-resolver.asp

depicts the configuration of a resolver, which consists of a stator, a rotor, and a rotary transformer [1].

The primary winding of a resolver, called reference winding, is located in the rotor and is excited through a rotary transformer. The two secondary windings, called SIN and COS Windings, are located in the stator and mechanically displaced 90° from each other. In the resolver operation, a shaft angle θ can be measured from signals induced in the secondary windings after injecting AC voltage signal into the primary winding as shown in Fig. 9.2.

When the primary winding is excited by an AC voltage V_r through a rotary transformer, voltages in the secondary windings are induced differently depending on the angle θ of the rotor shaft. The induced voltages vary as sine or cosine of the rotor angle, respectively. From an arctangent function of these signals, we can know the absolute angle θ of the rotor connected to the shaft as

$$\theta = \tan^{-1}\left(\frac{V_C}{V_S}\right) = \tan^{-1}\left(\frac{V_r \sin\theta}{V_r \cos\theta}\right) \tag{9.1}$$

The resolver usually uses a resolver-to-digital converter to provide resolver excitation and convert angular analogue signals of the resolver into a digital form (a serial binary output or pulses equivalent to an incremental encoder) that can be more easily used by digital controllers. The resolver is a rugged device, which can provide a reliable performance in high temperatures, vibration, and contaminated environments. However, resolvers are costly and require complex excitation and signal processing circuits that are susceptible to noise. In practice, a periodical position error may exist due to amplitude imbalance, inductive harmonics, reference phase shift, excitation signal distortion, and disturbance signals [2]. The position error causes a torque ripple with twice the electrical frequency, and thus should be corrected by a compensation method [3].

9.1.2 ROTARY ENCODER

A rotary encoder is a sensor of mechanical motion that generates digital signals in response to the rotational motion of the shaft. There are two main types of rotary encoder according to output forms: the *absolute encoder*, which can provide the absolute value of the rotation angle, and the *incremental encoder*, which provides only the incremental value of the rotation. Incremental encoders are the most widely used for most adjustable motor drive systems because absolute encoders are much more complicated and expensive. There are two sensor types used in encoders to generate digital signals such as magnetic and optical. The latter is more commonly used.

The incremental encoder generates a series of pulses as its shaft moves. These pulses can be used to measure the position, speed, and direction. The resolution of an incremental encoder is frequently described in terms of *pulse per revolution* (PPR), which is the total number of output pulses per complete revolution of the encoder shaft. An encoder with 512 or 1024 PPR is popular in the motor drive systems. The rotary encoders come in two configurations as shown in Fig. 9.3: a *shaft type* that connects the shaft of a rotor with a coupling, and a *hollow type* into which the shaft of a rotor is inserted into.

FIGURE 9.3

Incremental encoders.

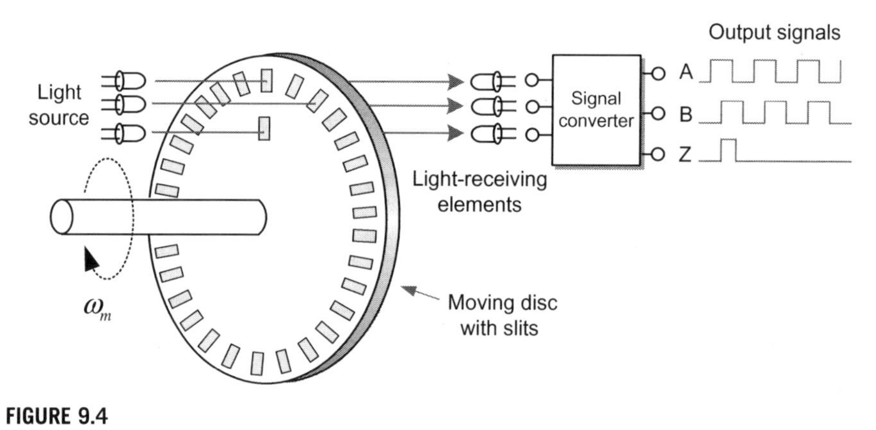

FIGURE 9.4

Simple configuration of an optical incremental encoder.

The encoders are a device that is sensitive to the surrounding environment such as temperature, shock, vibration, and contamination. To prevent noise from affecting encoder signals, it is preferable to use a shielded twisted pair cable for encoder output lines. Now we will explore how the incremental encoders work.

9.1.2.1 Operating principle of the optical incremental encoder

Fig. 9.4 shows the simple configuration of an optical incremental encoder, which consists of a moving disc mounted to the rotating shaft, light sources (LEDs), and light receivers (phototransistors). The moving disc has the same number of slits as PPR. The light of LEDs passing through the slits on the disc is transmitted to phototransistors, and in turn, is converted to square wave–shaped electric signals.

Commonly, the encoder has three outputs called A, B, and Z. The total number of A and B pulses per revolution is equal to PPR, with which the angular position and speed can be calculated. The A and B pulses are 90° out of phase, which allows the identification of the direction of rotation as shown in Fig. 9.5. For example, when rotating in the forward direction, pulse A is ahead of pulse B. There is another pulse Z known as the *index* or *reference pulse* besides pulses A and B. Pulse Z is generated once per revolution and can be used to set the reference position.

The absolute encoder is used when there is a need of an absolute position of the rotor. Since an absolute encoder generates a unique code (or multibit digital words) for each angular position of the rotor, we can find the actual position directly from the output signals. This output can be in binary code, binary coded decimal code, or gray code. The absolute encoder requires a complicated disc with many slits to generate the output code, resulting in a higher price.

Incremental encoders are the most widely used in motor drive applications. Since the incremental encoder produces a series of pulses as the rotor moves, we cannot measure the rotor speed directly from the encoders. Thus we will next discuss a method to estimate the speed from the output pulses of an encoder.

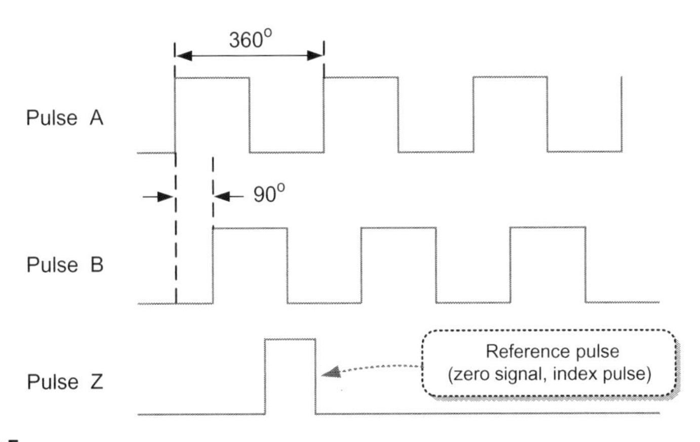

FIGURE 9.5

A, B, and Z pulses of incremental encoder.

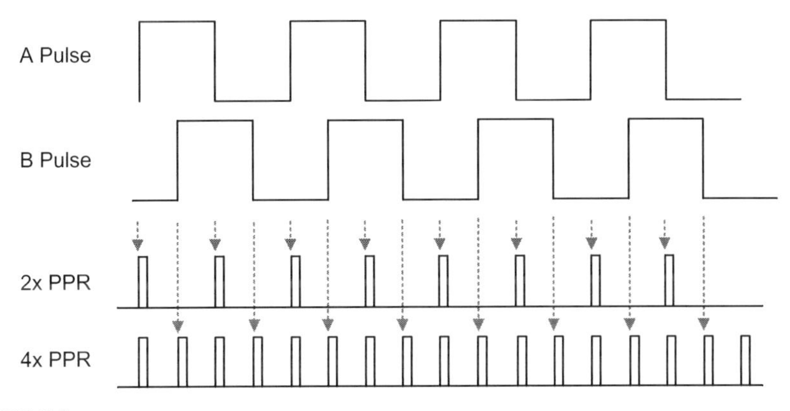

FIGURE 9.6

Pulse multiplication.

9.2 SPEED ESTIMATION USING AN INCREMENTAL ENCODER

We usually directly use the encoder pulses A or B themselves to calculate the speed. However, when more resolution is needed, we often increase the number of pulses by using a pulse multiplication. For a low-cost encoder with a low PPR, such multiplication can generate pulses more than the original pulse number, and thus results in a higher resolution. Fig. 9.6 shows that the PPR of an encoder can be doubled or quadrupled by counting the rising and falling edges of one or both pulses. In this way a 1000-PPR encoder can act like a 4000-PPR encoder by a 4× multiplication.

Now we will discuss how to calculate the speed of a rotor with pulses of an encoder mounted to the shaft of the rotor.

The angular velocity can be defined as the rate of change of the angular displacement X rad over the time interval T second as

$$\omega_m = \frac{X}{T} \quad \text{(rad/s)} \tag{9.2}$$

Here, the angular displacement X is obtained by counting the pulses produced from the encoder. The angular velocity can be often expressed in terms of revolution per minute unit as

$$N = \left(\frac{60}{2\pi}\right)\frac{X}{T} \quad \text{(r/min)} \tag{9.3}$$

There are three typical methods to obtain the angular velocity from the encoder pulses: M method, T method, and M/T method. In the M method, the angular velocity is obtained by measuring the displacement for a constant time interval. In the T method, the angular velocity is obtained by measuring the time interval for a constant displacement. Now, we will examine these methods in detail as follows.

9.2.1 M METHOD

In this method, the angular velocity is obtained by counting the pulses produced from the encoder during the constant sampling time interval T_c as shown in Fig. 9.7.

Assuming that the number of encoder pulses during the sampling time interval T_c is m, the angular displacement X over the T_c is given as

$$X = \frac{m}{PPR} \cdot 2\pi \quad \text{(rad)} \tag{9.4}$$

Thus the angular velocity is given from Eqs. (9.2) and (9.3) as

$$\omega_m = \frac{X}{T_c} = \frac{2\pi}{T_c}\frac{m}{PPR} \quad \text{(rad/s)} \tag{9.5}$$

$$N_f = \frac{60}{2\pi} \cdot \omega_m = \frac{60}{T_c} \cdot \frac{m}{PPR} \quad \text{(r/min)} \tag{9.6}$$

For example, if an encoder with a resolution of 1024 PPR produces 2048 pulses for 0.1 s, then the angular velocity can be calculated as

$$N = \frac{60 \cdot 2048}{0.1 \cdot 1024} = 1200\ \text{r/min} \tag{9.7}$$

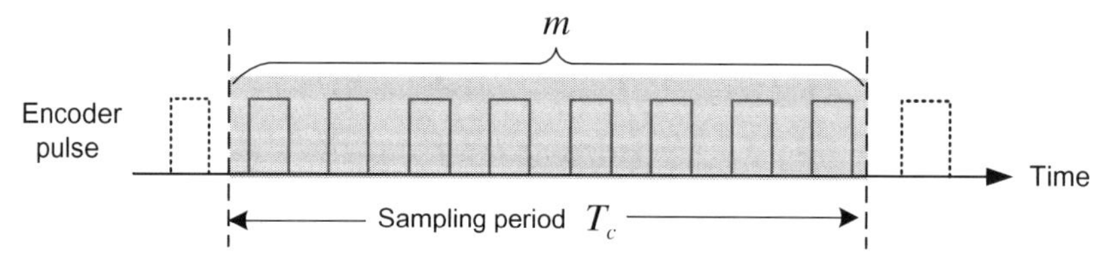

FIGURE 9.7

M method.

Since the M method is simple and easy to be implemented, it can be widely used in many applications where high precision in speed is not required. In addition, in this method, since the speed calculation time, i.e., the sampling time of the speed is constant, it is easy to design the speed controller where the speed is periodically controlled. The sampling period is commonly chosen within 1–3 ms according to the required control bandwidth. However, if the sampling time interval is not synchronous with pulses, as it is true in most cases, there exists a speed error. The maximum pulse error is one pulse, resulting in a maximum speed error of $60/(T_c\ \text{PPR})$ (r/min). For example, for PPR = 2000, 4 × multiplication, and $T_c = 1$ ms, the maximum speed error is $(60/(1\ \text{ms} \times 4 \times 2000)) = 7.5$ r/min. Thus in this case, it is impossible to identify a speed less than 7.5 r/min. In particular, this absolute pulse error of one pulse deteriorates the accuracy of the calculated speed in the low-speed region. It is because, in the low-speed region, the number of pulses becomes low as shown in Fig. 9.8. Thus this method is more effective in the high-speed region where there is are a large number of pulses. To reduce the speed error, it is necessary to use a large PPR or a long sampling time interval. However, a long time interval reduces the speed control bandwidth as we discussed in Chapter 2. Thus it is desirable to use an encoder with a large PPR, but it is an expensive solution.

9.2.2 T METHOD

In this method, the angular velocity is obtained by measuring the time interval T_c between two consecutive pulses to eliminate the pulse error as shown in Fig. 9.9.

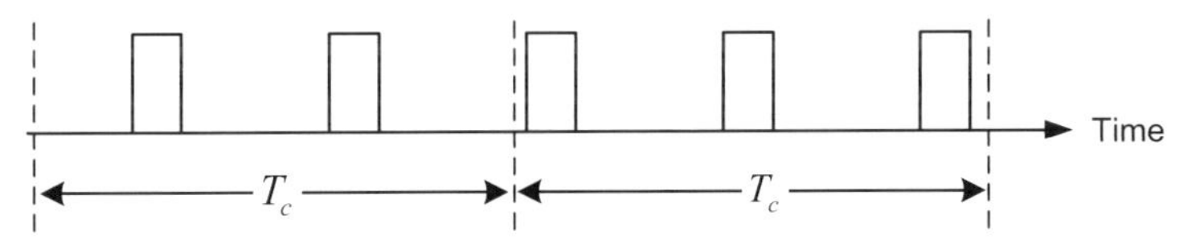

FIGURE 9.8

Pulse error in the low-speed region.

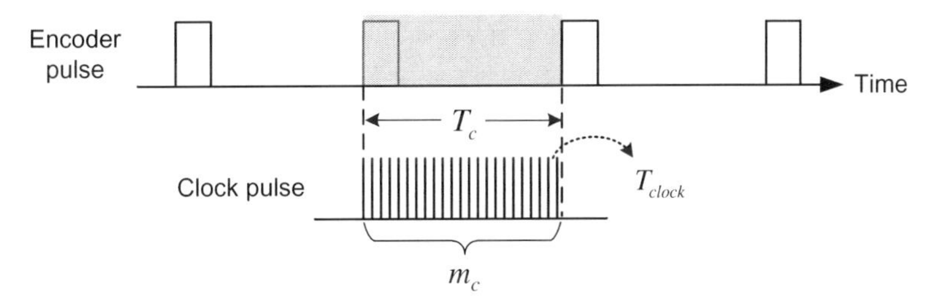

FIGURE 9.9

T method.

In the T method, the angular displacement X is always fixed as

$$X = \frac{2\pi}{PPR} \quad \text{(rad)} \tag{9.8}$$

The time interval T_c between two pulses is estimated by counting a reference clock, whose frequency is high enough compared to that of the encoder pulse. Assuming that the period of the reference clock is T_{clock} and the number of the reference clock generated for the time between two pulses is m_c, then the time T_c is given as

$$T_c = m_c \cdot T_{clock} \quad \text{(s)} \tag{9.9}$$

Thus the angular velocity is given as

$$\omega_m = \frac{X}{T} = \frac{2\pi}{PPR} \cdot \frac{1}{m_c T_{clock}} \quad \text{(rad/s)} \tag{9.10}$$

$$N = \frac{60}{2\pi} \cdot \omega_m = \frac{60}{PPR} \cdot \frac{60}{m_c \cdot T_{clock}} \quad \text{(r/min)} \tag{9.11}$$

Unlike the M method, the T method has an advantage of enabling an accurate calculation in the low-speed region because it has no pulse omitting. On the other hand, in the high-speed region, the frequency of the reference clock should be high enough to calculate the pulse period accurately. However, when such high-frequency reference clock is used, a clock counter with large bits is required for the operation in the low-speed region. Besides these, the calculation of Eq. (9.11) requires a division, which takes a long calculation time for a digital controller. Moreover, since the speed calculation time T_c varies with the speed, this method is not preferable for use for a speed controller, which carries out the speed control at each constant time interval.

9.2.3 M/T METHOD

The M/T method combines the two methods explained previously, which improves the accuracy of the speed calculation. The M/T method is now widely used in many motor drive applications for obtaining the speed information with a high resolution.

In the M/T method, similar to the M method, the encoder pulses are first counted in the constant sampling time interval T_c. However, if the sampling time does not synchronize with the last pulse, the extra time ΔT to the last pulse is additionally measured to eliminate the pulse error by adopting the T method as shown in Fig. 9.10. Thus in this method, the angular velocity can be calculated accurately by measuring the total time $T(=T_c + \Delta T)$ for the encoder pulses m_1.

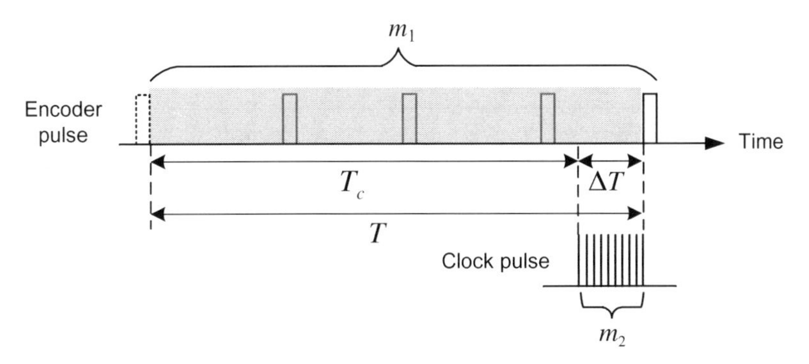

FIGURE 9.10

M/T method.

For the number of the encoder pulses m_1, the angular displacement X is given as

$$X = \frac{2\pi}{PPR} m_1 \quad \text{(rad)} \tag{9.12}$$

If the period of the reference clock is T_2 and counts of the number of the clock pulses is m_2, the total duration time for the pulses m_1 is given as

$$T = T_c + \Delta T = T_c + T_2 m_2 \quad \text{(s)} \tag{9.13}$$

Then the angular velocity is given as

$$\omega_m = \frac{X}{T} = \frac{2\pi}{PPR} \cdot \frac{m_1}{T_c + m_2 T_2} \quad \text{(rad/s)} \tag{9.14}$$

$$N = \frac{60}{2\pi} \cdot \omega_m = \frac{60}{PPR} \cdot \frac{m_1}{T_c + m_2 T_2} \quad \text{(r/min)} \tag{9.15}$$

The speed calculation based on the M/T method is more accurate than the other two methods, but its implementation is quite difficult. The M/T method has a problem in the very low-speed region. Since, in the very low-speed region, the number of the encoder pulses is too small, the extra time ΔT may be larger than the sampling time interval T_c. Thus the period for the speed calculation can vary with the speed.

The speed estimation methods introduced above cannot inherently estimate the true instantaneous speed, but only a discrete average speed over the sample interval. Since the actual speed within the sampling period interval cannot be identified, the speed control cannot work properly for a system with a long sampling period. In addition, a system with a high bandwidth of speed control may become unstable in the low-speed region, where the speed detection delay time is long. To overcome this problem, we need to estimate the instantaneous speed by using a *position observer* or a *state filter* with the discrete average speed obtained from the encoder pulses [4,5].

9.3 SENSORLESS CONTROL OF ALTERNATING CURRENT MOTORS

The vector control of AC motors requires the position of the rotor flux. The knowledge of the rotor position is used for identifying the position of the rotor flux. For this purpose, a position sensor such as an encoder or a resolver is installed at the rotor shaft as shown in Fig. 9.11.

A resolver is usually used for PMSM drives that require an absolute position of the rotor for start-up and vector control. On the other hand, an incremental encoder to measure the angular displacement of the rotor is commonly used for induction motor drives. When an incremental encoder is used for PMSMs, the initial rotor position should be provided [6].

Such position sensors cause several problems. Above all, these sensors increase the cost of the whole motor drive system. These sensors are also sensitive to the surrounding environment. The DC power lines and interface lines for a sensor increase the complexity of the system and are susceptible to noise. This may degrade the system reliability. In addition, the sensor attached to the shaft of a motor increases its size and requires maintenance periodically.

To solve this problem, a control technology without any position sensor, referred as *sensorless control*, has emerged and become an important research subject in the field of AC motor drives. The vector control using a sensor can

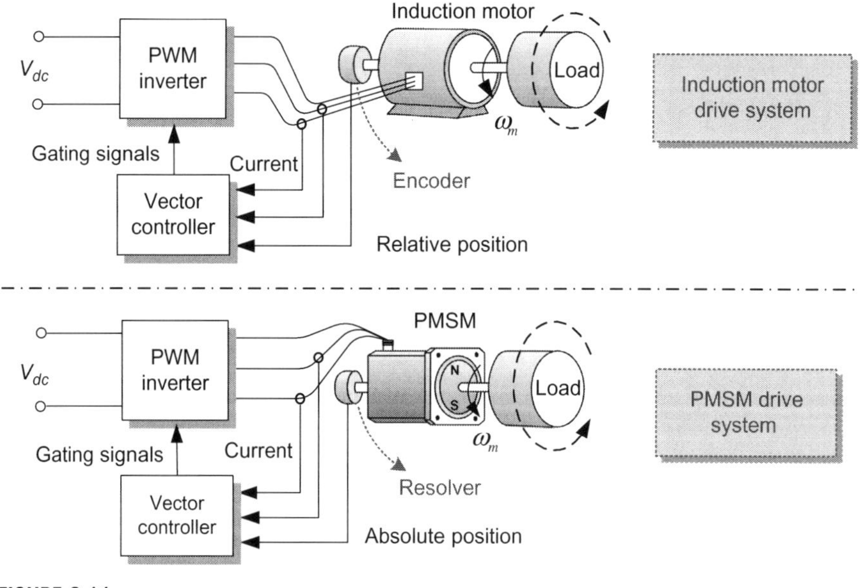

FIGURE 9.11

Position sensor for AC motor drive systems.

provide the speed accuracy of $\pm 0.01\%$ and the speed control range of 1:1000. These days, the sensorless vector control can provide a speed accuracy of $\pm 0.5\%$ and a speed control range of 1:150. The performance of the sensorless vector control exceeds that of the scalar control such as the *V/f* control, which has a speed accuracy of $\pm 1 \sim 2\%$. Nowadays, sensorless control techniques to achieve a performance comparable to that of the sensored vector control have been investigated continuously.

Understanding the sensorless control techniques requires a comprehensive knowledge of the motor control that was stated in the previous chapters 4–8. Now we will review briefly the sensorless control techniques of AC motors.

9.3.1 TYPES OF SENSORLESS CONTROL

For more than two decades, many efforts have been made to develop a sensorless drive of AC motors. These sensorless techniques are subdivided into two major groups according to their method of deriving the rotor position as shown in Fig. 9.12. One typical group is based on back-electromotive force (EMF) retaining information on the rotor speed [7–17]. The technique of this group is that the rotor position can be obtained through an estimator or observer using mathematical equations of a motor. Another group obtains the rotor position from the characteristics of a motor itself [18–24]. In this technique, a special signal to extract the rotor position-dependent characteristics from a motor is injected into the motor. High-frequency voltage is commonly used as the special signal.

9.3.1.1 Sensorless technique using the motor model

The basic concept of this sensorless technique is to use the back-EMF retaining information on the rotor speed to obtain the rotor flux position [7–17]. This sensorless technique usually uses an estimator or observer to obtain the rotor flux

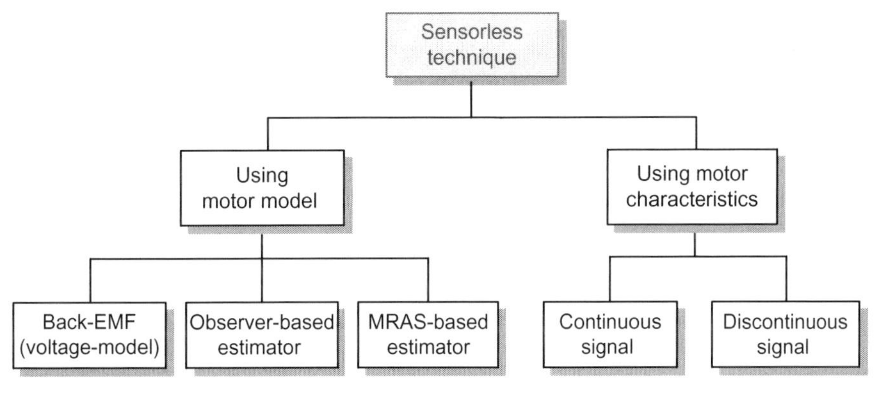

FIGURE 9.12

Different types of sensorless control methods.

position or the rotor flux linkage by using mathematical equations of a motor with its measured voltage and current.

We have already seen one example of the sensorless control of an induction motor based on this concept in Chapter 5. In the voltage model of Section 5.3.1, it was seen that the rotor flux angle could be estimated from the rotor flux linkage obtained by using the stator voltage equations. We will now review this process. The stator flux linkages are obtained first by the integral of back-EMFs, and in turn, the rotor flux linkages are estimated from the stator flux linkages as

$$\hat{\lambda}^s_{dqs} = \int (v^s_{dqs} - R_s i^s_{dqs})dt = \int e^s_{dqs}dt \rightarrow \hat{\lambda}^s_{dqr} = \frac{L_r}{L_m}(\hat{\lambda}^s_{dqs} - \sigma L_s i^s_{dqs}) \quad (9.16)$$

From the rotor flux linkages, the rotor flux angle $\hat{\theta}_e$ and the rotor speed $\hat{\omega}_r$ can be estimated as

$$\hat{\theta}_e = \tan^{-1}\left(\frac{\hat{\lambda}^s_{qr}}{\hat{\lambda}^s_{dr}}\right) \quad (9.17)$$

$$\hat{\omega}_e = \frac{d\hat{\theta}_e}{dt} = \frac{d}{dt}\left[\tan^{-1}\left(\frac{\hat{\lambda}^s_{qr}}{\hat{\lambda}^s_{dr}}\right)\right] \quad (9.18)$$

$$\hat{\omega}_r = \hat{\omega}_e - \omega_{sl} \quad (9.19)$$

Likewise, for a PMSM, the rotor flux angle $\hat{\theta}_r$ and the rotor speed $\hat{\omega}_r$ can be estimated by using the stator voltage and the stator flux linkage equations as

$$\begin{aligned} \hat{\lambda}^s_{dqs} &= \int (v^s_{dqs} - R_s i^s_{dqs})dt \\ &= \int e^s_{dqs}dt \quad (e^s_{dqs} = v^s_{dqs} - R_s i^s_{dqs}) \end{aligned} \quad (9.20)$$

$$\hat{\theta}_r = \tan^{-1}\left(\frac{\hat{\lambda}^s_{qr}}{\hat{\lambda}^s_{dr}}\right) = \tan^{-1}\left(\frac{\hat{\lambda}^s_{qs} - L_s i^s_{qs}}{\hat{\lambda}^s_{ds} - L_s i^s_{ds}}\right) \quad (9.21)$$

$$\hat{\omega}_r = \frac{d\hat{\theta}_r}{dt} \quad (9.22)$$

As we can see, the sensorless control technique based on the motor model is simple compared to the other methods using the characteristics of a motor and is capable of providing a satisfactory estimating performance in the medium- to high-speed range. However, since it is based on the back-EMF proportional to a rotor speed, its performance is inevitably limited in the low-speed range. A satisfactory performance cannot be obtained 10% below the rated speed. Especially, at zero speed where the back-EMF is equal to zero (or at zero stator frequency for an induction motor), no information on the rotor flux can be acquired, so this technique will fail to control the motor. Therefore this sensorless control technique is not an easy task to improve the performance in the low-speed range and at zero speed.

The accuracy of the estimation obtained from the sensorless control technique based on the motor model depends on the accuracy of motor parameters and input values used in the model as well as the accuracy of the model itself. The motor parameters can vary easily according to operating conditions such as winding temperature and flux level. The variation of parameters has influence on the accuracy of the estimation. For example, it can be readily seen that an accurate information on the stator resistance is required to estimate the stator flux linkage accurately from Eq. (9.16) or (9.20). Thus for an accurate sensorless control, an algorithm which provides real-time adaptation for motor parameters is necessary. The input values used in the model such as the current and voltage of the motor should also be measured precisely. Mostly, the motor currents are directly measured by using current sensors. However, since the voltages applied to the motor are generated as a pulse width modulation (PWM) waveform, it is hard to be directly measured. Thus voltage commands or output voltages calculated reversely by the switching pattern of an inverter are commonly used instead. In this case, as stated in Chapter 7, nonlinearity of an inverter due to a voltage drop on the switching devices and the dead time effect should be considered.

As a type of this sensorless control method for induction motors, the model reference adaptive control (MRAC) method, which estimates the rotor speed by comparing outputs of two models that estimate the rotor flux linkage, is a classic example [7,8]. Besides the MRAC, several techniques using the advanced control theory, such as the adaptive speed observer or Kalman filter, have been developed [9,10]. The stable operation region for such sensorless control methods is limited to 1–3 Hz due to insufficient back-EMF information in the low-speed range. However, some researches to further lower the available operating range have been proposed [11].

Several approaches used for induction motors are also adopted for the sensorless control of PMSMs such as a method estimating the rotor position from the flux obtained by an integral of back-EMF [13], a method adopting the MRAC using the error between outputs of two motor models [13,14], and a method using the advanced control theory such as the state observer or Kalman filter [15–17].

9.3.1.2 Sensorless technique using the characteristics of a motor

Sensorless control methods using the motor model mentioned previously cannot inherently avoid performance degradation in the low-speed range and at zero speed due to their dependency on back-EMF. Thus sensorless control methods using a different concept have been researched. These methods derive information on the rotor position from secondary effects of a motor such as eccentricity of a rotor, magnetic saliency, and slot harmonics [18,19]. Since these characteristics appear regardless of the rotor speed, they can be exploited for sensorless control even in the low-speed range and at zero speed.

Most of these sensorless methods exploit spatial magnetic saliency, which indicates the spatial variation of the inductance according to the rotor position

[20–25]. In this case, a special signal is injected into the motor to extract the magnetic saliency. A discontinuous pulse signal or modified PWM signal can be used as the injected signal, but a continuous high-frequency signal of a sinusoidal waveform is mainly used. Voltage rather than current has been mainly used as a type of high-frequency injection signal. The high-frequency injection voltage is added to the output of the current controller. The rotor position can be estimated from the motor's response to the injected voltage signal because the motor's response varies according to the magnetic saliency. The high-frequency signal enables us to estimate the magnetic saliency for a surface-mounted permanent magnet synchronous motor or an induction motor without magnetic saliency on its rotor as well as an interior permanent magnet synchronous motor with magnetic saliency on its rotor.

Fig. 9.13 shows a block diagram of a typical sensorless drive based on the high-frequency signal injection. The sensorless drive system consists of three parts. First one is a high-frequency signal injection part, which injects a special signal continuously to obtain the saliency information of the motor. Another is a signal processing part, which decomposes the current induced by the injected voltage and extracts the rotor position-related value. Finally, there is a position observer part, which estimates the rotor position and speed from the extracted rotor position-related value.

In the high-frequency signal injection, it is known that the voltage signal injection into the *d*-axis of the estimated rotor reference frame can result in a simpler signal processing and a better performance, even though the signal can be injected into any reference frame. The signal processing part often requires low-pass filters to extract the fundamental current and band-pass filters to extract the injected frequency component current from the measured motor current. Time delay by these filters limits the control bandwidth of the sensorless drives. The bandwidth of the speed controller is usually only about several Hertz.

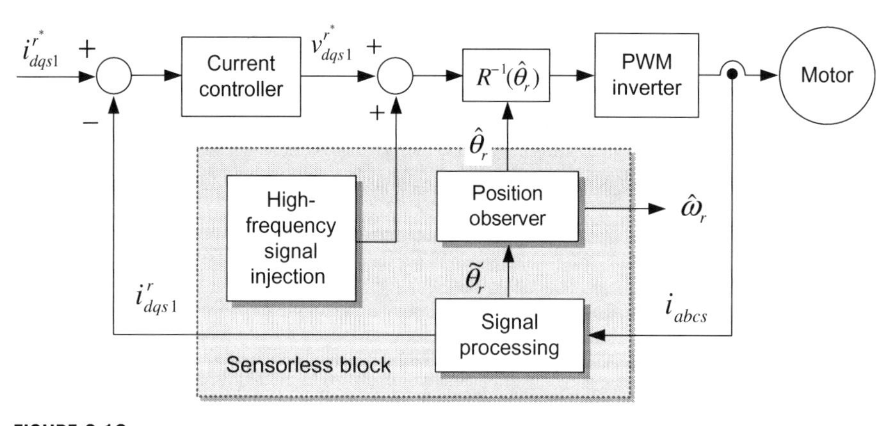

FIGURE 9.13

Typical sensorless drive system based on the high-frequency signal injection.

Recently, instead of the conventional sinusoidal-type signal injection, a square wave–type voltage injection has been proposed to eliminate the low-pass filters for the signal processing and enhance the control bandwidth of the sensorless drives. In these methods, the control bandwidth is very close to the value available in sensored drives [22–24].

The sensorless control methods using magnetic saliency and high-frequency voltage injection give a complex signal processing to estimate the rotor position information, requiring a high-performance microprocessor. In addition, the voltage injected to extract the rotor position produces an additional loss, acoustic noise, and torque ripple. Another problem is that there is a voltage shortage in the high-speed operation because an extra voltage is used to generate the injection signal. Thus this leads to a loss of high-speed operation. To overcome this problem, hybrid sensorless methods combing the two techniques mentioned above have been proposed for operation in the entire speed range [25]. The hybrid sensorless methods use the high-frequency signal injection technique at low-speed region and at zero speed but the back-EMF technique at a higher speed region, in which sufficient back-EMF information is available. In the hybrid methods, a transition between the two techniques is an important issue.

REFERENCES

[1] <http://www.amci.com/tutorials/tutorials-what-is-resolver.asp>.

[2] D.C. Hanselman, Resolver signal requirement for high accuracy resolver-to-digital conversion, IEEE Trans. Ind. Electron. 37 (6) (Dec. 1990) 556–561.

[3] H.-S. Mok, S.-H. Kim, Y.-H. Cho, Torque ripple reduction of PMSM caused by position sensor error for EPS application, Electron. Lett. 43 (11) (2007) 646–647.

[4] H.W. Kim, S.K. Sul, A new motor speed estimator using Kalman filter in low-speed range, IEEE Trans. Ind. Electron. 43 (4) (Aug. 1996) 498–504.

[5] R.D. Lorenz, K.V. Patten, High-resolution velocity estimation for all-digital AC servo drives, IEEE Trans. Ind. Appl. 27 (4) (Aug. 1991) 701–705.

[6] N.-C. Park, Y.-K. Lee, S.-H. Kim, Initial rotor position estimation for an interior permanent-magnet synchronous motor using inductance saturation, Trans. Korean Inst. Power Electron. 16 (4) (Aug. 2011) 374–381.

[7] C. Schauder, Adaptive speed identification for vector control of induction motors without rotational transducers, IEEE Trans. Ind. Appl. 28 (5) (Sep./Oct. 1992) 1054–1061.

[8] T. Ohtani, N. Takada, K. Tanaka, Vector control of induction motor without shaft encoder, IEEE Trans. Ind. Appl. 28 (1) (Jan./Feb., 1992) 157–164.

[9] H. Kubota, K. Matsuse, T. Nakano, DSP-based adaptive flux observer of induction motor, IEEE Trans. Ind. Appl. 29 (2) (Mar./Apr. 1993) 344–348.

[10] Y.R. Kim, S.K. Sul, M.H. Park, Speed sensorless vector control of induction motor using extended Kalman filter, IEEE Trans. Ind. Appl. 30 (5) (Sep./Oct. 1994) 1225–1233.

[11] J. Holtz, Sensorless control of induction motor drives, Proc. IEEE 90 (8) (Aug. 2002) 1359–1394.

[12] R. Wu, G.R. Slemon, A permanent magnet motor drive without a shaft sensor, IEEE Trans. Ind. Appl. 27 (5) (Sep./Oct. 1991) 1005−1011.

[13] N. Matsui, M. Shigyo, Brushless DC motor control without position and speed sensor, IEEE Trans. Ind. Appl. 28 (1) (1992) 120−127.

[14] N. Matsui, T. Takeshita, K. Yasuda, A new sensorless drive of brushless DC motor, in: Proc. 1992 Int. Conf. IECON, 1992, pp. 430−435.

[15] L.A. Jones, J.H. Lang, A state observer for the permanent-magnet synchronous motor, IEEE Trans. Ind. Electron. 36 (3) (1989) 374−382.

[16] Z. Chen, M. Tomita, S. Doki, S. Okuma, An extended electromotive force model for sensorless control of interior permanent magnet synchronous motors, IEEE Trans. Ind. Appl. 38 (2) (2003) 288−295.

[17] S. Bolognani, R. Oboe, M. Zigliotto, Sensorless full-digital PMSM drive with EMF estimation of speed and rotor position, IEEE Trans. Ind. Electron. 46 (2) (1999) 240−247.

[18] R. Blasco-Gimenez, G.M. Asher, M. Sumner, K.J. Bradley, Performance of FFT-rotor slot harmonic speed detector for sensorless induction motor drives, IEE Proc. Elect. Power Appl. 143 (3) (May 1996) 258−268.

[19] M. Schroedl, Sensorless control of AC machines at low speed and standstill based on "INFORM" method, in: Conf. Rec. IEEE IAS Annu. Meeting, 1996, pp. 270−277.

[20] J.-I. Ha, S.-K. Sul, Sensorless field-orientation control of an induction, IEEE Trans. Ind. Appl. 35 (1) (Jan./Feb. 1999) 45−51.

[21] J. Holtz, Sensorless control of induction machines-with or without signal injection?, IEEE Trans. Ind. Electron. 53 (1) (2006) 7−30.

[22] Y.-D. Yoon, S.-K. Sul, S. Morimoto, K. Ide, High-bandwidth sensorless algorithm for AC machines based on square-wave-type voltage injection, IEEE Trans. Ind. Appl. 47 (3) (2011) 1361−1370.

[23] S.-M. Kim, J.-I. Ha, S.-K. Sul, PWM switching frequency signal injection sensorless method in IPMSM, IEEE Trans. Ind. Appl. 48 (5) (2012) 1576−1586.

[24] N. Park, S.-H. Kim, A simple sensorless algorithm for IPMSMs based on high-frequency voltage injection method, IET Elect. Power Appl. 8 (2) (2014) 68−75.

[25] K. Ide, J.-I. Ha, M. Sawamura, A hybrid speed estimator of flux observer for induction motor drives, IEEE Trans. Ind. Electron. 53 (1) (2006) 130−137.

CHAPTER

Brushless direct current motors

10

As explained in Chapter 2, direct current (DC) motors need mechanical commutation devices consisting of brushes and commutators that change the direction of the current of conductors to produce an average torque for continuous rotation as shown in Fig. 10.1. However, this mechanical commutation causes an electromagnetic and acoustic noise. The commutators and brushes also need a periodical maintenance and replacement due to wear-out and flashover. Small DC motors with a radius of $\phi25-\phi34$ can operate about 1000 hours, whereas ones with a radius of $\phi37-\phi60$ can operate about 2000 hours. Specially designed DC motors are capable of operating about 3000 hours.

DC motors have been widely used for speed or position control applications because of their control simplicity when compared with alternating current (AC) motors. However, since they require periodic maintenance of brushes and commutators, their utilization has been reduced in many motor drive applications, which require continuous running and enhanced system reliability.

To overcome this problem of DC motors, a motor called *brushless direct current (BLDC) motor* was developed in 1962 [1]. This motor has similar electrical characteristics to a DC motor, but it has an enhanced reliability by replacing mechanical commutation with electronic commutation. To implement the electronic commutation, BLDC motors use sensors and driving circuits. The sensors detect the position of magnets on the rotor. By using the detected magnet position, the driving circuits excite a specific winding for continuous rotation.

In BLDC motors, to eliminate the brushes of DC motors, the armature windings are placed on the stator side and the magnets are placed on the rotor side. As a result, BLDC motors have a different configuration from that of DC motors. Since there is a degree of freedom in the motor configuration when designing to eliminate brushes, various BLDC motor designs to fit a wide application needs such as a smaller or thinner configuration are possible. The BLDC motors have many merits such as high efficiency, high-power density, high torque-to-inertia ratio, high-speed operation capability, simple drive method, and low cost. Thus, nowadays, they are widely used for cost-effective solution in many small and medium motor drive applications such as home appliances, industrial, office products, and light vehicles.

In this chapter, we will discuss the configuration, driving principle, mathematical model, speed and current control, and sensorless techniques of BLDC motors.

Electric Motor Control. DOI: http://dx.doi.org/10.1016/B978-0-12-812138-2.00010-6

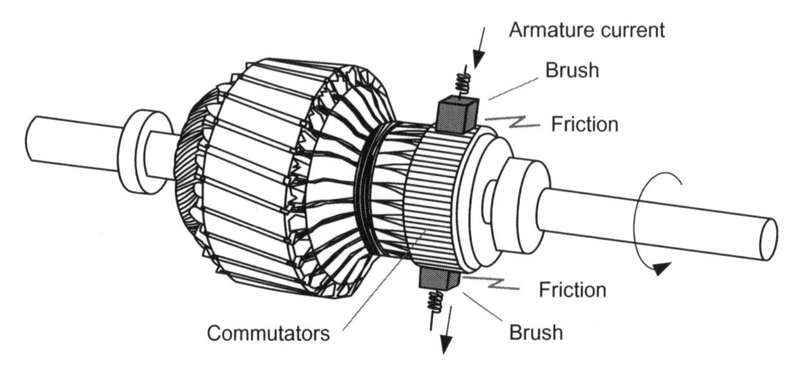

FIGURE 10.1

Mechanical commutation devices of a DC motor.

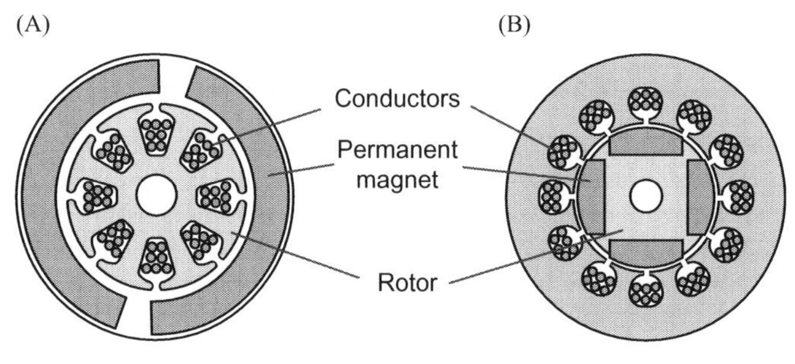

FIGURE 10.2

Configurations of (A) DC motor and (B) BLDC motor.

10.1 CONFIGURATION OF BRUSHLESS DIRECT CURRENT MOTORS

10.1.1 COMPARISON BETWEEN BRUSHLESS DIRECT CURRENT MOTORS AND DIRECT CURRENT MOTORS

BLDC motors do not have the crucial weakness of DC motors because the function of brushes and commutators is replaced with semiconductor switches operating based on the information from the rotor position. As a result of this replacement, the configuration of a BLDC motor becomes different from that of a DC motor, but similar to the configuration of a permanent magnet synchronous motor (PMSM). As shown in Fig. 10.2B, the BLDC motor has a configuration in which the windings are placed on the stator side and the magnets on the rotor side. This results in the reversed configuration of a DC motor as shown in Fig. 10.2A. Instead, its configuration resembles that of a PMSM, shown in Fig. 4.8.

However, similar to the current flowing in armature windings of a DC motor, the current flowing in the windings of a BLDC motor is a quasi-square waveform.

Such configuration of BLDC motors has the following merits over DC motors. Compared to the heavy rotor of DC motors consisting of many conductors, BLDC motors have a low inertia rotor. Thus BLDC motors can provide a rapid speed response. Moreover, windings placed on the stator side can easily dissipate heat, allowing BLDC motors to have a better attainable peak torque capability compared to DC motors whose maximum current is limited to avoid the demagnetization of magnets. In addition, BLDC motors can operate at a higher speed because of nonmechanical commutation devices.

10.1.2 COMPARISON BETWEEN BRUSHLESS DIRECT CURRENT MOTORS AND PERMANENT MAGNET SYNCHRONOUS MOTORS

Due to their structural similarity, BLDC motors are often confused with PMSMs. Commonly, BLDC motors can be distinguished from PMSMs by their shape of back-electromotive force (back-EMF). A BLDC motor is designed to develop a trapezoidal back-EMF waveform as shown in Fig. 10.3A. Thus the amplitude of the magnetic flux density generated by the rotor magnets remains constant along the air gap. This can be achieved by using magnets with parallel magnetization.

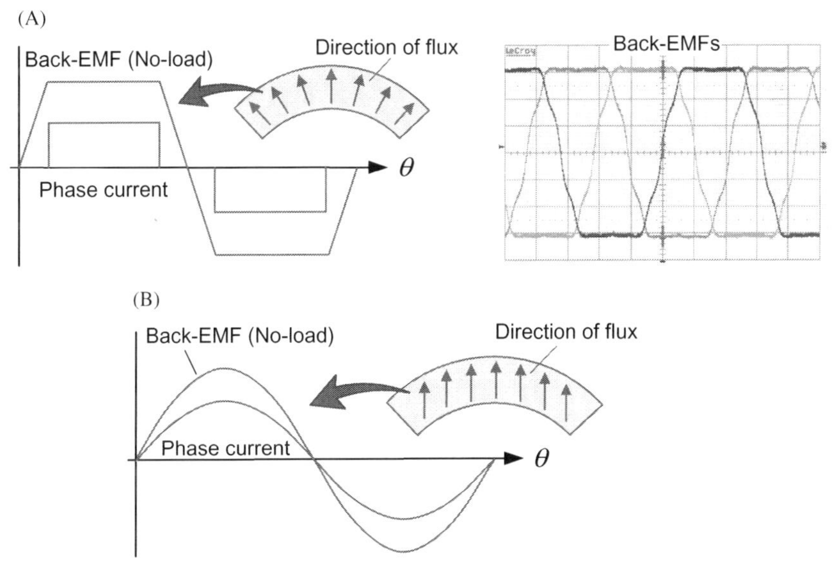

FIGURE 10.3

Comparison between (A) BLDC motors and (B) PMSMs.

When a BLDC motor with a trapezoidal back-EMF waveform is fed with a rectangular stator current, a constant torque can be developed. On the other hand, a PMSM is a type of AC motor with a sinusoidal back-EMF waveform, and thus its current should be a sinusoidal waveform for constant torque generation as shown in Fig. 10.3B.

A sinusoidal back-EMF waveform requires the magnetic flux density generated by the rotor magnets to be distributed sinusoidally along the air gap [2]. A PMSM is often referred to as *BLAC motor*, in contrast to BLDC motors.

Power density of a BLDC motor with a trapezoidal back-EMF waveform is 15% higher than that of a PMSM [3]. This is because the trapezoidal waveform has a higher fundamental component than the sinusoidal waveform, even though they have the same peak value. Besides the back-EMF waveform, there are several differences between BLDC motors and PMSMs. The comparison is listed in Table 10.1.

As shown in Fig. 10.4, a BLDC motor commonly uses concentrated stator windings, and quasi-square waveform current flows into them. In contrast, a PMSM usually uses distributed stator windings, into which the sinusoidal waveform current flows as shown in Chapter 3. However, recently, PMSMs often use concentrated windings due to their short end windings and simple structure suitable for automated manufacturing.

There is a clear difference in drive methods for these two motors. In the excitation of a three-phase BLDC motor, the phase currents flow only in two of the three-phase windings at a time and thus, each switch of the inverter always operates for a 120° conduction interval per fundamental operating cycle. By contrast, as described in Chapter 7, in a three-phase PMSM, the phase currents flow in all three-phase windings at a time and thus, each switch of the inverter always operates for a 180° conduction interval per fundamental operating cycle.

Table 10.1 Comparison Between BLDC Motors and PMSMs

	BLDC Motor	PMSM
Back-EMF	Trapezoidal waveform	Sinusoidal waveform
Stator winding	Concentrated winding	Distributed winding
Stator current	Quasi-square waveform	Sinusoidal waveform
Driving circuit	Inverter (120° conduction)	Three-phase inverter (180° conduction)
Drive method	Simple, using low-cost Hall effect sensors	Complex (using high-resolution position sensor such as an encoder or a resolver)
Torque ripple	Significant torque ripple	Nearly constant torque
System cost	Low cost	High cost

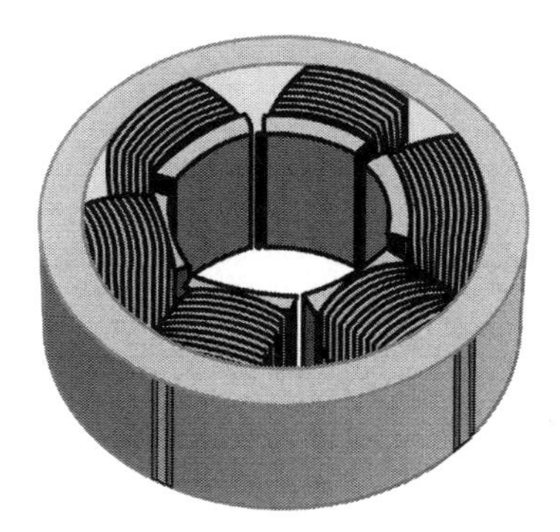

FIGURE 10.4

Concentrated stator windings of BLDC motors.

The BLDC motor drive is simple and inexpensive compared with the PMSM drive, which requires a complex method of the vector control as described in Chapter 5. However, the BLDC motors have a significant torque ripple during the phase commutation of changing an active switch, which will be explained later.

10.1.3 CONSTRUCTION OF BRUSHLESS DIRECT CURRENT MOTORS

BLDC motors can be categorized according to their number of stator windings: single-phase, two-phase, three-phase, and multi-phase.

Single-phase BLDC motors are widely used for appliances and small machines due to their simple structure, simple driving circuit, and low cost. However, such motors can rotate only in one direction. Furthermore, since they have a detent point where there is no starting torque, a motor design including an ancillary part is necessary for start-up. A reluctance torque is usually used as the starting torque, resulting in a large cogging torque. Single-phase BLDC motors are beneficial to small power applications below 10 W such as fans and blowers that require a low starting torque.

Multi-phase BLDC motors above four-phase can be mainly applied to aerospace and military applications requiring high reliability due to increased power density and fault-tolerance capability. Three-phase BLDC motors are the most widely used and our discussion will be limited to these motors in this book.

There are two BLDC motor designs which are classified in terms of magnetic flux direction: *radial-flux type*, where the flux from the rotor magnet crosses the air gap in a radial direction and *axial-flux type*, where the flux crosses the air gap in an axial direction as shown in Fig. 10.5.

Traditionally the radial-flux motors have been used almost exclusively. This type of motor can be either an *inner rotor type* or an *outer rotor type* as shown in Fig. 10.5A and B. As the most common type, the inner rotor design has an advantage of higher heat dissipating capacity, high torque-to-inertia ratio, and lower rotor inertia. Its common application is servo drives requiring a quick dynamic response.

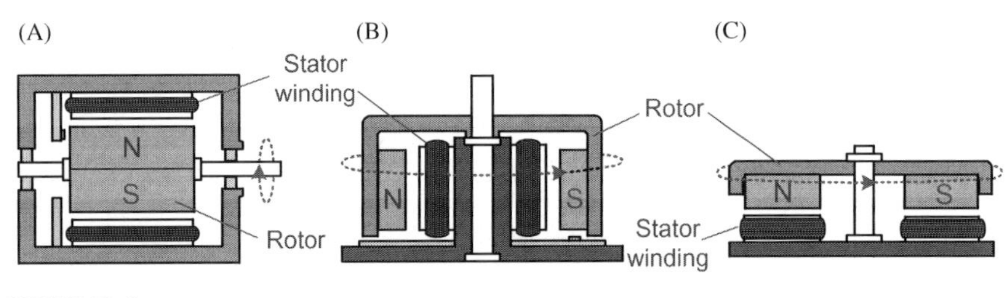

FIGURE 10.5

Classification of BLDC motors. (A) Radial-flux (inner rotor), (B) radial-flux (outer rotor), and (C) axial-flux.

Since the rotor is cylindrical in shape with a shaft on which the bearings are mounted, this type of motor can produce less vibration and acoustic noise. Outer rotor motors shown in Fig. 10.5B, in which the rotor magnets rotate around the stator windings located in the iron core of the motor, have relatively high rotor inertia. Thus this motor is favorable for systems requiring constant speed operation. The magnets affixed inside the yoke are beneficial to high-speed operation. The outer rotor motor can use more magnetic material than the inner rotor device, which means it is capable of more flux even though lower energy product magnets are used. These days, this type of motor is increasingly being used in many applications such as computer disk drives, cooling fans, and washing machines.

In the axial-flux motors shown in Fig. 10.5C, a disc rotor with magnets whose flux is in an axial direction rotates facing the stator. Permanent magnets are glued to the surface of the rotor. In this design topology small motors often have coreless stator windings mounted to a nonmagnetic substrate or slotless stator windings mounted to an iron core without slots. The primary advantage of such a technology is that it has a very low ripple torque and acoustic noise since the cogging torque associated with typical iron core motors can be eliminated due to no variations in reluctance. Their relatively high rotor inertia is favorable to constant speed operation. However, due to increased effective air gap, the available flux is somewhat low. Common applications are VCR and CD player drives. This axial-flux motor with a slimmer structure of shorter axial length is very suitable for applications in which the axial length of the motor is the limiting design parameter, or the motor is directly coupled to the driven load. Such applications include electrical vehicles in-wheel motors and elevator motors.

10.2 DRIVING PRINCIPLE OF BRUSHLESS DIRECT CURRENT MOTORS

The basic driving principle of the BLDC motor is to change the phase windings, which should be excited according to the position of permanent magnet on the

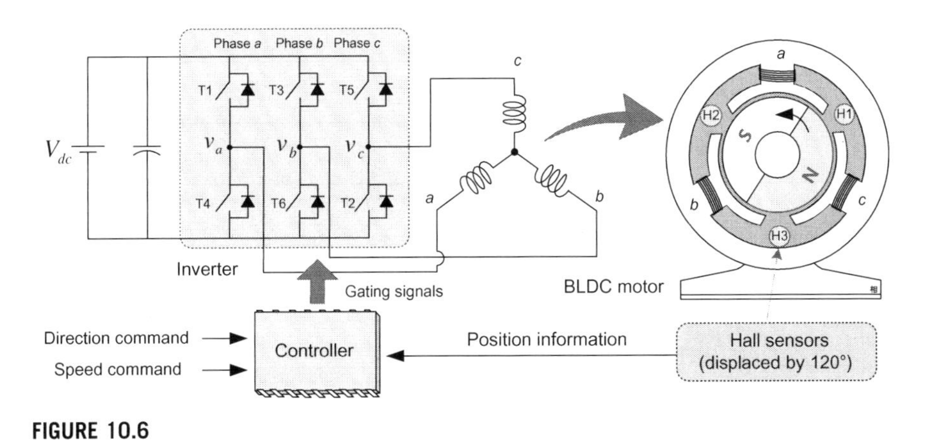

FIGURE 10.6

BLDC motor drive system.

rotor for producing a continuous torque. To implement this function, information on the rotor magnet position is indispensable. The position sensing is commonly achieved with Hall effect sensors.

A two-pole three-phase Y-connected BLDC motor drive system is shown in Fig. 10.6. Three Hall effect sensors are displaced from each other by 120 electrical degrees on the stator to detect the magnetic field flux produced from the rotor magnets. Three output signals of Hall effect sensors enable us to recognize the rotor position divided into six different sections. Accordingly, a basic drive (often called *six-step drive*) to complete one electrical cycle consists of six different sections.

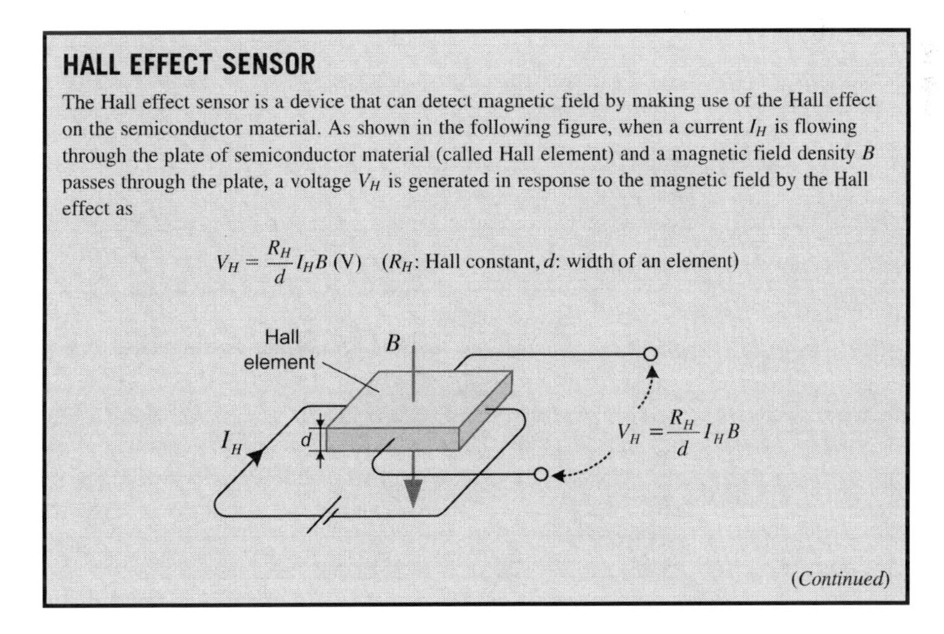

HALL EFFECT SENSOR

The Hall effect sensor is a device that can detect magnetic field by making use of the Hall effect on the semiconductor material. As shown in the following figure, when a current I_H is flowing through the plate of semiconductor material (called Hall element) and a magnetic field density B passes through the plate, a voltage V_H is generated in response to the magnetic field by the Hall effect as

$$V_H = \frac{R_H}{d} I_H B \text{ (V)} \quad (R_H\text{: Hall constant}, d\text{: width of an element})$$

(Continued)

HALL EFFECT SENSOR (CONTINUED)

From measuring the voltage V_H, the position and polarity of a magnetic field can be detected. Thus the Hall effect sensors are used to detect the rotor position of permanent magnet motors such as BLDC motors and PMSMs. They are also used for current sensors as shown in the following figure. From measuring the output voltage of the Hall effect sensor, the magnitude and the direction of the current, which produces the magnetic field, can be detected.

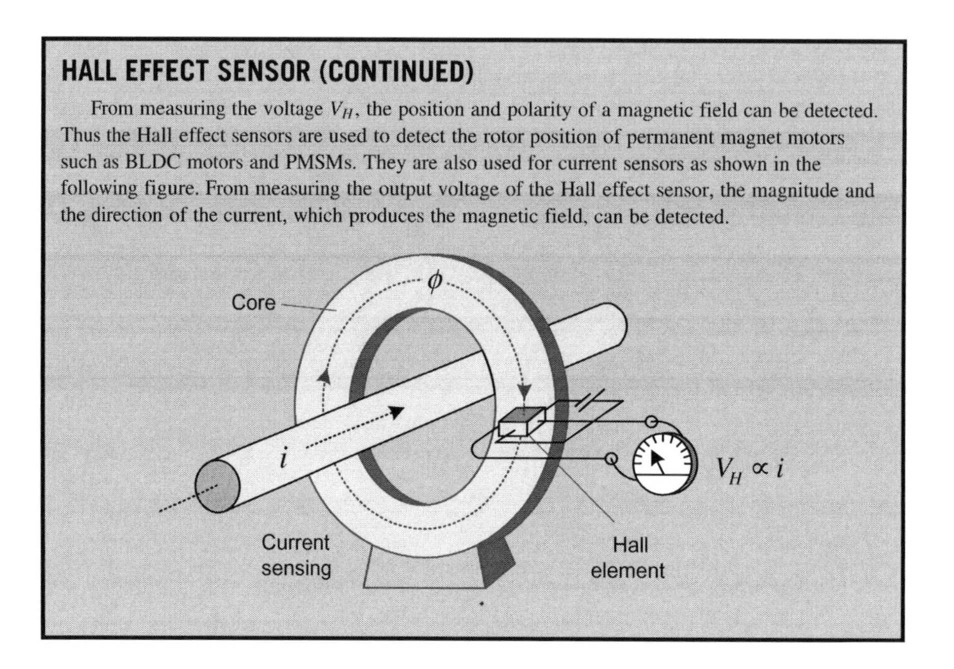

The switching sequence for the six-step drive is illustrated in Fig. 10.7. In the BLDC motor drive, only two of the three-phase windings are excited, while the other winding is left unexcited. This is the difference from the inverter drive method for AC motors, which is mentioned in Chapter 7. Rotor position feedback signals can be used to determine which two of the three-phase windings should be excited to produce the continuous torque at each instant. As a driving circuit, a three-phase inverter is used to flow the current into the required two-phase windings. In the inverter for a BLDCM drive as shown in Fig. 10.7, switching devices of only two phases work at any given instant. Accordingly, each switching device has a 120° conduction interval. In the six-step drive, a changeover of an active switch is done to the switch of the other phase, and thus a dead time is not required for short-circuit protection in the inverter.

Fig. 10.8 illustrates Hall effect sensor signals (H_1, H_2, H_3) with respect to back-EMFs of stator windings in the six-step drive as shown in Fig. 10.7, and the relationship between the sensor signals and the phase currents. Here, assume that each sensor outputs a digital high level for the north pole, whereas it outputs a low level for the south pole. From the sensor signals, the excited phase winding, i.e., an active phase winding, should be changed every 60 electrical degrees of rotation for producing a continuous torque. The transition of an active phase winding is called *commutation*. For the reverse rotation, the switching sequence with respect to Hall effect sensor signals should be altered.

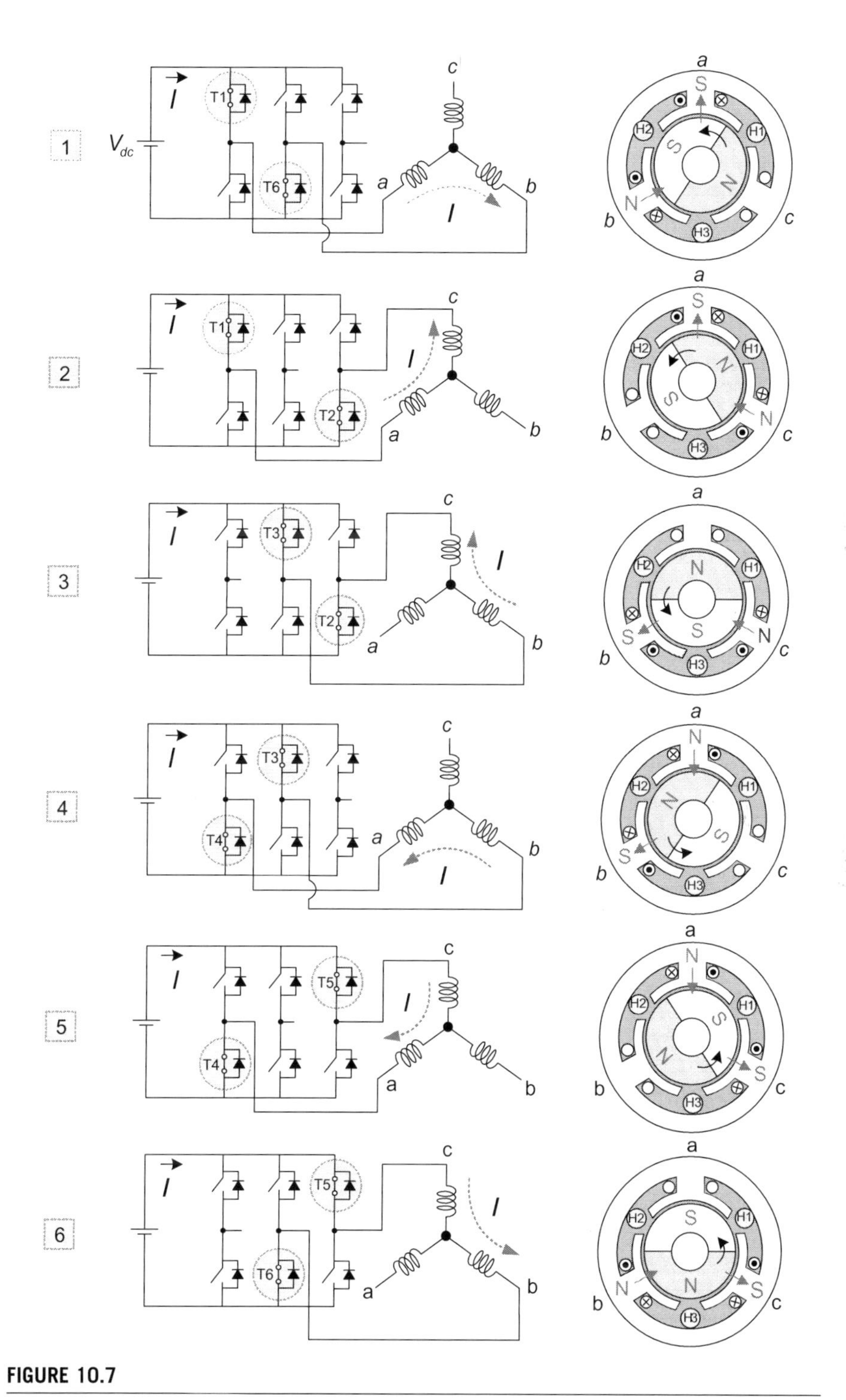

FIGURE 10.7

Switching sequence for two-pole three-phase BLDC motor.

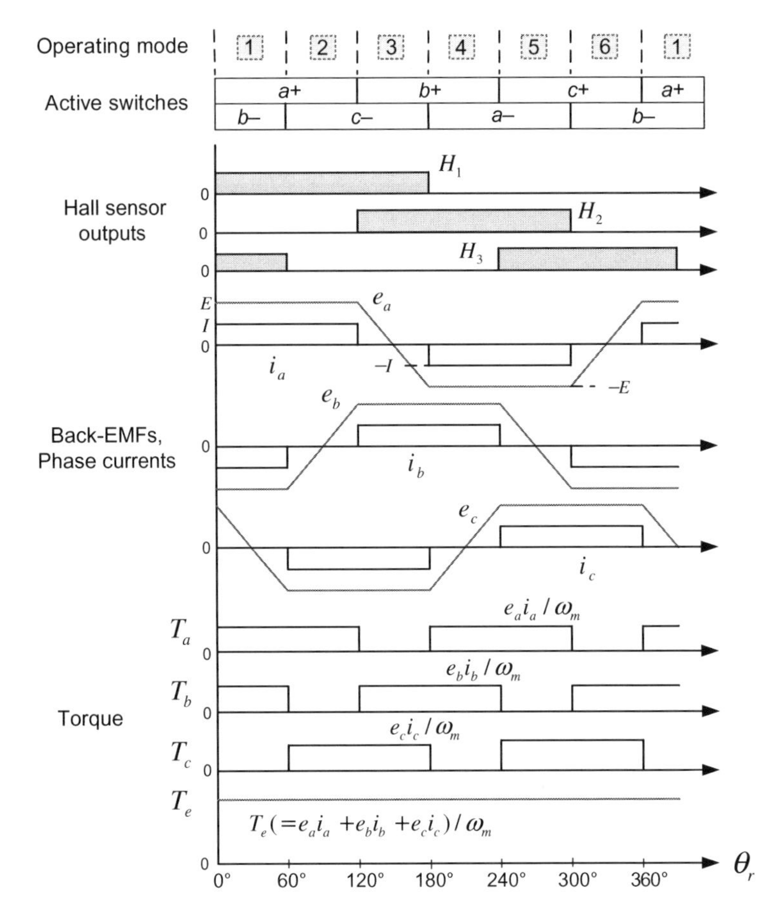

FIGURE 10.8

Driving principle of three-phase BLDC motor.

10.3 MODELING OF BRUSHLESS DIRECT CURRENT MOTORS

In this section, we will derive the mathematical model of a three-phase BLDC motor [3]. The model of a BLDC motor is similar to that of a PMSM due to their structural similarity.

10.3.1 VOLTAGE EQUATIONS

Let us consider a two-pole three-phase BLDC motor as shown in Fig. 10.9. Similar to AC motors examined in Chapter 4, the voltage equation for the stator windings of a BLDC motor can be expressed as

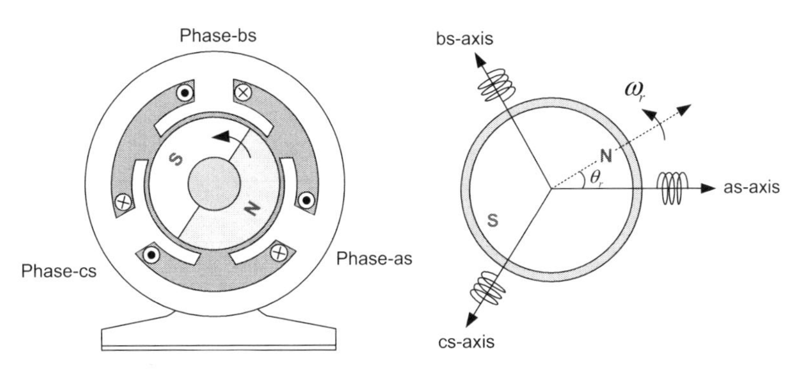

FIGURE 10.9

Windings of a two-pole three-phase BLDC motor.

$$\boldsymbol{v}_{abcs} = \boldsymbol{R}_s \boldsymbol{i}_{abcs} + \frac{d\boldsymbol{\lambda}_{abcs}}{dt} \quad (10.1)$$

where the stator voltage $\boldsymbol{v}_{abcs} = [v_{as} v_{bs} v_{cs}]^T$, the stator current $\boldsymbol{i}_{abcs} = [i_{as} i_{bs} i_{cs}]^T$, the stator flux linkage $\boldsymbol{\lambda}_{abcs} = [\lambda_{as} \lambda_{bs} \lambda_{cs}]^T$, and the stator resistance $\boldsymbol{R}_s = \begin{bmatrix} R_s & 0 & 0 \\ 0 & R_s & 0 \\ 0 & 0 & R_s \end{bmatrix}$.

The flux linkage $\boldsymbol{\lambda}_{abcs}$ of the stator windings consists of $\boldsymbol{\lambda}_{abcs(s)}$ due to the stator currents $\boldsymbol{i}_{abcs}$ and $\boldsymbol{\lambda}_{abcs(f)}$ due to the permanent magnet as

$$\boldsymbol{\lambda}_{abcs} = \boldsymbol{\lambda}_{abcs(s)} + \boldsymbol{\lambda}_{abcs(f)} \quad (10.2)$$

Substituting Eq. (10.2) into Eq. (10.1) gives the following stator voltage equation.

$$\boldsymbol{v}_{abcs} = \boldsymbol{R}_s \boldsymbol{i}_{abcs} + \frac{d\boldsymbol{\lambda}_{abcs}}{dt} = \boldsymbol{R}_s \boldsymbol{i}_{abcs} + \frac{d\boldsymbol{\lambda}_{abc(s)}}{dt} + \boldsymbol{e}_{abcs} \quad (10.3)$$

where the back-EMF due to the magnet flux is expressed as $\boldsymbol{e}_{abcs} = d\boldsymbol{\lambda}_{abc(f)}/dt$ and is also given as

$$\boldsymbol{e}_{abcs} = \begin{bmatrix} e_{as} \\ e_{bs} \\ e_{ac} \end{bmatrix} = \omega_m \begin{bmatrix} \lambda_{asf} \\ \lambda_{bsf} \\ \lambda_{csf} \end{bmatrix} = \omega_m \lambda_f \begin{bmatrix} f(\theta_r) \\ f(\theta_r - 120^\circ) \\ f(\theta_r - 240^\circ) \end{bmatrix} \quad (10.4)$$

where $\lambda_f (= N\phi_f)$ is the amount of the magnet flux ϕ_f linking N turns of the stator windings, $f(\theta_r)$ is a unit function representing the waveform of the back-EMF and θ_r is the rotor position.

The unit function for the trapezoidal back-EMF waveform of a BLDC motor can be expressed as

$$f(\theta_r) = \begin{cases} 6\theta_r/\pi & (0 \leqq \theta_r < \pi/6) \\ 1 & (\pi/6 \leqq \theta_r < 5\pi/6) \\ -6\theta_r/\pi & (5\pi/6 \leqq \theta_r < 7\pi/6) \\ -1 & (7\pi/6 \leqq \theta_r < 11\pi/6) \\ 6\theta_r/\pi - 12 & (11\pi/6 \leqq \theta_r < 2\pi) \end{cases} \quad \left(\theta_r = \int \omega_{rm}(t)\,dt\right) \tag{10.5}$$

The stator flux linkage $\boldsymbol{\lambda}_{abcs(s)}$ due to the stator currents is given by

$$\boldsymbol{\lambda}_{abcs(s)} = \boldsymbol{L}_s \boldsymbol{i}_{abcs} = \begin{bmatrix} L_{aa} & L_{ab} & L_{ac} \\ L_{ba} & L_{bb} & L_{bc} \\ L_{ca} & L_{cb} & L_{cc} \end{bmatrix} \begin{bmatrix} i_{as} \\ i_{bs} \\ i_{cs} \end{bmatrix} \tag{10.6}$$

As can be seen in Section 4.1.1, for symmetry three-phase windings, the self-inductances are all the same and the mutual-inductances are all the same as in the following

$$L_{aa} = L_{bb} = L_{cc} = L_s = L_{ls} + L_m \tag{10.7}$$

$$L_{ab} = L_{ac} = L_{ba} = L_{bc} = L_{ca} = L_{cb} = -\frac{1}{2}L_m = M \tag{10.8}$$

where $L_{\alpha\beta}(=\lambda/i_\beta)$ expresses the winding inductance, which is the ratio of the flux linkage λ of the winding α to the current i_β that produces the flux. Here, the values of leakage inductance L_{ls} and magnetizing inductance L_m are the same as those of a PMSM described in Chapter 4.

From Eqs. (10.6)–(10.8), the stator voltage equation is rewritten as

$$\boldsymbol{v}_{abcs} = \boldsymbol{R}_s \boldsymbol{i}_{abcs} + \boldsymbol{L}_{abcs} \frac{d\boldsymbol{i}_{abc(s)}}{dt} + \boldsymbol{e}_{abcs} \tag{10.9}$$

$$\begin{bmatrix} v_{as} \\ v_{bs} \\ v_{cs} \end{bmatrix} = \begin{bmatrix} R_s & 0 & 0 \\ 0 & R_s & 0 \\ 0 & 0 & R_s \end{bmatrix} \begin{bmatrix} i_{as} \\ i_{bs} \\ i_{cs} \end{bmatrix} + \begin{bmatrix} L_s & M & M \\ M & L_s & M \\ M & M & L_s \end{bmatrix} \frac{d}{dt} \begin{bmatrix} i_{as} \\ i_{bs} \\ i_{cs} \end{bmatrix} + \begin{bmatrix} e_{as} \\ e_{bs} \\ e_{cs} \end{bmatrix} \tag{10.10}$$

here, since $i_{as} + i_{bs} + i_{cs} = 0$, the mid-term of Eq. (10.10) is reduced as

$$\frac{d}{dt}[L_s i_{as} + M i_{bs} + M i_{cs}] = \frac{d}{dt}[L_s i_{as} - M i_{as}] \tag{10.11}$$

Thus Eq. (10.10) becomes the following voltage equations.

$$\begin{bmatrix} v_{as} \\ v_{bs} \\ v_{cs} \end{bmatrix} = \begin{bmatrix} R_s & 0 & 0 \\ 0 & R_s & 0 \\ 0 & 0 & R_s \end{bmatrix} \begin{bmatrix} i_{as} \\ i_{bs} \\ i_{cs} \end{bmatrix} + \begin{bmatrix} L_s - M & 0 & 0 \\ 0 & L_s - M & 0 \\ 0 & 0 & L_s - M \end{bmatrix} \frac{d}{dt} \begin{bmatrix} i_{as} \\ i_{bs} \\ i_{cs} \end{bmatrix} + \begin{bmatrix} e_{as} \\ e_{bs} \\ e_{cs} \end{bmatrix} \tag{10.12}$$

Voltage Equations of a Brushless Direct Current Motor

$$v_{as} = R_s i_{as} + (L_s - M)\frac{di_{as}}{dt} + e_{as} \tag{10.13}$$

$$v_{bs} = R_s i_{bs} + (L_s - M)\frac{di_{bs}}{dt} + e_{bs} \tag{10.14}$$

$$v_{cs} = R_s i_{cs} + (L_s - M)\frac{di_{cs}}{dt} + e_{cs} \tag{10.15}$$

Unlike AC motors that use the $d-q$ axes reference frame to facilitate their control, BLDC motors directly use three-phase abc quantities for their control. The $d-q$ transformation is necessary for the sinusoidal quantities. Since currents, flux, and back-EMFs of a BLDC motor are nonsinusoidal quantities, the $d-q$ transformation on the BLDC motor is meaningless.

10.3.2 TORQUE EQUATION

The output torque T_e of a three-phase motor is generally calculated from the output power P_e and the mechanical angular velocity ω_m as

$$P_e = e_{as}i_{as} + e_{bs}i_{bs} + e_{cs}i_{cs} \tag{10.16}$$

$$T_e = \frac{P_e}{\omega_m} = \frac{e_{as}i_{as} + e_{bs}i_{bs} + e_{cs}i_{cs}}{\omega_m} \tag{10.17}$$

here, the mechanical angular velocity $\omega_m = \omega_r/P$. In addition, ω_r is the electrical angular velocity and P is the number of pole.

From this torque equation, we can readily see that the phase current is needed to be in phase with the back-EMF to produce the maximum torque. The drive shown in Fig. 10.8 satisfies this requirement for maximum torque production.

Now we will discuss the output torque of a three-phase BLDC motor driven by the 120° conduction method. In a BLDC motor, the currents of two-phase windings are the same in magnitude but flow in the reverse direction except during the commutation interval. As an example, consider Section 1 shown in Fig. 10.8. In that section, $i_{as} = I$, $i_{bs} = -I$, $i_{cs} = 0$ and $e_{as} = E$, $e_{bs} = -E$. Since the phase current is in phase with the back-EMF, the torque of Eq. (10.17) can be simply expressed as the product of the magnitudes of the phase current and back-EMF as

$$T_e = \frac{e_{as}i_{as} + e_{bs}i_{bs} + e_{cs}i_{cs}}{\omega_m} = 2\frac{EI}{\omega_m} \tag{10.18}$$

For the remaining sections, we can obtain the same result.

If I and E are constant values, the output torque of Eq. (10.18) becomes a constant value. However, if the back-EMF is not an ideal trapezoidal waveform, there is a torque ripple even though the BLDC motor is fed with constant stator currents. For a BLDC motor having arbitrary back-EMF waveforms of Eq. (10.4), the output torque can be expressed as

$$T_e = \lambda_f[i_{as}f(\theta) + i_{as}f(\theta - 240°) + i_{as}f(\theta - 120°)] \tag{10.19}$$

10.3.2.1 Torque ripple during the commutation

Even though the back-EMF is an ideal trapezoidal waveform, a torque ripple may occur due to the current ripple introduced during the commutation of phase currents [4–8]. This torque ripple may be a major obstacle in applying a BLDC motor for high-performance motor drives. Below is a description on the causes of the torque ripple during the commutation of phase currents.

Consider the commutation of the phase current from Section 2 to Section 3 in Fig. 10.10. The stator current during the commutation is assumed to be constant and equal to I. The magnitude of back-EMF is also supposed to remain as a constant value E.

In Section 2, the stator current I flows from phase *as* to phase *cs*, while in Section 3, the current I flows from phase *bs* to phase *cs*. Thus the current is commutated from phase *as* to phase *bs* during the transition from Section 2 to Section 3. In this case the phase *as* current i_{as} decreases to zero, while the phase *bs* current increases to its final value I. Meanwhile, the phase *cs* current i_{cs}, i.e.,

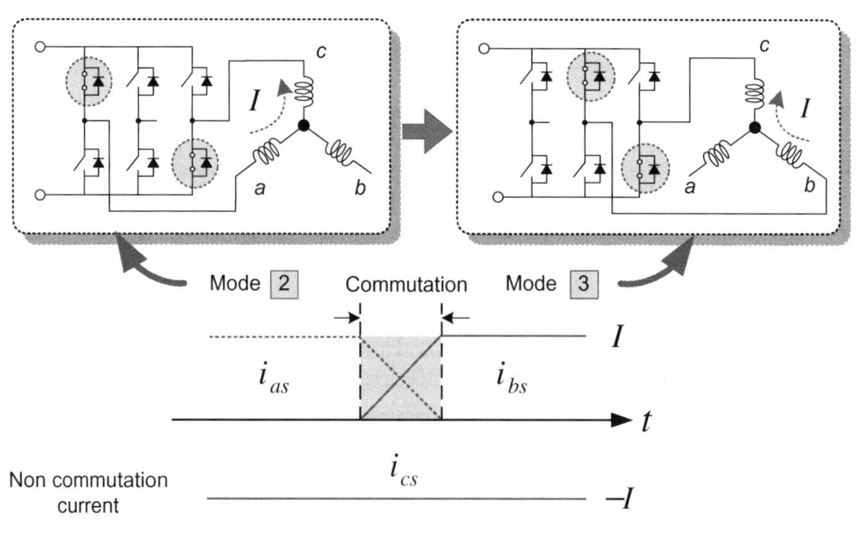

FIGURE 10.10

Commutation of phase currents.

noncommutation current, remains unchanged. During the transition from Section 2 to Section 3, $e_{as} = e_{bs} = E$, $e_{cs} = -E$. Thus the output torque of Eq. (10.18) can be expressed as

$$T_e = \frac{e_{as}i_{as} + e_{bs}i_{bs} + e_{cs}i_{cs}}{\omega_m} = \frac{E(i_{as} + i_{bs}) - Ei_{cs}}{\omega_m} = -\frac{2E}{\omega_m}i_{cs} \tag{10.20}$$

From Eq. (10.20), it can be readily seen that, during the commutation, the torque is proportional to the noncommutation current i_{cs}. If the noncommutation current i_{cs} remains a constant value during the commutation, then the torque also becomes a constant value.

During the commutation, it takes time to change the current due to the winding inductance. If the decreasing rate of the phase *as* current is equal to the increasing rate of the phase *bs* current, then $i_{as} + i_{bs} = I$, so the phase *cs* current i_{cs} remains at a constant value, $-I$. However, in practice, these two rates of change are usually not equal to each other due to the back-EMF, DC-link voltage, time constant of windings, etc. Accordingly, the noncommutation current cannot remain as a constant value, resulting in a torque ripple. For example, Fig. 10.11 shows that the rate of change of the current varies according to the back-EMF and DC-link voltage [4].

Fig. 10.11A shows that the phase *bs* current reaches the final value I before the phase *as* current reduces to zero. Hence, the noncommutation current, the phase *cs* current i_{cs}, becomes larger than I, resulting in an increase in the torque. On the other hand, Fig. 10.11C shows that the phase *as* current reduces to zero before the phase *bs* current reaches the final value, I. In this case the noncommutation current, phase *cs* current i_{cs}, becomes smaller than I, resulting in a decrease in the torque. Fig. 10.11B shows an ideal case where the phase *as* current reduces

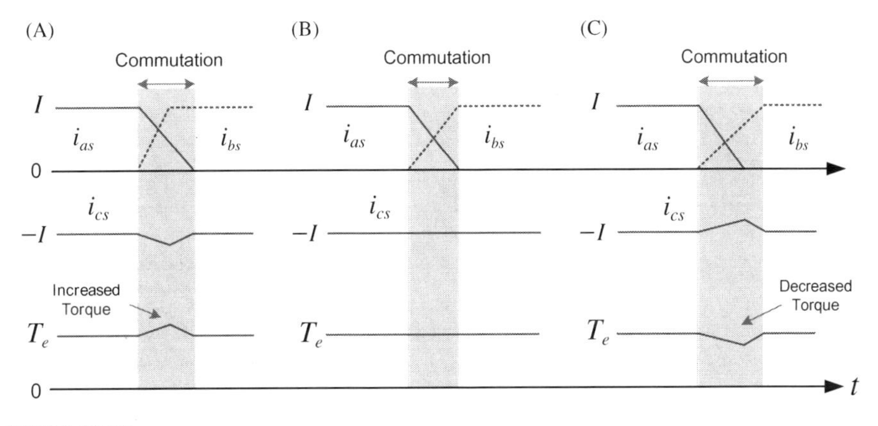

FIGURE 10.11

Phase currents during commutation in different cases. (A) low-speed ($V_{dc} > 4E$), (B) mid-speed ($V_{dc} = 4E$), and (C) high-speed ($V_{dc} < 4E$).

to zero at the same time the phase *bs* current reaches the final value *I*. In this case the noncommutation current remains constant, and thus results in the constant torque.

For this reason, whenever the winding current is commutated from one phase to another, a ripple in the output torque is generated. Thus this ripple of the torque occurs six times per cycle. The magnitude of the torque ripple depends on the operating current level and the operating speed. This is also different according to the pulse width modulation (PWM) techniques used in a driving inverter. Fig. 10.12A and B shows output torque and phase currents for the case shown in Fig. 10.11C. Commutation torque ripples produce noise and degrade speed control characteristics especially at low speeds. Thus many methods to reduce this torque ripple have been developed [4–8]. As an example, Fig. 10.12C shows that the ripple of the phase currents (thus, torque) is eliminated by using a commutation ripple compensation in which the motor input voltage is adjusted to equalize the rate of change of the current during the commutation [8]

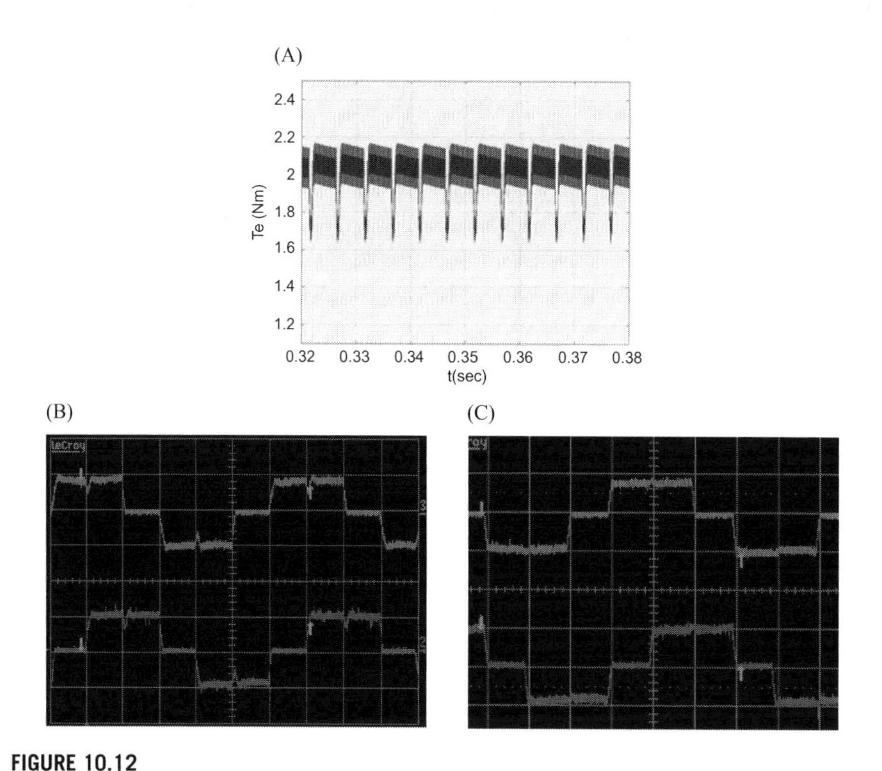

FIGURE 10.12

Ripple of output torque and phase currents. (A) output torque, (B) phase current (without compensation), and (C) with compensation.

MATLAB/SIMULINK SIMULATION: BRUSHLESS DIRECT CURRENT MOTOR

The six-step drive for 4-pole, 300-V BLDC motor is simulated by using a permanent magnet synchronous machine block in SimPowerSystems/Machines library and Mosfet blocks in SimPowerSystems/Power Electronics library.

- Overall diagram for simulation

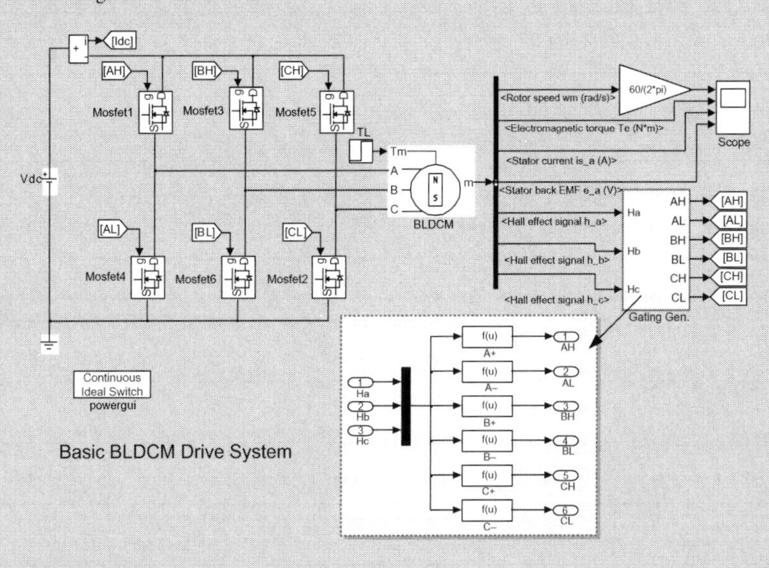

- Simulation results

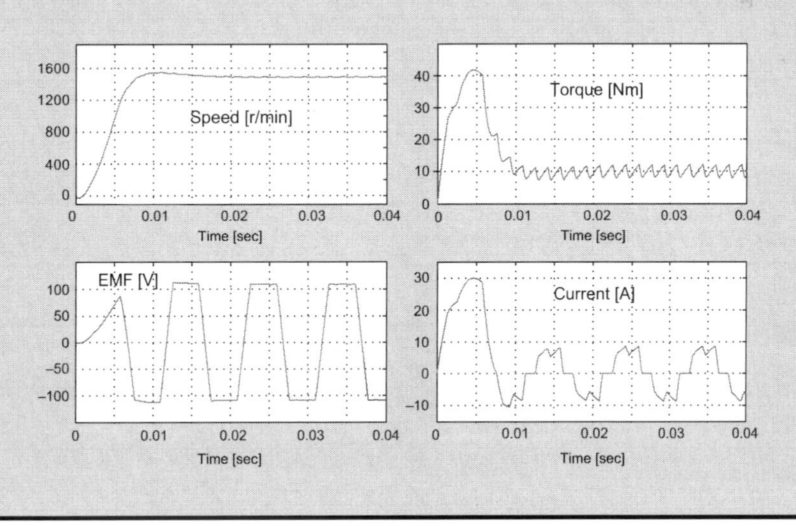

10.4 CONTROL OF BRUSHLESS DIRECT CURRENT MOTORS

As can be seen in the Section 10.2, we can operate a BLDC motor easily by a proper commutation of phase currents based on the information of the rotor position. Similar to a DC motor, the operating speed of a BLDC motor is proportional to the voltage applied to the motor, and thus its speed can be controlled by adjusting the applied voltage.

10.4.1 SPEED CONTROL

The simplest speed control system to control the speed of a BLDC motor is shown in Fig. 10.13. A proportional–integral (PI) controller as mentioned in Chapter 2, is commonly used for the speed control. This PI speed controller outputs the motor voltage reference (or PWM duty) as

$$V^* = \left(K_P + \frac{K_i}{s}\right)\cdot(\omega_m^* - \omega_m) \tag{10.21}$$

where K_P and K_i are the proportional and integral gains of the PI speed controller, respectively.

This reference voltage is generated by a PWM technique and then applied to the BLDC motor. This speed control system is simple but has a big problem. In this method, the motor current is hard to control within a proper range. This is because when a speed command is changed, the voltage reference may be changed largely. Thus we cannot expect to obtain a good dynamic response of the speed control. Moreover, this may incur a large transient current more than the rated current, which may lead to a shutdown of the drive system.

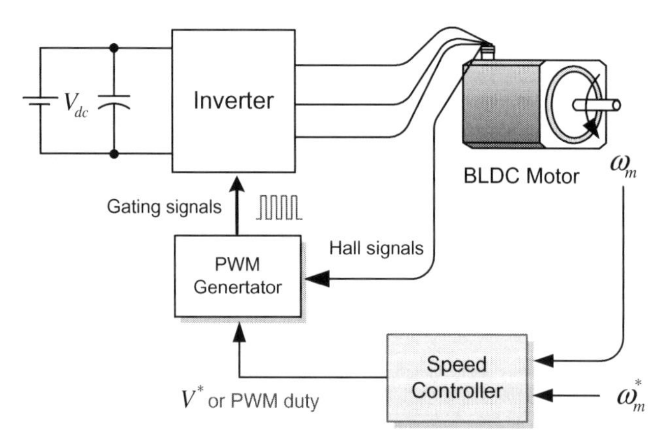

FIGURE 10.13

Speed control system of a BLDC motor.

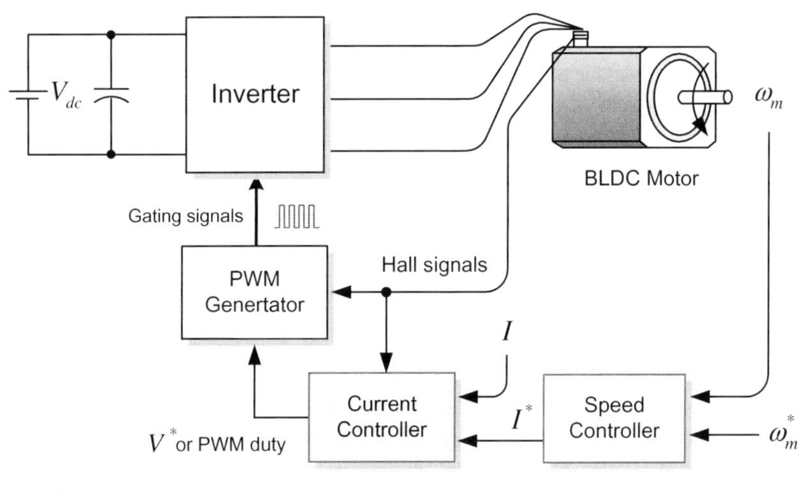

FIGURE 10.14

Speed control system based on the motor current.

To obtain a better dynamic response of the speed control, it is necessary to control the speed by controlling the torque or current of the BLDC motor. An enhanced speed control system based on the current control is shown in Fig. 10.14. Compared to the previous system of Fig. 10.13, this system includes a current controller to control the speed by regulating the current (thus, torque). In this case, the output of the PI speed controller becomes the motor current reference and a current sensor is needed to measure the actual motor current for the current control.

For AC motor drives, both the amplitude and phase of three-phase currents are instantaneously regulated to control the torque. However, for the BLDC motor drive, only the amplitude of phase currents needs to be regulated. This is because, as we can see in Eq. (10.18), the developed torque is proportional to the amplitude of the phase current. In addition, when controlling the amplitude of the current, we may regulate three-phase currents individually like in the current control of AC motors. However, since the amplitude of the phase current of a BLDC motor is proportional to the DC-link side current I_{dc}, the amplitude of the phase current is commonly controlled by regulating I_{dc}. In this case, the drive system needs only one current sensor at the DC-link side, and thus is more cost-effective.

Fig. 10.15 shows the speed control system using the regulation of the DC-link current. In this system, the speed controller produces the DC-link current reference as

$$I_{dc}^* = \left(K_{ps} + \frac{K_{is}}{s}\right) \cdot (\omega_m^* - \omega_m) \tag{10.22}$$

here, K_{ps} and K_{is} are the proportional and integral gains of the PI speed controller, respectively. These values can be determined from the gains selection procedure of a PI speed controller that was explained in Section 2.7.

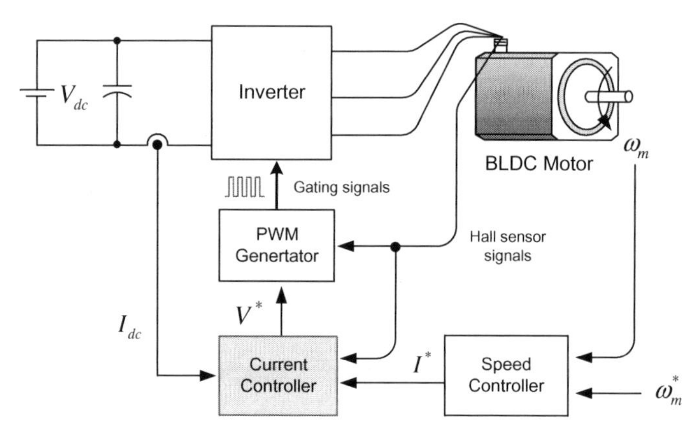

FIGURE 10.15

Speed control system using the regulation of the DC-link current.

To achieve the DC-link current reference I_{dc}^* produced by the speed controller, switching signals for an inverter are usually generated by a hysteresis regulation technique or a PWM technique.

10.4.2 CURRENT CONTROL

As can be seen from Chapter 6, the hysteresis technique is simply implemented and gives an excellent transient response because it directly determines the switching states from the current error. However, the hysteresis technique has a major drawback of varying the switching frequency with operating conditions, such as the back-EMF, load condition, etc. Thus, to have a constant switching frequency, a PWM technique is commonly used, but it is inferior to the hysteresis technique in performance.

In the case of using a PWM technique, a motor voltage reference (or PWM duty) is generated by the PI current controller from an error between the current command I_{dc}^* and the actual current as

$$V^* = \left(K_{pc} + \frac{K_{ic}}{s}\right) \cdot (I_{dc}^* - I_{dc}) \tag{10.23}$$

here, K_{pc} and K_{ic} are the proportional and integral gains of the current controller, respectively. These values can also be determined from the gains selection procedure of a PI current controller that was explained in Chapter 2.

Finally, the actual active switches are determined by combining the PWM switching signals with the operating mode signal decoded by using Hall effect sensor signals.

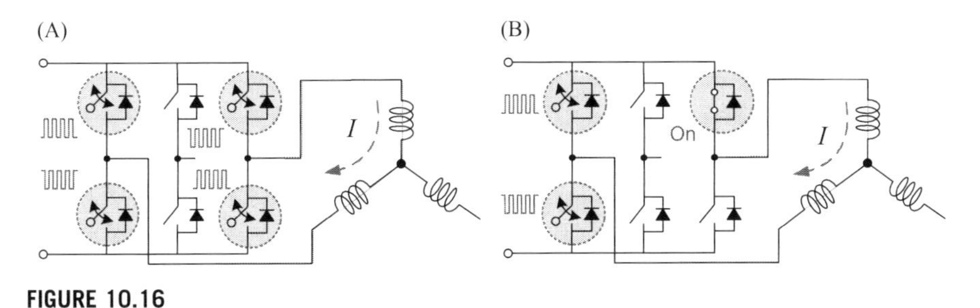

FIGURE 10.16

PWM techniques. (A) bipolar switching and (B) unipolar switching.

10.5 PULSE WIDTH MODULATION TECHNIQUES

In BLDC motor drives, there are two conventional PWM techniques for generating a motor applied voltage the as shown in Fig. 10.6: *bipolar switching method* and *unipolar switching method* [9–10]. These are similar to the PWM techniques of the H-bridge circuit for DC motor drives in Section 2.8. This similarity is due to the facts that switches of only two phases are being driven in the three-phase inverter for BLDC motor drives. However, there are several variations for the unipolar switching method. Now let us examine these PWM techniques.

10.5.1 BIPOLAR SWITCHING METHOD

In the bipolar switching method, a PWM signal is applied to all switches of two phases as shown in Fig. 10.16A. By contrast, in the unipolar switching method as shown in Fig. 10.16B, a PWM signal is applied to the switches of only one phase while one switch of the other phase is kept at an on-state. The bipolar method is simple and can give a better transient response because $+V_{dc}$ or $-V_{dc}$ is applied across the phase winding. However, the current ripple (thus, torque ripple) and switching losses are greater than those of the unipolar switching method. The PWM gating signals by the bipolar switching method are shown in Fig. 10.17.

10.5.2 UNIPOLAR SWITCHING METHOD

In the unipolar switching method, switching losses can be reduced because the PWM signal is applied to the switches of only one phase. In addition, since the applied voltage to the phase windings is 0 and $+V_{dc}$ or 0 and $-V_{dc}$, the current ripple is half of that of the bipolar switching method. Due to these advantages, the unipolar switching method is more widely used for BLDC motor drives. However, this method is complicated and has a slower response than the bipolar switching method. In addition, circulating currents may occur in the inactive phase windings. Thus this method is less favorable for precision servo drives.

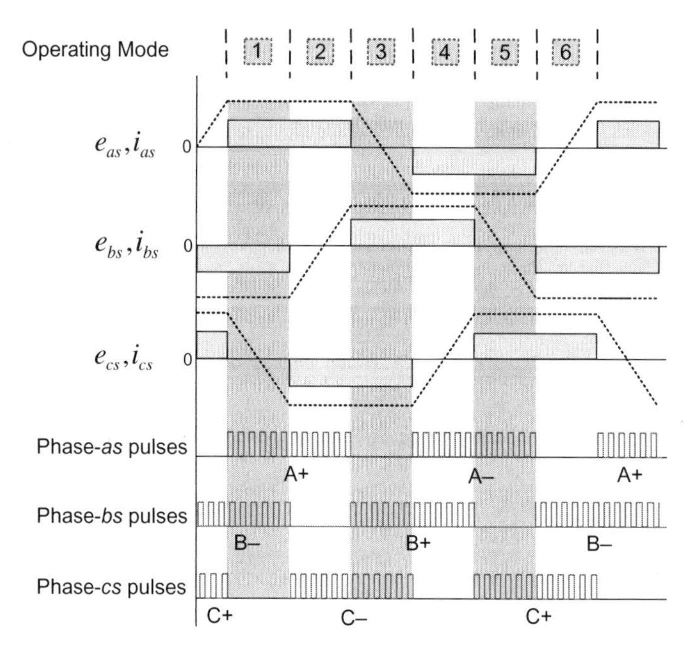

FIGURE 10.17

Bipolar switching method.

The unipolar switching method is also divided into several schemes according to the executing switch. In the *upper switch PWM scheme* in Fig. 10.18A, the PWM signal is applied to the upper switch only while the lower switch is kept at an on-state. On the other hand, in the *lower switch PWM scheme* in Fig. 10.18B, the PWM signal is applied to the lower switch only while the upper switch is kept at an on-state. In these two schemes, however, the utilization of switches and switching losses are biased because only one specific switch makes a continuous switching.

As types of improved schemes through uniform switching, there are *on-going PWM scheme* and *off-going PWM scheme* [9]. In the on-going PWM scheme shown in Fig. 10.18C, the PWM signal is applied to each switch during the front 60° part of the 120° conduction interval. In the off-going PWM scheme shown in Fig. 10.18D, the PWM signal is applied to each switch during the latter 60° part of the 120° conduction interval.

These unipolar PWM schemes have a disadvantage of having an increased ripple torque and a lowered efficiency because of ripple current, which occurs in the inactive phase windings due to diode freewheeling. The magnitude and direction of the ripple current depends on the PWM schemes. We can see, from Fig. 10.19, the ripple current during the inactive interval for unipolar PWM schemes that were explained above. In addition, the bipolar switching method has no ripple current due to no diode freewheeling. The PWM scheme, which is called

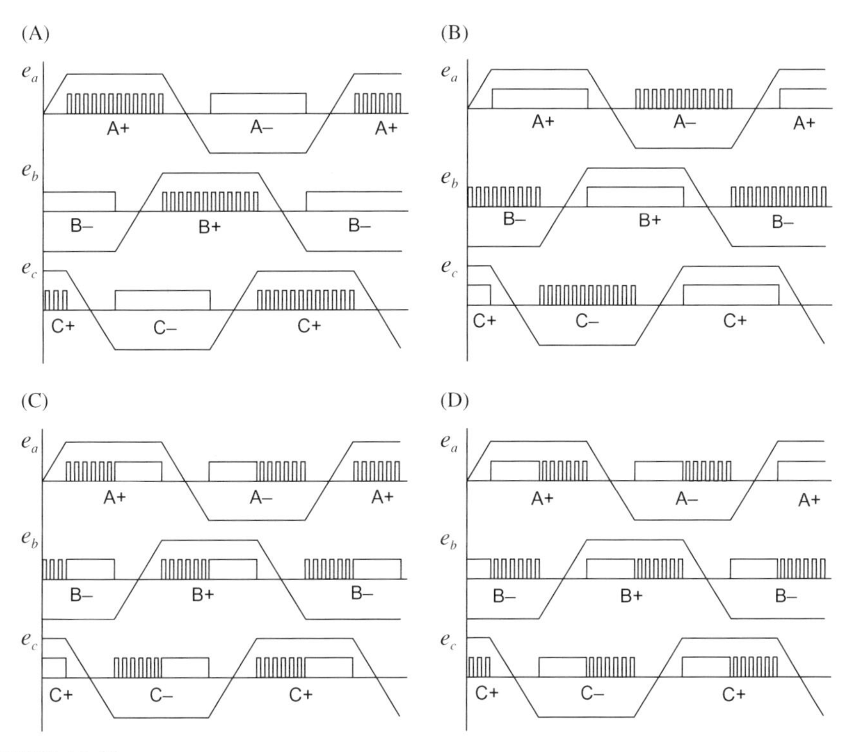

FIGURE 10.18

Different schemes of the unipolar switching method. (A) upper switch PWM, (B) lower switch PWM, (C) on-going PWM, and (D) off-going PWM.

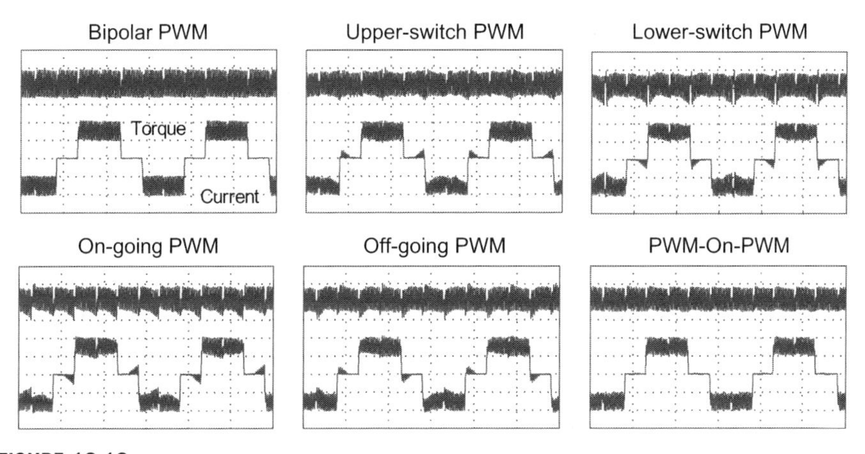

FIGURE 10.19

Comparison of phase and torque for PWM schemes.

PWM_ON_PWM scheme, to eliminate diode freewheeling in the inactive phase has been developed [10]. In this scheme the switches are in PWM mode in the beginning 30° and the last 30° zones, and in continuous on-state in the middle 60° zone of the 120° conduction interval.

As it can be seen from above, characteristics such as ripple torque during commutation, switching losses, and sensorless control performance vary according to different unipolar PWM schemes. The torque ripple comparison of PWM schemes is shown in Fig. 10.19.

10.6 SENSORLESS CONTROL OF BRUSHLESS DIRECT CURRENT MOTORS

As described in the Section 10.2, Hall effect sensors for obtaining the rotor position are indispensable for BLDC motor drives. However, since the position sensors increase the cost and size of the motor and reduce the reliability of a drive system, a BLDC motor drive without position or speed sensors is becoming more popular.

Various sensorless methods for BLDC motors have been seen in the literature [11–15]. One well-known method is the back-EMF-based method [11]. There is also a method based on the current of the freewheeling diodes of the noncommutation phase [13] and a method based on a flux observer. Among these, we will explore the back-EMF-based method, which is the most widely used for low-cost applications such as fan, pump, and compressor drives due to its easy principle and implementation.

10.6.1 SENSORLESS CONTROL BASED ON THE BACK-ELECTROMOTIVE FORCE

The back-EMF of a motor includes information on the magnetic flux. Thus the rotor position can be obtained by detecting the back-EMF. It is hard to measure the back-EMF of AC motors because all three windings are excited at all times. On the other hand, in three-phase BLDC motor drives, since only two of the three-phase windings are conducting at a time, the back-EMF appears in the open winding of the nonconducting phase. Thus the back-EMF can be detected by sensing the voltage of the nonconducting phase. In BLDC motors, we can consider that the nonconducting phase winding plays the role of a sensor to detect the position.

In this case, there is no need to detect the whole waveform of the back-EMF. Instead, the commutation instants can be identified by detecting only the zero crossing point (ZCP) of the back-EMF. From Fig. 10.20, we can easily see that the back-EMFs of a nonconducting phase always pass through zero (see the parts arrows are indicating). For example, in Section 2 where phase *as* and *cs* are conducting and phase *bs* is nonconducting, we can identify that the back-EMF of

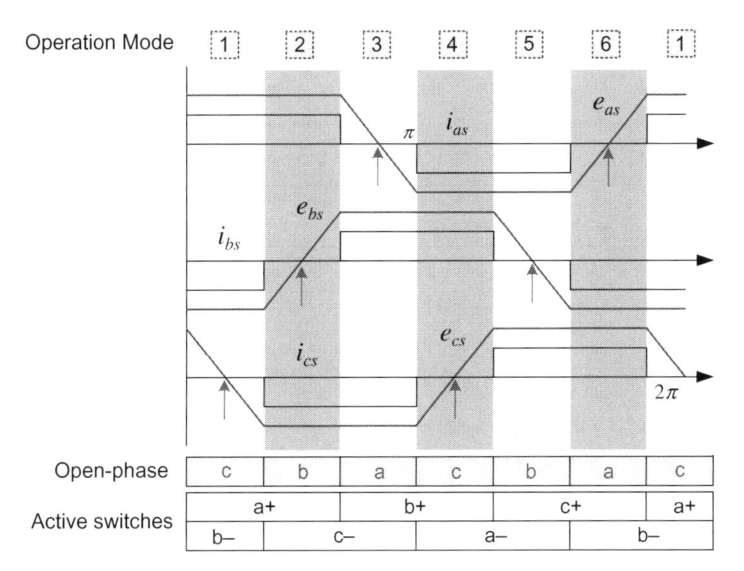

FIGURE 10.20

ZCPs of the back-EMF.

phase *bs* passes through zero. In addition, it can be seen that the commutations happen 30 electrical degrees after the ZCPs of the back-EMFs.

Therefore, without any position sensor, the phase commutation is made possible by detecting the ZCPs of the back-EMFs.

For three-phase Y-connected windings, the back-EMF across a phase can be obtained directly from measuring the phase terminal voltage referred to the neutral point of windings as shown in Fig. 10.21. However, in most cases, the neutral point of windings is not accessible. Thus the most commonly used method is to create a virtual neutral point by using the three-phase terminal voltages.

The back-EMF-based methods have the following problems. When measuring the terminal voltage, a large amount of electrical noise is induced on the sensed terminal voltage due to PWM switching signals driving the motor. Low-pass filters are usually used to remove unwanted switching noise, but the phase delay of the low-pass filter causes a commutation delay at high speeds. In addition, attenuation by a voltage divider will be required to lower the level of the sensed signal to an acceptable range of the control circuit. This lowers the signal-to-noise ratio at low speeds, resulting in degradation of low-speed operation performance. Moreover, the commutation happens 30 electrical degrees after the ZCPs of the back-EMFs. It is hard to obtain commutation instants precisely when the operating speed is changing. To improve these problems, the back-EMF integration the third harmonic voltage integration, and the method detecting the conducting current of the freewheeling diodes in the unexcited phase have been presented [13–14]. However, they still have a low accuracy problem at low speeds.

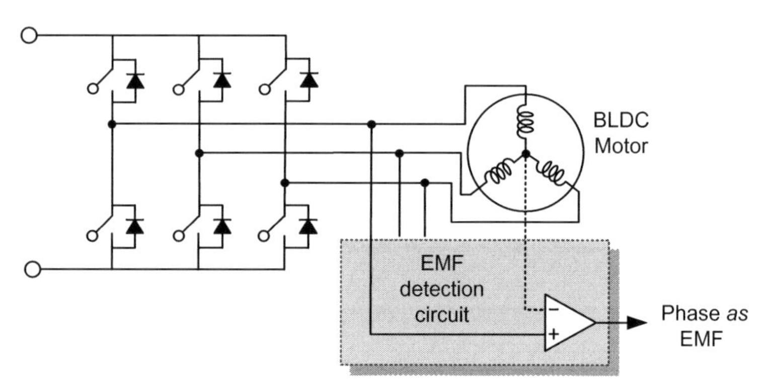

FIGURE 10.21

Back-EMF detection circuit.

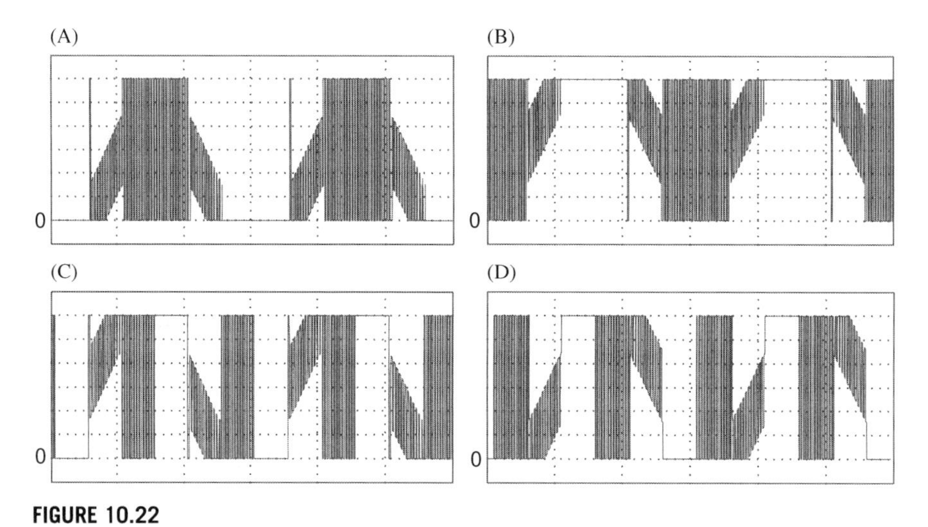

FIGURE 10.22

Back-EMFs according to unipolar PWM techniques: (A) upper switch PWM, (B) lower switch PWM, (C) on-going PWM, and (D) off-going PWM.

We have just now discussed the ZCPs of the back-EMF waveforms for a six-step drive in Fig. 10.20. However, BLDC motors are usually driven by using a PWM technique. In this case, PWM signals are superimposed on the back-EMF waveforms, as can be seen in Fig. 10.22. Thus, The ZPSs are obscured by the PWM signals. Several improved back-EMF sensing methods linked to PWM techniques, which require neither a virtual neutral voltage nor a great amount of filtering, have been introduced [14].

However, these back-EMF-based methods have intrinsic restrictions on low-speed operations where the back-EMF becomes insufficient. In addition, since there is no information on the back-EMF at start-up, an additional method for start-up is needed. Commonly, a sensorless BLDC motor is first started using initial rotor position detection method and brought up to a certain speed by an open-loop operation. When the motor reaches a speed where the back-EMF is sufficient to be sensed, the operation is transferred to the sensorless control.

REFERENCES

[1] G. Bauerlein, A brushless DC motor with solid-state commutation, IRE Natl. Conv. Rec. (1962) 184–190.

[2] T.M. Jahns, W.L. Soong, Pulsating torque minimization technique for permanent magnet AC motor drives—a review, IEEE Trans. Ind. Electron. 43 (2) (Apr. 1996) 321–330.

[3] P. Pillay, R. Krishnan, Modeling, simulation, and analysis of permanent-magnet motor drives, Part II. The brushless DC motor drive, IEEE Trans. Ind. Appl. 25 (2) (Mar./Apr. 1989) 274–279.

[4] R. Carlson, M. L-Mazenc, J. Fagundes, Analysis of torque ripple due to phase commutation in brushless DC machines, IEEE Trans. Ind. Appl. 28 (3) (1992) 441–450.

[5] C. Berendsen, G. Champenois, A. Bolopion, Commutation strategies for brushless DC motors: influence on instant torque, IEEE Trans. Power Electron. 8 (2) (1993) 231–236.

[6] K-W Lee, et al., Current control algorithm to reduce torque ripple in brushless DC motors, in: Conf. Rec. ICPE'98, vol. 1, Seoul, Korea, October 1998, pp. 380–385.

[7] J.H. Song, I. Choy, Commutation torque ripple reduction in brushless DC motor drives using a single DC current sensor, IEEE Trans. Power Electron. 19 (2) (Mar., 2004) 312–319.

[8] K.-J. Kwun, S.-H. Kim, A current control strategy for torque ripple reduction on brushless DC motor during commutation, Trans. Korean Inst. Power Electron. 9 (4) (Jun. 2002) 195–202.

[9] Z. Xiangjun. C. Boshi, The different influences of four PWM modes on the commutation torque ripples in sensorless brushless DC motors control system, in: Proc. the Fifth International Conference Electrical Machines and Systems, vol. 1, 2001, pp. 575–578.

[10] U. Vinatha, S. Pola, K.P. Vittal, A novel PWM scheme to eliminate the diode freewheeling in the inactive in BLDC motor, in: Conf. Rec. IEEE PESC'2004, pp. 2282–2286.

[11] K. Iizuka, et al., Microcomputer control for sensorless brushless motor, IEEE Trans. Ind. Appl. 27 (May/Jun., 1985) 595–601.

[12] P.P. Acarnley, J.F. Watson, Review of position-sensorless operation of brushless permanent-magnet machines, IEEE Trans. Ind. Electron. 53 (2) (Apr. 2006) 321–330.

[13] T. Kim, H.-W. Lee, M. Ehsani, Position sensorless brushless DC motor/generator drives: review and future trends, IET Electr. Power, Appl. 1 (4) (2007) 557–564.

[14] S. Ogasawara, H. Akagi, An approach to position sensorless drive for brushless DC motors, IEEE Trans. Ind. Appl. 27 (5) (Sep./Oct., 1991) 928–933.
[15] Y.-S. Lai, Y.-K. Linl, A unified approach to back-emf detection for brushless dc motor drives without current and hall sensors, in Conf. Rec. IEEE IECON 2006, pp. 1293–1298.

Index

Note: Page number followed by "*f*," "*t*," and "*b*" refer to figures, tables, and boxes, respectively.

B

C

I

K

L

M

N

O

R

S

T

Printed in the United States
by Baker & Taylor Publisher Services